Analysis of Variance for High-Dimensional Data

Analysis of Variance for High-Dimensional Data

Applications in Life, Food, and Chemical Sciences

Age K. Smilde
Swammerdam Institute for Life Sciences
University of Amsterdam
The Netherlands

and

Department of Plant and Environmental Sciences
University of Copenhagen
Denmark

Federico Marini
Department of Chemistry
University of Rome "La Sapienza"
Italy

Johan A. Westerhuis
Swammerdam Institute for Life Sciences
University of Amsterdam
The Netherlands

Kristian H. Liland
Faculty of Science and Technology
Norwegian University of Life Sciences
Norway

WILEY

This edition first published 2025

Registered Office(s)
John Wiley & Sons, Inc., 111 River Street, Hoboken, NJ 07030, USA
John Wiley & Sons Ltd, New Era House, 8 Oldlands Way, Bognor Regis, West Sussex, PO22 9NQ, UK

For details of our global editorial offices, customer services, and more information about Wiley products visit us at www.wiley.com.

The manufacturer's authorized representative according to the EU General Product Safety Regulation is Wiley-VCH GmbH, Boschstr. 12, 69469 Weinheim, Germany, e-mail: Product_Safety@wiley.com.

Wiley also publishes its books in a variety of electronic formats and by print-on-demand. Some content that appears in standard print versions of this book may not be available in other formats.

Library of Congress Cataloging-in-Publication Data applied for:

Hardback ISBN: 9781394211210

Cover Design: Wiley
Cover Image: Courtesy of Age K. Smilde, Federico Marini, Johan A. Westerhuis, and Kristian H. Liland

Set in 10/12pt WarnockPro by Straive, Chennai, India

Contents

About the Authors

Age K. Smilde is an Emeritus Professor of Biosystems Data Analysis at the Swammerdam Institute for Life Sciences, University of Amsterdam. He also holds a part-time position at the Department of Plant and Environmental Sciences, University of Copenhagen. His research interests include modeling complex biological systems using multiway, multiset, and latent-path models.

Federico Marini is a Professor of Analytical Chemistry at the Department of Chemistry, University of Rome “La Sapienza”. His main research interest includes model development, particularly for classification, variable selection, non-linear modeling, and data fusion, with applications in spectroscopy and imaging, food science, and various omics.

Johan A. Westerhuis is an Assistant Professor of Biosystems Data Analysis at the Swammerdam Institute for Life Sciences, University of Amsterdam. His main research is in the development and validation of chemometric and machine learning tools and their application to high-dimensional metabolomics and microbiome data.

Kristian H. Liland is a Professor of Statistics at the Faculty of Science and Technology, Norwegian University of Life Sciences. His main research interests include model development and applications using multivariate statistics and machine learning within spectroscopy, bioinformatics, various omics, health data, and industrial applications.

Foreword

The reason for studying a biological system is the desire to acquire an understanding of that system. The classical scientific royal road proceeds via experimental designs, accompanied by statistical analyses of variance. By realizing that manipulations can affect the system in different ways, the necessity of studying experimental effects on multiple aspects jointly became eminently clear. With the recent major developments in measurement instruments, it is within common reach to collect high-quality data on very many variables, which are supposed to be indicative of different aspects of a system.

Yet, then the key challenge becomes to unravel the nature of the experimental effects from the observed high-dimensional data. This is exactly where analysis of variance for high-dimensional data comes in. As simple as its basic idea is – combine analysis of variance with dimension reduction – as intricate are the possible decisions to get meaningful results. This book equips the reader with the necessary knowledge to do so. It brings a unified treatment of the methods, carefully built upon the mathematical and conceptual foundations introduced. It discusses the properties and their implications for empirical use, supported by clarifying figures. Further, it treats the similarities and differences between the various methods and their implications. It discusses all essential aspects required for insightful modeling, including preprocessing the data, model diagnostics, statistical testing, validation, and visualization.

The empirical use of the methods is illustrated with examples of high-dimensional experimental data from various fields in the life, food, and chemical sciences. These inspiring examples provide the reader with a good sense of the line of reasoning guiding the specific choices to be made and show what can be learned from high-dimensional experimental data. Further inspiration can be gained from the references provided in the review of currently available applications. The kind invitation to the reader to apply and profit from the methods is made complete by the overview of the available software implementations of the HD-ANOVA methods, and specifically the gentle introduction to the R package HDANOVA.

Though the book is clearly rooted in and focused on the life, food, and chemical sciences, it is also of great value to empirical scientists in other fields, like the behavioral and social sciences and economics. After all, the principles of and methods for studying a system based on multivariate data resulting from an experimental design are generally applicable and provide a rapid path to knowledge about the system. The book provides a state-of-the-art overview and thereby clearly indicates the open issues that deserve attention. I do express the hope that the book will also serve as a starting point for method developers. Specifically, the framework of restricted multivariate multiple regression offers ample opportunities for further useful developments. This will provide method developers with a deeper understanding of the methods themselves, as well as enrich our method toolbox – with the ultimate aim to help us understand our world and the ways we can fruitfully contribute to it.

Groningen, The Netherlands
September 2024

Marieke Timmerman

Preface

In many cases in the life, food, and chemical sciences, measurements are performed of a system under various experimental conditions generated according to an experimental design, which provides samples with a grouping structure. Traditionally, such data is analyzed by analysis of variance (ANOVA) and generalizations thereof, such as linear mixed models. These methods are, however, no longer at par with the increasing complexity of the measurements: in many cases, a large number of variables are measured, and there is even the tendency to measure not just one set of related variables, but multiple sets pertaining to the same problem, such as gene expression and metabolomics measurements on the same samples.

Analyzing such complex data requires methods that go beyond ANOVA. In statistics, a generalization to ANOVA for multivariate data is Multivariate ANOVA. This method, however, breaks down when the number of variables exceeds the number of samples which is often the case in modern measuring devices. About 20 years ago, a method was proposed, called Analysis of Variance Simultaneous Component Analysis (ASCA), which attempts to remedy this problem and can be used to analyze high-dimensional data with a grouping structure. This method has undergone many changes and generalizations in the last two decades, which urgently calls for a book with an overview of these methods.

This book offers such an overview and summarizes the most important developments in the field of high-dimensional ANOVA. We try to accommodate our readers at different levels. In this book, we explain the theory and show when to use which method underpinned by many examples. This should give the practitioners an idea of which method to use when and how to use it and interpret the results. In elaborations, we give more detail for the readers who like to go in more depth and the algorithms serve the method developers to encourage them to critically see if they can improve on these methods and corresponding algorithms.

A minimum read of the book should be Chapters 1, 4, and 5 for those who are already familiar with matrix algebra and linear models. For those who need to refresh the latter, we propose to include reading Chapters 2 and 3 as well. We also provide software (R-scripts) that contain most of the presented methods.

We are very grateful to all our friendly colleagues, Fentaw Abegaz, Jasper Engel, Fred van Eeuwijk, Peter Filzmoser, Anna Heintz-Buschart, Margriet Hendriks, Ulf Indahl, Tormod Næs, Jean-Michel Roger, Edoardo Saccenti, Joakim Skogholt, and Raffaele Vitale, who took the time to proofread the chapters of our book. We are also grateful to Marieke Timmerman who wrote the foreword. Needless to say, all remaining errors are our responsibility!

Utrecht, The Netherlands — *Age K. Smilde*
September 2024

Rome, Italy — *Federico Marini*
September 2024

Dronten, The Netherlands — *Johan A. Westerhuis*
September 2024

Ås, Norway — *Kristian H. Liland*
September 2024

1
Introduction

1.1 Types of Data

The amount of data in the life, food, and chemical sciences is exploding. This is mainly due to the rapid improvement and automation of measuring instruments and development of new measuring devices. This has also raised the expectations of scientists to receive answers to increasingly more complex questions. To fill the gap between the large amount of data generated on the one hand, and the complex questions on the other hand, advanced data analysis methods are needed.

Without claiming to be exhaustive, we will give a short overview of what kind of data structures are presently being generated in the life sciences. We will generally use the term life sciences, although also applications from food and chemical sciences will be provided. The first important difference in data sets to be made is shown in Figure 1.1. Tall or skinny data sets – $\mathbf{X}_1$ in Figure 1.1a – are characterized by having many samples or cases (I) and relatively few variables (J). Data from cohort studies are examples of this, and these are also sometimes called multivariate data. Contrarily to that, there are short or fat data sets such as $\mathbf{X}_2$ in Figure 1.1b. In those cases, the number of variables (far) exceeds the number of samples. These data sets are called high-dimensional. In the life sciences, the latter types of data are far in the majority since collecting samples is usually costly or time-consuming, and modern sequencing techniques and mass-spectrometry instruments easily generate thousands of features. In this book, we will discuss methods for the analysis of multivariate data and for high-dimensional data. We will always use the convention of having the samples in the rows of matrices and the variables or features in their columns.

In many cases in the life sciences, there are multiple data blocks being measured simultaneously on the same or on comparable systems. This can result in different data set configurations, as shown in Figure 1.2. If different types of variables are measured for the same samples, then the data can be fused in the common sampling mode (Figure 1.2a). If the same variables are measured on different (but comparable) sets of samples, then the data can be fused in the common variable mode (Figure 1.2b). The methods to perform such data fusions are manifold, and some of these are explained in Smilde *et al.* (2022).

Yet other data structures can be envisaged, such as shown in Figure 1.3. The separate data blocks may occur in a certain topology as can be inferred from substantive arguments (Figure 1.3a). This calls for approaches using pathway analysis such as path-models (Kaplan, 2008). In other cases, the data can be arranged in a three-way array (Figure 1.3b) and should then be analyzed with three-way methods (Smilde *et al.*, 2005; Kroonenberg, 2008). An example is the measurement of metabolites at different time points on the same individuals during a challenge test (Vis *et al.*, 2015).

Analysis of Variance for High-Dimensional Data: Applications in Life, Food and Chemical Sciences,
First Edition. Age K. Smilde, Federico Marini, Johan A. Westerhuis, and Kristian H. Liland.

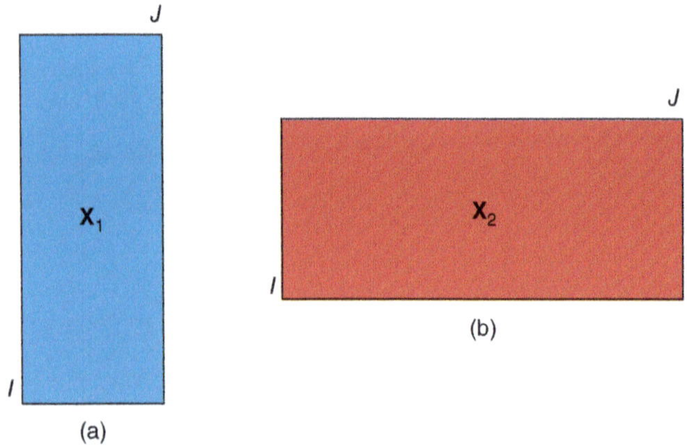

Figure 1.1 Two different types of data sets: (a) long, tall, or skinny, (b) short or fat.

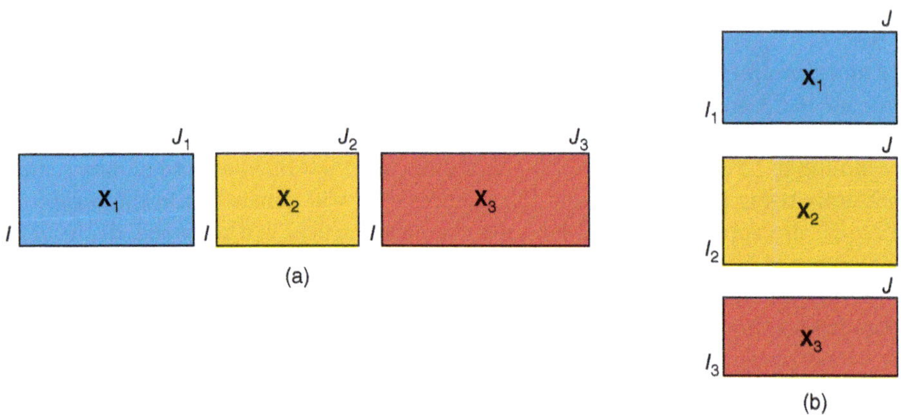

Figure 1.2 Two different types of data fusion: (a) along the sample mode or (b) along the variable mode.

In science, data are always embedded in an environment of prior knowledge. This is visualized in Figure 1.4 where an example from life science is chosen. Suppose that for a group of people, samples are taken and their microbiome, metabolome, and transcriptome are measured. For the microbiome, there may be information available regarding the phylogeny of the bacteria and viruses involved; this information can be available at the genus or species level. For the metabolome, there may be information present about metabolic pathways and networks or knowledge on metabolites that are disease-related. For the transcriptome, there may be information available on gene regulation, pathways, and also disease-related gene-expression patterns. On a general level, databases may be available, related cohort data, and of course biochemical knowledge.

Often there is also prior knowledge regarding a structure across the samples when analyzing multivariate or high-dimensional data. This can be knowledge regarding a certain grouping structure (*e.g.*, family relations) or meta-data (disease status, lifestyle variables, diet information) or a clustering of the samples as a result of a preliminary cluster analysis. Another type of knowledge can be whether groups of samples are related to certain treatments. This can be in the form of a design that is used to generate a sampling structure. There are very many types of

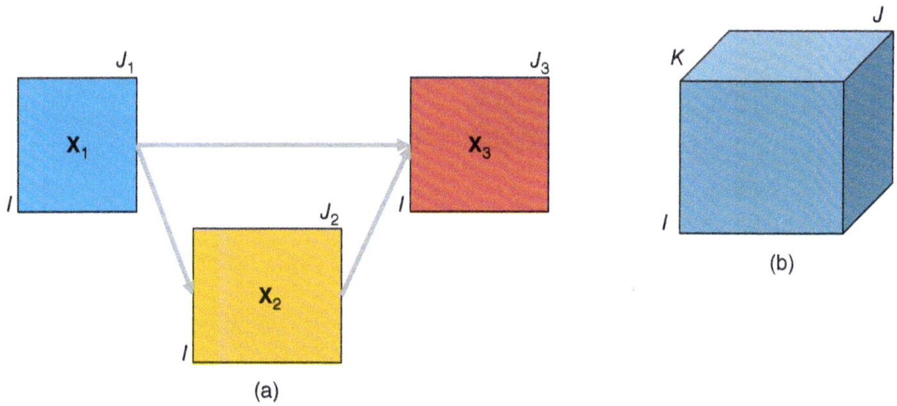

Figure 1.3 Data blocks arranged in a pathway topology (a) or in a three-way array (b).

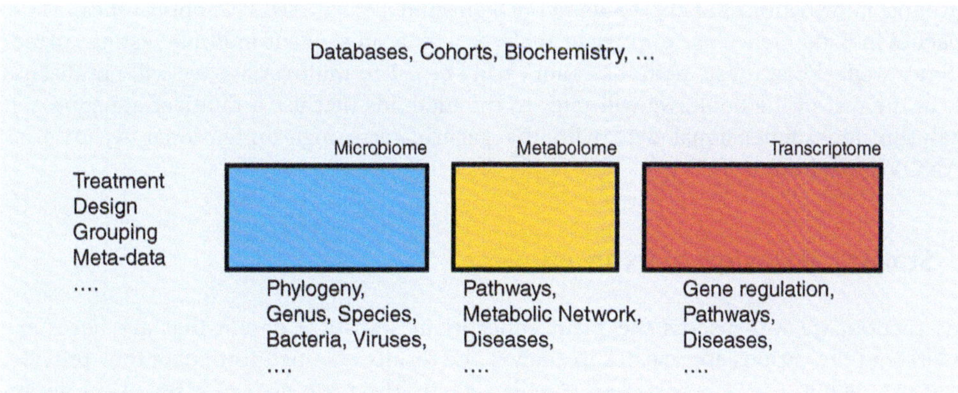

Figure 1.4 Prior knowledge available for a data set.

designs in use in the life sciences, and all these introduce a structure in the sampling mode of the data. The main goal of this book is to introduce, discuss, and compare data analysis methods that are able to estimate high-dimensional treatment effects from such data where various treatment factors have been used to induce differences between groups of subjects. These methods are focused on partitioning the treatment-induced variation according to the different factors used in the design, while explorative methods such as Principal Component Analysis (PCA) (Pearson, 1901), t-distributed Stochastic Neighbor Embedding (tSNE) (Van der Maaten and Hinton, 2008), or Uniform Manifold and Projections (UMAP) (McInnes *et al.*, 2018) analyze the total variation as a whole and are not able to partition the variation to specific causes.

We are going to discuss methods that can take into account a possible design or grouping structure underlying the sampling mode. Such information on design or grouping is usually encoded in the design matrix with zeros and ones (see Figure 1.5). Different types of codings exist, and these will be discussed in much detail in Chapter 3. This holds for encoding qualitative variables; for quantitative variables this works differently, as discussed in detail in Draper and Smith (1998). There may be a whole variety of designs underlying the sampling mode. In terms of classical statistical design, these can be crossed designs based on fixed effects, cross-over

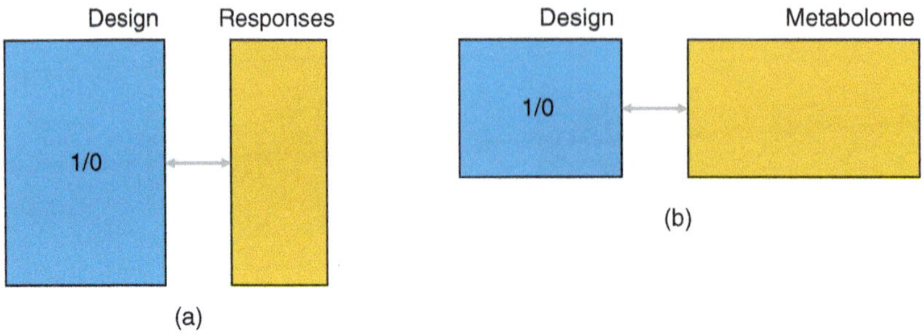

Figure 1.5 Design information about the samples: (a) encoding for skinny data and (b) encoding for fat data.

designs, and nested designs to name a few. All these types of designs are used in the life sciences. We will also discuss random effect and mixed effect models, since these are also used in these fields.

In this book we will introduce univariate Analysis of Variance (ANOVA) models as background information, but the focus is on high-dimensional ANOVA approaches. Many approaches in data science use univariate analyses with a subsequent multiple testing correction. Since we feel that these methods cannot truly be called multivariate, we will not discuss those. In the rest of the book, we will refer to the methods that use a multivariate approach to analyzing high-dimensional data with the generic term high-dimensional ANOVA or HD-ANOVA for short.

1.2 Statistical Design of Experiments

In this section, we will discuss the main concepts in statistical design that are necessary to understand the various approaches to analyze the results obtained from data they provide. We will use definitions as reported in Casella (2008). The main goal of a statistical design is to systematically investigate the effect size of different levels of one or more treatment factors on the experimental units and compare this effect size to the natural variation between experimental units. The *experimental unit* is the unit that is randomly assigned to a treatment. This can be, *e.g.*, a subject, an animal, a plant, or a cage. The experimental unit is not always the same as the sampling or *observational unit*. Multiple observations can be obtained from the experimental unit, but they depend on the experimental unit and, therefore, special care needs to be taken into account during the statistical analysis. As an example, multiple neuronal cells (observational unit) can be measured in the same mouse (experimental unit) to evaluate the effect of the different levels of treatment that are applied to the mice.

The variation between the experimental units is expected to be homogeneous. This means that if none of the treatments has an effect on the response, all response values come from the same distribution. However, the use of different batches, different plots, measurements on different days, or unknown confounding variables can lead to systematic differences between the experimental units. To reduce bias and maximize precision of treatment comparisons, three key design principles are being used.

- **Replication**: The separate levels of the treatment factor are provided to more than one experimental unit to average out differences between experimental units. Taking multiple measurements (observational units) from each experimental unit, can lead to pseudo-replication, and give the impression that the variation between experimental units is smaller than what it actually is.

- **Randomization**: The treatment levels are randomly applied to the experimental units to ensure a fair assessment of the different levels of the treatment factor without bias. Randomization corrects for unknown confounders such as lifestyle or different amounts of light in a greenhouse.
- **Blocking**: If there are known sources of bias, such as the use of different microarrays, blocks of experimental units can be defined in which the variation between experimental units within a block is homogeneous. This is also referred to as local control. As an example, take an experiment in which three types of fertilizer are compared. Each fertilizer is added to 10 randomly selected pots of soil and each pot has 3 plants. The 30 pots are randomly assigned to 2 greenhouse benches, in such a way that an equal number of pots with each fertilizer are on both benches. In this example, the 2 benches are the blocks, and the pots are the experimental units, while the plants are observational units.

The total variation between the pots is due to the different types of fertilizer, due to the specific greenhouse bench, and due to the natural variation between the pots. The part of the design that takes care of how the different levels of the treatment factor are distributed over the experimental units is called the treatment design. The part of the design that takes care of the blocking, randomization, and replication structure is called the experimental design. In most omics experiments, samples of each experimental unit are obtained and stored. The actual measurement of the samples takes place later and can suffer from the same biases as explained above. To reduce bias and variation between experimental units, blocking and randomization are used to deal with measurements taken in different batches over different days. The definition of when and how the measurements are performed (sometimes called the measurement design) is part of the experimental design. The combination of the treatment design and experimental design is called the statistical design.

The treatment design defines how the different levels of each treatment factor are assigned to the experimental units. If there are two or more treatment factors, these treatment factors can be crossed or nested. For crossed treatment factors the exact same levels of one factor are repeated for each level of the other factor. If plants are subjected to different temperatures and to different light conditions the effect of these crossed treatment factors can be observed independently, as well as their interaction. Nested treatment factors describe hierarchical relationships between treatments. Treatments are nested if different levels of Factor `B` appear within the levels of Factor `A`. The different levels of `A` can be tested separately, but comparing the levels of `B` can only be done within a specific level of Factor `A`. Nested factors therefore have no interaction. If a treatment (Factor `A`) is applied to a plant and 4 leaves (Factor `B`) are measured, then Factor `B` is nested in `A` as the variation can be described as having two sources: due to treatment level and to the plant-specific leaf level. A general leaf effect cannot be observed as it depends on which plant (Factor `A`) was used.

Often the experimental unit is also considered as a treatment factor. This factor is usually nested inside another treatment factor. For example, the pot experimental unit is nested inside the fertilizer treatment factor as for each type of fertilizer different pots are used. Sometimes the experimental unit factor can be crossed, for example in a cross-over design where each subject (experimental unit) takes all different diets (levels of the diet treatment factor) in different order, to correct for individual variation.

In the fertilizer example there was randomization of pots inside the blocks, but not over the blocks. This means that within each block, it is made sure that each type of fertilizer has the same number of pots. The blocking structure thus puts a restriction on the randomization structure and, therefore, it changes the way of estimating the experimental variability. In models that describe the effects of different treatments on a response, usually, a distinction is made between factors that describe how the intended variation, *e.g.*, the different types of fertilizer, is distributed over the experimental units, and factors that describe the blocking and randomization structure that is needed to obtain homogeneous variation for the experimental units. The distinction between these factors is further discussed in Chapter 3.

For each design, the different factors of the experimental part and the treatment part will have a crossed or nested structure. This needs to be clarified and considered when analyzing the data they provide. Many predefined designs exist. The simplest design is the Completely Randomized Design (CRD) (Casella, 2008). For this design, no blocking structure is present, and the random allocation of treatments to the experimental units is not restricted in any way. An easy extension to situations where blocks are needed to correct for heterogeneity in the experimental units is a Randomized Complete Block design (RCB) (Casella, 2008) where all treatments are applied in equal amounts within each block. In this book, we are mainly dealing with quantifying the effect of different treatments in the treatment design. If also the part that deals with the blocking and replication structure is taken into account, this will be mentioned specifically.

The unbiased estimation of treatment effects depends on how the experiments are performed with respect to blocking, randomization, and replication as well as whether treatments are crossed or nested and whether their effects are considered fixed or random. Each combination of these selections gives rise to a different model, and the effects can be calculated with ANOVA or linear models. In ANOVA, F-tests are used to calculate the significance of treatment effects, which are ratios of the variation explained by the treatments in the numerator and the residual variation in the denominator. The treatment design gives information on the degrees of freedom (related to the number of levels used for each factor), while the experimental design gives information on the expected mean squares of the residual variation and its degrees of freedom. More details will be provided in Chapter 3.

1.3 High-Dimensional Data

Already the title of this book refers to the biggest hurdle to take: high-dimensional data. There are very many ways that measurements can become high-dimensional in the life sciences. In omics research, it is not unusual to have thousands of variables (or features) measured per sample. This happens in RNAseq (orders of 10,000–30,000), in methylation (orders of 50,000–100,000), and in metabolomics, proteomics, and microbiomics (orders of 1000–10,000). Also vibrational spectroscopy generates many variables. One of the first problems of high-dimensional data is the curse of dimensionality, see Elaboration 1.1.

ELABORATION 1.1

Curse of dimensionality

The curse of dimensionality refers to the fact that given a certain sample size, increasing the number of measurements per sample will decrease the density of the sample points in the row-space (in which the samples are points) dramatically. This is illustrated in Figure 1.6. Suppose that we have 20 samples and consider the number of samples in a unit bin. In Figure 1.6a, we see that the unit bin contains half of the samples on average. By adding a dimension (Figure 1.6b), now a quarter of the samples are in a unit bin. Going up to three dimensions decreases the number of samples in a unit bin to about one-eighth (Figure 1.6c). Hence, in the limit, most unit bins contain one or even zero samples.

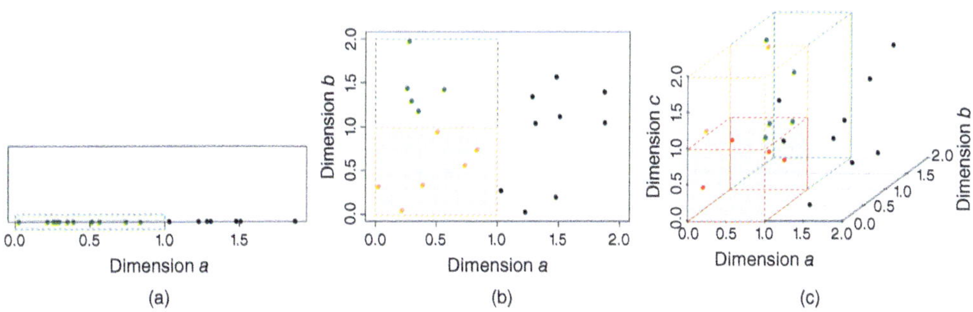

Figure 1.6 The curse of dimensionality. For an explanation, see the text. Source: Reproduced from Parsons *et al.* (2004)/with permission of ACM (The Association for Computing Machinery).

The curse of dimensionality, as presented above, represents the worst-case scenario since it does not account for correlation. Suppose for example that we perform near-infrared (NIR) measurements on a chemical system where around 200 variables are measured per sample. Assume that the chemical system contains three absorbing constituents and that Beer's law holds (Christian and O'Reilly, 1988). Then the data set has rank three, apart from some noise which is low in NIR spectroscopy. Stated otherwise, there is a high amount of correlation between the variables in such data.

Obviously, the characteristics of a data set in terms of the curse of dimensionality depend on the system and the type of measurements performed. Usually, omics data (RNAseq, metabolomics, proteomics and microbiomics) are not of low rank, have relatively high noise levels, and are measured on complex systems. Chemical systems and spectroscopic measurements are less complex in terms of rank and measurement error.

Another aspect to consider is the relevance of the measured variables. In omics data, it is not unusual that only a small number of the variables are really important for the research question. This calls for variable selection, which is not an easy task in high-dimensional data. Another route to take is using regularization approaches that invoke sparseness of the solution, thereby reducing the number of variables to (hopefully) the most relevant ones (Tibshirani, 1996). This selection of relevant variables is exacerbated when there is grouping in the samples and for each group a different set of variables is relevant (Friedman and Meulman, 2004). Yet, variable selection or regularization approaches are not trivial for certain kinds of high-dimensional data.

One of the mainstream methods for analyzing multivariate responses obtained through a statistical design is multivariate analysis of variance (MANOVA) (Mardia *et al.*, 1979). Although it may work for skinny data (see Figure 1.5a) this method breaks down for high-dimensional data, see Elaboration 1.2.

ELABORATION 1.2

Covariance matrix of high-dimensional data

Consider an experiment where one factor is varied at K levels and J measurements are performed on I_k samples at each level k. Many data analysis methods calculate a pooled within-sum-of-squares matrix $\mathbf{W}$, and this matrix needs to be non-singular since it is used in subsequent calculations in the form of $\mathbf{W}^{-1}$. This pooled within-sum-of-squares matrix is obtained as a weighted average of the within-sum-of-squares matrices per group which are indicated by $\mathbf{S}_k(J \times J), k = 1, \ldots, K$.

If $J > I_k$ each matrix $\mathbf{S}_k$ has maximum rank $I_k - 1$.[1] An upper bound on the rank of $\mathbf{W}$ is then $\sum_k (I_k - 1)$ (Schott, 2016). For high-dimensional data, this rank is still (much) lower than the number of variables J so that $\mathbf{W}$ is rank-deficient. Solutions for this problematic issue are found in the use of low-dimensional approximations of the original data or shrinkage approaches for the covariance matrix (Schäfer and Strimmer, 2005).

1.4 Examples

In this section, we will give some examples of high-dimensional data from different fields that are collected according to an underlying treatment design. This serves to show the growing abundance of such data structures. A more detailed review of papers using an HD-ANOVA method is given in Chapter 7.

1.4.1 Metabolomics

Metabolomics is a field of growing interest in the life sciences. It concerns the measurements of the level of small biochemical compounds in biological systems. The term metabolomics may refer to the measurements themselves, mostly performed by advanced instrumental analysis (see Elaboration 1.3), but also to the subsequent data analysis and biological interpretation. In our case, we are concerned with the processing of the data generated in a metabolomics experiment. Typically, these data need to be preprocessed to arrive at so-called clean data. This preprocessing is meant to remove batch effects, noise, and other artifacts that affect the data and depend on the setup of the experiment and the instrument used. There is a host of literature on data preprocessing of metabolomics data, and we will not discuss this in this book. Hence, we will assume that we have clean data albeit that these data may still have problems (see Section 1.5).

ELABORATION 1.3

Metabolomic measurements

Depending on the type of application, typically metabolomics samples (*e.g.*, blood plasma or plant material) need some sample work-up. After that, they can be analyzed by an instrumental technique. The three techniques most in use are NMR, LC-MS, and GC-MS. Nuclear Magnetic Resonance (NMR) is the fastest and easiest to perform, but it has a low sensitivity and limited selectivity. With a limit of detection of about 1–5 μM, the concentrations of the metabolites need to be reasonably high, and thus the number of separate metabolites that can be distinguished is relatively low (in the order of 100–200) (Wishart, 2008). Furthermore, the amount of sample needed for the measurement is rather large (~500 μL). Liquid-chromatography (LC) and gas-chromatography (GC) coupled with mass spectrometry (LC-MS and GC-MS) are much more sensitive and selective. Usually, these methods are either geared toward measuring a predefined class of compounds (targeted analysis) or are used to cover a large range of metabolites (untargeted), of which most are then unidentified. In the latter case, preprocessing the raw data to arrive at clean data is really cumbersome. Also the LC-MS and GC-MS analyses need accurate sample work-up and are more elaborate to perform than NMR.

1 These matrices are calculated on centered data.

Metabolomics is used in a wide variety of applications in the life sciences; for a review see Chapter 7. It is important to distinguish the different organismal levels that can be subjected to a metabolomics experiment. Such experiments can be performed at the cellular level, such as in biotechnology and human cell-lines. In such cases, biological interpretation of the results may be phrased in terms of the metabolic network of the cellular systems. In most cases, in human and animal studies, the metabolome of body fluids is measured, such as in blood, urine, saliva, fecal fluid, or cerebrospinal fluid (CSF). Clearly, in such cases, biological interpretation is more difficult since there are no direct relationships of such biological compartments with cellular metabolic networks. An intermediate level at which measurements can be performed is in tissue. This can be muscle tissue from humans and animals or specific tissues from plants, *e.g.*, the leaves or roots; this may also include a spatial dimension to the data. An example of metabolomics using the analysis of variance simultaneous component analysis (ASCA) method (see Chapter 4) is given in Example 1.1.

Example 1.1: Metabolomics analysis of cell differentiation

In the field of cell differentiation, specifically, the differentiation of types of stem cells into hepatocytes, metabolomics measurements can be used to quantify the different stages of the differentiation (Moreno-Torres *et al.*, 2022). Figure 1.7 shows the treatment design of the study. Briefly, during a time-course, stem cells are stimulated in different ways and turn into hepatocytes. Along the way, they are also subjected to different renewed *Medium compositions*. Metabolites are obtained at different *Days* and measured with LC-MS.

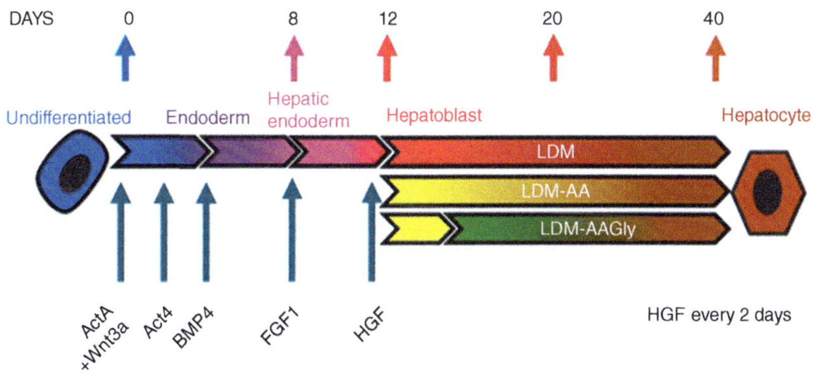

Figure 1.7 The treatment design of the hepatocyte differentiation study. Details of the cell differentiation strategy and assessment of the influence of transcription factors on the cell's metabolome. Cells are treated with different *Medium compositions* (*LDM*, *LDM-AA*, and *LDM-AAGly*) and analyzed at different *Days* during the differentiation process. For detailed explanation, see the text. Source: Reproduced from Moreno-Torres *et al.* (2022)/with permission of American Chemical Society.

An ASCA model (see Chapter 4) was used to explore the metabolite differences due to the different *Cell types*, the different *Time points*, and the different renewed *Medium compositions*. The results are shown in Figure 1.8.

The figure shows spider score plots that connect the experiments of the same treatment level. In spider plots, the center of the spider shows the overall effect size and the arms represent

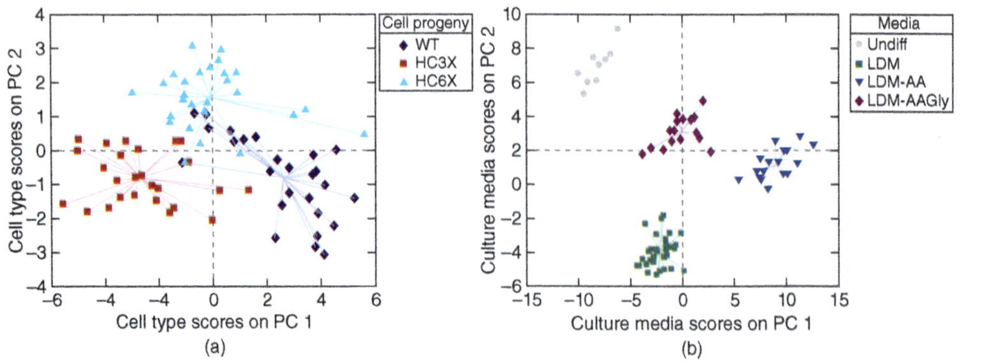

Figure 1.8 The ASCA results of the hepatocyte study: (a) cell type scores plot with different types of cells, *WT*, *HC3X*, and *HC6X*; (b) culture media scores plot with different *Media*, *Undiff*, *LDM*, *LDM-AA*, and *LDM-AAGly*. These are spider plots in which each center of the spider shows the overall effect size and the arms represent individual measurements. Source: Reproduced from Moreno-Torres *et al.* (2022)/with permission of American Chemical Society.

individual measurements, (see Section 8.6.1.3). The position of a measurement represents a summary of the levels of all metabolites. Dots in the same color represent replicates for the level of the treatment indicated. In Figure 1.8a the three cell types are indicated by color, and a clear difference can be observed between the three cell types. In Figure 1.8b we can see that the metabolite levels are rather different at the different time points. The larger the separation between the levels, the more different they are. Besides the main factors shown in Figure 1.8, significant interactions between the *Cell type* and time as well as between *Medium compositions* and *Time* were reported.

1.4.2 Genomics

Genomics is a huge field and it encompasses many aspects of the genome such as the makeup of the DNA, epigenetics (*e.g.*, methylation patterns of the DNA), and gene expression. HD-ANOVA analyses are mainly applied to gene-expression measurements, which are performed with micro-arrays and modern sequencing methods, especially RNAseq. These types of measurements can easily produce thousands of variables for a single sample. One of the complications with modern RNAseq measuring devices is that they need to be preprocessed prior to analysis because there are differences in library sizes, or sometimes referred to as sequencing depth, *i.e.*, the number of sequenced bases in the samples. A normalization step is necessary to make the samples comparable (Evans *et al.*, 2018) (see Section 1.5.1). There are many options and no clear consensus on how to do this. Normalization is one remedy for a type of incomparability which will be explained in more detail in Section 1.5.5, and indeed these types of problems are also encountered in other omics measurements such as in metabolomics (Uh *et al.*, 2020) and microbiome data analysis (Lin and Peddada, 2020). An example of the use of ASCA for RNAseq data is given in Example 1.2. For more applications on the use of HD-ANOVA methods, see Chapter 7.

Example 1.2: Gene-expression analysis of apples

Soft-scald is an injury often observed for "Honeycrisp" apples (Leisso *et al.*, 2016). Apples of two different categories of *Risk* for developing soft-scald were stored for different periods of *Time*. This results in two factors that can be studied: *Risk* category and storage *Time*. RNAseq measurements were performed on the fruit, and an ASCA model was used to study the effect of the factors involved. Those factors did not only affect the soft-scald independently, but there appeared to be a strong interaction of those two factors as shown in Figure 1.9 indicating an early response and a delayed response.

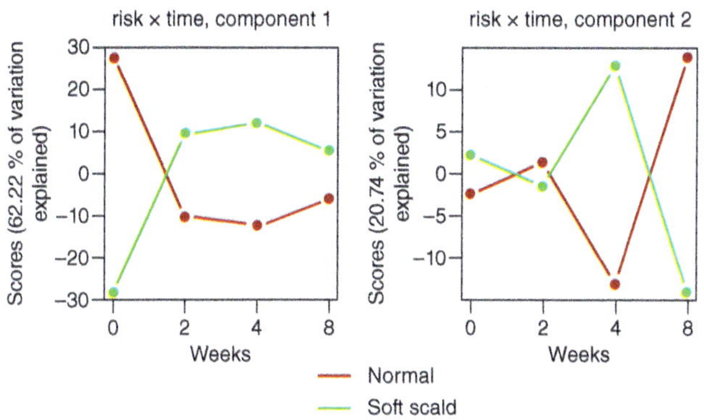

Figure 1.9 The ASCA results of the (*Risk:Time*) interaction effect of the apple study. Source: Leisso *et al.* (2016)/Springer Nature/CC BY 4.0.

Apart from the RNAseq measurements, metabolomics measurements were performed which were also subjected to an ASCA analysis. For more details regarding this study, we refer to the original publication (Leisso *et al.*, 2016).

1.4.3 Microbiome

A microbiome is a community of microbes located in a certain niche. These microbes may contain, *a.o.*, bacteria, fungi, and viruses. Microbiome research has been exploding in the last decade. The topic is also very broad since microbiomes are literally everywhere. Most papers deal with the human gut-microbiome, but there are also papers about the oral microbiome and skin microbiome. Also in other areas of biology, microbiomes are being studied, *e.g.*, in plants' roots, soil, and water. There is a growing awareness of the importance of these microbiomes, *e.g.*, in their relation to health in general.

As in the case of metabolomics, a strong driver in microbiome research has been the development of measuring devices. Whereas in the early days, HITChips were being used (Rajilić-Stojanović *et al.*, 2009), nowadays advanced sequencing techniques are available to measure the microbiome such as 16S rRNA sequencing (amplicon sequencing) and metagenome sequencing (Janda and Abbott, 2007). The measured microbiome variables can be expressed in Operational Taxonomic Units (OTUs) or in Amplicon Sequence Variants

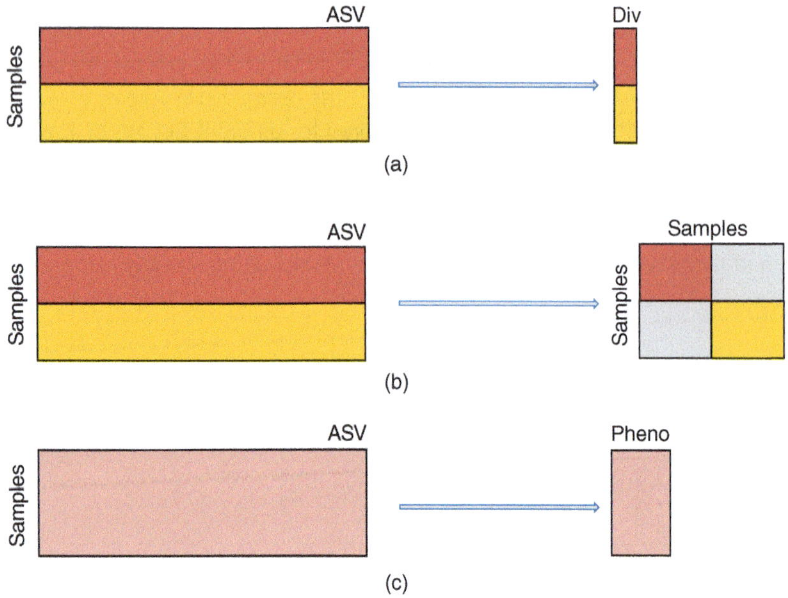

Figure 1.10 Three different ways to deal with microbiome data: (a) calculate the alpha diversity of the whole microbiome (all ASVs) and express this in a single number per sample (Div); (b) calculate distances between the microbiome samples and work with a sample to sample distance matrix. The different colors in the data blocks indicate a grouping structure or statistical design. The gray boxes indicate the distances between samples of different groups which are usually larger than the distances between samples of the same group (indicated by the red and yellow boxes); (c) work directly with the ASV data to predict a phenotype.

(ASVs), which are counts for a given observed DNA sequence. Microbiome data are intrinsically different from metabolomics data, because the first represents compositions whereas the second represents concentrations, see Elaboration 1.4. In some microbiome applications, there is a natural grouping in the subjects, and in some cases, *e.g.*, in the plant sciences, there is a statistical design underlying the samples. Hence, in these cases, the use of HD-ANOVA methods is warranted. In the majority of cases, there are three different routes followed for microbiome data analysis. These are shown in Figure 1.10.

ELABORATION 1.4

Composition versus concentrations

Microbiome data are conceptually different from metabolomics data. In metabolomics, we measure concentrations that only describe a very small part of the whole sample: the amount of water or other fluid in which the metabolites are measured is considered as a background and not quantified. In contrast, in microbiome data, the whole composition of the sample is measured, not only a part. This makes microbiome data by definition compositional, which means that if the level of one of the species increases, other species have to decrease due to the compositional constraint that the total number of species counts is constant. This has repercussions for subsequent analysis of the data. This is discussed in more detail in Section 6.1.3.

In the first route (see Figure 1.10a), alpha diversity calculates a property for each sample based on how many species are present (richness) or how equal the abundances of the different species are (evenness). Differentiating between the two groups (indicated with different colors in the figure) then translates into the problem of comparing two vectors of values which can easily be done with a regular t-test between the means of both groups (or a non-parametric form of this).

The often-selected second route (see Figure 1.10b) is to use beta diversity, which focuses on the between-community differences resulting in a distance matrix that can then be subjected to a PERMANOVA (see Chapter 6) or Principal Coordinate Analysis (see Elaboration 1.5). Note that the data are now expressed in terms of a distance matrix between the samples. Such an approach is called an indirect analysis in the context of multivariate and high-dimensional data analysis.

ELABORATION 1.5

Principal Coordinate Analysis

Principal Coordinate Analysis (PCoA) is often used for ordination analysis in ecology and also in microbiome data analysis (Legendre and Legendre, 2012). PCoA uses a sample-to-sample distance matrix, where the specific distance measure chosen relates to the type of data and type of interpretation, *e.g.*, whether the abundance of species is relevant or only its presence. The use of the Euclidean distance is often discouraged for ecological data as microbiome count data has a complex distribution with many zeros and variances that depend on the number of counts. Therefore, multivariate explorative analyses of sample-to-sample distances in microbiome analyses often use distance measures that take the presence/absence aspect as well as the quantitative aspect into account.

The third route takes the data as is and tries to relate the ASV's directly to the phenotype. This opens up possibilities to use regression techniques or sparse versions of those. In this case, some kind of normalization of the data is required because the number of reads per sample (*i.e.*, the total measured ASV abundances) are different across samples; a similar problem is encountered in RNAseq data for gene expression (see Section 1.4.2).

All routes are extensively used in microbiome and ecological research (Legendre and Legendre, 2012), and they all have advantages and disadvantages. Expressing a whole profile in one number may result in losing a lot of information. In the second route, a distance measure has to be chosen and there are many of those. Although often the Bray–Curtis distance is chosen (see Section 2.1.4), this may be more a matter of custom than based on strong scientific grounds. Moreover, the second approach is a so-called indirect analysis (see Section 1.6). A major disadvantage of indirect analysis is that the importance of individual features is not easily established since distances are calculated across features. The third approach may be hampered by an excessive amount of zeros and needs some kind of normalization or transformation. An example of using ASCA in microbiome research is given in Example 1.3. The microbiome data analysis field is still dominated by approaches from ecology, and ASCA and related methods are not yet used that much. For a review, see Chapter 7.

Example 1.3: Microbiome analysis of bioreactor fermentations

Microbiota of several subjects are used as starting material for fermentations at two inoculation conditions (*Dilutions* C and D), and samples were taken at several *Time* points (Raimondi *et al.*, 2021). From these samples, the microbiome sequences were measured with 16S-rRNA gene profiling, and also the volatile organic compounds (VOCs) in the headspace of the cultures were measured using GC-MS. A scheme of the experimental setup is shown in Figure 1.11.

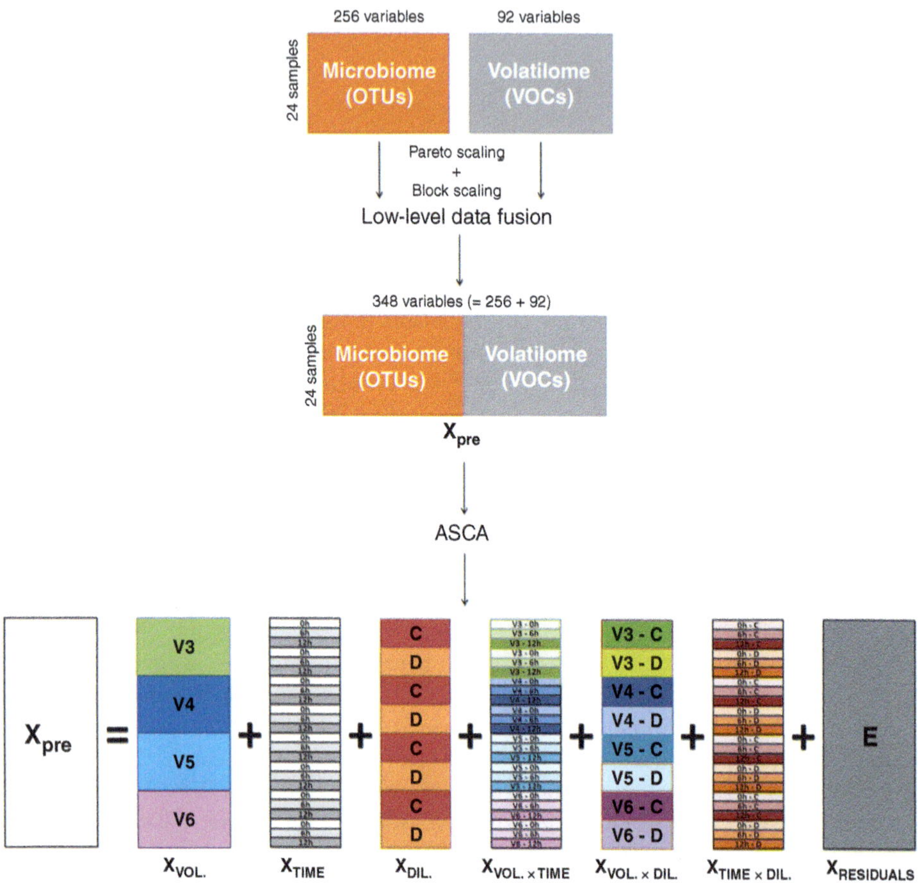

Figure 1.11 Schematic representation of low-level data fusion of microbiome and volatilome data, and ASCA decomposition of the fused dataset. Source: Raimondi *et al.* (2021)/Frontiers Media S.A/CC BY 4.0.

This is an advanced application where apart from using ASCA also a fusion between microbiome and metabolome data is performed. The results show significant effects of the factors *Subjects* (or volunteers), *Time*, *Dilution* and their interactions on the microbiome only, see Table 1.1. Note that the *Subjects* are treated as a fixed effect in this analysis, and an alternative might be to treat the subjects as random effects.

The main conclusion that can be seen from this table is that the most contributing factor to the variation in the data is the subject factor. This is typical for human-based studies; the individual variation is usually large.

Table 1.1 Effect sizes and significance in the microbiome example.

Factor	Effect (%)	*p*-value
Subject	30.8	0.001
Time	21.2	0.001
Dilution	6.6	0.001
Subject:Time	20.5	0.001
Subject:Dilution	8.5	0.001
Time:Dilution	4.2	0.007
Error	8.2	

Source: Raimondi *et al.* (2021)/Frontiers Media S.A/CC BY 4.0.

1.4.4 Proteomics

Proteomics deals with measuring proteins in biological systems. As with metabolomics, proteomics has taken advantage of many developments in instrumentation, notably in mass-spectrometry techniques. Similar to metabolomics, proteomics encompasses not only the measurements of the levels and identity of many proteins, but also the study of the role and behavior of these proteins. An example of using ANOVA–PCA (APCA) (see Chapter 4) is given in Example 1.4, and more references are given in Chapter 7.

Example 1.4: Proteomics measurements of Escherichia coli strains

Four *Strains* of *E. coli* were used at four different *Growth ages* to study the reproducibility of MALDI-MS measurements (Chen *et al.*, 2008). The data was analyzed using APCA. In APCA, the total variation in the data is partitioned into sources defined by the treatments in the treatment design. This provides a decomposition of the data matrix into its separate contributions, similar to the decomposition visualized in Figure 1.11. The variation due to a single treatment is called the effect matrix for that treatment. In APCA, a PCA is performed on the addition of an effect matrix and the matrix of residuals, generating plots like in Figure 1.12. The residual matrix is obtained after removing the explained variation due to the factors and interaction and represents the within-group variation between the replicates of each group.

In Figure 1.12a, it is clear that *Strain B − 14948* behaves differently from the other *Strains*. In Figure 1.12b we can see that the *A − D1 Growth age* is separated from the other levels. The average difference between the groups per PC, *e.g.*, 5.31 and 0.752, is calculated as the projected difference resolution and represents a geometric mean of the resolution of all pairs of groups per factor.

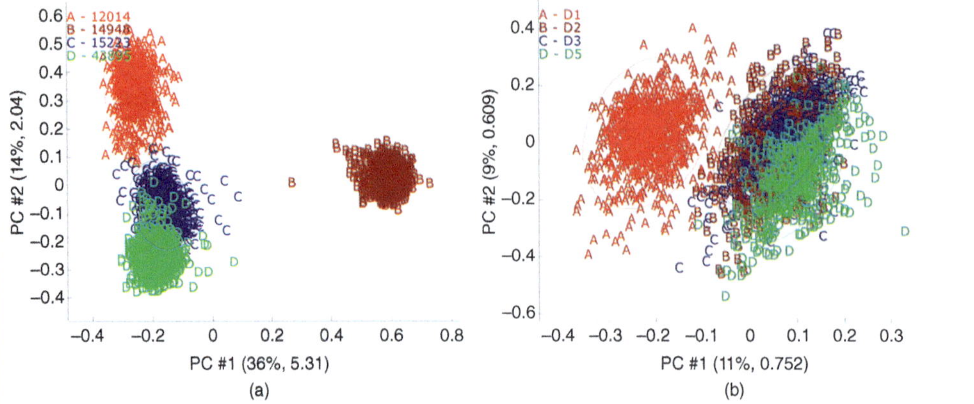

Figure 1.12 The APCA results of the *E. coli* study: (a) score plot of the matrix resulting from the addition of the residuals to the effect matrix of the *Strains* factor (see labels); (b) score plot of the matrix resulting from the addition of the residuals to the effect matrix of the *Growth age* factor (see labels). Source: Reproduced from Chen *et al.* (2008)/with permission of American Chemical Society.

1.4.5 Food Science

Food science is a very broad field where traditionally many measurements are performed, and sometimes high-dimensional data are produced, *e.g.*, using NIR. Moreover, also in many cases, there is an underlying statistical design in such studies. Hence, HD-ANOVA methods are increasingly being used in this field (see also Chapter 7). An example of ASCA in the field of food science is given in Example 1.5.

Example 1.5: Analysis of pasta

The effect of *Drying time, Drying temperature*, and *Egg concentration* on the quality of pasta was explored by Bevilacqua *et al.* (2013). NIR spectroscopy was used to measure the pasta produced at different experimental conditions, and ASCA was used to analyze the data. Figure 1.13 shows the results for the different levels of the factor *Drying temperature*. Two approaches of validation are visualized. Figure 1.13a visualizes the result of a permutation test. In the ASCA permutation test (see Section 4.1.3) the samples are randomly assigned to a level of the *Drying temperature*, and the sum-of-squares of explained variation is recalculated. The histogram shows that the original data (red dot) has a much higher sum-of-squares (1200) than any of the permuted results indicating that a significant *Drying temperature* effect exists. Figure 1.13b uses back-projection (see Section 4.1.4 and Zwanenburg *et al.* (2011)) to give an indication of whether the between-group variation is large compared to the within-group variation. Here, a PCA is applied to the sum of the effect matrix and residual matrix, and the scores indicate that the variation between the groups is large compared to the spread of the samples from the same group.

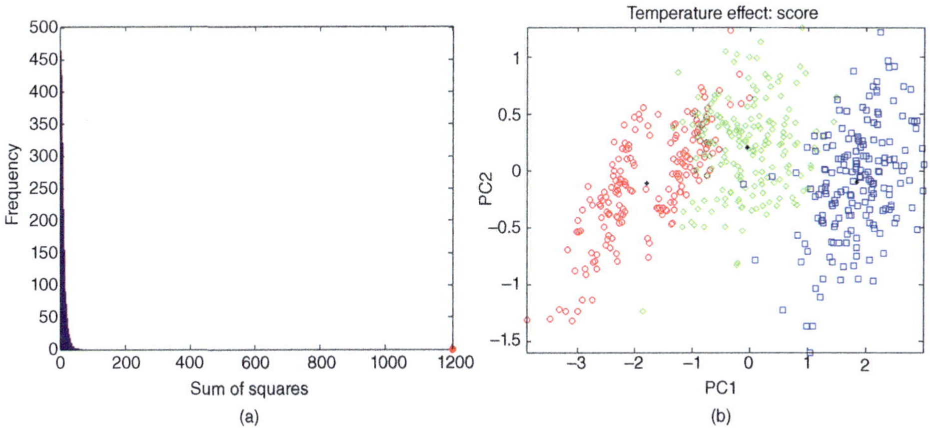

Figure 1.13 The ASCA results of the pasta study: (a) the permutation results of the factor *Drying time*. The blue histogram shows the distribution of the sum-of-squares under the null hypothesis (no effect), and the red dot at a sum-of-squares of 1200 indicates the estimated effect in the data; (b) score plot of the effect of *Drying temperature* after back-projection of the residuals. The *Drying temperature* levels of *40 °C* (red), *52 °C* (green), and *65 °C* (blue). Source: Reproduced from Bevilacqua *et al.* (2013)/with permission of Elsevier.

1.4.6 Sensory Science

In sensory science, the measurements can take on many forms, *e.g.*, assessors scoring attributes of various products or chemical measurements of food products. Sometimes, there is also an underlying statistical design where the factors can be the products, the assessors, or factors changed during the processing of the food products. An example of ASCA in sensory analysis is given in Example 1.6 based on Luciano and Næs (2009); this paper also discusses the difference between ASCA and PC-ANOVA (see Section 5.4.1.2). A review of applications is given in Chapter 7.

Example 1.6: Sensory analysis of candy

Luciano and Næs (2009) explored the sensory differences between five *Candies* judged by 11 *Assessors* on 9 sensory attributes. Besides looking at the differences between the *Candies*, such a study also provides information on the similarities and differences between the assessors. This is visualized in Figure 1.14.

From the score plot, we can see that *Assessors 5* and *7* are clearly deviating probably due to their assessment of acid and sweet (see Figure 1.14b). Assessors *2*, *4*, and *10* are close together, and also the remaining *Assessors* seem to form a cluster. The interpretations of such ASCA score and loading plots are the same as for "normal" PCA.

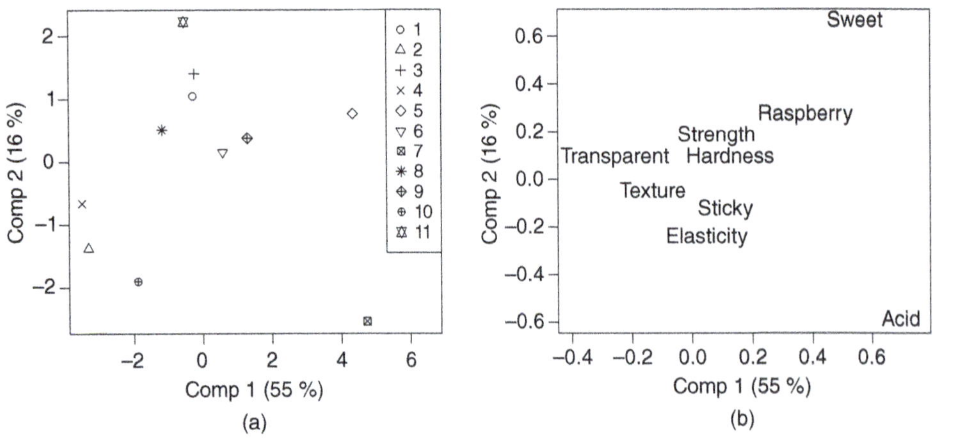

Figure 1.14 The ASCA results of the sensory study: (a) visualizes the scores of a PCA on the *Assessor* effect matrix; (b) shows the corresponding loadings of the sensory attributes used. Source: Luciano and Næs (2009)/with permission of Elsevier.

1.4.7 Chemistry

Most of the applications of HD-ANOVA methods can be found in analytical and environmental chemistry (see Chapter 7), because in those fields it is customary to perform high-dimensional measurements using MS, NMR, and vibrational spectroscopy. An illustration from the field of analytical chemistry is provided in Example 1.7.

Example 1.7: Analysis of stationary phases for HILIC chromatography

Hydrophilic interaction chromatography (HILIC) is an important method for analyzing polar compounds in the field of analytical chemistry. Navarro-Reig *et al.* (2017) investigated different types of *Stationary phases* for HILIC chromatography. They used a treatment design for measuring metabolite concentrations, using different *Stationary* and *Mobile phases* (with different pHs and ionic strengths). The resulting effects of the *Stationary phases* are shown in the loading and score plots in Figure 1.15.

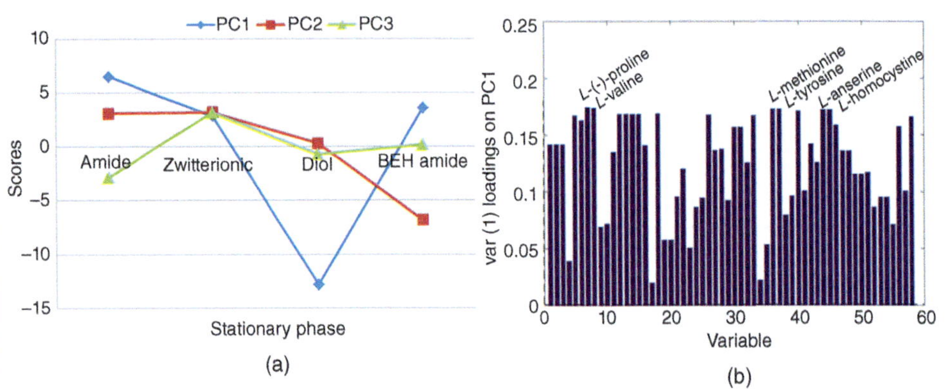

Figure 1.15 The ASCA results of the HILIC study: (a) the scores of a PCA of the effect matrix corresponding to the *Stationary phases;* (b) the corresponding loading for PC1 of the PCA analysis. Source: Navarro-Reig *et al.* (2017)/MDPI/CC BY 4.0.

The scores of PC1 in Figure 1.15a show that the *Diol stationary phase* was dissimilar from the other ones. This was mostly due to the effect of the *Diol stationary phase* on six indicated amino acid concentrations as presented in the PC1 loading plot in Figure 1.15b.

Another chemical example is presented in Example 1.8. Here, ASCA is used in the context of environmental and pollution science to support the development of a phytoremediation strategy.

Example 1.8: Microbially assisted phytoremediation of soils

The assessment of the applicability of a "toolbox" for microbially assisted phytoremediation of soils contaminated by mining activities was reported by Sprocati *et al.* (2014). A field trial was conducted, using *Euphorbia pithyusa* as a pioneer plant and investigating the effect of different treatments used singly or in combination (bioaugmentation with *Bacterial consortia, Mycorrhizal fungi*, and *Viromine*, a commercial mineral amendment) on soil and plant parameters. A set of around 50 variables, including total and diethylenetriaminepentaacetic acid-extractable amounts of selected metals, concentration of specific metabolites, activities of three hydrolytic enzymes (arylesterase, acid and alkaline phosphatase), pH, carbon and nitrogen content, and average well color development, were measured. Only the effect of the supplementation of *Viromine*, and its interaction with either *Bacteria* or *Mycorrhizal fungi* were found to be significant. The scores and loadings on the first component for the matrix resulting from the combination of the main effect of *Viromine* and its interaction with *Bacteria* are shown in Figure 1.16.

The results shown in Figure 1.16 indicate that *Viromine* promotes an increase of pH, organic carbon, and total carbon content in the soil together with the capacity of metabolizing xylose and glucose-1-phosphate; average well color development (AWCD) also increases while metal mobility decreases.

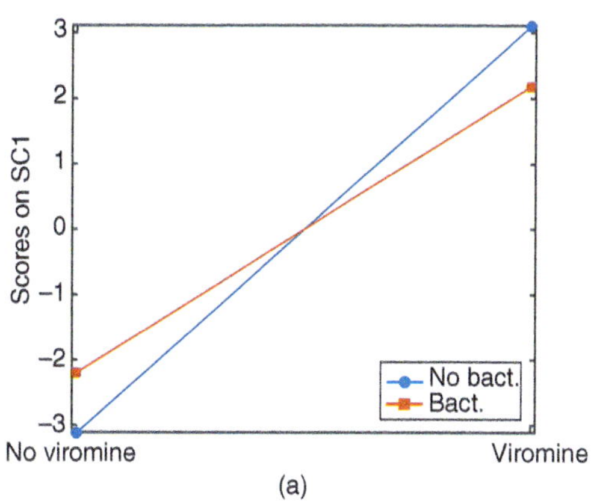

Figure 1.16 The ASCA results of the phytoremediation study: (a) scores of PC1 for the combined effect of *Viromine* and its interaction with *Bacteria* on the y-axis vs the level of *Viromine;* (b) the corresponding loadings on PC1 together with their confidence intervals obtained by bootstrap analysis (y-axis) of the variables. Source: Adapted from Sprocati *et al.* (2014).

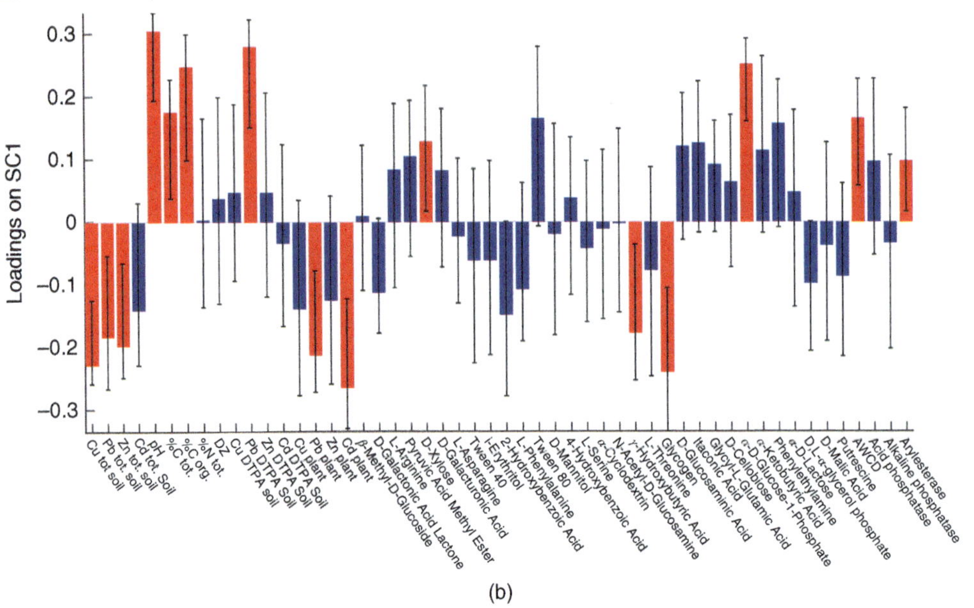

Figure 1.16 *(Continued)*

1.5 Complexities

The data measured in the life sciences often have specific properties that do not match the assumptions of HD-ANOVA methods. We already discussed the problem of high-dimensionality, but there are many others as well. In this section, we will discuss these complex properties and provide some solutions.

1.5.1 Normalization

In many of the previous examples, issues arise about the comparability of samples. In metabolomics, the concentration of urine samples can be very different at different times of the day, and this makes it difficult to compare metabolite concentrations of such different samples. In genomics and microbiome analysis or metagenomics, RNAseq measurements can have different numbers of total counts over the whole sample, which also makes the number of counts per feature incomparable between samples. Normalization is a data pretreatment step to make the values of variables for different samples comparable. Several reasons can make samples incomparable, *e.g.*, different dilution of urine samples as a result of how much individuals drank or the time of the day that the sample has been taken, different library sizes in RNAseq analysis, differences due to the use of different measurement batches.

Several approaches can be used to correct for these differences:

- An analyte that is considered to be present in equal amounts in each sample can be used as an internal standard (*e.g.*, creatinine in urine samples, or household genes in RNASeq samples). Note that this internal standard is already present in the sample and can be used to correct for dilution differences in urine samples.
- Batch effects can be corrected for by using it as a cofactor in linear models. In metabolomics, a fixed amount of an internal standard is usually added to each sample before sample workup

which (partly) corrects for differences. Note that since this internal standard is added to the sample it cannot be used to correct for dilution differences (see the first item on this list).
- Total-sum-normalization where the sum of the intensities of all variables in the samples are made equal across samples.
- Probabilistic Quotient Normalization in which the median or the most common (mode) ratio between a sample and a reference for all features is used as a dilution correction factor for that sample (Dieterle *et al.*, 2006).

Which method to use depends on the data characteristics such as the number of zeros, or whether or not it is expected that only a few features are important (Boshuizen and Te Beest, 2023). Normalization in RNAseq and microbiome sequencing data analysis is often used to remove any bias caused by differences in library sizes between samples. For microbiome studies (see Section 1.4.3), the problems are even more severe due to zero counts for many features in most samples. Various approaches are taken, but the underlying assumptions of these normalization approaches often make them not appropriate to be applied to such data (Boshuizen and Te Beest, 2023).

1.5.2 Different Measurement Scales

There is a whole multitude of different types of data that can be analyzed by HD-ANOVA-type methods. Reference should be given to Stevens (1946) who was the first to systematically treat the topic of scales of measurements. Most readers will be familiar with the typology as shown in Table 1.2. As an example, length can be measured in centimeters or in meters. Thus the permissible transformation is $x' = \alpha x$ $(\alpha > 0)$, where x refers to the original measurements (*e.g.*, in centimeters) and x' to the transformed measurement (*e.g.*, in meters). If all measurements of a system are changed accordingly, then the description of the system in terms of those measurements remains essentially the same: we can change the unit of the measurement. Of special importance is the last column of this table showing the limitations in terms of statistics to be used for the different scales. This has, of course, also repercussions for the HD-ANOVA method to use. This will be exemplified in the following chapters where needed.

1.5.3 Different Distributions

Apart from possibly having to deal with different measurement scales, there is another property of the data that needs to be considered: their distribution. There are two aspects to this: (i) the variability between replicate samples and (ii) the variability due to measurement error.

Table 1.2 Overview of types of measurement scales and their permissible transformations and statistics.

Scale-type	Example	Permissible transformations	Permissible statistics
Nominal	Categories	One-to-one	Number of cases, frequencies
Ordinal	Survey data	Monotonic	Median, IQR
Interval	Degree celsius	Positive linear transformation	Mean, Standard deviation
	Calendar time	$x' = \alpha x + \beta$ $(\alpha > 0)$	
Ratio	Length	Similarity transformation	Coefficient of variation
	Mass	$x' = \alpha x$ $(\alpha > 0)$	
Absolute	Counts	$x' = x$	All previous

Source: Reproduced from Smilde *et al.* (2022)/with permission of John Wiley & Sons.

The latter will be discussed in Section 1.5.4. Most measurements in the life sciences are non-negative; an exception is a difference measurement resulting from comparing a measurement with a reference which can result in (small) negative values. Apart from that, the distribution of replicate measurements of protein or metabolite levels tends to be skewed to the right and have log-normal-like distributions (Lu and King, 2009).

In many cases where measured (relative) concentrations are involved, transformations are used to make the data more symmetrically or normally distributed. For metabolomics, proteomics, and micro-array data, the logarithmic transformation is very popular because it solves the skewness of the distribution and corrects for the heteroscedasticity, but it runs into problems for values between one and zero. Here, the log-transformation transforms small concentration differences into large transformed values, *e.g.*, concentration values of 0.01 and 0.001 become −2 and −3). Thus log-transformation tends to blow up the measurement error (Durbin *et al.*, 2002). There are alternatives, *e.g.*, taking a square root transformation (Waaijenborg *et al.*, 2018).

RNAseq data tends to follow a Poisson or Negative Binomial (NB) distribution. Amplicon sequencing or metagenomics data of the microbiome tend to follow a Zero-Inflated Negative Binomial (ZINB) distribution (Yau *et al.*, 2003). Both types of data are sometimes subjected to a Centered Log-Ratio transformation (CLR) (Aitchison, 1994) to account for compositionality. However, after the CLR transformation, the transformed counts of the different features still depend on the geometric mean of the original counts, which also partly depends on the evenness of the composition, *i.e.*, how similar the abundances are in the composition (see Section 6.1.3). Values in different samples that are equal after CLR transformation might not have been equal in their original composition if the evenness of the samples was different. Thus the compositionality is not fully removed (Boshuizen and Te Beest, 2023). More on compositionality is discussed in Chapter 6. Table 1.A.1 in Section 1.7.1 gives an overview of the distribution of many of the discussed measurements. Obviously, when applying HD-ANOVA methods to the data the way of estimating the model parameters must be in line with the assumed sampling distribution.

1.5.4 Heteroscedastic Error

Apart from the distribution of the samples, we also have to consider the distribution of the measurements errors, since that may have repercussions for the estimation of the parameters in the model. Ordinary least squares (LS) estimation, *e.g.*, assumes errors to be homogeneously distributed around zero with equal variance (*i.e.*, homoscedastic).[2] The measurement error variance should not depend on the measured value. This is very rarely the case in practice. In many cases where chemical analysis measurements are used (*e.g.*, in metabolomics and proteomics), the measurement error variance tends to depend on the size of the instrumental read-out (Van Batenburg *et al.*, 2011). Stabilizing transformations can help alleviate this problem. Different approaches exist; *e.g.*, for micro-array data, arcsinh transformed intensities between the sample and a reference are used (Huber *et al.*, 2002), while in the case of RNAseq or 16S rRNA sequencing in microbiomics, which is considered to have a Poisson or Negative Binomial distribution, the variance stabilization is performed using a log transform (Love *et al.*, 2014).

Whereas problems related to the issues of distributional properties and heteroscedasticity of the measurement error hold in general for data in the life sciences, there is an additional issue at stake in the context of this book. Since this book deals with data collected according to a statistical design, there will be different experimental groups in the data according to the levels

2 It is often stated that LS estimation assumes normality. This is not true.

of the design factors. Most of the HD-ANOVA-type methods assume a similar distribution of the (repeated) samples within each experimental group. This is an assumption that should be checked in the data or has to be assumed *a priori*.

1.5.5 Comparability

When dealing with data, a fundamental question to ask is whether two numbers in a data set can be compared. Suppose, *e.g.*, that we have measured a 2 and a 4 in a data set. Then we can ask whether the ratio 4/2 has meaning, or whether their difference $4 - 2$ has meaning. This goes back to the problem of measurement scales. But are we even allowed to compare the numbers 2 and 4 anyway? Such fundamental questions are studied in the field of data theory (Coombs, 1964).

The question of comparability can be split into two concepts: (i) conditionality and (ii) lack-of-invariance. Conditionality concerns the aspect of whether the numerical values of two numbers can be compared. A good example is from RNAseq analysis where two counts for the same gene in two samples cannot be compared directly because of differences in library size. The data has to be normalized first, in order to make these counts comparable (see Section 1.5.1). Similar examples can be found in metabolomics measurements of urine; if the urine is more diluted, all the concentrations will be lower. Hence, this also needs normalization. In sensory analysis, different assessors may use the scale for measuring a sensory attribute differently leading also to conditionality problems. Data that can be compared within a sample are called row-conditional data. Similarly, data that can be compared across features (columns in a data matrix) are called column-conditional; and data that can be compared across a whole matrix are called matrix-conditional.

An even more fundamental problem arises when a variable across the samples changes its *meaning*. If a variable has a different meaning across different samples, there is a lack-of-invariance. An example is a measurement performed at a certain time point for a set of individuals. These can easily be put in a matrix with the individuals as samples and the time points as variables, but why should there be a relation across individuals for the same time point? Obviously, this depends on the setup of the study. It is important to realize that decisions regarding lack-of-invariance are not statistical decisions but domain-substantive decisions. For a more detailed explanation, see Van Mechelen and Smilde (2011) and Smilde *et al.* (2022).

1.5.6 Sparseness, Non-detects, and Missing Values

Depending on the type of data, there can be many different characteristics making those data complex. The first complication is sparseness. Some types of data can be very sparse (with very many zeros or below detection values). An example is microbiome data that can easily contain 80–90% zeros (Lutz *et al.*, 2022). These zeros can be the result of different phenomena and thus could have different interpretations (Silverman *et al.*, 2020) and thus should be treated with care (Jiang *et al.*, 2022). Oftentimes some filtering is necessary to remove ASVs with very many zeros (Abegaz *et al.*, 2024). It is sometimes argued that indirect analysis of microbiome data circumvents the problem of the zeros (since distances are calculated across ASVs), but this depends on the type of distance measure chosen. Another alternative that reduces the number of zeros is to aggregate the ASVs in a larger family, *e.g.*, at the genus level. But this hampers interpretation at the species level.

A problem often encountered in metabolomics is non-detects, which are sometimes reported as zero or as missing. Such non-detects can also be the result of different situations: (i) they can represent a real concentration of zero (*i.e.*, the metabolite is not present), (ii) they can represent a very small concentration below the limit of detection, or (iii) they can be artifacts due to the pre-processing. The last reason can sometimes be discovered by carefully checking the pre-processing of the data. The first two reasons cannot be distinguished, and biological reasoning has to be used to make a decision, *e.g.*, this metabolite can never be present in this sample. Methods that can deal with the problem of zero or missing data problem stem mostly from methods dealing with left-censored data.

The final complication to be discussed in this section is the existence of missing data. This is quite often encountered in real-life data, and missingness can have different backgrounds. Missings can be considered to be "not at random," "at random" or "completely at random." Excellent textbooks exist explaining these different types of missing data in detail (Little and Rubin, 2019). Depending on the type of missingness, various methods exist that can deal with this. Wherever encountered in this book, we will explain how the missingness is treated. Sometimes the zeros in microbiome data or in metabolomics data are treated as missing data, which is questionable. A zero value means the level measured is zero or below the detection limit, while missing means the level is unknown due to whatever reason (Silverman *et al.*, 2020).

1.5.7 Unbalancedness

Since this book deals with data collected according to an underlying statistical design, we also have to deal with the issue of unbalancedness, *i.e.*, when the number of samples in the different experimental groups differ. Already for standard ANOVA, this is not trivial (Searle, 1971), let alone for the high-dimensional case. For mild unbalancedness, the standard methods can still be used but, for severe unbalancedness, different approaches have to be used (Miller Jr, 1997). One of the ad hoc remedies is to subsample (or downsample) the "richest occupied" experimental cell in the design, to the level of samples in the "poorest occupied" cell, which is sometimes repeated several times to study the variability induced by this step.

Another method to use is imputation, where the "poorly occupied" groups are enriched with imputed samples to make them more similar to the "richly occupied" groups in terms of available replicates (Little and Rubin, 2019). The quality of this method depends crucially on the quality of the imputation which in turn has to be based on distributional assumptions which may or may not hold.

Yet another method is to work with cell-means where the replicates within each experimental group are averaged to obtain one single observation. Then in the subsequent analysis, the differences in variances of these group-means (as a result of averaging across a different number of replicates) can to a certain extent be compensated by assigning weights in an estimation procedure (Searle, 2006).

All these methods are not very satisfactory, so the problem of unbalancedness should be avoided whenever possible. Of course, in practice, there is always fall-out in experiments causing (unwanted) unbalancedness. There are some methods to deal with such issues, and these are discussed in Chapter 3.

1.6 Direct Versus Indirect Methods

Within the field of machine learning and statistics, there are different ways to analyze data. Suppose that we have measured the microbiome (J OTUs) of a set of I samples resulting in the matrix $\mathbf{X}(I \times J)$. We can use these data directly and perform our favorite methods on $\mathbf{X}$. This is called a direct method of data analysis, and PCA and ASCA are examples of direct methods.

Another route to take (and this is done frequently in ecology and microbiomics) is to calculate distances between all sample pairs in $\mathbf{X}$ and collect these in a distance matrix $\mathbf{D}(I \times I)$. Then, analysis can be done on this matrix $\mathbf{D}$. This is called an indirect method of data analysis. There are different ways to come from a matrix $\mathbf{X}$ to a matrix $\mathbf{D}$, *e.g.*, there are many types of distances that can be considered. Also kernel methods, which can be used to model nonlinear relationships, belong to this category (Murphy, 2012). Examples of indirect methods are principal covariate analysis (PCoA) and PERMANOVA, which both work on a sample-by-sample distance matrix.

There are advantages and disadvantages of both methods. Direct methods have the advantage of staying close to the original data, thereby also showing the importance of the variables which facilitates interpretation. The influences of the individual variables are lost in the process of making the data indirect although there are some remedies, *e.g.*, pseudo-samples (Engel *et al.*, 2015). Distance-based methods tend to alleviate the problem of many zeros in the original data, but that depends also on the type of distance chosen. More of this will be explained in Chapter 6 which deals exclusively with indirect methods.

1.7 Some History

Methods for analysis of high-dimensional data obtained from studies in which various treatments were tested in a structured approach using a statistical design are based on the marriage of PCA (Pearson, 1901) and ANOVA (Fisher and Mackenzie, 1923). ANOVA methods provide tools for significance analysis of treatment effects from simple to complex statistical designs for a single response. With the birth of new analytical instruments that are able to measure many compounds simultaneously and produce thousands of variables, new data analysis methods had to be developed as the natural extension, MANOVA (Stahle and Wold, 1990), was not able to deal with these large numbers of variables (see Elaboration 1.2). Redundancy analysis (RDA) (Van Den Wollenberg, 1977) was extended to Principal Response Curves (PRC) (Van den Brink and Ter Braak, 1999) that can be used for analysis of treatment effects in experiments with a repeated measures design. In 2005, ANOVA-Simultaneous Component Analysis (ASCA) (Smilde *et al.*, 2005) was introduced, and since then many HD-ANOVA methods have been developed and applied in many fields. Figure 1.17 shows how all these new data analysis methods were built on the strong shoulders of PCA and ANOVA.

1.A Appendix

1.A.1 Types of Measurements

In this section of the appendix, we give a bird's eye view of some of the most common types of measurements generating the data as discussed in this book. This is not intended to be an exhaustive list, but to give a flavor of the types of measurements.

1.A.2 Notation and Terminology

In the introductory chapters (1–3) we will use the symbols $\mathbf{x}$ and $\mathbf{y}$ for general vectors and the symbols $\mathbf{X}$ and $\mathbf{Y}$ for general matrices. In the later chapters, these symbols receive a special meaning. There we will reserve $\mathbf{Y}(I \times J)$ for response variables; the symbol $\mathbf{X}(I \times K_f)$ for the design matrix of fixed effects and $\mathbf{Z}(I \times K_r)$ for the design matrix of random effects. The matrix

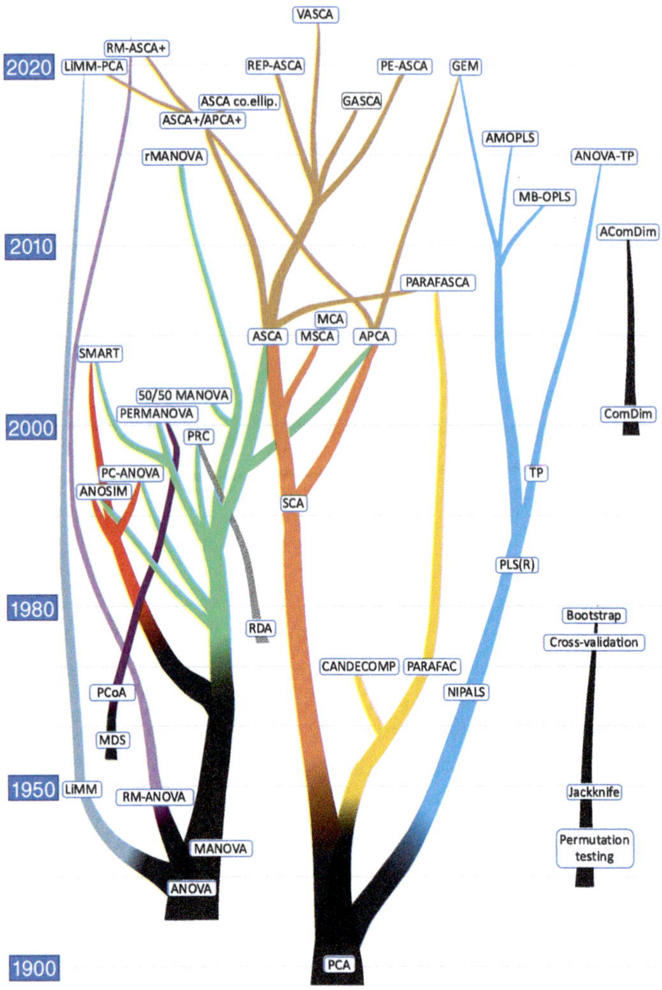

Figure 1.17 A graphical depiction of the history of the development of HD-ANOVA methods.

D is reserved to indicate a distance matrix unless stated otherwise. Score matrices are denoted by **T** and loading matrices by **P** or **Q**. The matrices $\mathbf{E}_X$ and $\mathbf{E}_Y$ are used to indicate error or residual matrices. The notational conventions below in Table 1.A.2 are phrased in terms of a general matrix **X**. We will use the generic term "sample" for rows of a matrix; this may pertain to subjects, objects, plants, or other units on which the treatment is applied. Columns will be called variables, and all methods that try to deal with high-dimensional designed or grouped data will be termed HD-ANOVA. A combination of factor settings will be referred to as a cell in the design matrix.

Apart from these notational conventions, we will also make use of the convenient ANOVA notations for means and sums. The sum of values of y_{abn} across index $n = 1, \ldots, N$ will be denoted as $y_{ab.}$ and likewise for double or triple indices, $y_{...}$ will be used for the summation of the values across all indices. Similarly, the mean of the values across an index will be notated with an overline: $\overline{y}_{ab.}$ is the mean of the values y_{abn} across index $n = 1, \ldots, N$. Analogous definitions hold for $\overline{y}_{.b.}$ and $\overline{y}_{...}$. Concatenation of matrices will be indicated by a |, so [**X**|**Y**] is the

Table 1.A.1 Overview of types of measurements in different fields.

Field	Number of variables	Replicates distribution	Instruments	Section
Metabolomics	≈1000	Lognormal	NMR, LC-MS, GC-MS	1.4.1
Genomics	≈20,000	Poisson, NB	RNAseq	1.4.2
Microbiome	≈1000	Poisson, NB	RNAseq	1.4.3
Proteomics	≈1000	Lognormal	LC-MS	1.4.4
Food/chemistry	≈300	Normal, Lognormal	Spectroscopy, LC, GC	1.4.5
				1.4.7
Sensory	≈30	Various distributions	Panel assessment	1.4.6

NB means negative binomial.

concatenation of the matrices **X** and **Y**. The size of a matrix with I rows and J columns will be indicated by $\mathbf{X}(I \times J)$.

1.A.3 Some Definitions

In this book, we will use some basic definitions from matrix algebra which will be discussed briefly here. For a full account, see Schott (2016).

Norms of vectors and matrices measure their "size" in some way, and there are formal definitions. The norm we will be using mostly is the Frobenius (a.k.a. Euclidean) norm. For a vector $\mathbf{x}(I \times 1)$ with typical elements x_i this is defined as:

$$||\mathbf{x}|| = (\mathbf{x}^t\mathbf{x})^{\frac{1}{2}} = \sqrt{\sum_{i=1}^{I} x_i^2} \tag{1.A.1}$$

and it is sometimes more convenient to take the square of this norm $||\mathbf{x}||^2$. This will be the notation we will use in our book: $||\mathbf{x}||$ will always mean this Frobenius norm unless stated otherwise. This norm generalizes easily to matrices, *e.g.*, for the matrix $\mathbf{X}(I \times J)$ with typical elements x_{ij} the squared Frobenius norm is

$$||\mathbf{X}||^2 = \sum_{i=1}^{I}\sum_{j=1}^{J} x_{ij}^2 \tag{1.A.2}$$

The trace of a square matrix $\mathbf{X}(I \times I)$ is defined as

$$tr(\mathbf{X}) = \sum_{i=1}^{I} x_{ii} \tag{1.A.3}$$

which is simply the sum of its diagonal elements. A very convenient property of traces is cyclic permutation:

$$tr(\mathbf{ABC}) = tr(\mathbf{CAB}) = tr(\mathbf{BCA}) \tag{1.A.4}$$

if the product of the matrices involved is defined given their sizes. A property relating Frobenius norms to traces is

$$||\mathbf{X}||^2 = tr(\mathbf{X}^t\mathbf{X}) \tag{1.A.5}$$

which is easily verified.

Table 1.A.2 Overview of notation as used in this book.

x	A scalar
$\mathbf{x}$	Column vector: bold lowercase
$\mathbf{X}$	Matrix: bold uppercase
$\mathbf{X}^t$	Transpose of $\mathbf{X}$
$\underline{\mathbf{X}}$	Three-way array: bold uppercase underlined
$\mathbf{1}_I$	Column vector of ones of length I
$\mathbf{0}_{I \times J}$	Matrix of zeros of size $(I \times J)$
$\mathbf{I}_I$	Identity matrix of size $(I \times I)$
$diag(\mathbf{S})$	Column vector containing the diagonal of $\mathbf{S}$
$i = 1, \ldots, I$	Index for subjects
$j = 1, \ldots, J$	Index for variables
$n = 1, \ldots, N$	Index for replicates
$m = 1, \ldots, M$	Index for blocks in a multiblock dataset
$a = 1, \ldots, A$	Levels for the first experimental factor
$b = 1, \ldots, B$	Levels for the second experimental factor
$c = 1, \ldots, C$	Levels for the third experimental factor
$l = 1, \ldots, L$	Index for (principal) components
$t = 1, \ldots, T$	Index for timepoints
K_f, K_r	Number of columns in the design matrix related to fixed and random effects
$\otimes$	Kronecker product
$\odot$	Khatri–Rao product (column-wise Kronecker product)
$*$	Hadamard or Element-wise product
$\oplus$	Direct sum of spaces

Two vectors $\mathbf{x}$ and $\mathbf{y}$ are called orthogonal if $\mathbf{x}^t\mathbf{y} = 0$. Likewise, two matrices $\mathbf{X}$ and $\mathbf{Y}$ are called orthogonal if $\mathbf{X}^t\mathbf{Y} = \mathbf{0}$. A matrix $\mathbf{X}$ is called orthogonal if $\mathbf{X}^t\mathbf{X} = \mathbf{X}\mathbf{X}^t = \mathbf{I}$ (with $\mathbf{I}$ of the proper size). A matrix $\mathbf{X}$ is symmetric if $\mathbf{X}^t = \mathbf{X}$, while it is called idempotent if $\mathbf{X}^2 = \mathbf{X}$. A matrix $\mathbf{X}$ is called positive definite if $\mathbf{y}^t\mathbf{X}\mathbf{y} > 0; \forall \mathbf{y} \neq 0$. Likewise, it is called positive semi-definite if $\mathbf{y}^t\mathbf{X}\mathbf{y} \geq 0; \forall \mathbf{y}$.

1.A.4 Abbreviations

In this book we will use a lot of abbreviations. Table 1.A.3 provides abbreviations used including the chapter(s) in which they appear.

Table 1.A.3 Abbreviations of the different methods.

Abbreviation	Full description	Chapter
AComDim	ANOVA Common Dimensions	5
AMOPLS	ANOVA Multiblock Orthogonal Partial Least Squares	5
ANOVA	Analysis of Variance	3
ANOVA-TP	ANOVA-Target Projection	5
APCA	ANOVA-PCA	4
ASCA	ANOVA-Simultaneous Component Analysis	4
ASCA+	ANOVA-Simultaneous Component Analysis+	4
FFMANOVA	Fifty–Fifty Multivariate ANOVA	5
GASCA	Group-wise ASCA	4
GEM	General Effect Modeling	5
GLM	Generalized Linear Model	3
LiMM–PCA	Linear Mixed Model–PCA	5
LMM	Linear Mixed Model	3
MANOVA	Multivariate ANOVA	1,3
MSCA	Multilevel Simultaneous Component Analysis	4
OPLS	Orthogonal Projections to Latent Structures	5
PCoA	Principal Coordinate Analysis	1,6
PC-ANOVA	Principal Component ANOVA	5
PE-ASCA	Penalized ASCA	5
PERMANOVA	Permutational MANOVA	6
PLS	Partial Least Squares	2
PRC	Principal Response Curves	4
RDA	Redundancy Analysis	2
REP-ASCA	Replication ASCA	4
rMANOVA	Regularized MANOVA	5
RM-ASCA+	Repeated-Measures ASCA+	5
SMART	Scaled-to-Maximum and Aligned Reduced Trajectories	4

2

Basic Theory and Concepts

In this chapter, we are giving some background in mathematics and statistics needed to understand the high-dimensional analysis of variance (HD-ANOVA) methods as presented in this book. This will only be a brief survey for two reasons. First, to save space we only discuss the most relevant topics as a refresher and, secondly, many books exist that explain these topics in much more depth. For more details regarding the mathematics involved, see Searle (1982) and Schott (2016) and for more details about statistics, see Searle (1971) and Searle *et al.* (1992).

2.1 Mathematical Background

2.1.1 Vector Spaces and Subspaces

A basic concept in all data analysis methods is the notion of a vector. A vector is a collection of I numbers organized in a single column.[1] Such a vector is then an element of an I-dimensional space, denoted as $\mathbb{R}^I$. We can make linear combinations of vectors, such as $\mathbf{x}_{LC} = w_1\mathbf{x}_1 + w_2\mathbf{x}_2$ (where w_1 and w_2 are scalars and LC stands for Linear Combination) which is also an element of $\mathbb{R}^I$. If we have J vectors $\mathbf{x}_j$ then the span of $\mathbf{x}_j$; $j = 1, \dots, J$ is the set of all possible linear combinations of $\mathbf{x}_j$ and this generates a subspace of $\mathbb{R}^I$ (a subspace has the property that if $\mathbf{x}_1$ and $\mathbf{x}_2$ are elements of that space then so is $w_1\mathbf{x}_1 + w_2\mathbf{x}_2$ for all values of w_1, w_2). This subspace has a certain dimension which can never be higher than I.

In this book, we will present all data in the form of matrices (or three-way arrays when needed) where the samples form the rows and the variables are in the columns. We will start with the simple example of a matrix $\mathbf{X}(I \times J)$. Each sample is thus a row in this matrix and the ith sample can be represented by $\mathbf{x}_i^t(1 \times J)$; $i = 1, \dots, I$. Since this sample has J numbers, it lives in the J-dimensional space denoted as $\mathbb{R}^J$. Similarly, a column $\mathbf{x}_j$ lives in the I-dimensional space denoted by $\mathbb{R}^I$. The span of $\mathbf{x}_i^t$ is called the row-space of $\mathbf{X}$ and the span of $\mathbf{x}_j$ is called the column-space. The dimensionality of both spaces is equal[2] and is called the *rank* of $\mathbf{X}$. We will denote the row-space of $\mathbf{X}$ as $R(\mathbf{X}^t)$ and similarly the column-space as $R(\mathbf{X})$.

The above reasoning also shows that we can represent a matrix in two ways: (i) the rows as points in the row-space and (ii) the columns as points in the column-space. There are I points in the row-space of $\mathbf{X}$ which is embedded in $\mathbb{R}^J$, and there are J points in the column space of $\mathbf{X}$ which is embedded in $\mathbb{R}^I$. Both representations have their value, and a small example is given in Elaboration 2.1.

---

1 We will notate all vectors as column vectors.

2 This property does not hold for three-way arrays.

Analysis of Variance for High-Dimensional Data: Applications in Life, Food and Chemical Sciences,
First Edition. Age K. Smilde, Federico Marini, Johan A. Westerhuis, and Kristian H. Liland.

ELABORATION 2.1

Representations of data

To show the different types of representations of a data matrix consider the matrix:

$$\mathbf{X} = \begin{bmatrix} 3 & 1 \\ 1 & 3 \\ 5 & 2 \end{bmatrix} \tag{2.1}$$

and there are two ways of plotting the data. The samples or, more generally, the objects may be plotted in the row-space, or conversely the variables may be plotted in the column-space, see Figure 2.1.

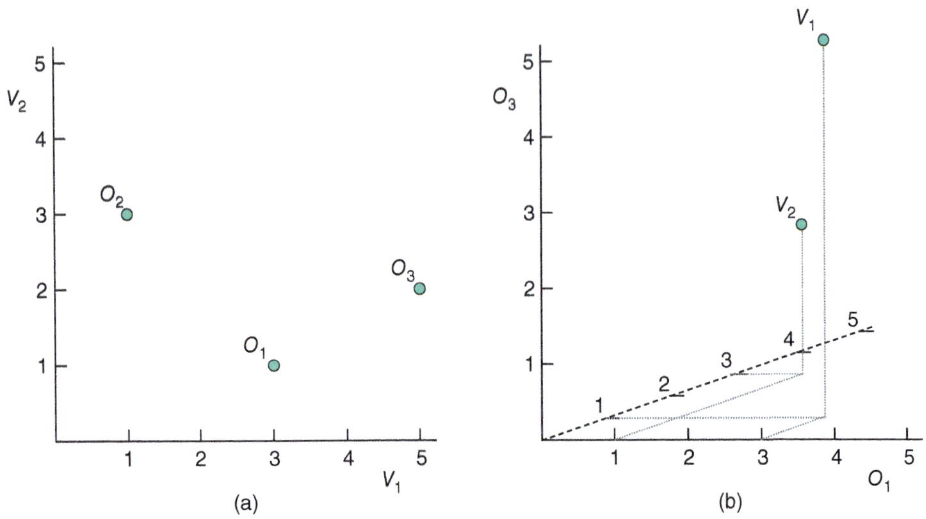

Figure 2.1 Row-space (a) and column-space (b) representation of the raw data given in Equation 2.1; O = objects and V = variables.

For the row-space representation of $\mathbf{X}$, the objects (rows of $\mathbf{X}$) are seen as points in the space. This is the most used representation of data and is also sometimes called a scatter plot. The usefulness of the column-space representation becomes apparent when considering centered data. If the matrix $\mathbf{X}$ is column-centered then this results in

$$\mathbf{X}_C = \begin{bmatrix} 0 & -1 \\ -2 & 1 \\ 2 & 0 \end{bmatrix} \tag{2.2}$$

and this matrix can again be represented in two different ways, see Figure 2.2.

When the data have been centered, the row-space representation (Figure 2.2a), shows that the coordinate system is translated, so the origin is now in the center of the data, but the distances between objects (rows) are kept intact. In the column-space (Figure 2.2b), the distances between the points change. The cosine of the angle between the centered variables is also the correlation between variables. Moreover, the squares of the lengths of the vectors representing the variables in this plot are the variances of those variables. Hence, both plots have their added value. In regression situations, mostly the column-spaces are used for illustrating the machinery of regression models which will become clear in the following chapters.

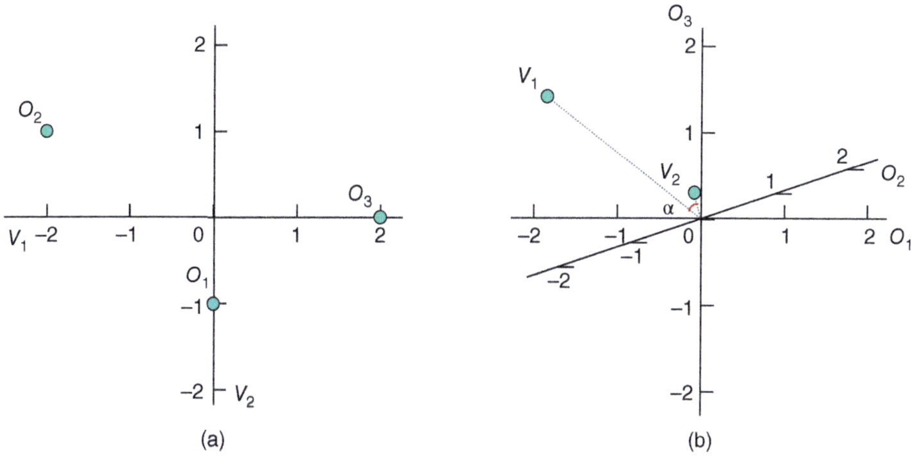

Figure 2.2 Row-space (a) and column-space (b) representation of the centered data given in Equation 2.2; O = objects and V = variables.

In the sequel we need a few definitions regarding subspaces. The intersection of the two subspaces $R(\mathbf{X})$ and $R(\mathbf{Y})$, both embedded in $\mathbb{R}^I$, is the subspace containing all the vectors which are both in $R(\mathbf{X})$ and in $R(\mathbf{Y})$ (notation: $R(\mathbf{X}) \cap R(\mathbf{Y})$). The union of two subspaces contains all the vectors that are either in $R(\mathbf{X})$ or in $R(\mathbf{Y})$ (notation: $R(\mathbf{X}) \cup R(\mathbf{Y})$). This is not necessarily a subspace; if $\mathbf{x} \in R(\mathbf{X})$ and $\mathbf{y} \in R(\mathbf{Y})$ then $\mathbf{x} + \mathbf{y}$ is not necessarily in either $R(\mathbf{X})$ or $R(\mathbf{Y})$ thereby violating one of the definitions of linear spaces. The sum of two subspaces $R(\mathbf{X})$ and $R(\mathbf{Y})$ contains all the vectors that can be written as a sum of vectors of $R(\mathbf{X})$ and $R(\mathbf{Y})$. This is indeed again a subspace and is denoted as $R(\mathbf{X}) + R(\mathbf{Y})$. If it also holds that $R(\mathbf{X}) \cap R(\mathbf{Y}) = \{0\}$,[3] then this sum is called the direct sum, denoted as $R(\mathbf{X}) \oplus R(\mathbf{Y})$. In the latter case, every vector $\mathbf{z} \in R(\mathbf{X}) \oplus R(\mathbf{Y})$ can be written uniquely as $\mathbf{z} = \mathbf{x} + \mathbf{y}$ where $\mathbf{x} \in R(\mathbf{X})$ and $\mathbf{y} \in R(\mathbf{Y})$. Note that the direct sum does not imply that $R(\mathbf{X})$ and $R(\mathbf{Y})$ are orthogonal.

2.1.2 Matrix Decompositions

There are several important matrix decompositions that are used very frequently in data science. The first one is the Singular Value Decomposition (SVD) which holds for any matrix $\mathbf{X}(I \times J)$ with rank R. The SVD of this matrix is

$$\mathbf{X} = \mathbf{USV}^t \tag{2.3}$$

with $\mathbf{U}(I \times I)$; $\mathbf{U}^t\mathbf{U} = \mathbf{UU}^t = \mathbf{I}_I$ and $\mathbf{V}(J \times J)$; $\mathbf{V}^t\mathbf{V} = \mathbf{VV}^t = \mathbf{I}_J$ and $\mathbf{S}(I \times J)$ with a diagonal block of size $R \times R$ with positive elements and zero elsewhere in its first R rows and R columns and the rest of $\mathbf{S}$ is also zero. We can write this decomposition in a more convenient way by distinguishing two cases corresponding to tall or fat matrices (see Figure 1.1). The first case pertains to tall matrices in which $I \geq J$ and gives matrices $\mathbf{U}(I \times J)$; $\mathbf{U}^t\mathbf{U} = \mathbf{I}_J$; $\mathbf{V}(J \times J)$; $\mathbf{V}^t\mathbf{V} = \mathbf{VV}^t = \mathbf{I}_J$ and $\mathbf{S}(J \times J)$. The second case pertains to fat matrices with $I \leq J$ and generates matrices $\mathbf{U}(I \times I)$; $\mathbf{U}^t\mathbf{U} = \mathbf{UU}^t = \mathbf{I}_I$; $\mathbf{V}(J \times I)$; $\mathbf{V}^t\mathbf{V} = \mathbf{I}_I$ and $\mathbf{S}(I \times I)$.

3 Note that subspaces always contain the zero-vector.

The SVD has many useful properties. The number of non-zeros on the diagonal of $\mathbf{S}$ is equal to the rank of $\mathbf{X}$. If the rank of $\mathbf{X}$ is $R < J$ then a more compact way of writing the SVD is

$$\mathbf{X} = \mathbf{U}_R\mathbf{S}_R\mathbf{V}_R^t \tag{2.4}$$

with $\mathbf{U}_R(I \times R)$; $\mathbf{U}_R^t\mathbf{U}_R = \mathbf{I}_R$; $\mathbf{V}_R(J \times R)$; $\mathbf{V}_R^t\mathbf{V}_R = \mathbf{I}_r$ and the diagonal of $\mathbf{S}_R(R \times R)$ has positive elements. This decomposition is sometimes called the economy or parsimonious version of the SVD. The matrix $\mathbf{U}_R$ gives an orthonormal basis for $R(\mathbf{X})$, and the matrix $\mathbf{V}_R$ gives an orthonormal basis for $R(\mathbf{X}^t)$. There is a very useful relation between least squares and the SVD (see Section 2.1.5). If the values on the diagonal of $\mathbf{S}$ are all different, then the SVD is unique.

Another very useful decomposition holds for (square) symmetric matrices and is called the Spectral Decomposition (SD):

$$\mathbf{X} = \mathbf{V}\mathbf{D}\mathbf{V}^t \tag{2.5}$$

where $\mathbf{X}(J \times J)$ is symmetric; the matrix $\mathbf{V}(J \times J)$; $\mathbf{V}^t\mathbf{V} = \mathbf{V}\mathbf{V}^t = \mathbf{I}_J$ contains the eigenvectors of $\mathbf{X}$ and the matrix $\mathbf{D}(J \times J)$ is diagonal and contains the eigenvalues. Hence, $\mathbf{V}^t = \mathbf{V}^{-1}$ (see Section 2.1.3) and Equation 2.5 can be written accordingly. Moreover, Equation 2.5 can also be written as

$$\mathbf{V}^t\mathbf{X}\mathbf{V} = \mathbf{D} \tag{2.6}$$

showing that $\mathbf{X}$ is diagonalizable (Schott, 2016). For Gramian matrices – which can always be written as $\mathbf{X} = \mathbf{A}^t\mathbf{A}$ and are positive semi-definite – it additionally holds that $\mathbf{D}$ has nonnegative diagonal elements, and these are usually sorted in decreasing size. In many regression situations, matrices of the form $\mathbf{Z} = \mathbf{X}^t\mathbf{X}$ are obtained, and these thus allow for a very simple SD of $\mathbf{Z}$. If all eigenvalues are different, then the SD is unique (Schott, 2016).

The SD has many nice properties and one which is especially useful is that the problem

$$\max_{\mathbf{p}^t\mathbf{p}=1} \mathbf{p}^t\mathbf{X}\mathbf{p} \tag{2.7}$$

is solved by taking $\mathbf{p} = \mathbf{v}_1$ which is the eigenvector[4] of $\mathbf{X}$ corresponding to the largest eigenvalue. This eigenvalue is then also the maximum as sought in Equation 2.7.

The relationship between the SVD and the SD becomes clear when writing out $\mathbf{X}^t\mathbf{X}$ in terms of the SVD of $\mathbf{X}$:

$$\mathbf{X}^t\mathbf{X} = \mathbf{V}\mathbf{S}^t\mathbf{U}^t\mathbf{U}\mathbf{S}\mathbf{V}^t = \mathbf{V}\mathbf{S}^t\mathbf{S}\mathbf{V}^t = \mathbf{V}\mathbf{S}^2\mathbf{V}^t \tag{2.8}$$

using the diagonality of $\mathbf{S}$ which also shows that the matrix $\mathbf{D}$ for Gramian matrices is nonnegative.

2.1.3 Inverses and Generalized Inverses

An important concept in matrix algebra in the context of HD-ANOVA and regression models is the inverse of a matrix. A full-rank matrix $\mathbf{X}(I \times I)$ has an inverse $\mathbf{X}^{-1}$ such that $\mathbf{X}\mathbf{X}^{-1} = \mathbf{X}^{-1}\mathbf{X} = \mathbf{I}$, and this matrix is unique. Clearly, for a matrix to have an inverse it needs to be square and of full rank. The inverse has many applications; *e.g.*, in solving a linear system of equations such as $\mathbf{Y} = \mathbf{X}\mathbf{B}$ where $\mathbf{X}$ has size $(I \times I)$ then $\mathbf{B} = \mathbf{X}^{-1}\mathbf{Y}$ if the inverse of $\mathbf{X}$ exists.[5]

---

4 Eigenvectors usually are normalized to have length one.

5 In practice this solution is not recommended due to numerical instabilities.

The concept of inverses can be generalized to non-square matrices and matrices not of full rank. The most used generalized inverse is the Moore–Penrose (MP) inverse $\mathbf{X}^+$ which is defined for any general matrix and has the following properties:

1) $\mathbf{XX}^+\mathbf{X} = \mathbf{X}$
2) $\mathbf{X}^+\mathbf{XX}^+ = \mathbf{X}^+$
3) $(\mathbf{XX}^+)^t = \mathbf{XX}^+$
4) $(\mathbf{X}^+\mathbf{X})^t = \mathbf{X}^+\mathbf{X}$

which are called the Moore–Penrose conditions. The MP inverse is unique and has many applications, *e.g.*, in regression theory. An explicit equation for the MP-inverse of a matrix $\mathbf{X}(I \times J)$ using the SVD of $\mathbf{X}$ is

$$\mathbf{X}^+ = \mathbf{V}_R\mathbf{S}_R^{-1}\mathbf{U}_R^t \tag{2.9}$$

where we have used the parsimonious version of the SVD and $\mathbf{S}_R^{-1}$ is the inverse of $\mathbf{S}$ (all elements on the diagonal of $\mathbf{S}$ are inverted).

Another useful generalized inverse is the $\{1\}$-inverse or g-inverse that only obeys the first of the Moore–Penrose conditions and is sometimes simply called the generalized inverse and notated as $\mathbf{X}^-$. Also this inverse is useful in regression situations (see Section 2.2.2). This inverse is not unique and can also be expressed explicitly using the SVD of $\mathbf{X}$:

$$\mathbf{X}^- = \begin{bmatrix} \mathbf{S}_R^{-1} & \mathbf{A} \\ \mathbf{B} & \mathbf{C} \end{bmatrix} \tag{2.10}$$

where $\mathbf{A}$, $\mathbf{B}$, and $\mathbf{C}$ are arbitrary matrices of conformable sizes (Schott, 2016) and $\mathbf{S}_R^{-1}$ is as before. This shows that the generalized inverse is not unique.

An inverse that plays a role in least squares problems is the least squares inverse (LS-inverse) denoted as $\mathbf{X}^L$ which is in fact a $\mathbf{X}^{\{1,3\}}$ inverse obeying the first and third MP-conditions. For any least squares inverse, it holds that $\mathbf{XX}^L = \mathbf{XX}^+$ and $(\mathbf{X}^t\mathbf{X})^-\mathbf{X}^t$ is an LS-inverse of $\mathbf{X}$ for any choice of generalized inverse of $\mathbf{X}^t\mathbf{X}$.

2.1.4 Distances and Projections

Distances play a crucial role in HD-ANOVA methods; the most prominent being PERMANOVA (see Chapter 6). There is a close relationship between distances and projections, and thus it is worthwhile to give a more formal introduction to distances and give some background on the ones mostly used in this book. This is then followed by a description of projections.

2.1.4.1 Formal Description of Distances

The formal definition of a distance function or a distance $d(\mathbf{x}, \mathbf{y}) : \mathbb{R}^J \times \mathbb{R}^J \rightarrow \mathbb{R}$ between two vectors $\mathbf{x}$ and $\mathbf{y}$ is (Schott, 2016):

a) $d(\mathbf{x}, \mathbf{y}) \geq 0$ (2.11)

b) $d(\mathbf{x}, \mathbf{y}) = 0$ if and only if $\mathbf{x} = \mathbf{y}$

c) $d(\mathbf{x}, \mathbf{y}) = d(\mathbf{y}, \mathbf{x})$

d) $d(\mathbf{x}, \mathbf{z}) \leq d(\mathbf{y}, \mathbf{x}) + d(\mathbf{y}, \mathbf{z})$

where the last inequality is called the triangle inequality (and **z** is also a vector). We will see that not all "distances" in this book (and used in the life sciences) are proper distances. Some of them do not obey the triangle inequality, and some of them (*e.g.*, the Kullback–Leibler divergence) are not symmetric and thus disobey requirement (c) in Equation 2.11. The distances obeying all four requirements are sometimes called metric; if only requirements (a), (b), and (c) are met, then these distances are called semi-metric.

We will discuss the distances in terms of two samples $\mathbf{x} = (x_1, \ldots, x_J)$ and $\mathbf{y} = (y_1, \ldots, y_J)$ each containing measurements on the same J variables. The most basic distance is the Euclidean distance which is defined as

$$d_E(\mathbf{x}, \mathbf{y}) = \sqrt{\sum_{j=1}^{J} (x_j - y_j)^2} = ||\mathbf{x} - \mathbf{y}|| \tag{2.12}$$

which is a metric distance. This distance is abundantly used in multivariate analysis and has many nice properties of which simplicity is the most prominent. It has no fixed upper limit. A disadvantage of the Euclidean distance is that it does not account for the correlation between the J variables (see Example 2.1). A distance that does take this into account is the Mahalanobis distance which is defined as:

$$d_M(\mathbf{x}, \mathbf{y}) = \sqrt{(\mathbf{x} - \mathbf{y})^t \mathbf{S}^{-1} (\mathbf{x} - \mathbf{y})} \tag{2.13}$$

where **S** is the covariance matrix of the J variables. The Mahalanobis distance does also not have an upper limit. Note that if $\mathbf{S} = \mathbf{I}$, then the Mahalanobis distance equals the Euclidean distance.

Example 2.1: Euclidean versus Mahalanobis distance

As mentioned in the main text, the Euclidean distance does not take into account the correlation structure of the data. This is shown with a small example in Figure 2.3. In this figure, the Euclidean distance between points (samples) A and B is as large as the Euclidean distance between points A

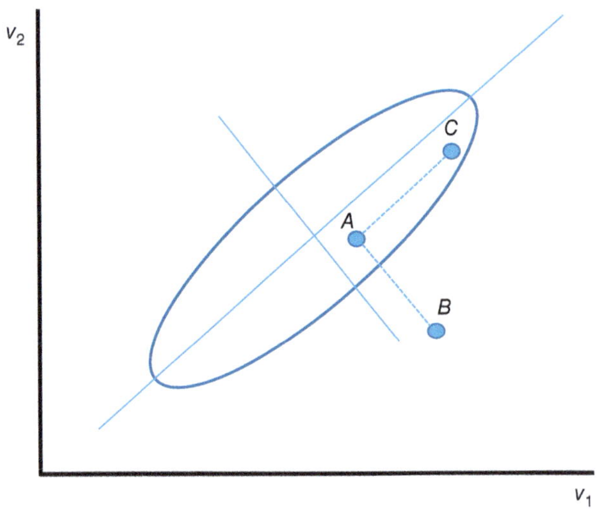

Figure 2.3 Two variables v_1 and v_2 are correlated as shown by the 95% contour of the data ellipsoid. For more explanation: see the text.

and C. Yet, in terms of the correlation structure of the data (correlation between the variables v_1 and v_2) it can be argued that points A and C are more alike than points A and B. This is accounted for by the Mahalanobis distance: $d_M(A, C) < d_M(A, B)$.

All points on the 95% contour ellipse have the same Mahalanobis distance to the centroid.

Especially in the field of ecology and microbiome data analysis other distances are in use. The most prominent ones are the Hellinger and Canberra distances and the Bray–Curtis dissimilarity. The Hellinger distance is defined as:

$$d_H(\mathbf{x}, \mathbf{y}) = \sqrt{\sum_{j=1}^{J} \left[\sqrt{\frac{x_j}{x_+}} - \sqrt{\frac{y_j}{y_+}}\right]} \tag{2.14}$$

where $x_+ = \sum_j x_j$ and y_+ likewise. Note that both x_j and y_j are nonnegative since they usually refer to count data. The Hellinger distance is nonnegative and has upper limit $\sqrt{2}$. It calculates the differences between the relative abundances of the species in both samples. The Canberra distance is defined as:

$$d_C(\mathbf{x}, \mathbf{y}) = \sum_{j=1}^{J} \left[\frac{|x_j - y_j|}{(|x_i| + |y_i|)}\right] \tag{2.15}$$

which has no fixed upper limit. Both the Hellinger and Canberra distances are metric distances. The Bray–Curtis dissimilarity is defined as

$$d_{BC}(\mathbf{x}, \mathbf{y}) = 1 - \frac{2C_{xy}}{x_+ + y_+} = \frac{D_{xy}}{x_+ + y_+} \tag{2.16}$$

where C_{xy} is the sum of the lowest value of x or y in which they have common non-zeros, and $D_{xy} = \sum_{j=1}^{J} |x_j - y_j|$. The Bray–Curtis dissimilarity is a ratio of the sum of all differences and the sum of all totals in both samples. The Bray–Curtis dissimilarity is bounded between zero and one, where a zero means that the composition of both samples is the same and a one means that none of the species present in sample x is present in sample y and vice versa.

Euclidean, Hellinger, Canberra distances obey the triangle inequality (metric), whereas the Bray–Curtis does not (semi-metric). None of these measures are affected by double zeros (see Section 6.1.1). All distances are affected by features that are only present in one of the samples. However, the Canberra distance is equally affected by features that have a single count in one sample and zero counts in the other. This, however, appears a lot in microbiome data due to reading errors that mistakenly add a feature to the data due to a wrongly identified species. Examples of these distances are given in Example 2.2.

Example 2.2: Distances and dissimilarities

A simple example of the Bray–Curtis dissimilarity is the following. For the samples $\mathbf{x} = (6, 7, 4)$ and $\mathbf{y} = (10, 0, 6)$ the nonzero values in common are for variables 1 and 3 and thus $d_{BC}(\mathbf{x}, \mathbf{y}) = 1 - \frac{2(6+4)}{(17+16)} = \frac{(4+7+2)}{(17+16)} = 0.39$ where the 6 in the numerator comes from sample $\mathbf{x}$ and the 4 also comes from sample $\mathbf{x}$.

To show the differences between several of these distances and dissimilarity measures assume we have the data as shown in Table 2.1 which are count data of amplicon sequence variants (ASVs) that are common in microbiome data. Note the large number of zeros which are always present in such data.

Table 2.1 ASV values of three samples.

Samples	ASV1	ASV2	ASV3	ASV4	ASV5	ASV6	ASV7	ASV8	ASV9	ASV10	ASV11
1	10	20	40	100	50	20	0	0	0	0	0
2	100	2	50	20	50	10	0	0	0	0	0
3	11	21	39	90	45	10	5	6	10	10	15

The different distances and dissimilarities are shown in Table 2.2.

Table 2.2 Distances and dissimilarities between the samples.

Samples	$D(1,2)$	$D(1,3)$	$D(2,3)$
Euclidean	123	27	117
Bray–Curtis	0.44	0.15	0.49
Hellinger	0.61	0.44	0.71
Canberra	2.75	5.52	7.44

For the Euclidean distance, samples 1 and 3 are the most similar. For the Canberra distance, samples 1 and 2 are the most similar. This is due to the pattern of zeros which is more similar between samples 1 and 2. Clearly, when using a distance or dissimilarity measure, domain-specific knowledge is important to make a specific choice.

2.1.4.2 Projections

The concept of projections is crucial in multivariate analysis. In this section we will describe the basics, starting with simple projections and later expanding this to a more formal treatment.

Simple Projections

We will first start with a simple projection using geometry and later also give the matrix algebra machinery. In its most basic form, we can project vector $\mathbf{y}$ orthogonally onto $\mathbf{x}$ where both vectors are in $\mathbb{R}^I$. This is illustrated in Figure 2.4. Clearly, the vector $\hat{\mathbf{y}} = b\mathbf{x}$, and the only unknown is the scalar b. The cosine rule as applied to the vectors $\mathbf{x}$ and $\mathbf{y}$ is

$$cos(\alpha) = \frac{\mathbf{x}^t\mathbf{y}}{||\mathbf{x}|| \cdot ||\mathbf{y}||} \tag{2.17}$$

but the cosine can also be written differently as:

$$cos(\alpha) = \frac{||\hat{\mathbf{y}}||}{||\mathbf{y}||} = \frac{||b\mathbf{x}||}{||\mathbf{y}||} = \frac{b||\mathbf{x}||}{||\mathbf{y}||} \tag{2.18}$$

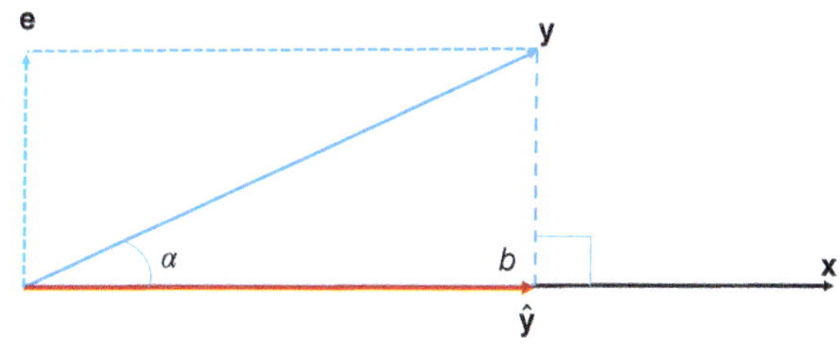

Figure 2.4 The vector $\mathbf{y}$ is projected onto the vector $\mathbf{x}$ giving $\hat{\mathbf{y}}$. For more explanation, see the main text.

since b is positive in this case. Combining Equations 2.17 and 2.18 then gives after some arrangement:

$$b = \frac{\mathbf{x}^t\mathbf{y}}{\mathbf{x}^t\mathbf{x}} = (\mathbf{x}^t\mathbf{x})^{-1}\mathbf{x}^t\mathbf{y} \tag{2.19}$$

using the equality $\mathbf{x}^t\mathbf{x} = ||\mathbf{x}||^2$. A similar derivation can be made for the case that b is negative. Note that the residual vector $\mathbf{e}$ in Figure 2.4 is orthogonal to the vector $\mathbf{x}$ which is a property of regression. This residual vector also has the shortest Euclidean distance between $\mathbf{y}$ and all vectors on the line generated by the vector $\mathbf{x}$. This shows the connection between projections and distances. The vector $\hat{\mathbf{y}}$ is called the least squares (LS) estimate of $\mathbf{y}$ given $\mathbf{x}$ (see Section 2.2.1.1). Now the vector $\mathbf{y}$ can be written as $\mathbf{y} = b\mathbf{x} + \mathbf{e} = \hat{\mathbf{y}} + \mathbf{e}$ which is the most basic regression equation.

Orthogonal projections are not the only possibility. It is also possible to define oblique projections or non-orthogonal projections, and a simple example is given in Elaboration 2.2.

ELABORATION 2.2

Oblique projections

A simple example of an oblique projection is shown in Figure 2.5 where now $\mathbf{y}$ is projected onto $\mathbf{x}$ along the direction $\mathbf{z}$.

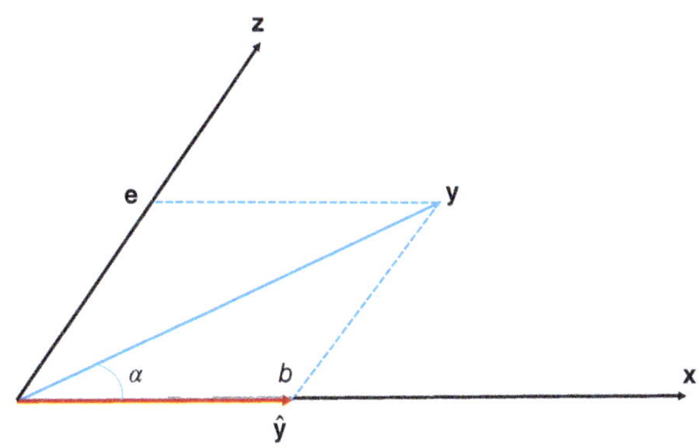

Figure 2.5 The vector $\mathbf{y}$ is projected onto the vector $\mathbf{x}$ along the vector $\mathbf{z}$ giving $\hat{\mathbf{y}}$. For more explanation, see the main text.

In this case, the calculations for b are a bit more elaborate. We need to extend Figure 2.5 with a help vector $\mathbf{u}$ which is orthogonal to $\mathbf{z}$, see Figure 2.6.

Let $\mathbf{v}$ be the vector obtained by an orthogonal projection of $\mathbf{y}$ onto $\mathbf{u}$ and likewise $\mathbf{e}_x$ is the vector obtained by an orthogonal projection of $\mathbf{x}$ onto $\mathbf{u}$, then the triangle formed by $\mathbf{x}$, $\mathbf{e}_x$, and the origin is similar to the triangle formed by $\hat{\mathbf{y}}$, $\mathbf{v}$, and the origin. Thus, b can be phrased in terms of $\mathbf{x}$ (proportion

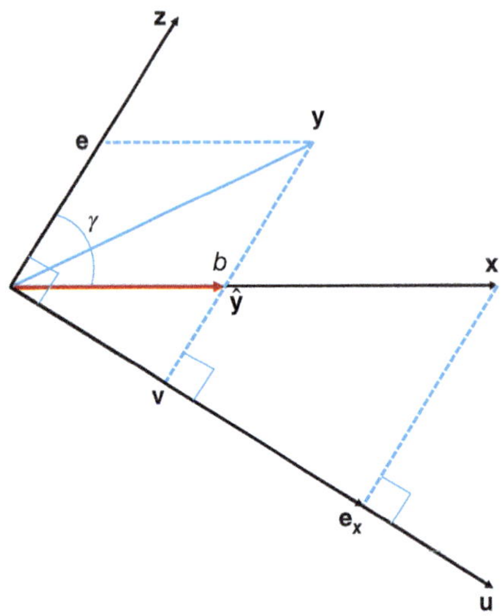

Figure 2.6 Extended figure for oblique projections.

of its length) as well as in terms of $\mathbf{e}_x$ (proportion of its length). This goes as follows. First, project $\mathbf{x}$ onto $\mathbf{z}$ and obtain the residual:

$$\mathbf{e}_x = \mathbf{x} - \mathbf{z}(\mathbf{z}^t\mathbf{z})^{-1}(\mathbf{z}^t\mathbf{x}) \tag{2.20}$$

using Equation 2.19. In the next step, $\mathbf{y}$ can be projected onto $\mathbf{e}_x$ to give b:

$$b = (\mathbf{e}_x^t\mathbf{e}_x)^{-1}\mathbf{e}_x^t\mathbf{y} \tag{2.21}$$

using again Equation 2.19. After some rearrangement (and using the cosine rule several times), b can be expressed in the original variables $\mathbf{x}$, $\mathbf{y}$, and $\mathbf{z}$:

$$b = \frac{(\mathbf{y}^t\mathbf{x})}{(\mathbf{x}^t\mathbf{x})}\frac{1}{(1 - cos^2\gamma)} - \frac{(\mathbf{z}^t\mathbf{y})cos\gamma}{||\mathbf{x}|| \cdot ||\mathbf{z}||(1 - cos^2\gamma)} \tag{2.22}$$

where γ is the angle between $\mathbf{x}$ and $\mathbf{z}$. This is not an easy expression anymore, but for the orthogonal case, it holds that $\mathbf{z}^t\mathbf{x}$ is zero and thus $cos\gamma = 0$. Then Equation 2.22 reduces to 2.19.

It is also possible to project vectors onto subspaces, as shown in Figure 2.7. Then the vector $\hat{\mathbf{y}}$ can be expressed as $b_1\mathbf{x}_1 + b_2\mathbf{x}_2$. This is the most basic multiple regression problem, and the machinery will be explained in Section 2.2.2. The vectors $\mathbf{x}_1$ and $\mathbf{x}_2$ are linearly independent.

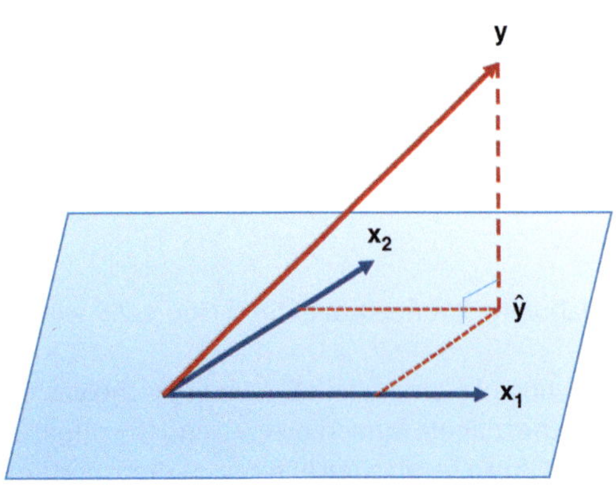

Figure 2.7 The vector $\mathbf{y}$ is projected onto the subspace spanned by the vectors $\mathbf{x}_1$ and $\mathbf{x}_2$ giving $\hat{\mathbf{y}}$. For more explanation, see the main text.

For a set of dependent vectors, the projection on the subspace generated by these vectors is more complicated, see Elaboration 2.3.

ELABORATION 2.3

Projection on the span of a dependent set of vectors

When the three vectors $\mathbf{x}_1$, $\mathbf{x}_2$, and $\mathbf{x}_3$ span a two-dimensional subspace on which a vector $\mathbf{y}$ is projected, then the result is shown in Figure 2.8. This figure has a close resemblance with Figure 2.7 with the crucial difference that now three vectors $\mathbf{x}$ are involved instead of two.

Figure 2.8 Projection on a set of dependent vectors.

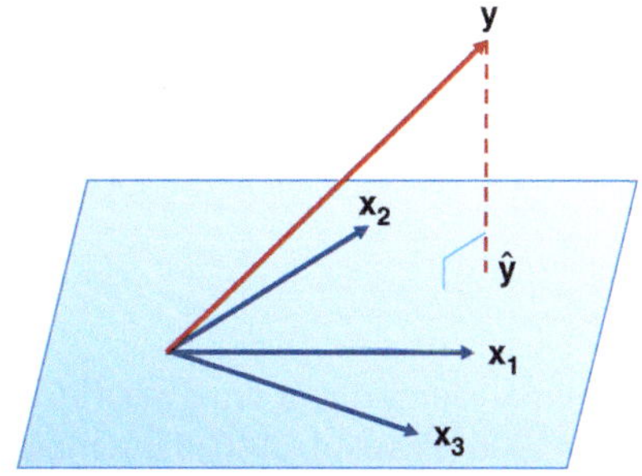

This means that $\hat{\mathbf{y}}$ in this case cannot be expressed uniquely as a linear combination of $\mathbf{x}_1$, $\mathbf{x}_2$, and $\mathbf{x}_3$; in fact, there is an infinite set of linear combinations of $\mathbf{x}_1$, $\mathbf{x}_2$, and $\mathbf{x}_3$ that can be used to define $\hat{\mathbf{y}}$. It is important to note that whereas such a linear combination is not unique, the projection $\hat{\mathbf{y}}$ *itself* is unique.

There is a nice analogy between generalized inverses and the two projection problems of Figures 2.7 and 2.8. The uniqueness of the MP-inverse is linked to the projection problem of Figure 2.7, and the generalized inverse (which is not unique) is linked to the problems as shown in Figure 2.8. This will become apparent in Chapters 3 and 4.

FORMAL DESCRIPTION OF PROJECTIONS

The formal definition of a projection goes as follows. Suppose S is an R-dimensional subspace of $\mathbb{R}^I$. Then every vector $\mathbf{x} \in \mathbb{R}^I$ can be expressed uniquely as $\mathbf{x} = \mathbf{x}_1 + \mathbf{x}_2$ where $\mathbf{x}_1 \in S$ and $\mathbf{x}_2$ is orthogonal to all vectors in S. The vector $\mathbf{x}_1$ is the orthogonal projection of $\mathbf{x}$ onto S. If the set of vectors $\mathbf{z}_1, \ldots, \mathbf{z}_R$ forms an orthonormal basis of S, then $\mathbf{ZZ}^t$ (with $\mathbf{Z} = [\mathbf{z}_1 | \ldots | \mathbf{z}_R]$) is the matrix that performs this projection:

$$\begin{aligned} \mathbf{ZZ}^t\mathbf{x} &= \mathbf{x}_1 \\ \mathbf{ZZ}^t\mathbf{x}_2 &= 0 \\ \mathbf{ZZ}^t\mathbf{x}_1 &= \mathbf{x}_1 \end{aligned} \tag{2.23}$$

and it can be checked easily that $\mathbf{ZZ}^t\mathbf{ZZ}^t$ equals $\mathbf{ZZ}^t$ and $\mathbf{ZZ}^t$ is symmetric. A matrix for which it holds that $\mathbf{AA} = \mathbf{A}$ is called idempotent. A matrix is a projection matrix if and only if it is idempotent and symmetric. For a general basis $\mathbf{Q}$ of S, not necessarily orthonormal, it can be shown that the projection matrix is $\mathbf{Q}(\mathbf{Q}^t\mathbf{Q})^{-1}\mathbf{Q}^t$ which is again symmetric and idempotent.

There exists a nice relationship between projections and MP-inverses. For any matrix $\mathbf{X}$, the matrix $\mathbf{XX}^+$ projects onto $R(\mathbf{X})$ and the matrix $\mathbf{X}^+\mathbf{X}$ projects onto $R(\mathbf{X}^+)$. If $\mathbf{X}$

has full column-rank then $\mathbf{X}\mathbf{X}^{+} = \mathbf{X}(\mathbf{X}^t\mathbf{X})^{-1}\mathbf{X}^t$. If $\mathbf{X}$ does not have full column-rank then $\mathbf{X}\mathbf{X}^{L} = \mathbf{X}(\mathbf{X}^t\mathbf{X})^{-}\mathbf{X}^t = \mathbf{X}\mathbf{X}^{+}$ for any generalized inverse (see Section 2.1.3) which also projects onto $R(\mathbf{X})$.

The formal description of oblique projections goes as follows (Schott, 2016). Suppose S_1 and S_2 are subspaces of $\mathbb{R}^I$ and $S_1 \oplus S_2 = \mathbb{R}^I$. Then every vector $\mathbf{x}$ can be written as $\mathbf{x}_1 + \mathbf{x}_2$ where $\mathbf{x}_1 \in S_1$ and $\mathbf{x}_2 \in S_2$. This is clearly a generalization of the orthogonal projection, and $\mathbf{x}_1$ is called the projection onto S_1 along S_2. If $\mathbf{Z}_1(I \times R)$ is a basis of S_1 and $\mathbf{Z}_2(I \times R)$ is a basis of $S_2^{\perp}$ (which is the space orthogonal to S_2), respectively, then the projection matrix $\mathbf{H}$ onto S_1 along S_2 is:

$$\mathbf{H} = \mathbf{Z}_1(\mathbf{Z}_2^t\mathbf{Z}_1)^{-1}\mathbf{Z}_2^t \tag{2.24}$$

and it can be verified that $\mathbf{H}$ is idempotent ($\mathbf{H}^2 = \mathbf{H}$) and not symmetric (Takeuchi *et al.*, 1982). These oblique projections are used in partial least squares (PLS) (see Elaboration 2.6). There exists a nice relationship between a generalized inverse $\mathbf{X}^-$ and oblique projections. The matrix $\mathbf{X}\mathbf{X}^-$ projects onto $R(\mathbf{X})$ along the vector space $R(\mathbf{I} - \mathbf{X}\mathbf{X}^-)$. This is only an orthogonal projection if $\mathbf{X}\mathbf{X}^-$ is symmetric (Schott, 2016).

2.1.5 Principal Component Analysis

Principal Component Analysis (PCA) is the workhorse of data analysis and dates back to the early 1900 (Pearson, 1901; Hotelling, 1933), and good monographs on this topic exist (Jolliffe, 2010). In its most basic form following Pearson (1901) it goes like this. Suppose a data matrix $\mathbf{X}(I \times J)$, then – as explained in Elaboration 2.1 – the I rows of $\mathbf{X}$ can be considered as points in $\mathbb{R}^J$. These points show a certain configuration in that space which may be approximated by subspaces of a dimension lower than $\min(I, J)$. To quantify approximation, the concept of projections is used. The formal reasoning is then to choose lines or planes – which are defined by directions $\mathbf{p}_r$; $r = 1, \dots, R$ in $\mathbb{R}^J$ – in such a way that the projections of the points (*i.e.*, the rows of $\mathbf{X}$) are as close as possible to the original points. This leads then to orthogonal projections.

The projection of point i or the ith row of $\mathbf{X}$ (which is $\mathbf{x}_i^t$) onto $\mathbf{p}_1$ is shown already in Figure 2.4 (with different nomenclature), and thus the corresponding projection equals

$$\hat{\mathbf{x}}_i = \frac{\mathbf{x}_i^t\mathbf{p}_1}{\mathbf{p}_1^t\mathbf{p}_1}\mathbf{p}_1 = t_{i1}\mathbf{p}_1 \tag{2.25}$$

with t_{i1} implicitly defined, and thus the residual of this projection is

$$\mathbf{e}_i = \mathbf{x}_i - \hat{\mathbf{x}}_i = \mathbf{x}_i - t_{i1}\mathbf{p}_1 \tag{2.26}$$

which has sum-of-squares $\sum_{j=1}^{J} e_{ij}^2$ where e_{ij} is a typical element of $\mathbf{e}_i$. This is then also the minimum sum-of-squares which can be attained due to the use of an orthogonal projection. It is convenient to choose $\mathbf{p}_1$ to have length one since it simplifies notation because then $\mathbf{p}_1^t\mathbf{p}_1 = 1$. The vector $\mathbf{p}_1$ is called a loading and the scalar t_{i1} which is the coordinate of the projected point on $\mathbf{p}_1$ is called the score of $\mathbf{x}_i$ on $\mathbf{p}_1$. When all rows of $\mathbf{X}$ are thus projected onto $\mathbf{p}_1$, then these projected points can be described by

$$\hat{\mathbf{X}} = \mathbf{t}_1\mathbf{p}_1^t \tag{2.27}$$

where the vector $\mathbf{t}_1$ contains the elements t_{i1} and the corresponding residuals are

$$\mathbf{E} = \mathbf{X} - \hat{\mathbf{X}} = \mathbf{X} - \mathbf{t}_1\mathbf{p}_1^t \tag{2.28}$$

where $\mathbf{E}$ contains the vectors $\mathbf{e}_i^t$ in its rows. The question is now how to select $\mathbf{p}_1$ in such a way that the total sum of squares of $\mathbf{E}$ is minimized. This seems like a complicated problem, but this problem is actually easy to solve. Using the fact that $\mathbf{t} = \mathbf{Xp}$ (assuming $\mathbf{p}$ of length one and post-multiplying Equation 2.28 by $\mathbf{p}$ and using $\mathbf{Ep} = 0$ due to the regression step, suppressing the subscripts) the problem becomes

$$\min_{\|\mathbf{p}\|=1} \|\mathbf{E}\|^2 = \min_{\|\mathbf{p}\|=1} \|\mathbf{X} - \mathbf{Xpp}^t\|^2 \tag{2.29}$$

and using the relationship between the Frobenius norm and the trace (see Section 1.7.3) this can be written as

$$\min_{\|\mathbf{p}\|=1} \|\mathbf{X} - \mathbf{Xpp}^t\|^2 = \min_{\|\mathbf{p}\|=1} tr(\mathbf{X} - \mathbf{Xpp}^t)^t(\mathbf{X} - \mathbf{Xpp}^t) \tag{2.30}$$

which after using cyclic permutations can be written as

$$\min_{\|\mathbf{p}\|=1} tr(\mathbf{X}^t\mathbf{X}) - tr(\mathbf{p}^t\mathbf{X}^t\mathbf{Xp}) \tag{2.31}$$

and since $tr(\mathbf{X}^t\mathbf{X})$ is fixed the problem of finding the optimal $\mathbf{p}$ can be phrased as

$$\max_{\|\mathbf{p}\|=1} tr(\mathbf{p}^t\mathbf{X}^t\mathbf{Xp}) \tag{2.32}$$

which is an easy problem since $\mathbf{X}^t\mathbf{X}$ is Gramian thus $\mathbf{p} = \mathbf{v}_1$ where $\mathbf{v}_1$ is the first eigenvector of $\mathbf{X}^t\mathbf{X}$ corresponding to the largest eigenvalue of $\mathbf{X}^t\mathbf{X}$, see Equation 2.7.

When one dimension is not enough to approximate $\mathbf{X}$ adequately, then more components can be selected. It is convenient to select the loadings $\mathbf{p}_r, r = 1, \ldots, R$ orthogonal to each other and of length one. These vectors are collected in the loading matrix $\mathbf{P}(J \times R)$ with $\mathbf{P}^t\mathbf{P} = \mathbf{I}$. This then results in the following model for $\mathbf{X}$:

$$\mathbf{X} = \mathbf{TP}^t + \mathbf{E} = \hat{\mathbf{X}} + \mathbf{E} \tag{2.33}$$

with $\mathbf{T} = \mathbf{XP}$ and residuals collected in $\mathbf{E}$ with minimum sum-of-squares. A geometrical summary of PCA is given in Figure 2.9. In this figure, two axes $\mathbf{p}_1$ and $\mathbf{p}_2$ are used to span a two-dimensional subspace.

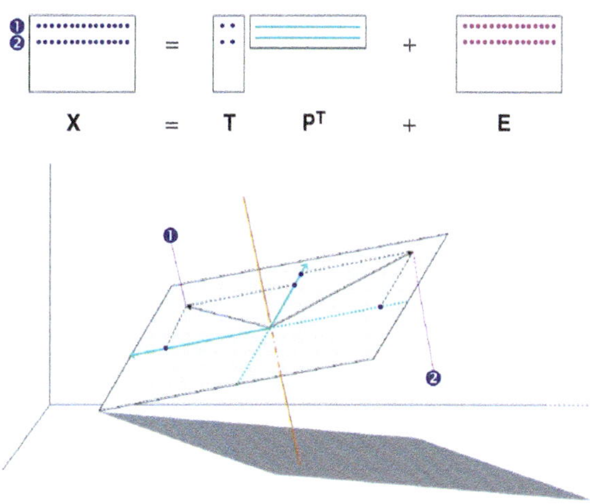

Figure 2.9 The two green lines are the loadings, and the blue dots numbered 1 and 2 are two rows of X. These rows are projected onto the gray model plane resulting in projected points with scores on the axes spanning the model plane. This model plane is spanned by the loadings. Source: Reproduced with permission from H.J. Ramaker (TIPb, The Netherlands).

It is customary to judge the quality of the approximation of **X** by a PCA model using the diagnostic

$$ExplVar_R = 100\left(1 - \frac{||\mathbf{E}||^2}{||\mathbf{X}||^2}\right)\% \tag{2.34}$$

where $ExplVar_R$ is the abbreviation of explained variance after having used R principal components. By increasing the number of components R, the sum-of-squares of the residuals will become smaller and the $ExplVar_R$ becomes higher. There are very many ways to select the number of components such as using the $ExplVar_R$ as a function of R (with the risk of overfitting), using cross-validation (Bro *et al.*, 2008; Camacho and Ferrer, 2012, 2014) or permutation testing (Endrizzi *et al.*, 2013; Vitale *et al.*, 2017). There is a lot of discussion about how to choose the optimal number of components, but this is situation dependent.

Some remarks are appropriate:

- Calculating subsequent principal components can be done by repeating the whole analysis using the residuals **E** from Equation 2.28; a procedure known as deflation or by using the SVD of **X**. The latter rests on a theorem of Eckart–Young stating that the SVD has least squares properties: the SVD of **X** truncated to R components (*i.e.*, $\mathbf{U}_R\mathbf{S}_R\mathbf{V}_R^t$) is the best rank R approximation of **X** in the LS sense (Eckart and Young, 1936).
- By working out $\mathbf{T}^t\mathbf{T} = \mathbf{P}^t\mathbf{X}^t\mathbf{X}\mathbf{P} = \mathbf{D}$ it is clear that the columns of **T** are orthogonal (see also Equation 2.6). Hence, this is not an active constraint but a mathematical property of the SD.
- The columns of **T** are in the column-space of **X** as is clear from the equation $\mathbf{T} = \mathbf{X}\mathbf{P}$. It can also be shown that **P** is in the row-space of **X** (see Ten Berge, 1993). Considering Equation 2.33 for fixed regressors **T** the **P** is solved as $\mathbf{P} = \mathbf{X}^t\mathbf{T}(\mathbf{T}^t\mathbf{T})^{-1}$, see Section 2.2.1.1. Hence, **P** is in the row-space of **X**.
- It is sometimes claimed that PCA assumes normality. Nothing in the above derivation which followed the original ideas uses that fact, so this is not true.
- PCA uses projections and consequently sums of squared residuals. From a modeling perspective, it may, therefore, seem reasonable that the residuals have a mean zero and are of comparable size (*i.e.*, are homoscedastic). This means that for heteroscedastic data, PCA has to be adapted to those situations, *e.g.*, by using weighted least squares as a criterion (Wentzell *et al.*, 1997).
- An alternative view of PCA is due to Hotelling (1933) and starts with the problem of finding a linear combination of **X** with the highest variance. It can be shown that this leads to exactly the same solution as the one shown above (Smilde *et al.*, 2022).

2.2 Statistical Background

This section discusses some basic statistical concepts and issues that are needed in the rest of this book. For a full description of the different topics, we will give references to appropriate textbooks where needed.

2.2.1 Estimation Methods

There are many ways to estimate parameters in a model. In this section, we will describe the most used methods namely least squares and maximum likelihood. Sometimes these two methods give the same results but not always.

2.2.1.1 Least Squares

The least squares (LS) method of estimating parameters goes back to Gauss and Legendre at around 1800. We already encountered the principle of LS when discussing PCA and showed that it was closely related to orthogonal projection. The basic equation is:

$$\mathbf{X} = \mathbf{M}(\boldsymbol{\beta}) + \mathbf{E} \tag{2.35}$$

and the related problem:

$$\min_{\boldsymbol{\beta}} ||\mathbf{X} - \mathbf{M}(\boldsymbol{\beta})||^2 = \min_{\boldsymbol{\beta}} ||\mathbf{E}||^2 \tag{2.36}$$

where $\mathbf{X}$ is an $(I \times J)$ matrix of measured values and $\mathbf{M}(\boldsymbol{\beta})$ is an $(I \times J)$ matrix of values according to a model parametrized by $\boldsymbol{\beta}$.

The simplest example of an LS-solution is the mean of a set of values x_i which can be obtained by solving

$$\min_{\mu} \sum_{i=1}^{I} (x_i - \mu)^2 \tag{2.37}$$

and differentiating the term $\sum_{i=1}^{I} (x_i - \mu)^2$ with respect to μ, and equating this to zero gives

$$\sum_{i=1}^{I} (x_i - \mu) = 0 \tag{2.38}$$

which gives the familiar estimate for the mean of x_i:

$$\hat{\mu} = \bar{x} = \frac{1}{I} \sum_{i=1}^{I} x_i \tag{2.39}$$

which is indeed minimizing problem 2.37 since the second derivative of $\sum_{i=1}^{I} (x_i - \mu)^2$ with respect to μ is 2 which is positive.

In Section 2.2.2 we will discuss the familiar regression equation

$$\mathbf{y} = \mathbf{X}\boldsymbol{\beta} + \epsilon \tag{2.40}$$

which has to be solved for $\boldsymbol{\beta}$. We can again formulate this as an LS problem by solving:

$$\min_{\boldsymbol{\beta}} ||\mathbf{y} - \mathbf{X}\boldsymbol{\beta}||^2 \tag{2.41}$$

and differentiating the term $||\mathbf{y} - \mathbf{X}\boldsymbol{\beta}||^2$ with respect to $\boldsymbol{\beta}$ and equating this to zero gives:

$$(\mathbf{X}^t\mathbf{X})\mathbf{b} = \mathbf{X}^t\mathbf{y} \tag{2.42}$$

which are called the normal equations. If $\mathbf{X}$ has full column-rank then the solution is

$$\mathbf{b} = (\mathbf{X}^t\mathbf{X})^{-1}\mathbf{X}^t\mathbf{y} \tag{2.43}$$

and the second derivative of $||\mathbf{y} - \mathbf{X}\boldsymbol{\beta}||^2$ with respect to $\boldsymbol{\beta}$ is $(\mathbf{X}^t\mathbf{X})$ which is a (semi-)positive definite matrix. Hence, the solution of Equation 2.42 is indeed a minimum.

It can be shown that this LS estimator for the regression coefficient $\mathbf{b}$ is BLUE: the Best Linear Unbiased Estimator. This means that in the class of linear estimators (*i.e.*, linear functions of the data $\mathbf{y}$) and under the assumption that the model is correct, this estimator is unbiased and has the lowest variance (Johnston, 1963). Note that also in this case (as in the case of PCA) we do not need to assume normality.

2.2.1.2 Maximum Likelihood

Maximum likelihood (ML) is another workhorse for parameter estimation. In this case, we need to assume a statistical distribution of the data. The example of estimating the mean of x_i goes as follows, assuming that x_i follows a normal distribution $N(\mu, \sigma^2)$. For a single observation x_i it holds that its probability density is

$$f_X(x_i) = \frac{1}{\sqrt{2\pi}\sigma} e^{-\frac{1}{2}\left(\frac{x_i-\mu}{\sigma}\right)^2} \tag{2.44}$$

which is a function of x_i given the parameters μ and σ^2. Yet, this function can also be seen differently, namely a function of μ given x_i, σ^2 and it is customary to call this the likelihood function. Since we assume that all $x_i, i = 1, \dots, I$ are random drawings, the likelihood for μ given the vector $\mathbf{x} = (x_1, \dots, x_I)^t$ is

$$L(\mu|\mathbf{x}, \sigma^2) = \prod_{i=1}^{I} \frac{1}{\sqrt{2\pi}\sigma} e^{-\frac{1}{2}\left(\frac{x_i-\mu}{\sigma}\right)^2} \tag{2.45}$$

which is the function that needs to be maximized with respect to μ. Maximizing this function is the same as maximizing the logarithm of this function which is

$$logL(\mu|\mathbf{x}, \sigma^2) = I \cdot log\frac{1}{\sqrt{2\pi}\sigma} - \frac{1}{2}\sum_{i=1}^{I}\left(\frac{x_i - \mu}{\sigma}\right)^2 \tag{2.46}$$

and since maximizing $logL$ is the same as minimizing $-logL$ the problem can be written as

$$\min_{\mu} \ - logL(\mu|\mathbf{x}, \sigma^2) = \min_{\mu} \frac{1}{2}\sum_{i=1}^{I}\left(\frac{x_i - \mu}{\sigma}\right)^2 \tag{2.47}$$

where the constant term $log\frac{1}{\sqrt{2\pi}\sigma}$ has been dropped since it is immaterial for the minimization (assuming a fixed σ^2). This is seen to be exactly the same problem as in the LS estimation of the mean in Section 2.2.1.1, and the solution is the same: $\hat{\mu} = \frac{1}{I}\sum_{i=1}^{I} x_i$. Indeed, under the assumption of normality, this LS estimator is also ML (Johnston, 1963).

The ML estimators have certain desirable properties but are not by definition unbiased. The ML estimator for the variance of a normally distributed variable is biased downward (see Elaboration 2.4). The ML estimator does not take into account that one degree of freedom is lost by using $\bar{x}$ in the estimator. A method often used in the context of linear mixed models is Restricted Maximum Likelihood or REML, see Elaboration 2.4.

ELABORATION 2.4

Restricted Maximum Likelihood

Restricted Maximum Likelihood (REML) is a procedure to estimate parameters related to Maximum Likelihood (ML). Where ML estimates parameters using a fit of all available data, REML uses a likelihood function calculated from a transformed set of data. As an example take the estimation of the variance of a normally distributed variable x.

The log-likelihood function is

$$logL(\mu, \sigma^2, x_1, \dots, x_I) = -\frac{I}{2}log(2\pi\sigma^2) - \sum_{i=1}^{I}\frac{(x_i - \mu)^2}{2\sigma^2} \tag{2.48}$$

In general the expected value μ will not be known, and $\bar{x} = \frac{1}{I}\sum_{i=1}^{I} x_i$ is its estimate. The maximum likelihood estimator (MLE) is then defined as:

$$\hat{\sigma}^2 = \frac{1}{I}\sum_{i=1}^{I}(x_i - \bar{x})^2 \tag{2.49}$$

The $E(\hat{\sigma}^2) = \frac{I-1}{I}\sigma^2$ which is biased downward. Hence, the familiar estimator $s^2 = \frac{1}{I-1}\sum_i (x_i - \bar{x})^2$ is indeed unbiased. To obtain such an unbiased estimate of σ^2 we should consider that $\mathbf{x} \sim N(\mu\mathbf{1}_I, \sigma^2 I_I)$, where $\mathbf{1}_I$ is an I-dimensional vector of ones and $\mathbf{I}_I$ is the identity matrix $(I \times I)$. $\mathbf{A}(I \times (I-1))$ is a matrix of $(I-1)$ columns that are all orthogonal to $\mathbf{1}_I$. Then $\mathbf{u} = \mathbf{A}^t\mathbf{x}$ is a vector of error contrasts and $\mathbf{u} \sim N(\mathbf{0}, \sigma^2\mathbf{A}^t\mathbf{A})$. The MLE for σ^2 then yields $\hat{\sigma}^2 = \mathbf{x}\mathbf{A}(\mathbf{A}^t\mathbf{A})^{-1}\mathbf{A}^t\mathbf{x}/(I-1)$. This estimate for σ^2 is called the restricted maximum likelihood (REML) estimate, as it is restricted to $(I-1)$ error contrasts and is unbiased. Intuitively, the influence of estimating μ is accounted for. Note that any matrix $\mathbf{A}$ that satisfies the condition of having $(I-1)$ linearly independent columns orthogonal to $\mathbf{1}_I$ will lead to the same REML estimate (Verbeke *et al.*, 2009).

Whereas in many cases it is sufficient to use the normal distribution, this is not always the only option, see Elaboration 2.5.

ELABORATION 2.5

Different types of distributions

In some cases in the life sciences, we need other distributions than the normal distribution. For instance, RNAseq data or microbiome data are count data and need specialized distributions such as the Poisson distribution:

$$p_X(x) = \begin{cases} e^{-\mu}\frac{\mu^x}{x!}, & x = 0, 1, 2, \ldots \\ 0, & \text{otherwise} \end{cases} \tag{2.50}$$

which has both expected value and variance equal to μ.

Another important distribution for count data is the Negative Binomial (NB) distribution:

$$p_X(x) = \frac{\Gamma(x+\alpha^{-1})}{x!\Gamma(\alpha^{-1})}\left(\frac{\alpha^{-1}}{\mu+\alpha^{-1}}\right)^{\alpha^{-1}}\left(\frac{\mu}{\mu+\alpha^{-1}}\right)^{x} \tag{2.51}$$

where Γ is the Gamma-function and α, μ are parameters. The expected value is μ and the variance is $\mu(1+\alpha\mu)$.

In microbiome data there are usually many zeros and this can, to some extent, be treated with specialized distributions such as the Zero-Inflated Poisson distribution or the Zero-Inflated Negative Binomial distribution.

2.2.2 Regression Methods

Almost all methods described in this book can be put under the umbrella of regression methods. This section discusses the most important ones. For a more detailed analysis, see, *e.g.*, Draper and Smith (1998).

2.2.2.1 Multiple Linear Regression: Full Rank Case

The basic regression model was already briefly introduced in Section 2.2.1.1 and is repeated here:

$$\mathbf{y} = \mathbf{X}\boldsymbol{\beta} + \boldsymbol{\epsilon} \tag{2.52}$$

where $\mathbf{y}$ is a vector of size $(I \times 1)$ containing the measured response or dependent variable; the matrix $\mathbf{X}(I \times J)$ contains the regressors or independent variables; $\boldsymbol{\beta}$ is the vector of unknown regression weights and $\boldsymbol{\epsilon}$ is a vector of residuals. Upon assuming that $\mathbf{X}$ has full column-rank (which implies that $J \leq I$), then the LS solution for $\boldsymbol{\beta}$ is $\mathbf{b} = (\mathbf{X}^t\mathbf{X})^{-1}\mathbf{X}^t\mathbf{y}$ which can also be written as $\mathbf{b} = \mathbf{X}^{+}\mathbf{y}$. The fitted values $\hat{\mathbf{y}}$ can now be written as $\hat{\mathbf{y}} = \mathbf{X}\mathbf{b} = \mathbf{X}(\mathbf{X}^t\mathbf{X})^{-1}\mathbf{X}^t\mathbf{y}$, and the matrix $\mathbf{X}(\mathbf{X}^t\mathbf{X})^{-1}\mathbf{X}^t$ is called the hat matrix: it "puts the hat on $\mathbf{y}$."

Upon assuming that the model is correct, this estimator is unbiased (*i.e.*, $E(\mathbf{b}) = \boldsymbol{\beta}$) and has variance/covariance matrix $\hat{\sigma}^2(\mathbf{X}^t\mathbf{X})^{-1}$ where $\hat{\sigma}^2 = \frac{1}{I-J}\sum_i e_i^2$. All this is standard Multiple Linear Regression (MLR) theory. The geometry of MLR is shown in Figure 2.7. The regression coefficients are sometimes indicated with $\mathbf{b}_{MLR}$.

From the equation of $\mathbf{b}$ and its variance/covariance matrix, it becomes clear that problems can occur when the columns in $\mathbf{X}$ are (highly) collinear. This can be shown by using the SVD of $\mathbf{X}$ ($\mathbf{X} = \mathbf{U}\mathbf{S}\mathbf{V}^t$). Then

$$(\mathbf{X}^t\mathbf{X})^{-1} = \mathbf{V}\mathbf{S}^{-2}\mathbf{V}^t = \sum_{j=1}^{J} \frac{1}{s_j^2}\mathbf{v}_j\mathbf{v}_j^t \tag{2.53}$$

where $\mathbf{V} = [\mathbf{v}_1, \ldots, \mathbf{v}_J]$ and s_j is the jth diagonal element of $\mathbf{S}$. When the columns of $\mathbf{X}$ are highly collinear then some singular values are very small. Equation 2.53 shows that such singular values can inflate the variance/covariance of the estimator $\mathbf{b}$ considerably since the squared reciprocal of such singular values appears in this equation.

There are many ways to combat the consequences of highly collinear $\mathbf{X}$ matrices. Some of these will be described in the following sections.

2.2.2.2 Multiple Linear Regression Using Dummy Variables

ANOVA models can be written as multiple linear regression models once the treatment design is encoded in a design matrix $\mathbf{X}$. The model then becomes simply $\mathbf{y} = \mathbf{X}\boldsymbol{\theta} + \epsilon$ which can be solved by LS.[6] The advantage of writing an ANOVA model in terms of a regression equation is that it provides a general framework for different ANOVA models and also facilitates estimating effects in the case of unbalanced data. Examples of different coding schemes and subsequent regression models are given in Section 3.4. This can easily be extended to the multivariate case through multivariate multiple regression models such as $\mathbf{Y} = \mathbf{X}\boldsymbol{\Theta} + \mathbf{E}$.

2.2.2.3 Multiple Linear Regression: Rank Deficient Case

When the standard ANOVA model is described in terms of an MLR problem, there are exact linear relationships between the columns of $\mathbf{X}$. This means that standard MLR theory cannot be applied since the inverse of $\mathbf{X}^t\mathbf{X}$ does not exist. Solving the LS problem gives indeed the normal equations:

$$(\mathbf{X}^t\mathbf{X})\hat{\boldsymbol{\theta}} = \mathbf{X}^t\mathbf{y} \tag{2.54}$$

but these equations have an infinite number of solutions for $\hat{\boldsymbol{\theta}}$, as illustrated by Figure 2.8. The traditional solution to this problem in the ANOVA literature is to impose restrictions on the

6 We will use the symbol $\boldsymbol{\theta}$ to notate the parameters to be estimated in ANOVA-regression models.

estimated parameters which translates into using a specific generalized inverse to solve the normal equations. Alternatively, certain coding schemes can be used that generate full-rank design matrices. This is also linked to the type of constraints put on the factor estimates and the type of generalized inverse. This is explained in more detail in Section 3.4.

2.2.2.4 Penalized Regression

A way to deal with a rank-deficient or very collinear regression matrix **X** is by using penalized regression methods. A brief explanation is given here and based on Hastie *et al.* (2009).

One of the oldest ideas to regularize ill-posed regression problems is ridge regression (Hoerl and Kennard, 1970). Ridge regression solves the problem

$$\min_{\beta} ||\mathbf{y} - \mathbf{X}\boldsymbol{\beta}||^2 + \lambda||\boldsymbol{\beta}||^2 \tag{2.55}$$

where λ is a small positive value. The term $||\boldsymbol{\beta}||^2$ in this minimization problem is the sum of squared regression coefficients which is thus bounded. This is easier to see when rewriting the ridge problem as:

$$\min_{\beta} ||\mathbf{y} - \mathbf{X}\boldsymbol{\beta}||^2; \quad s.t. \quad ||\boldsymbol{\beta}||^2 \leq \kappa \tag{2.56}$$

where κ is now the meta-parameter and is one-to-one related to λ. It can be shown that the $\boldsymbol{\beta}$ solving this problem is

$$\mathbf{b} = (\mathbf{X}^t\mathbf{X} + \lambda\mathbf{I})^{-1}\mathbf{X}^t\mathbf{y} \tag{2.57}$$

which clearly shows that the collinearity problem is solved by adding small values to the diagonal of $\mathbf{X}^t\mathbf{X}$ giving this matrix a better condition (*i.e.*, having better numerical stability properties).

Ridge regression tends to make the regression coefficient smaller but not necessarily zero, where the latter would simplify the regression model considerably. The LASSO method does exactly that and solves:

$$\min_{\beta} ||\mathbf{y} - \mathbf{X}\boldsymbol{\beta}||^2; \quad s.t. \quad ||\boldsymbol{\beta}||_1 \leq \kappa \tag{2.58}$$

where $||\boldsymbol{\beta}||_1$ is the L_1 norm which can also be written as $\sum_j|\beta_j|$ which simply sums the absolute values of the regression coefficients. This can generate zeros in these regression coefficients, thereby performing an implicit variable selection (Tibshirani, 1996). There have been many advancements reported based on the LASSO idea, and one of them is the Group LASSO (see Elaboration 5.1). The LASSO has certain disadvantages that are solved by the Elastic Net which solves the problem:

$$\min_{\beta} ||\mathbf{y} - \mathbf{X}\boldsymbol{\beta}||^2 + \lambda_1||\boldsymbol{\beta}||^2 + \lambda_2||\boldsymbol{\beta}||_1 \tag{2.59}$$

which combines an L_1 and L_2 norm (Zou and Hastie, 2003). In all these cases, the meta-parameters λ_1 and λ_2 need to be optimized; typically this is done by cross-validation.

2.2.2.5 Principal Component Regression

For cases of highly collinear or even rank-deficient **X** matrices, another route to solve the MLR problem of Equation 2.52 is to start with a PCA-model of **X**:

$$\mathbf{X} = \mathbf{T}\mathbf{P}^t + \mathbf{E}_X \tag{2.60}$$

where R principal components are selected and $R \ll \min(I, J)$. Subsequently, **y** is regressed on **T**:

$$\mathbf{y} = \mathbf{T}\mathbf{b}_{PCR} + \mathbf{e}_y \tag{2.61}$$

and since $\mathbf{T} = \mathbf{XP}$ (assuming orthonormal columns in $\mathbf{P}$) this can be written as

$$\mathbf{y} = \mathbf{XPb}_{PCR} + \mathbf{e}_y = \mathbf{X}\widetilde{\mathbf{b}}_{PCR} + \mathbf{e}_y \tag{2.62}$$

where $\widetilde{\mathbf{b}}_{PCR} = \mathbf{Pb}_{PCR}$ which resembles the MLR problem. It is important to stress that Equation 2.62 is not the LS solution for the MLR problem of Equation 2.52, and the residual sum-of-squares of Equation 2.62 is larger than that of Equation 2.52 using an LS-based estimator. The method described above is called Principal Component Regression (PCR).

If all the principal components of $\mathbf{X}$ are selected in the subsequent PCR regression model, then this solution coincides with the Moore–Penrose solution $\mathbf{b}_{MP} = \mathbf{X}^+\mathbf{y}$.

2.2.2.6 Partial Least Squares

Whereas PCR is a two-step method to solve the MLR problem for collinear data, it is also possible to use a one-step method called Partial Least Squares or PLS for short. This method is abundantly used in chemometrics and beyond and goes back to the 1980s (Wold *et al.*, 1984). The basic machinery for PLS is based on maximizing the covariance of a linear combination of the columns of $\mathbf{X}$, expressed as $\mathbf{t} = \mathbf{Xw}$, and $\mathbf{y}$:

$$\max_{\mathbf{w}} cov(\mathbf{t}, \mathbf{y}) \tag{2.63}$$

under the constraint that $||\mathbf{w}|| = 1$, to make this maximization meaningful. After having found this first PLS component, the matrix $\mathbf{X}$ is deflated by solving:

$$\mathbf{X} = \mathbf{tp}^t + \mathbf{E}_1 \tag{2.64}$$

for the loadings $\mathbf{p}$ which is an ordinary least squares problem. The PLS model is now continued using the residuals $\mathbf{E}_1$ which are orthogonal to $\mathbf{t}$ giving the second PLS component. This process is continued until a proper number of components is available (*e.g.*, selected by cross-validation). The final PLS model can now be phrased as:

$$\begin{aligned}
\mathbf{X} &= \mathbf{TP}^t + \mathbf{E}_X \\
\mathbf{T} &= \mathbf{XR} = \mathbf{XW}(\mathbf{P}^t\mathbf{W})^{-1} \\
\mathbf{y} &= \mathbf{Tq} + \mathbf{e}_y = \mathbf{XRq} + \mathbf{e}_y = \mathbf{Xb}_{PLS} + \mathbf{e}_y
\end{aligned} \tag{2.65}$$

where $\mathbf{T} = [\mathbf{t}_1, \ldots, \mathbf{t}_R]$, $\mathbf{W} = [\mathbf{w}_1, \ldots, \mathbf{w}_R]$, $\mathbf{P} = [\mathbf{p}_1, \ldots, \mathbf{p}_R]$, $\mathbf{R}$ are defined implicitly and $\mathbf{q}$ is a vector of regression coefficients.

The advantage of PLS over PCR is that the PLS finds components in the column-space of $\mathbf{X}$ that are also related to $\mathbf{y}$; this is not necessarily the case for PCR. PLS can also be seen as a solution of the regression Equation 2.52 but similar to PCR, does not provide an LS estimator. Since both PLS and PCR take successive components of $\mathbf{X}$, they both efficiently deal with multicollinearity of ill-posed regression problems. It can be shown that the PLS solution is an example of an oblique projection, see Elaboration 2.6.

ELABORATION 2.6

PLS and oblique projections

The PLS regression vector can be written as (Phatak and De Jong, 1997):

$$\mathbf{b}_{PLS} = \mathbf{W}(\mathbf{P}^t\mathbf{W})^{-1}\mathbf{P}^t\mathbf{b}_{MLR} \tag{2.66}$$

and upon examining the oblique projections (see Section "Formal Description of Projections") this shows that the vector $\mathbf{b}_{MLR}$ is projected onto $R(\mathbf{W})$ along the orthogonal complement of $R(\mathbf{P})$.

The rationale of this is that in case of a collinear matrix **X** the MLR regression vector tends to become unstable and has high variance. This is circumvented by projecting onto the stable subspace $R(\mathbf{W})$, and PLS is shrinking the MLR regression coefficients (De Jong, 1995) thereby stabilizing these coefficients. This is effectuated by the oblique projection.

Another way to understand and interpret PLS is by using the concept of target projections, see Elaboration 2.7 (Kvalheim and Karstang, 1989; Kvalheim, 2010).

ELABORATION 2.7

Target projection

Target projection was introduced in the context of PLS and bilinear modeling, in general, to facilitate interpretation as it results in a single predictive component for each response constituting the dependent block. When considering a generic regression problem that can be solved by PLS regression, according to:

$$\mathbf{y} = \hat{\mathbf{y}} + \mathbf{e}_y = \mathbf{X}\mathbf{b}_{PLS} + \mathbf{e}_y = \mathbf{T}\mathbf{q} + \mathbf{e}_y \tag{2.67}$$

the target projection is obtained by projecting the matrix **X** onto the PLS regression vector $\mathbf{b}_{PLS}$:

$$\mathbf{t}_{TP} = \mathbf{X}\frac{\mathbf{b}_{PLS}}{||\mathbf{b}_{PLS}||} = \frac{\hat{\mathbf{y}}}{||\mathbf{b}_{PLS}||} \tag{2.68}$$

Given these scores, the corresponding target projection loadings can be calculated as:

$$\mathbf{p}_{TP} = \frac{\mathbf{X}^t\mathbf{t}_{TP}}{\mathbf{t}_{TP}^t\mathbf{t}_{TP}} \tag{2.69}$$

Then, the response-related fraction of the X variability can be reconstructed as:

$$\mathbf{X}_{TP} = \mathbf{t}_{TP}\mathbf{p}_{TP}^t \tag{2.70}$$

When the regression problem involves more than one response to be predicted, target projection scores and loadings can be calculated for each of the columns of the dependent matrix according to the same procedure.

The main advantage of TP lies in the possibility of obtaining a single predictive component for each modeled response, even in cases where the regression model involves more than one latent variable. It has been shown how the TP scores are proportional to the predicted response, while the amount of variance associated with a response explained by each predictor is given by the square of the corresponding TP loadings. Interpretation is then facilitated since only a single component has to be inspected for each investigated response.

2.2.2.7 Redundancy Analysis

A regression method that is also useful in the context of HD-ANOVA approaches is called redundancy analysis (Davies and Tso, 1982) or RDA. The problem statement is as follows:

$$\mathbf{Y} = \mathbf{X}\mathbf{B} + \mathbf{E}_1;\ s.t.\ rank(\mathbf{B}) \leqslant s \tag{2.71}$$

which is multivariate multiple regression with a constraint on the rank of the matrix of the regression coefficients **B**, *i.e.*, a meta-parameter of the method and needs to be specified *a priori*. The sum of squared residuals in Equation 2.71 is now minimized subject to (*s.t.*) the rank constraint. We will denote the unconstrained residuals in the regression problem of Equation 2.71

with the symbol **E**. The rank constraint has several advantages which will become clear in the rest of this book. An easy way to solve the problem is given in Algorithm 2.1.

Algorithm 2.1

SOLVING THE RDA PROBLEM

The RDA problem can be solved in the following way:

1) Calculate the unrestricted matrix $\hat{\mathbf{B}} = \mathbf{X}^{+}\mathbf{Y}$ minimizing the unrestricted residuals $\mathbf{E}$ ($||\mathbf{E}||^2$) of Equation 2.71.
2) Calculate the fitted values $\hat{\mathbf{Y}} = \mathbf{X}\hat{\mathbf{B}} = \mathbf{X}\mathbf{X}^{+}\mathbf{Y}$.
3) Perform an SVD on $\hat{Y}$ and truncate to R components: $\hat{\mathbf{Y}}_R = \mathbf{U}_R\mathbf{S}_R\mathbf{V}^t_R$.
4) Then $\tilde{\boldsymbol{B}} = \hat{\mathbf{B}}\mathbf{V}_R\mathbf{V}^t_R$ are the required regression coefficients

where clearly the rank of $\tilde{\mathbf{B}} \leqslant R$. Hence, the final regression equation can be written as $\mathbf{Y} = \mathbf{X}\tilde{\mathbf{B}} + \mathbf{E}_1 = \hat{\mathbf{Y}}_R + \mathbf{E}_1$. Some properties are the following:

1) $\mathbf{E}^t_1\hat{\mathbf{Y}}_R = 0$.
2) $||\mathbf{E}_1||^2 \geqslant ||\mathbf{E}||^2$

where the last inequality is due to the fact that an active restriction on a model will typically reduce the fit. For proofs: see Section 2.3.2.

A PCA of the fitted values of an RDA analysis ($\hat{\mathbf{Y}}_R$) can thus be seen as a PCA on $\mathbf{Y}$ where the components are restricted to be in the column-space of $\mathbf{X}$.

A useful extension of RDA is partial RDA (Davies and Tso, 1982) which goes as follows. The model considered in that case is

$$\mathbf{Y} = \mathbf{X}_1\mathbf{B}_1 + \mathbf{X}_2\mathbf{B}_2 + \mathbf{E}_1;\ s.t.\ rank(\mathbf{B}_2) \leqslant R \tag{2.72}$$

which represents a mixture of concomitant variables in $\mathbf{X}_1$ (additional regressors such as covariates) and a reduced rank part ($\mathbf{X}_2\mathbf{B}_2$). Algorithm 2.2 shows how to calculate the regression coefficients $\mathbf{B}_1$ and $\mathbf{B}_2$ using partial RDA.

Algorithm 2.2

PARTIAL RDA

The partial RDA model can be solved in the following way (Davies and Tso, 1982):

1) Project $\mathbf{X}_2$ on $\mathbf{X}_1$ and define the matrix $\tilde{X}_2 = (\mathbf{I} - \mathbf{X}_1\mathbf{X}_1^{+})\mathbf{X}_2$. By construction $\tilde{X}_2 \perp \mathbf{X}_1$.
2) Rewrite the problem $\mathbf{Y} = \mathbf{X}_1\mathbf{B}_1 + \mathbf{X}_2\mathbf{B}_2 + \mathbf{E}_1$ as $\mathbf{Y} = \mathbf{X}_1(\mathbf{B}_1 + \mathbf{X}_1^{+}\mathbf{X}_2\mathbf{B}_2) + (\mathbf{I} - \mathbf{X}_1\mathbf{X}_1^{+})\mathbf{X}_2\mathbf{B}_2 + \mathbf{E}_1 = \mathbf{X}_1(\mathbf{B}_1 + \mathbf{X}_1^{+}\mathbf{X}_2\mathbf{B}_2) + \tilde{X}_2\mathbf{B}_2 + \mathbf{E}_1$.
3) Perform an RDA of $\mathbf{Y}$ on $\tilde{X}_2$ which gives $\tilde{\boldsymbol{B}}_2$.
4) Regress $\mathbf{Y}$ on $\mathbf{X}_1$ giving $\tilde{\boldsymbol{B}}_1 = \mathbf{X}_1^{+}\mathbf{Y} = \hat{\boldsymbol{B}}_1 + \mathbf{X}_1^{+}\mathbf{X}_2\tilde{\boldsymbol{B}}_2$.
5) $\hat{\boldsymbol{B}}_1 = \tilde{\boldsymbol{B}}_1 - \mathbf{X}_1^{+}\mathbf{X}_2\tilde{\boldsymbol{B}}_2$

The required regression coefficients are now $\hat{\boldsymbol{B}}_1$ and $\tilde{\boldsymbol{B}}_2$. For proofs: see Section 2.3.2.

2.2.3 Significance Tests

In univariate ANOVA, there are tests to decide whether effects and interactions are significant. In this section, we will describe three approaches to this end. Of these approaches, the ones using permutations and likelihood ratio tests are generalized for the high-dimensional case, and the one using normality is used to build confidence ellipsoids (see Section 4.3.1).

2.2.3.1 Classical Tests

The classical tests for ANOVA require normality of the residuals. Taking a two-way ANOVA model with fixed effects as an example (see Chapter 3), this means the following. Suppose a response is measured as a function of a treatment design consisting of two crossed Factors A and B with treatment levels $a = 1, \ldots, A$ and $b = 1, \ldots, B$, respectively, and $n = 1, \ldots, N$ replicates per treatment group.[7] The ANOVA model is then:

$$y_{abn} = \mu + \alpha_a + \beta_b + (\alpha\beta)_{ab} + \epsilon_{abn} \tag{2.73}$$

where α_a represents the treatment effect of Factor A at level a; β_b represents the treatment effect of Factor B at level b; $(\alpha\beta)_{ab}$ represents the interaction effect and ϵ_{abn} are the residuals. These residuals are assumed to be independently, normally distributed with mean zero with variance σ^2. Sums-of-squares (SS) can now be calculated, and this results in the (classical) ANOVA table (see Table 2.3).

Testing proceeds by comparing the F-ratios with the proper critical values of the F-distribution at a chosen significance level. Alternatively, a p-value can be calculated. The null hypothesis of no effect is tested against the alternative that there is an effect. The assumption of normality is needed to arrive at the F-distribution. This is the basic principle of classical significance testing in ANOVA. For other types of designs, the principle is the same but the implementation can be different (Searle, 1971). It is customary to first test the significance of the interactions. If these are significant, then the main effects are always included in the model to aid interpretation.

Table 2.3 ANOVA table for a crossed two-way design.

Source	DF	SS	MS	*F*-ratio
Mean	1	SS_M	MS_M	
Factor A	$(A-1)$	SS_A	$MS_A = \dfrac{SS_A}{(A-1)}$	$\dfrac{MS_A}{MS_R}$
Factor B	$(B-1)$	SS_B	$MS_B = \dfrac{SS_B}{(B-1)}$	$\dfrac{MS_B}{MS_R}$
Interaction	$(A-1)(B-1)$	SS_{AB}	$MS_{AB} = \dfrac{SS_{AB}}{(A-1)(B-1)}$	$\dfrac{MS_{AB}}{MS_R}$
Error	$AB(N-1)$	SS_R	$MS_R = \dfrac{SS_R}{AB(N-1)}$	
Total	ABN	SS_T		

DF is degrees of freedom, SS is sum-of-squares, MS is mean sum-of-squares.

7 We now change to the ANOVA notation using the index n for replicates.

2.2.3.2 Permutation Tests

The idea of permutation tests dates back to Fisher (1935). Nowadays, permutation tests are most often used when there are no mathematical distributions known for the statistic being tested or when the data does not obey the assumptions of the statistical test. We will explain the basic idea by using the two-sample t-test for the difference between the means of the samples. Suppose that two groups of measurements are available and we want to test whether there is a difference between the means of the groups. For simplicity assume that the number of replicates in both groups is the same: $(x_{11}, \ldots, x_{1N})$ and $(x_{21}, \ldots, x_{2N})$, respectively. A reasonable test statistic for the null-hypothesis that $\mu_1 = \mu_2$, is the squared difference in sample means $(\bar{x}_{1.} - \bar{x}_{2.})^2$. To arrive at a test, we could assume that the samples in both groups came from the same normal distribution with $\mu_1 = \mu_2$ and variance σ^2 and continue with the two-sample t-test.

If we cannot assume the samples to be normally distributed, the type 1 error cut-off value cannot be obtained from the t-distribution and must be obtained in another way. An alternative approach is developed by recognizing that under the null-hypothesis of no difference between the groups, the individual measurements x_{1n} and x_{2n} are *exchangeable*: it does not matter from which group they are taken under this null-hypothesis. Hence, we can arbitrarily exchange (*i.e.*, permute) numbers of groups one and two and calculate again the difference between the mean values for the groups. This can be done multiple times, and these differences can be plotted in a histogram, see Figure 2.10. The observed value of the original statistic $(\bar{x}_{1.} - \bar{x}_{2.})^2$ can also be plotted in the same figure (see the red line), and the fraction of the area to the right of this red line under the histogram is then an estimate of the p-value related to the above-mentioned null-hypothesis.

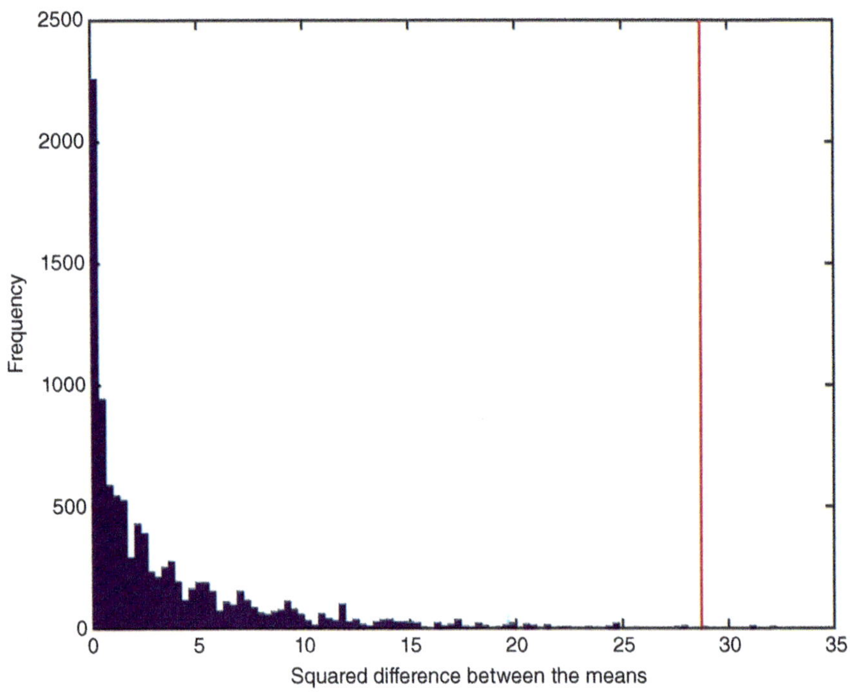

Figure 2.10 The value of the test statistic (squared difference between the means; red vertical bar) is compared to its null distribution non-parametrically evaluated through a permutation test with 10,000 randomizations (blue histogram). A p-value of 0.0023 can be estimated as described in the text.

Some remarks are appropriate:

- This is an example of Fisher's exact test since this type of permutation test gives exactly the correct Type I error (Anderson and Ter Braak, 2003).
- The assumption of exchangeability means that it is assumed that errors are independent and identically distributed; however, the assumption of homoscedasticity remains since a found effect can also be caused by heteroscedasticity (Hayes, 1996; Anderson, 2017).

A more extensive explanation of permutation tests in complex ANOVA models (and HD-ANOVA models) will follow in Chapters 3, 4, and 6.

2.2.3.3 Likelihood Ratio Tests

The Likelihood Ratio (LR) test makes use of likelihood functions and thus depends on an assumed distribution of the data. The basic principle will be explained using the example of testing $H_0 : \mu = \mu_0$ against $H_1 : \mu \neq \mu_0$ in a normally distributed variable. Suppose that a sample $\mathbf{x} = (x_1, \dots, x_N)$ is available from $N(\mu, \sigma^2)$ where σ^2 is assumed to be known. The likelihood is then

$$L(\mu|\mathbf{x}, \sigma^2) = \prod_{n=1}^{N} f(x_n|\mu, \sigma^2) \tag{2.74}$$

where $f(x_n|\mu, \sigma^2)$ is the probability distribution function of the normal distribution. This likelihood under H_0 is now $L(\mu_0|\mathbf{x}, \sigma^2)$. Under H_1 we have to estimate the μ using maximum likelihood, and the estimate is $\bar{x}$. The corresponding likelihood is $L(\bar{x}|\mathbf{x}, \sigma^2)$. Now the ratio of the two likelihoods (the likelihood ratio or LR) can be formed:

$$LR = \frac{L(\mu_0|\mathbf{x}, \sigma^2)}{L(\bar{x}|\mathbf{x}, \sigma^2)} \tag{2.75}$$

and the smaller this LR is, the less credible is H_0. The critical region for the test statistic is therefore $LR \leq c_\alpha$ where α is the pre-selected Type I error and c_α is obtained from the distribution of LR. This distribution is not always known or available, but under H_0, the value $-2lnLR$ has asymptotically a χ^2 distribution (Stuart *et al.*, 2004).

2.3 Association Measures

Multivariate analysis methods and methods for HD-ANOVA make use of association measures between variables. The most prominent ones are the correlation coefficient: either the Pearson correlation or the Spearman rank-correlation. The use of such association measures is not without difficulties as will be shown in this section.

2.3.1 Pearson and Spearman Correlation Coefficients

The Pearson correlation between two variables x and y with values $x_i, i = 1, \dots, I$ and $y_i, i = 1, \dots, I$ is defined as

$$r_P(x, y) = \frac{cov(x, y)}{\sqrt{var(x)}\sqrt{var(y)}} = \frac{\sum_i (x_i - \bar{x})(y_i - \bar{y})}{\sqrt{\sum_i (x_i - \bar{x})^2}\sqrt{\sum_i (y_i - \bar{y})^2}} \tag{2.76}$$

where $\bar{x}$ and $\bar{y}$ are the means of the variables x and y, respectively. The Pearson correlation is a measure of the linear relationship between the variables x and y and takes on values between

−1 and 1. If the vectors $\mathbf{x}_c$ and $\mathbf{y}_c$ collect the centered x and y values, respectively, then the correlation coefficient between x and y is equal to the cosine of the angle between the vectors $\mathbf{x}_c$ and $\mathbf{y}_c$ (see Elaboration 2.1).

Whereas the Pearson correlation assumes continuous data, sometimes data comes in terms of rank-order. If we define rank-order variables r and s such that the values $r_i, i = 1, \ldots, I$ are the rank-orders for the values of x_i and $s_i, i = 1, \ldots, I$ the rank-orders of the corresponding elements of y_i, then the Spearman correlation between x and y is the Pearson correlation between r and s. Such Spearman correlations are also sometimes used when the data is not normally distributed. Both the Pearson as well as the Spearman correlation coefficients are very useful summaries of the relationship between variables and are abundantly used, *e.g.*, in association networks.

The Pearson correlation coefficient measures the linear relationship between variables. There also exist association measures for nonlinear data such as the Mutual Information (Murphy, 2012). This measure is sometimes used for estimating nonlinear relationships but needs a substantial amount of data points to be reasonably stable.

2.3.2 Problems with Correlations

There are several issues regarding the use of correlation coefficients. Some of these are nicely illustrated by Saccenti (2023). The first issue concerns data that are very heterogeneously distributed. An extreme example of this is shown in Figure 2.11 where the correlation almost exclusively depends on a single observation. This may be an extreme case, but it shows that interpreting correlation coefficients should be done with care, and it is always good to plot the data. Of course, for high-dimensional data, the amount of such plots would explode but then other visualization tools should be used.

Another example of problematic correlations is highly relevant for the topic of this book. Suppose that data are collected according to a treatment design. For simplicity, we assume one factor at two levels. Within each treatment group, several samples are collected of which certain measurements are obtained. The data for two of these measurements x and y is shown in Figure 2.12. The overall correlation between all data points is dominated by the group averages (shown as the dotted line), but this does not reflect the essential feature of this data configuration. The relationship between the two variables depends on the level of the treatment factor which becomes clearly visible in this plot. Such patterns are by no means exceptional, see, *e.g.*, Caldana *et al.* (2011). The reason for this problematic situation is that the treatment

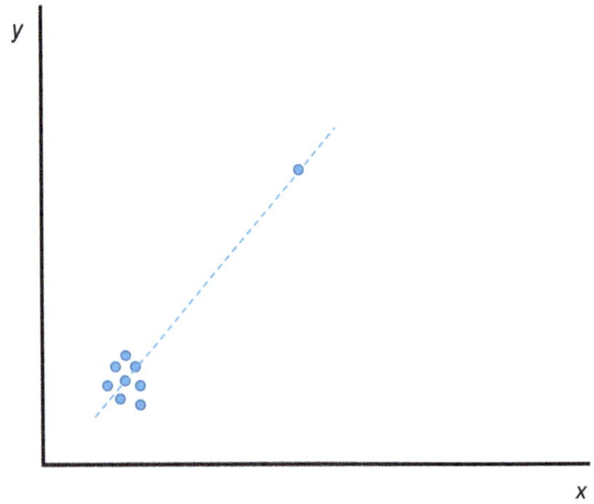

Figure 2.11 A problematic correlation depending completely on only one observation. The dotted line shows the linear relationship measured by the correlation.

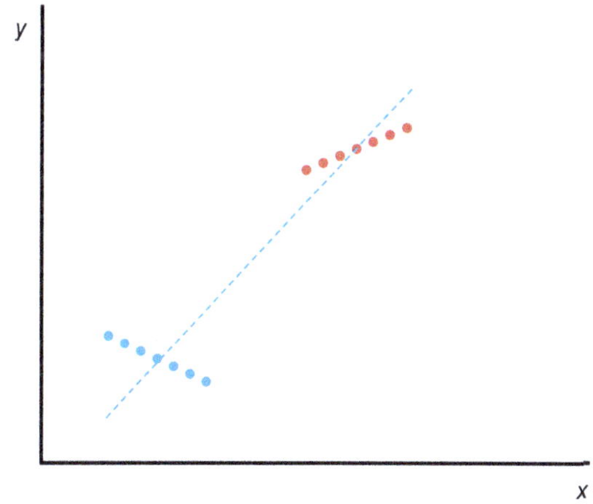

Figure 2.12 Data of two measurements obtained for a treatment factor at two levels. Each dot is a measurement. The blue dots are measured at the low level of the treatment factor and the red dots are measured at the high level of this factor. The dotted line shows the linear relationship measured by the correlation.

effect (between-group variation) is mixed with variation at the individual level (within-group variation). This is an instance of a phenomenon known as Simpson's paradox (Blyth, 1972), and this is exactly the reason for using methods like ASCA.

Correlation coefficients are easily calculated but as a statistical measure of a linear relationship, we should also care about their reliability: how accurate is a correlation coefficient? Elaboration 2.8 shows some results. One of the remedies for unstable correlations that are used in high-dimensional data is to penalize correlation matrices and use shrunken versions of those (Schäfer and Strimmer, 2005; Friedman *et al.*, 2008). This seems to work well in practice (Nørgaard *et al.*, 2021).

ELABORATION 2.8

Reliability of correlations

Some simple simulations can reveal the stability of correlation coefficients. Data are generated using a standard normal bivariate distribution with true correlation $r = 0.4$ or $r = 0.8$, and this is performed for different sample sizes ($n = 10, 50, 100$). For each situation, 1000 drawings are made and the results for the $r = 0.4$ case are shown in Figure 2.13. It is clear that for a sample size of 10, anything between −0.4 and 0.8 can be obtained. This improves a little when moving to larger sample sizes but even

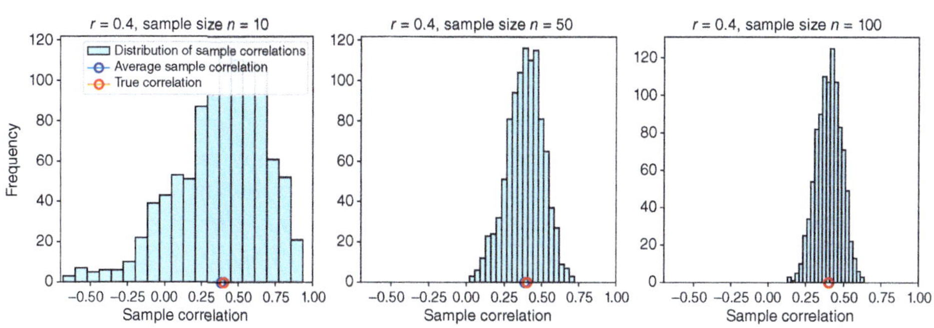

Figure 2.13 Histograms of the distribution of the Pearson correlation between two variables with a true correlation of $r = 0.4$ for different sample sizes based on 1000 drawings. Some blue circles are beneath the red circles.

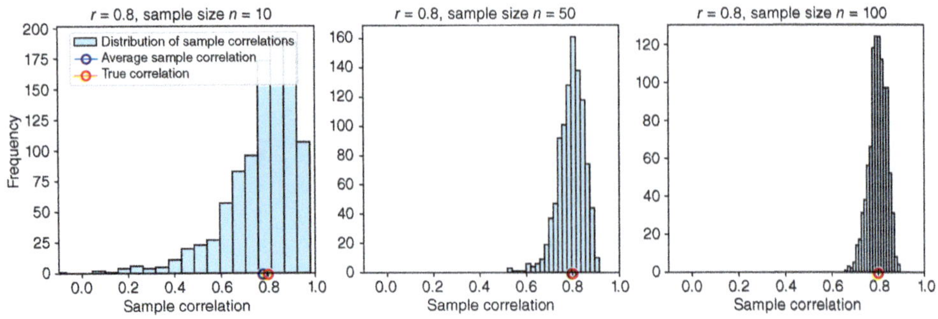

Figure 2.14 Histograms of the distribution of the Pearson correlation between two variables with a true correlation of $r = 0.8$ for different sample sizes based on 100 drawings. Some blue circles are beneath the red circles.

for a sample size of 100, the distribution is still rather wide. There is a bit of improvement when the true correlation is higher ($r = 0.8$; see Figure 2.14), but the conclusion is the same: for a reasonable estimate of a correlation coefficient we need many data points.

The final issue to discuss around correlations plays especially a role in comprehensive omics measurements such as metabolomics and serves as a warning about taking (biological) correlations too seriously. In such measurements, all metabolite abundances are measured in the same sample in "one shot." This means that the measurement errors (that are always present) are no longer independent: a measurement error in metabolite A is correlated with the measurement error in metabolite B (Saccenti *et al.*, 2020). The same phenomenon can be expected for other types of omics measurements such as proteomics and gene-expression. The measurement error structure can be quite complicated and in the case of measurement error correlation, this can severely obscure the "true" biological correlation. To estimate such measurement correlation, the same sample has to be measured multiple times. The differences between these measurements are now solely due to the measurement and not the biology. Such repeated measurements can thus be used to calculate measurement error correlation. Since this amounts to a heavy measurement burden, this approach is never used.

2.A Appendix

Some results for RDA are proved in this appendix. The problem of minimizing $||\mathbf{Y} - \mathbf{XB}||^2$ can be broken down in two parts:

$$||\mathbf{Y} - \mathbf{XB}||^2 = ||\mathbf{Y} - \widehat{\mathbf{Y}} + \widehat{\mathbf{Y}} - \mathbf{XB}||^2 = ||\mathbf{Y} - \widehat{\mathbf{Y}}||^2 + ||\widehat{\mathbf{Y}} - \mathbf{XB}||^2 \tag{2.A.1}$$

where $\widehat{\mathbf{Y}} = \mathbf{X}\widehat{\mathbf{B}}$ (the unrestricted LS estimate) and this holds for any matrix $\mathbf{B}$. This can be shown when writing out the terms $|| \cdot ||^2$ using traces and noting that the cross-product $(\mathbf{Y} - \widehat{\mathbf{Y}})^t(\widehat{\mathbf{Y}} - \mathbf{XB})$ is zero because using $\widehat{\mathbf{Y}} = \mathbf{XX}^+\mathbf{Y}$ gives the cross-product as $(\mathbf{Y} - \widehat{\mathbf{Y}})^t\mathbf{X}(\mathbf{X}^+\mathbf{Y} - \mathbf{B})$ which is zero since the residuals $(\mathbf{Y} - \widehat{\mathbf{Y}})$ are orthogonal to $\mathbf{X}$. Hence, this also holds for the rank-constrained matrix $\widetilde{\mathbf{B}}$:

$$||\mathbf{Y} - \mathbf{X}\widetilde{\mathbf{B}}||^2 = ||\mathbf{Y} - \widehat{\mathbf{Y}}||^2 + ||\widehat{\mathbf{Y}} - \mathbf{X}\widetilde{\mathbf{B}}||^2 \tag{2.A.2}$$

or

$$||\mathbf{E}_1||^2 = ||\mathbf{E}||^2 + ||\mathbf{E}_2||^2 \tag{2.A.3}$$

which shows that $||\mathbf{E}_1||^2 \geqslant ||\mathbf{E}||^2$ where $\mathbf{E}$ is the residual of the unrestricted LS solution.

Concentrating on the part $||\hat{\mathbf{Y}} - \mathbf{X}\tilde{\mathbf{B}}||^2$, it is important to realize that $\hat{\mathbf{Y}} = \mathbf{X}\hat{\mathbf{B}}$ is in the column space of $\mathbf{X}$. Hence, minimizing this expression can be done by performing a PCA on $\hat{\mathbf{Y}}$ since this will automatically give scores in the column-space of $\mathbf{X}$ (see Section 2.1.5). Now suppose that the SVD of $\hat{\mathbf{Y}}$ is $\mathbf{USV}^t$ then truncating the SVD to R components gives $\hat{\mathbf{Y}}\mathbf{V}_R = \mathbf{U}_R\mathbf{S}_R$ and thus $\hat{\mathbf{Y}}\mathbf{V}_R\mathbf{V}_R^t = \mathbf{U}_R\mathbf{S}_R\mathbf{V}_R^t = \hat{\mathbf{Y}}_R$. This results in $\hat{\mathbf{Y}}_R = \mathbf{X}\hat{\mathbf{B}}\mathbf{V}_R\mathbf{V}_R^t = \mathbf{X}\tilde{\mathbf{B}}$ where now the $\tilde{\mathbf{B}}$ are the reduced rank regression coefficients (assuming that $\mathbf{X}$ has full column-rank).

It holds that $\mathbf{E}_1 = \mathbf{E} + \mathbf{E}_2$; $\mathbf{E}^t\mathbf{X}\tilde{\mathbf{B}} = \mathbf{E}^t\hat{\mathbf{Y}}_R = 0$ due to the regression step and $\mathbf{E}_2^t\hat{\mathbf{Y}}_R$ is zero due to the PCA step. Hence, $\mathbf{E}_1^t\hat{\mathbf{Y}}_R = 0$ as stated in the main text.

For partial RDA it holds that $\tilde{\mathbf{B}}_1 = \mathbf{X}_1^+\mathbf{Y} = \mathbf{X}_1^+(\mathbf{X}_1\hat{\mathbf{B}}_1 + \mathbf{X}_2\tilde{\mathbf{B}}_2 + \mathbf{E}_1) = \mathbf{X}_1^+\mathbf{X}_1\hat{\mathbf{B}}_1 + \mathbf{X}_1^+\mathbf{X}_2\tilde{\mathbf{B}}_2 + \mathbf{X}_1^+\mathbf{E}_1 = \hat{\mathbf{B}}_1 + \mathbf{X}_1^+\mathbf{X}_2\tilde{\mathbf{B}}_2$ where $\mathbf{X}_1^+\mathbf{E}_1 = 0$ because of the regression step and which assumes that $\mathbf{X}_1$ has full column-rank. Thus, $\hat{\mathbf{B}}_1 = \tilde{\mathbf{B}}_1 - \mathbf{X}_1^+\mathbf{X}_2\tilde{\mathbf{B}}_2$

3

Linear Models

3.1 Introduction

In this chapter, we will introduce the concept of linear models, which is the basic machinery of many statistical methods and also of HD-ANOVA methods. There are many good textbooks about linear models (*e.g.*, Searle, 1971; Casella, 2008; Zimmerman, 2020), and thus we are not attempting to be complete but only to give an overview necessary for the rest of this book. Because of the rapid expansion of HD-ANOVA methods, we need to cover a lot of ground: fixed versus random effect models, nested models, and linear mixed models to name a few. The distinction between fixed and random effects is crucial and will be explained first by giving some examples:

1) To study the effect of different fertilizers (Factor A), plants are grown on different plots. Each plot receives a randomly chosen fertilizer.
2) Repeated batch fermentations (Factor A) are run under different process conditions (Factor B), and samples are taken in time (Factor C) to assess the progress of the batch-run.
3) In a study of between-day variability (Factor A) of hormone levels, these hormone levels are measured in a group of individuals (Factor B) on different days.

In example one, the interest is in studying the growth of the plant as a function of the type of fertilizer. Hence, the Factor *Fertilizer* is a fixed factor since the interest is in these specific fertilizers. These fertilizers are not considered as a random sample from all available fertilizers. The plants are grown on plots and each plot is thus an experimental unit, and the design is a one-way CRD. If different plants are grown on each plot, then the individual plants are the observational units, not to be confused with the experimental units (Casella, 2008), see Section 1.2.

In example two, the interest is in studying the behavior of the batch fermentation as a function of process conditions. This is an example of a split-plot design with batches as random Factor A (whole plot) since the specific batches are not of interest but the variability among batches. Process condition is fixed Factor B, and time points (fixed Factor C) is the split-plot treatment. In this case, there are two types of experimental units: batch and measurements at each time point.

In example three, both effects are random since the interest is not the specific day nor the specific individual but simply the variability of hormone levels among individuals. Factor A is time, nested in Factor B individual. The experimental unit is the individual.

Example one will lead to a fixed effect model; example three will result in a variance component model and example two will lead to a linear mixed effect model.

Analysis of Variance for High-Dimensional Data: Applications in Life, Food and Chemical Sciences,
First Edition. Age K. Smilde, Federico Marini, Johan A. Westerhuis, and Kristian H. Liland.

3.2 Simple ANOVA Models

We will begin our exposition of ANOVA models with the simplest case: fixed effects and balanced data. In the following sections, we will expand this to more complicated situations.

3.2.1 One-Way ANOVA

The most basic linear model is a one-way fixed effect model with a balanced design. Suppose that sets of measurements are available that are collected according to a treatment design with one Factor $\mathtt{A}$ at multiple levels ($a = 1, \ldots, A$) assuming a CRD (see Section 1.2). The layout of this design is shown in Figure 3.1. The most basic model is the so-called *means* model which has the following specification:

$$y_{an} = \mu_a + \epsilon_{an} \tag{3.1}$$

where μ_a is the factor level effect and these are estimated by the means of the respective group, hence the name means model. Although this model is simple to use, it is not very flexible and cannot be used for higher-order ANOVA models with interactions. This will be shown later.

A more flexible approach is by using the model:

$$y_{an} = \mu + \alpha_a + \epsilon_{an} \tag{3.2}$$

where μ is a general mean; α_a is the effect of the Factor $\mathtt{A}$ at level a; and ϵ_{an} is the residual. This is a linear model since the effects are linearly combined with a mean. The basic assumption of this model is that the residuals are independently and identically distributed (*iid*), thus with the same variance σ_ϵ^2, which does not depend on the levels of the factor. Usually, the assumed distribution of the residuals is the normal distribution. It is important to realize that the model of Equation 3.2 is over-parametrized. This is explained in Elaboration 3.1.

ELABORATION 3.1

The ANOVA model has too many parameters

Apart from σ_ϵ^2, the model as stated in Equation 3.2 has parameters $\mu, \alpha_1, \ldots, \alpha_A$ which are $(A + 1)$ parameters. The observations are all the values $y_{an}; a = 1, \ldots, A; n = 1, \ldots, N$ but only the group means $\bar{y}_{a.}$ are informative for estimating those parameters since the replicate variation will end up in the residuals and will be used to estimate σ_ϵ^2. Hence, we have only A informative values. Thus, there are too many parameters. This is solved in ANOVA models by using constraints or, stated differently, by using generalized inverses to solve the associated regression problem. This will be explained at length in the rest of this chapter.

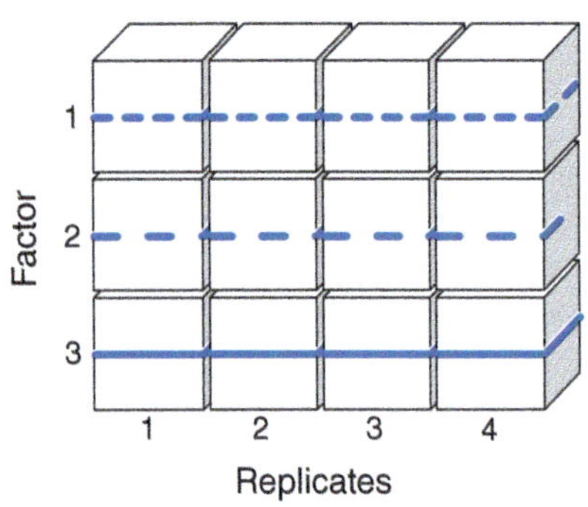

Figure 3.1 Pictorial representation of a one-way balanced design with three levels and four replicates. The drawn/dotted lines in the figure indicate the different effects at the different levels of the Factor $\mathtt{A}$.

Table 3.1 Overview of one-way ANOVA model for the balanced case.

Source	SS	df	MS	F
Treatment	SS_{A}	$A-1$	$MS_{\mathrm{A}} = SS_{\mathrm{A}}/(A-1)$	MS_{A}/MS_E
Error	SS_E	$I-A$	$MS_E = SS_E/(I-A)$	
Total	SS_T	$I-1$		

One way of dealing with the over-parametrization of the ANOVA model is by applying the so-called sum-to-zero constraints on the solution, which in this case amounts to $\sum_{a=1}^{A} \hat{\alpha}_a = 0$, and since this reduces the number of free parameters to A, the model is solvable. The estimated model gives the decomposition of the measurements y_{an} in the following way:

$$y_{an} = \bar{y}_{..} + (\bar{y}_{a.} - \bar{y}_{..}) + (y_{an} - \bar{y}_{a.}) \tag{3.3}$$

where $\bar{y}_{..}$ is the overall mean (both over a and n, see also the Notation in Chapter 1); $\bar{y}_{a.}$ is the mean of all measurements of factor level a; and $(\bar{y}_{a.} - y_{an})$ is the residual. Note that Equation 3.3 holds always (irrespective of a model), but in this special case $(\bar{y}_{a.} - \bar{y}_{..})$ is an estimate of the effect of Factor A at level a under the constraint $\sum_{a=1}^{A} \hat{\alpha}_a = 0$. Bringing the term $\bar{y}_{..}$ to the left-hand-side (LHS) of the equation and then taking sums-of-squares gives:

$$\sum_{a=1}^{A}\sum_{n=1}^{N}(y_{an} - \bar{y}_{..})^2 = \sum_{a=1}^{A}\sum_{n=1}^{N}(\bar{y}_{a.} - \bar{y}_{..})^2 + \sum_{a=1}^{A}\sum_{n=1}^{N}(y_{an} - \bar{y}_{a.})^2 \tag{3.4}$$

where the cross products of the two last terms on the right-hand-side (RHS) of Equation 3.3 have disappeared. Equation 3.4 is often written schematically as $SS_T = SS_{\mathrm{A}} + SS_E$ where SS_T represents the total sum-of-squares (relative to the mean), SS_{A} the sum-of-squares attributed to Factor A, and SS_E the residual sum-of-squares (or error). This is the basic decomposition of the sum-of-squares (SS) in ANOVA, and to arrive at mean sums-of-squares (MS) we need to divide the SS by the corresponding degrees of freedom. This is summarized in Table 3.1 where the index I is used to denote the total sample size which is AN. Due to the constraint, the degrees of freedom for the *Treatment* is $A-1$; for SS_T this number is $I-1$ (due to subtraction of the mean) leaving $I-A$ degrees of freedom for the error. For such CRDs, these calculations are easy. A more extensive treatment is given in Section 3.7.1.

The model in Equation 3.2 is called a one-way ANOVA model since there is only one factor involved. Testing the null hypothesis of no effect ($H_0 : \alpha_a = 0, \forall a$) can be done by comparing the F-value with the critical F-value at a chosen level of significance or by calculating a p-value; in both cases using an F-distribution with $(A-1), (I-A)$ degrees of freedom. In case of no effect, the values of MS_{A} and MS_E should be of similar size generating a ratio of around one. In case of an effect, the value of MS_{A} tends to be much higher than MS_E; hence, this is a one-sided test. Note that the assumption of normality of the residuals is crucial for testing purposes, since this leads to an F-distribution for the test statistic, but this assumption is not necessary for the decomposition of Equation 3.3. An example of such a model is shown in Example 3.1.

Example 3.1: Fixed effect one-way ANOVA

As an example of the application of fixed effect one-way ANOVA with a balanced design, a subset of the data collected in a metabolomic and transcriptomic study of Caldana *et al.* (2011) will be used. Overall, the study investigates the effect of the specific environmental (light/temperature) conditions and time on the metabolome and transcriptome of *Arabidopsis thaliana*. The treatment design for the overall study has three fixed factors (*Light*, *Temperature*, *Time*) which are all crossed. For the present example, we will focus on the effect of exposing the plants to different light conditions (*Dark*, *Low Light*, *Light*, and *High Light*) for 160 minutes at 21 °C. This part of the data corresponds to a one-factor CRD, where the fixed Factor *Light* is controlled at four levels. We focus on one metabolite (isoleucine) and five replicates for each design point were collected. Of interest is how the concentration of isoleucine responds to the different light conditions after 160 minutes at 21 °C. The data are graphically represented in Figure 3.2. Sometimes we need transformations to homogenize the variance across replicates but, in this case, we did not deem it necessary.

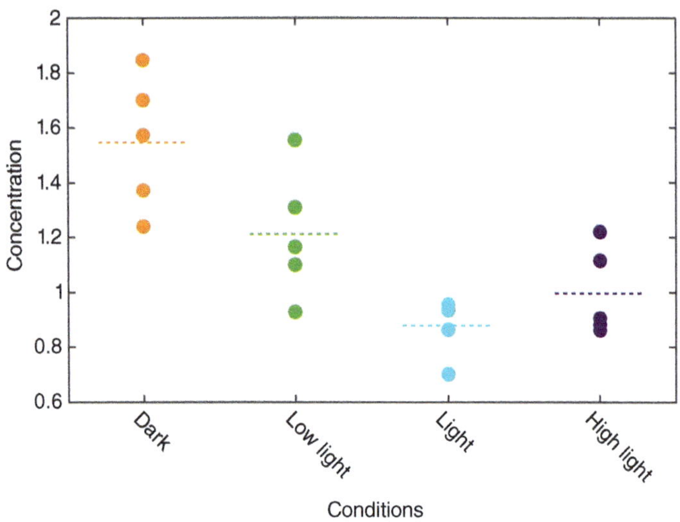

Figure 3.2 Concentration of isoleucine after 160 minutes exposure to different light conditions at 21 °C. Horizontal lines indicate averages of the five replicates.

By carrying out the ANOVA analysis as described in Table 3.1, the following results are obtained (see Table 3.2):

Table 3.2 Overview of one-way ANOVA model for isoleucine example.

Source	SS	df	MS	*F*
Light	1.293	3	0.4309	11.33
Error	0.608	16	0.038	
Total	1.901	19		

Since the critical F-value for 3 and 16 degrees of freedom and $\alpha = 0.05$ is 4.77, the effect of the different light conditions on isoleucine concentration of *Arabidopsis thaliana* after 160 minutes of exposure is found to be statistically significant. We could also interpret the results in Table 3.2 by calculating the corresponding p-value which is found to be equal to 0.0003.

3.2.2 Two-Way ANOVA

One-way ANOVA models can be extended to multiple factors, and we will sketch the models with two factors. These come in two flavors: based on crossed designs and based on nested designs (see Section 3.2.2.2). We will discuss them separately.

3.2.2.1 Crossed Designs

A natural extension of one-way ANOVA is to extend the model with more factors. Hence, assume that we have two factors A and B with levels $a = 1, \ldots, A$ and $b = 1, \ldots, B$, respectively, and assume again a balanced design with N replicates per group of factor levels. The layout of this design is shown in Figure 3.3. The extension of the model in Equation 3.2 is

$$y_{abn} = \mu + \alpha_a + \beta_b + \epsilon_{abn} \tag{3.5}$$

where we can again use the sum-to-zero constraints which are now $\sum_{a=1}^{A} \hat{\alpha}_a = 0$ and $\sum_{b=1}^{B} \hat{\beta}_b = 0$. Note that this model is introduced here for didactic purposes, but in general the model includes an interaction term which will be introduced later. Then the model in terms of the estimated parameters becomes:

$$y_{abn} = \bar{y}_{...} + (\bar{y}_{a..} - \bar{y}_{...}) + (\bar{y}_{.b.} - \bar{y}_{...}) + (y_{abn} - \bar{y}_{a..} - \bar{y}_{.b.} + \bar{y}_{...}) \tag{3.6}$$

where, *e.g.*, the term $(\bar{y}_{a..} - \bar{y}_{...})$ is an estimate of α_a under the sum-to-zero constraint similarly as in the one-way ANOVA case (also here, the decomposition of Equation 3.6 holds in general). Upon moving the term $\bar{y}_{...}$ to the LHS of the equation and taking sums-of-squares amounts to

$$\begin{aligned}\sum_{a=1}^{A}\sum_{b=1}^{B}\sum_{n=1}^{N}(y_{abn} - \bar{y}_{...})^2 = &\sum_{a=1}^{A}\sum_{b=1}^{B}\sum_{n=1}^{N}(\bar{y}_{a..} - \bar{y}_{...})^2 + \sum_{a=1}^{A}\sum_{b=1}^{B}\sum_{n=1}^{N}(\bar{y}_{.b.} - \bar{y}_{...})^2 + \\ &\sum_{a=1}^{A}\sum_{b=1}^{B}\sum_{n=1}^{N}(y_{abn} - \bar{y}_{a..} - \bar{y}_{.b.} + \bar{y}_{...})^2\end{aligned} \tag{3.7}$$

which leads to Table 3.3 where the index I is again used to denote the total sample size which is ABN and the terms SS_{A}, SS_{B}, and SS_E are self-explaining. Note that we have to assume that the interaction is not significant.

There are now two null hypotheses to test: one for Factor A and one for Factor B. They both require their own comparison with critical values given a level of Type I error.

When moving from a one-way ANOVA model to a two-way ANOVA model there is an important new concept involved: interaction between the factors. Simply stated, an interaction takes place if the effect of a factor depends on the level of the other factor. This is illustrated in Figure 3.4 where plants are grown under different conditions (yes/no fertilizer and amount

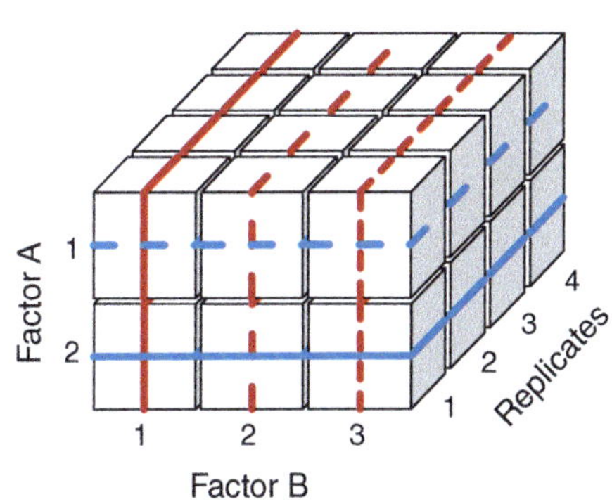

Figure 3.3 Pictorial representation of a balanced crossed two-way design. The drawn/dotted lines in the figure indicate the different effects at the different levels of the factors.

Table 3.3 Overview of a two-way ANOVA model for the balanced case.

Source	SS	df	MS	F
Factor A	SS_A	$A-1$	$MS_A = SS_A/(A-1)$	MS_A/MS_E
Factor B	SS_B	$B-1$	$MS_B = SS_B/(B-1)$	MS_B/MS_E
Error	SS_E	$I-A-B+1$	$MS_E = SS_E/(I-A-B+1)$	
Total	SS_T	$I-1$		

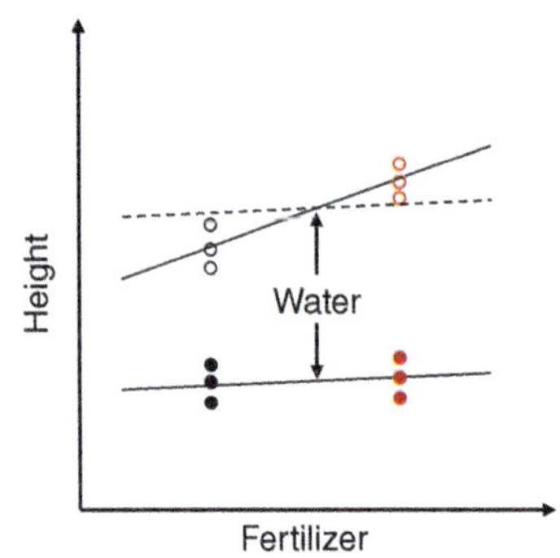

Figure 3.4 An example of interaction. The height of plants in a greenhouse is influenced by no/yes fertilizer (black versus red dots) and the amount of watering (open versus closed dots). The effect of the fertilizer depends on the level of watering, and there are three replicates per combination of factor levels. The dashed line represents the case where there would not have been an interaction.

of water applied). Clearly, the effect of the fertilizer depends on the level of the watering applied; thus, the factors interact in their effect on the response (the height of the plants). The ANOVA model to account for interaction can be written as

$$y_{abn} = \mu + \alpha_a + \beta_b + (\alpha\beta)_{ab} + \epsilon_{abn} \tag{3.8}$$

where the term $(\alpha\beta)_{ab}$ should be understood as a separate term and not the multiplication of α_a and β_b. Also this ANOVA model has too many parameters, and the sum-to-zero constraints on the estimated parameters are the same as for the two-way model without interaction with additional constraints $\sum_{a=1}^{A} \widehat{(\alpha\beta)}_{ab} = 0;\ \forall b$ and $\sum_{b=1}^{B} \widehat{(\alpha\beta)}_{ab} = 0;\ \forall a$. In terms of estimates, this becomes:

$$y_{abn} = \bar{y}_{...} + (\bar{y}_{a..} - \bar{y}_{...}) + (\bar{y}_{.b.} - \bar{y}_{...}) + (\bar{y}_{ab.} - \bar{y}_{a..} - \bar{y}_{.b.} + \bar{y}_{...}) + (y_{abn} - \bar{y}_{ab.}) \tag{3.9}$$

where each term again estimates a factor effect or an interaction (but the decomposition is again general). This leads to the decomposition of the sum-of-squares:

$$\begin{aligned}\sum_{a=1}^{A}\sum_{b=1}^{B}\sum_{n=1}^{N}(y_{abn} - \bar{y}_{...})^2 &= \sum_{a=1}^{A}\sum_{b=1}^{B}\sum_{n=1}^{N}(\bar{y}_{a..} - \bar{y}_{...})^2 + \sum_{a=1}^{A}\sum_{b=1}^{B}\sum_{n=1}^{N}(\bar{y}_{.b.} - \bar{y}_{...})^2 + \\ &\sum_{a=1}^{A}\sum_{b=1}^{B}\sum_{n=1}^{N}(\bar{y}_{ab.} - \bar{y}_{a..} - \bar{y}_{.b.} + \bar{y}_{...})^2 + \sum_{a=1}^{A}\sum_{b=1}^{B}\sum_{n=1}^{N}(y_{abn} - \bar{y}_{ab.})^2\end{aligned} \tag{3.10}$$

and the ANOVA Table 3.4.

Now there are three null hypotheses to test: Factor A, Factor B, and their interaction. This is again done by comparing F-values to their critical values given a level of Type I error. Example 3.2 illustrates a two-way ANOVA model with interaction.

Table 3.4 Overview of a two-way ANOVA model with interaction for the balanced case.

Source	SS	df	MS	F
Factor A	SS_A	$A-1$	$MS_A = SS_A/(A-1)$	MS_A/MS_E
Factor B	SS_B	$B-1$	$MS_B = SS_B/(B-1)$	MS_B/MS_E
Interaction AB	SS_{AB}	$(A-1)(B-1)$	$MS_{AB} = SS_{AB}/((A-1)(B-1))$	MS_{AB}/MS_E
Error	SS_E	$I-AB$	$MS_E = SS_E/(I-AB)$	
Total	SS_T	$I-1$		

Example 3.2: Fixed effect two-way ANOVA with interaction

As an example of the application of fixed effect two-way ANOVA with interactions, we will further exploit the data from (Caldana *et al.*, 2011). More specifically, by once again limiting to the set of data corresponding to the temperature of 21 °C, the effect of both time and light conditions will be examined, by selecting as response variable the concentration of alanine. The resulting statistical design is a balanced crossed two factors (*Light* and *Time*, both fixed) CRD. The interest is in studying the dynamics of the concentration of alanine as a function of the different light conditions. As in Example 3.1, five replicates for each group of factor levels were collected, and new plants were used at each time point. Factor *Light* is controlled at four levels, while Factor *Time* at seven levels. The data are graphically represented in Figure 3.5. Again, we did not consider variance-stabilizing transformations.

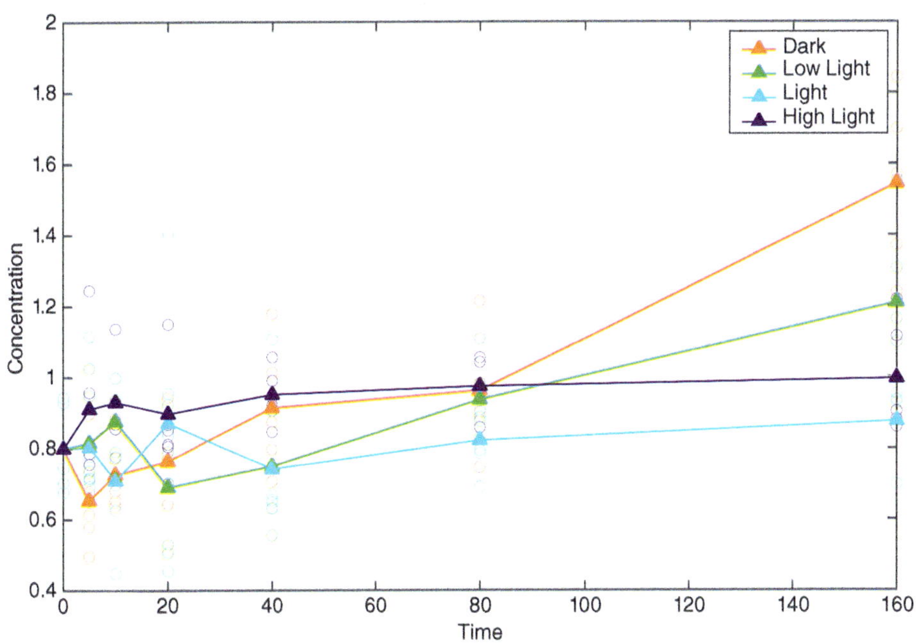

Figure 3.5 Concentration of alanine as a function of the exposure to different time and light conditions at 21 °C (empty circles). Horizontal lines and filled triangles indicate averages of the five replicates.

By carrying out the ANOVA analysis as described in Table 3.4, the following results are obtained (see Table 3.5):

Table 3.5 Overview of the two-way ANOVA model with interaction for the alanine example.

Source	SS	df	MS	F
Light	2.518	3	0.8392	30.46
Time	0.583	6	0.0972	3.53
Interaction	3.748	18	0.2082	7.56
Error	3.086	112	0.0275	
Total	9.935	139		

By comparing the F in Table 3.5 (30.46) with the critical value ($\alpha = 0.05$) for 3 and 112 degrees of freedom (2.686), the effect of light is found to be statistically significant ($p = 1.8 \times 10^{-14}$). Analogously, also the effect of time ($p = 0.0031$) and of the interaction ($p = 1.9 \times 10^{-12}$) is found to be significant. These results can be confirmed by taking a close look at Figure 3.5.

The interpretation of main and interaction effects should be done with care. Especially, if an interaction is found to be significant, then a main effect should be incorporated into the model, and this model should be interpreted as a whole. Hence, in practice the interactions are tested and, subsequently, if the interactions are found to be not significant, the main effects. It is always useful to make plots of the found effects. Good discussions of these issues can be found elsewhere (Ott and Longnecker, 2015).

Generalizations to more than two factors are straightforward. Three-way ANOVA has three factors and thus three main effects; three two-way interactions and one three-way interaction. The machinery to calculate effects in such models generalize from the two-way ANOVA models but interpretations may be harder: interpreting three-way interactions is not easy.

All ASCA (ANOVA Simultaneous Component Analysis; see Section 4.1) models are based on these simple ANOVA models for balanced data, and thus ASCA models can be calculated using the means from the ANOVA decompositions. Hence, the original ASCA model is a very simple extension of such ANOVA models.

3.2.2.2 Nested Designs

In nested designs, it is not possible to change both factors independently (see Section 3.1). An example is from baking bread; suppose that a design is made for preparing dough to bake bread but depending on the composition of the dough, different ranges of temperature should be applied. Then the factor temperature cannot be independently set to a certain value, because this depends on the composition of the dough. The layout of a nested design is shown in Figure 3.6. This gives rise to nested ANOVA models:

$$y_{abn} = \mu + \alpha_a + \beta_{b(a)} + \epsilon_{abn} \tag{3.11}$$

where we have assumed that Factor B (*Temperature*) is nested in Factor A (*Composition of the dough*), and each cell has the same number of replicates (N). The nesting is indicated by the term $\beta_{b(a)}$ which should not be confused with $(\alpha\beta)_{ab}$ which indicates an interaction in a two-way

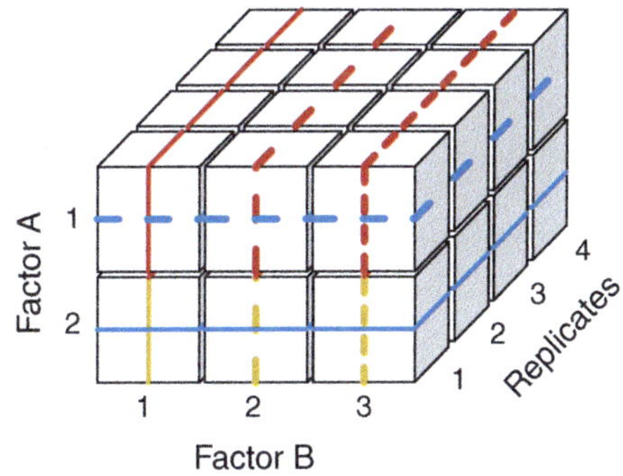

Figure 3.6 Pictorial representation of a balanced nested two-way design. The drawn/dotted lines in the figure indicate the different effects at the different levels of the factors. Factor B is nested in Factor A. Note the different colors for the levels of Factor B depending on the level of Factor A.

crossed-design ANOVA model. Similarly as for the other ANOVA models, a decomposition can be made as:

$$y_{abn} = \bar{y}_{...} + (\bar{y}_{a..} - \bar{y}_{...}) + (\bar{y}_{ab.} - \bar{y}_{a..}) + (y_{abn} - \bar{y}_{ab.}) \tag{3.12}$$

and similarly as in two-way ANOVA sums-of-squares can be calculated and subsequently mean sums-of-squares and F-values for testing hypotheses. Example 3.3 provides a demonstration of the analysis of a nested model.

Example 3.3: Fixed effect nested ANOVA

As an example of fixed effect nested ANOVA, we consider the case of a pharmaceutical batch process for the production of a formulation, where the drug could be produced in three different plants. We could then be interested in evaluating whether there are differences in the amount of the API depending on the production *Site* (Factor A, which can be considered as fixed), but also if there is a significant variability due to manufacturing *Protocols* (Factor B). Here, manufacturing protocols should in principle be the same, but, in practice, they could be slightly modified at each plant, to meet the specific characteristics of the plant itself. Accordingly, by definition, the Factor *Protocol* can be treated as a factor nested within the individual production plant (Factor *Site*). When considering investigating three manufacturing protocols per production site and collecting four samples from each manufacturing protocol at each production site, the overall design is depicted schematically in Figure 3.7, where the final API concentration is also displayed.

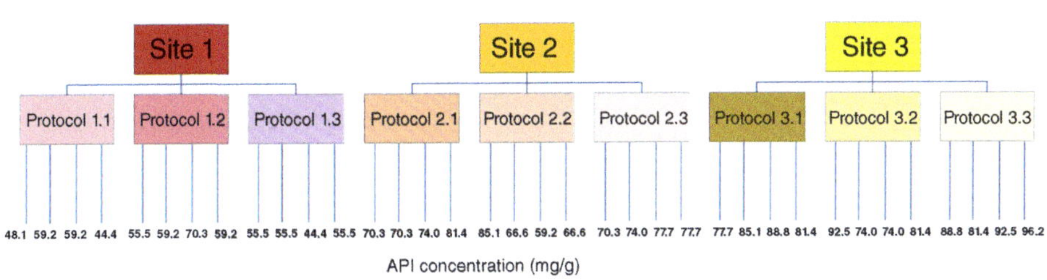

Figure 3.7 Scheme of the two-factor nested ANOVA example, where Factor A is the production site and Factor B are the manufacturing protocols, nested in the individual sites; four samples are taken from each manufacturing protocol at each site.

By carrying out the ANOVA analysis as described in Equation 3.12, the following results are obtained:

Table 3.6 Overview of the ANOVA model for the two-factor nested example.

Source	SS	df	MS	F
Site	5101.8	2	2551.9	53.24
Protocol (Site)	435.8	6	72.6	1.52
Error	1293.7	27	47.9	
Total	6831.3	35		

By comparing the F in Table 3.6 (53.24) with the critical value for 2 and 27 degrees of freedom (3.354), the effect of the production site on the API concentration is found to be statistically significant ($p = 4.2 \times 10^{-10}$). The effect of the Factor *Protocol* is found to be not significant, since its F-value (1.52) is lower than the 95th percentile of the null distribution with 6 and 27 degrees of freedom (2.459), with a p-value of 0.21.

When the number of levels of Factor B is the same for all levels of Factor A; and if the number of replicates per cell in the design is the same (*i.e.*, when the nested design is balanced), then the factor estimates can be obtained again by calculating proper means as in the ANOVA model for crossed-designs (Searle, 1971).

3.2.3 Unbalanced Designs

3.2.3.1 One-Way ANOVA

In the case of unbalanced designs for the one-way ANOVA model, the idea remains the same but the calculations become more complicated. Having A different levels for Factor A, in principle each level a may have its own number of replicates $N_a; a = 1, \ldots, A$ with running index $n = 1, \ldots, N_a$ in group a (see Figure 3.8). Note that for simplicity the index n is used for replicates in all the cells, but the cardinality (*i.e.*, the N_a values) of this index depends on the level of Factor A which makes indexing unambiguous. The decomposition remains the same as in the balanced case (see Equation 3.3), but it does not generate the proper factor effect estimates, see Elaboration 3.2.

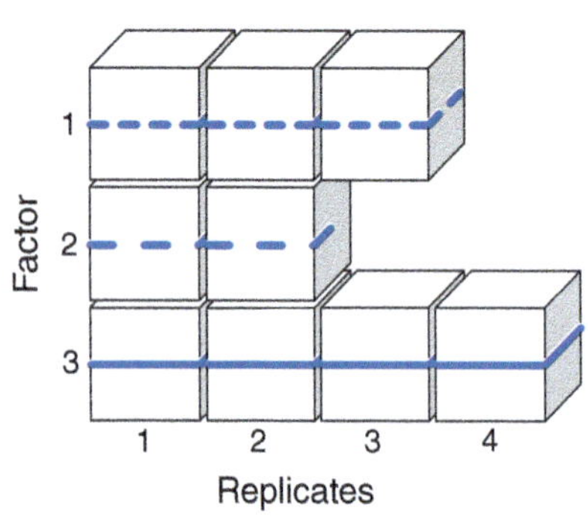

Figure 3.8 Pictorial representation of a one-way unbalanced design. The drawn/dotted lines in the figure indicate the different effects at the different levels of the factors.

ELABORATION 3.2

Standard ANOVA does not retrieve the correct factor estimates (1)

A simple numerical example is given to show the problem of using standard ANOVA calculations in unbalanced designs using factor estimates based on means. Suppose we have a one-way layout with one factor at two levels and per level two replicates. The y-values are generated in the following way:

$$\begin{aligned} y_{11} &= \mu + \alpha_1 + \epsilon_{11} = 2 + 1 + 0 = 3 \\ y_{12} &= \mu + \alpha_1 + \epsilon_{12} = 2 + 1 + 0 = 3 \\ y_{21} &= \mu + \alpha_2 + \epsilon_{21} = 2 - 1 + 0 = 1 \\ y_{22} &= \mu + \alpha_2 + \epsilon_{22} = 2 - 1 + 0 = 1 \end{aligned} \tag{3.13}$$

where we have assumed zero error for simplicity. Upon using the simple ANOVA calculations based on the means (see Equation 3.3) we retrieve the true values for μ, α_1, α_2 (*e.g.*, the value $\bar{y}_{..} = \frac{8}{4} = 2$ which can easily be checked).

Suppose now that trial y_{12} is missing, hence we have unbalanced data. The simple ANOVA calculations based on means now generate the solution $\hat{\mu} = \frac{5}{3}$, $\hat{\alpha}_1 = \frac{4}{3}, \hat{\alpha}_2 = -\frac{2}{3}$ (*e.g.*, the value $\bar{y}_{..} = \frac{5}{3}$ which can easily be checked). To retrieve the true values we need to use the regression approach to be explained later in this chapter.

The break-down of sum-of-squares after moving $\bar{y}_{..}$ to the LHS and taking squares on both sides becomes now

$$\sum_{a=1}^{A}\sum_{n=1}^{N_a}(y_{an} - \bar{y}_{..})^2 = \sum_{a=1}^{A}\sum_{n=1}^{N_a}(\bar{y}_{a.} - \bar{y}_{..})^2 + \sum_{a=1}^{A}\sum_{n=1}^{N_a}(y_{an} - \bar{y}_{a.})^2 \tag{3.14}$$

where again the cross-products vanish and the total sample size is now $I = \sum_{a=1}^{A} N_a$. Table 3.1 also holds for this situation.

3.2.3.2 Two-Way ANOVA for Crossed Designs

We also cannot use the simple ANOVA calculations for estimating the factor effects in the case of unbalanced two-way data (see Figure 3.9). This is shown in Elaboration 3.3.

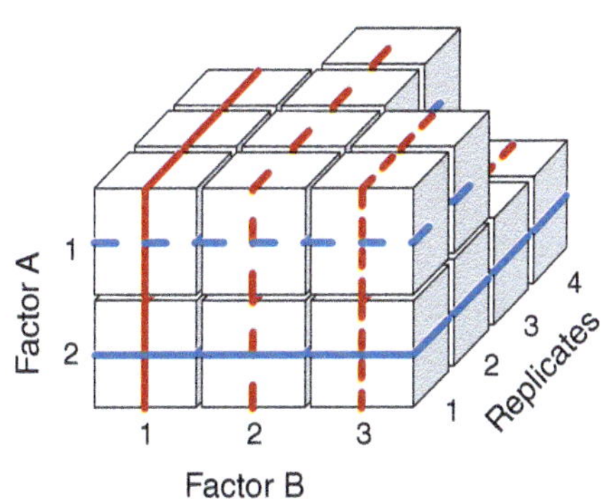

Figure 3.9 Pictorial representation of a two-way crossed unbalanced design. The drawn/dotted lines in the figure indicate the different effects at the different levels of the factors.

ELABORATION 3.3

Standard ANOVA does not retrieve the correct factor estimates (2)

A simple numerical example is given to show the problem of using standard ANOVA calculations using factor estimates based on means (Thiel *et al.*, 2017). Suppose we have an ANOVA model with parameters $\mu = 10$, $\alpha_1 = 2, \alpha_2 = -2$, $\beta_1 = 1, \beta_2 = 0.5, \beta_3 = -1.5$, no interaction, $\sigma^2 = 0$, and two replicates per cell. The vector of responses is then $(y_{111}, y_{112}, y_{121}, y_{122}, y_{131}, y_{132}, y_{211}, y_{212}, y_{221}, y_{222}, y_{231}, y_{232}) = (13.0, 13.0, 12.5, 12.5, 10.5, 10.5, 9.0, 9.0, 8.5, 8.5, 6.5, 6.5)$ which is a nicely balanced design. We can calculate the ANOVA factor estimates using the means as is done in Equation 3.6, and these ANOVA calculations retrieve the true factor estimates.

Suppose now that trials y_{222} and y_{232} are missing, hence we have unbalanced data. The simple ANOVA calculations based on means now generate the solution $\hat{\mu} = 10.45$, $\hat{\alpha}_1 = 1.5, \hat{\alpha}_2 = -2.25$, $\hat{\beta}_1 = 0.5, \hat{\beta}_2 = 0.667, \hat{\beta}_3 = -1.333$ which is obviously wrong. Moreover, it also generates non-zero values for the interactions. Hence, these estimators are wrong and even generate non-zero interactions that were not present in the underlying data. To retrieve the true values, we need to use the regression approach to be explained later in this chapter.

A problem of the two-way unbalanced ANOVA model can be illustrated in the following way. Each cell in the design has N_{ab} replicates. The ANOVA decomposition for the case without interaction is

$$y_{abn} = \bar{y}_{...} + (\bar{y}_{a..} - \bar{y}_{...}) + (\bar{y}_{.b.} - \bar{y}_{...}) + (y_{abn} - \bar{y}_{a..} - \bar{y}_{.b.} + \bar{y}_{...}) \tag{3.15}$$

which holds in general (also for the balanced case). After moving the term $\bar{y}_{...}$ to the LHS and taking squares on both sides does not generate a nice separation of sum-of-squares. For example, it can be shown (see Section 3.A.1) that

$$\sum_{a=1}^{A} \sum_{b=1}^{B} \sum_{n=1}^{N_{ab}} (\bar{y}_{a..} - \bar{y}_{...})(\bar{y}_{.b.} - \bar{y}_{...}) = \sum_{a=1}^{A} \sum_{b=1}^{B} N_{ab}\bar{y}_{a..}\bar{y}_{.b.} - I(\bar{y}_{...})^2 \tag{3.16}$$

which is not equal to zero in general. If $N_{AB} = N$ (*i.e.*, a balanced design) then the term $\sum_{a=1}^{A} \sum_{b=1}^{B} N_{ab}\bar{y}_{a..}\bar{y}_{.b.}$ boils down to $I(\bar{y}_{...})^2$, so for $N_{AB} \approx N$ this holds approximately and thus the decomposition in terms of sums-of-squares holds then approximately. In the more extreme case, some of the values N_{AB} may be zero, *i.e.*, these cells are empty. That case is more involved, and sometimes not all interactions can be estimated (Searle, 1971). Such data should be handled with extreme care. The break up of SS is more involved in the case of unbalanced designs, see Elaboration 3.4. Note that this problem also holds when the factor effect estimates are obtained using a regression approach to be explained later in this chapter.

ELABORATION 3.4

Types of SS in unbalanced designs

As shown in Equation 3.16, there is no easy break-up of the SS in unbalanced ANOVA models with at least two factors (Searle, 1971; Langsrud, 2003) since the cross-products do not disappear. This has led to different options to analyze unbalanced data and different ways of calculating SS coined Type I, Type II, and Type III analysis and corresponding SS. These types are also related to certain hypotheses to test.

Type I analysis (and SS) corresponds to adding terms in the model sequentially and testing the significance of these terms. This depends on the order in which the terms are evaluated due to the unbalancedness of the design. In Type II analysis each effect is adjusted for all other terms except for those that contain the one being tested: in a three-way analysis, the main effect of Factor A is not adjusted for the interactions AB, AC, and ABC. Stated otherwise, in a three-way analysis testing for the effect of Factor A is performed by comparing the residual sum-of-squares of the model containing Factors A, B, C and the interaction BC with a model containing only the terms B, C, and BC. Finally, in Type III analysis each effect is adjusted for all other effects including interactions. So, for finding the Type III sum-of-squares for Factor A we need to correct for all other factors and interactions. All these methods give different results. There has been considerable debate in the literature about which analysis to use, but there seems to be some consensus on using Type II or Type III analysis depending on the circumstances. For a more detailed exposition, see Nelder (1994) and Langsrud (2003).

3.2.3.3 Nested ANOVA

Also for nested models in unbalanced designs (see Figure 3.10), simple means cannot be used anymore for factor estimates, and we have to resort to a regression approach to be explained later in this chapter.

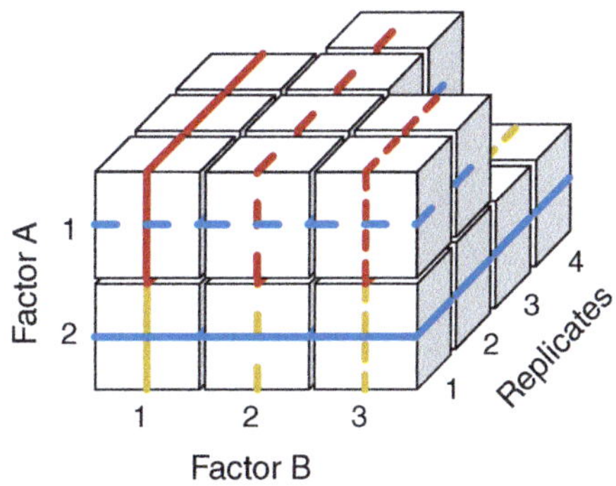

Figure 3.10 Pictorial representation of a two-way nested unbalanced design. The drawn/dotted lines in the figure indicate the different effects at the different levels of the factors.

3.3 Regression Formulation, Estimability, and Contrasts

In Chapter 2 we already introduced regression models including the way to solve such models using LS. Such regression models are very useful for formulating ANOVA models whereby the factors are introduced in the model using coding variables which will be explained later. There is one problem to solve, however, before doing so and that has to do with the fact that the resulting regression models are not of full rank; *i.e.*, the design matrix $\mathbf{X}$ is not of full rank; this is related to ANOVA models being over-parametrized and not all parameters can be estimated, see Elaboration 3.1.

For regression models not of full rank, the concept of *estimability* is very relevant. In the model $\mathbf{y} = \mathbf{X}\boldsymbol{\beta} + \epsilon$ a linear function of $\mathbf{c}^t\boldsymbol{\beta}$ is called estimable if a linear estimator $\mathbf{t}^t\mathbf{y}$ exists that estimates $\mathbf{c}^t\boldsymbol{\beta}$ unbiasedly (Zimmerman, 2020). In other words, we should be able to find a linear combination of the observed values that estimates this function of the parameters. Note that only existence is required, not uniqueness. A necessary and sufficient condition for estimability is that $\mathbf{c}$ is in the row-space of $\mathbf{X}$ (Zimmerman, 2020).

Considering the normal equations sheds some light on the estimability problem. The normal equation $(\mathbf{X}^t\mathbf{X})\mathbf{b} = \mathbf{X}^t\mathbf{y}$ is a consistent set of equations (Searle, 1971) and thus can always be solved for $\mathbf{b}$. Yet, in the case of a rank-deficient matrix $\mathbf{X}$, also the matrix $(\mathbf{X}^t\mathbf{X})$ is rank deficient, and thus the normal equations have an infinite number of solutions for $\mathbf{b}$. The general solution of the normal equations is $\mathbf{b} = (\mathbf{X}^t\mathbf{X})^-\mathbf{X}^t\mathbf{y} + (\mathbf{I} - (\mathbf{X}^t\mathbf{X})^-\mathbf{X}^t\mathbf{X})\mathbf{d}$ where $\mathbf{d}$ is an arbitrary vector (see Elaboration 3.5). Considering now the function $\mathbf{c}^t\mathbf{b}$ and using the fact that $\mathbf{c}$ is in the row-space of $\mathbf{X}$, *i.e.*, there exists a vector $\mathbf{a}$ such that $\mathbf{c}^t = \mathbf{a}^t\mathbf{X}$, gives upon substitution in the general solution: $\mathbf{c}^t\mathbf{b} = \mathbf{a}^t\mathbf{X}\mathbf{b} = \mathbf{a}^t\mathbf{X}(\mathbf{X}^t\mathbf{X})^-\mathbf{X}^t\mathbf{y} + \mathbf{a}^t\mathbf{X}\mathbf{d} - \mathbf{a}^t\mathbf{X}(\mathbf{X}^t\mathbf{X})^-\mathbf{X}^t\mathbf{X}\mathbf{d}$. Now the matrix $\mathbf{X}(\mathbf{X}^t\mathbf{X})^-\mathbf{X}^t$ projects orthogonally on the column-space of $\mathbf{X}$ for any generalized inverse, and since $\mathbf{Xd}$ is in that column space (by definition) the last term equals $\mathbf{a}^t\mathbf{Xd}$. Hence, this term cancels and thus $\mathbf{c}^t\mathbf{b} = \mathbf{a}^t\mathbf{Xb} = \mathbf{a}^t\mathbf{X}(\mathbf{X}^t\mathbf{X})^-\mathbf{X}^t\mathbf{y} = \mathbf{t}^t\mathbf{y}$. Thus, for such vectors $\mathbf{c}$, the function $\mathbf{c}^t\mathbf{b}$ is estimable.

ELABORATION 3.5

Solving a system of consistent equations

The general solution of a consistent set of equations $\mathbf{Ax} = \mathbf{f}$ is $\mathbf{x} = \mathbf{A}^-\mathbf{f} + (\mathbf{I} - \mathbf{A}^-\mathbf{A})\mathbf{w}$ where $\mathbf{w}$ is an arbitrary vector (Schott, 2016). This holds for any generalized inverse.

A special class of estimable functions are called *contrasts*. A contrast estimates an interpretable outcome of an ANOVA model. An example is in the case of a one-way ANOVA model with a Factor A varied at A levels. A convenient set of contrasts is then $\hat{\alpha}_i - \hat{\alpha}_{i'}; \forall i \neq i'$ which estimates (uniquely) all the differences between the outcomes of the factor levels.

3.4 Coding Schemes

To use regression models for fitting ANOVA models we need to encode the factor levels in the treatment design. This can be done in different ways, and these will be explained in this section. Quantitative covariates can be used as such and do not need encoding.

3.4.1 Codings for Balanced Designs

3.4.1.1 One-Way Layout

To set the scene for this section, we will start by explaining coding schemes for the balanced one-way ANOVA model. We will use the example of the Factor A at levels $a = 1, 2, 3$ and $N = 2$ replicates per level. The easiest model for this case is

$$y_{an} = \mu_a + \epsilon_{an} \tag{3.17}$$

with $\mathbf{y} = (y_{11}, y_{12}, y_{21}, y_{22}, y_{31}, y_{32})^t$. The design matrix $\mathbf{X}_{MM}$ becomes:

$$\mathbf{X}_{MM} = \begin{bmatrix} 1 & 0 & 0 \\ 1 & 0 & 0 \\ 0 & 1 & 0 \\ 0 & 1 & 0 \\ 0 & 0 & 1 \\ 0 & 0 & 1 \end{bmatrix} \tag{3.18}$$

which is of full rank. This is called the "means" model, hence the subscript MM. The cross-product matrix is now:

$$\mathbf{X}_{MM}^t\mathbf{X}_{MM} = \begin{bmatrix} 2 & 0 & 0 \\ 0 & 2 & 0 \\ 0 & 0 & 2 \end{bmatrix} \tag{3.19}$$

and the normal equations are

$$\begin{aligned} 2\widehat{\alpha}_1 &= y_{1.} \\ 2\widehat{\alpha}_2 &= y_{2.} \\ 2\widehat{\alpha}_3 &= y_{3.} \end{aligned} \tag{3.20}$$

leading to the solutions $\widehat{\mu}_1 = \overline{y}_{1.}$, $\widehat{\mu}_2 = \overline{y}_{2.}$ and $\widehat{\mu}_3 = y_{3.}$.

Another specification of the model would be:

$$y_{an} = \mu + \alpha_a + \epsilon_{an} \tag{3.21}$$

where now the mean is explicitly taken along in the model. The resulting coding is now:

$$\mathbf{X}_F = \begin{bmatrix} 1 & 1 & 0 & 0 \\ 1 & 1 & 0 & 0 \\ 1 & 0 & 1 & 0 \\ 1 & 0 & 1 & 0 \\ 1 & 0 & 0 & 1 \\ 1 & 0 & 0 & 1 \end{bmatrix} \tag{3.22}$$

where the subscript F indicates the full-coding design matrix. This matrix is clearly rank deficient (the rank is three) since the last three columns add up to the first one. The resulting matrix $\mathbf{X}_F^t\mathbf{X}_F$ is:

$$\mathbf{X}_F^t\mathbf{X}_F = \begin{bmatrix} 6 & 2 & 2 & 2 \\ 2 & 2 & 0 & 0 \\ 2 & 0 & 2 & 0 \\ 2 & 0 & 0 & 2 \end{bmatrix} \tag{3.23}$$

which has size (4×4) and thus is also rank deficient (rank is three also). The other relevant matrix is $\mathbf{X}_F^t\mathbf{y}$ which is:

$$\mathbf{X}_F^t\mathbf{y} = \begin{bmatrix} 1 & 1 & 1 & 1 & 1 & 1 \\ 1 & 1 & 0 & 0 & 0 & 0 \\ 0 & 0 & 1 & 1 & 0 & 0 \\ 0 & 0 & 0 & 0 & 1 & 1 \end{bmatrix} \begin{bmatrix} y_{11} \\ y_{12} \\ y_{21} \\ y_{22} \\ y_{31} \\ y_{32} \end{bmatrix} = \begin{bmatrix} y_{..} \\ y_{1.} \\ y_{2.} \\ y_{3.} \end{bmatrix} \tag{3.24}$$

and thus the normal equations $(\mathbf{X}_F^t\mathbf{X}_F)\widehat{\theta} = \mathbf{X}_F^t\mathbf{y}$ with $\widehat{\theta} = (\widehat{\mu}, \widehat{\alpha}_1, \widehat{\alpha}_2, \widehat{\alpha}_3)^t$ are

$$\begin{aligned} 6\widehat{\mu} + 2\widehat{\alpha}_1 + 2\widehat{\alpha}_2 + 2\widehat{\alpha}_3 &= y_{..} \\ 2\widehat{\mu} + 2\widehat{\alpha}_1 &= y_{1.} \\ 2\widehat{\mu} + 2\widehat{\alpha}_2 &= y_{2.} \\ 2\widehat{\mu} + 2\widehat{\alpha}_3 &= y_{3.} \end{aligned} \tag{3.25}$$

which is a rank-deficient system. The general solution to this problem is:

$$\begin{bmatrix} \hat{\mu} \\ \hat{\alpha}_1 \\ \hat{\alpha}_2 \\ \hat{\alpha}_3 \end{bmatrix} = \frac{1}{N}\begin{bmatrix} 0 \\ y_{1.} \\ y_{2.} \\ y_{3.} \end{bmatrix} - \lambda \begin{bmatrix} -1 \\ 1 \\ 1 \\ 1 \end{bmatrix} \tag{3.26}$$

and if we adopt the sum-to-zero constraint $\hat{\alpha}_1 + \hat{\alpha}_2 + \hat{\alpha}_3 = 0$ gives the solution $\lambda = y_{..}/N$, and this results in:

$$\begin{aligned} \hat{\mu} &= \bar{y}_{..} \\ \hat{\alpha}_1 &= \bar{y}_{1.} - \bar{y}_{..} \\ \hat{\alpha}_2 &= \bar{y}_{2.} - \bar{y}_{..} \\ \hat{\alpha}_3 &= \bar{y}_{3.} - \bar{y}_{..} \end{aligned} \tag{3.27}$$

which is the solution for the sum-to-zero constraint (Dobson and Barnett, 2018).

Sum-coding (also called effect-coding) uses the same model as the full-coding where again the constraint $\hat{\alpha}_1 + \hat{\alpha}_2 + \hat{\alpha}_3 = 0$ is used but now immediately invoked in the design matrix by using

$$\mathbf{X}_S = \begin{bmatrix} 1 & -1 & -1 \\ 1 & -1 & -1 \\ 1 & 1 & 0 \\ 1 & 1 & 0 \\ 1 & 0 & 1 \\ 1 & 0 & 1 \end{bmatrix} \tag{3.28}$$

which is full rank (the rank is three). The parameter vector is now $\hat{\theta} = (\hat{\mu}, \hat{\alpha}_2, \hat{\alpha}_3)^t$ where the parameter $\hat{\alpha}_1$ is skipped because it can be inferred from the other estimated parameters. The cross-product matrix is

$$\mathbf{X}_S^t\mathbf{X}_S = \begin{bmatrix} 6 & 0 & 0 \\ 0 & 4 & 2 \\ 0 & 2 & 4 \end{bmatrix} \tag{3.29}$$

which is full rank. The RHS matrix in the normal equations is

$$\mathbf{X}_S^t\mathbf{y} = \begin{bmatrix} 1 & 1 & 1 & 1 & 1 & 1 \\ -1 & -1 & 1 & 1 & 0 & 0 \\ -1 & -1 & 0 & 0 & 1 & 1 \end{bmatrix} \begin{bmatrix} y_{11} \\ y_{12} \\ y_{21} \\ y_{22} \\ y_{31} \\ y_{32} \end{bmatrix} = \begin{bmatrix} y_{..} \\ y_{2.} - y_{1.} \\ y_{3.} - y_{1.} \end{bmatrix} \tag{3.30}$$

resulting in the normal equations $(\mathbf{X}_S^t\mathbf{X}_S)\hat{\theta} = \mathbf{X}_S^t\mathbf{y}$:

$$\begin{aligned} 6\hat{\mu} &= y_{..} \\ 4\hat{\alpha}_2 + 2\hat{\alpha}_3 &= y_{2.} - y_{1.} \\ 2\hat{\alpha}_2 + 4\hat{\alpha}_3 &= y_{3.} - y_{1.} \end{aligned} \tag{3.31}$$

and this system can be solved by using again $y_{..} = y_{1.} + y_{2.} + y_{3.}$. This then gives exactly the same solution as for the full-coding case, see Equation 3.27. Hence, the sum-to-zero constraint can be invoked on the normal equations or built into the design matrix using sum-coding. Another way to understand the working of sum-coding is explained in Elaboration 3.6.

ELABORATION 3.6

Sum-coding explained differently

We will explain sum-coding in a bit more detail using again the example of a one-way ANOVA with Factor A at three levels. The constraint $\alpha_1 + \alpha_2 + \alpha_3 = 0$ can also be written as $\alpha_1 = -\alpha_2 - \alpha_3$[1] and using this constraint will result in the model:

$$
\begin{aligned}
y_{11} &= \mu - \alpha_2 - \alpha_3 + \epsilon_{11} \\
y_{12} &= \mu - \alpha_2 - \alpha_3 + \epsilon_{12} \\
y_{21} &= \mu + \alpha_2 + \epsilon_{21} \\
y_{22} &= \mu + \alpha_2 + \epsilon_{22} \\
y_{31} &= \mu + \alpha_3 + \epsilon_{31} \\
y_{32} &= \mu + \alpha_3 + \epsilon_{32}
\end{aligned} \tag{3.32}
$$

and this results in a regression equation with design matrix $\mathbf{X}_S$ and a parameter vector where α_1 is skipped.

Another way to understand sum-coding is as follows. The constraint $\alpha_1 + \alpha_2 + \alpha_3 = 0$ can also be written as $\mathbf{d}^t\theta = 0$ with $\mathbf{d} = (0, 1, 1, 1)^t$ and $\theta = (\mu, \alpha_1, \alpha_2, \alpha_3)^t$. The original regression equation for the model using full-coding can now be augmented to also include the constraint:

$$
\begin{bmatrix} \mathbf{y} \\ 0 \end{bmatrix} = \begin{bmatrix} 1 & 1 & 0 & 0 \\ 1 & 1 & 0 & 0 \\ 1 & 0 & 1 & 0 \\ 1 & 0 & 1 & 0 \\ 1 & 0 & 0 & 1 \\ 1 & 0 & 0 & 1 \\ \hdashline 0 & 1 & 1 & 1 \end{bmatrix} \theta + \begin{bmatrix} \epsilon \\ 0 \end{bmatrix} \tag{3.33}
$$

where the last line in this regression equation reflects the constraint. The augmented design matrix is now full rank due to the augmentation with the row $\mathbf{d}^t$. Hence, also the resulting normal equations will be of full rank and thus exactly solvable, thereby generating the results of the sum-coding regression equation.

For reference-coding (also called treatment-coding) the details are as follows. The assumed model is the same as model Equation 3.21, but now it is assumed that $\hat{\alpha}_1 = 0$. This means that we have to remove the column associated with this parameter from $\mathbf{X}_F$ and thus:

$$
\mathbf{X}_R = \begin{bmatrix} 1 & 0 & 0 \\ 1 & 0 & 0 \\ 1 & 1 & 0 \\ 1 & 1 & 0 \\ 1 & 0 & 1 \\ 1 & 0 & 1 \end{bmatrix} \tag{3.34}
$$

which is of full rank. Then the cross-product matrix is:

$$
\mathbf{X}_R^t\mathbf{X}_R = \begin{bmatrix} 6 & 2 & 2 \\ 2 & 2 & 0 \\ 2 & 0 & 2 \end{bmatrix} \tag{3.35}
$$

1 There is a subtle difference in applying the constraints on the parameters instead of on its estimates, but for simplicity we ignore this for the moment.

and the normal equations are

$$
\begin{aligned}
6\hat{\mu} + 2\hat{\alpha}_2 + 2\hat{\alpha}_3 &= y_{..} \\
2\hat{\mu} + 2\hat{\alpha}_2 &= y_{2.} \\
2\hat{\mu} + 2\hat{\alpha}_3 &= y_{3.}
\end{aligned} \tag{3.36}
$$

and this can be solved to give the estimates:

$$
\begin{aligned}
\hat{\mu} &= \bar{y}_{1.} \\
\hat{\alpha}_1 &= 0 \\
\hat{\alpha}_2 &= \bar{y}_{2.} - \bar{y}_{1.} \\
\hat{\alpha}_3 &= \bar{y}_{3.} - \bar{y}_{1.}
\end{aligned} \tag{3.37}
$$

which shows that now the factor estimates are expressed relative (or as reference) to factor level one. Another way to understand the working of reference-coding is explained in Elaboration 3.7. Note that the parameters for the reference-coded model have a different interpretation from the ones in the sum-coded model (see Elaboration 3.7); for simplicity, we used the same symbols.

ELABORATION 3.7

Reference-coding explained differently

We will explain reference-coding in a bit more detail using again the example of a one-way ANOVA with Factor A at three levels. Invoking the constraint that $\alpha_1 = 0$ results in the following model:

$$
\begin{aligned}
y_{11} &= \mu + \epsilon_{11} \\
y_{12} &= \mu + \epsilon_{12} \\
y_{21} &= \mu + \alpha_2 + \epsilon_{21} \\
y_{22} &= \mu + \alpha_2 + \epsilon_{22} \\
y_{31} &= \mu + \alpha_3 + \epsilon_{31} \\
y_{32} &= \mu + \alpha_3 + \epsilon_{32}
\end{aligned} \tag{3.38}
$$

where the parameters μ, α_2, and α_3 change meaning compared to the sum-coded ones as is evident from Equation 3.38. This now results in a regression equation with design matrix $\mathbf{X}_R$ and a parameter vector where α_1 is removed.

The constraint $\alpha_1 = 0$ can also be written as $\mathbf{d}^t\theta = 0$ with $\mathbf{d} = (0, 1, 0, 0)^t$ and $\theta = (\mu, \alpha_1, \alpha_2, \alpha_3)^t$. The original regression equation for the model using full-coding can now again be augmented to also include the constraint:

$$
\begin{bmatrix} \mathbf{y} \\ 0 \end{bmatrix} = \begin{bmatrix} 1 & 1 & 0 & 0 \\ 1 & 1 & 0 & 0 \\ 1 & 0 & 1 & 0 \\ 1 & 0 & 1 & 0 \\ 1 & 0 & 0 & 1 \\ 1 & 0 & 0 & 1 \\ \hdashline 0 & 1 & 0 & 0 \end{bmatrix} \theta + \begin{bmatrix} \epsilon \\ 0 \end{bmatrix} \tag{3.39}
$$

where the last line in this regression equation reflects the constraint. The augmented design matrix is now full rank due to the augmentation with the row $\mathbf{d}$. Hence, also the resulting normal equations will be of full rank and thus exactly solvable thereby generating the results of the reference-coding regression equation.

These simple one-way examples show that there are different codings possible which are all related to a model specification and constraints on the estimated parameters. All contrasts which are relevant to this one-way ANOVA model (*e.g.*, $\hat{\alpha}_1 - \hat{\alpha}_2$) are independent of the coding schemes used to formulate the ANOVA model in a regression framework and will thus generate the same results.

3.4.1.2 Two-Way Crossed Designs

For two-way crossed treatment designs, the means models of Equation 3.17 cannot be used anymore (Dobson and Barnett, 2018), see Elaboration 3.8.

ELABORATION 3.8

Means model for two-factor ANOVA model without interaction

For an ANOVA model with Factor A at three levels and Factor B at two levels and only one replicate, the means model would lead to the design matrix:

$$\mathbf{X}_{MM} = \begin{bmatrix} 1 & 0 & 0 & 1 & 0 \\ 1 & 0 & 0 & 0 & 1 \\ 0 & 1 & 0 & 1 & 0 \\ 0 & 1 & 0 & 0 & 1 \\ 0 & 0 & 1 & 1 & 0 \\ 0 & 0 & 1 & 0 & 1 \end{bmatrix} \tag{3.40}$$

and the corresponding cross-product matrix is

$$\mathbf{X}^t_{MM}\mathbf{X}_{MM} = \begin{bmatrix} 2 & 0 & 0 & 1 & 1 \\ 0 & 2 & 0 & 1 & 1 \\ 0 & 0 & 2 & 1 & 1 \\ 1 & 1 & 1 & 3 & 0 \\ 1 & 1 & 1 & 0 & 3 \end{bmatrix} \tag{3.41}$$

and this matrix is rank-deficient since the first three columns have the same sum as the last two columns. Hence, the normal equations have no unique solution, and extra restrictions on the estimated parameters are necessary.

As for one-way ANOVA models, there are also different ways to encode an ANOVA model for a two-way crossed treatment design to enable writing those models as regression models. We will explain different coding schemes by using the simple crossed fixed-effect ANOVA (balanced) model with two factors and interaction:

$$y_{abn} = \mu + \alpha_a + \beta_b + (\alpha\beta)_{ab} + \epsilon_{abn} \tag{3.42}$$

and we will assume that $A = 2, B = 3$, and $N = 2$. We will discuss three types of coding schemes and will start with the most obvious one, namely full-coding. This means that we simply use dummy variables to indicate the levels of the factors. The model can thus be written as

$$\mathbf{y} = \mathbf{X}_F\theta + \epsilon \tag{3.43}$$

where $\mathbf{y} = (y_{111}, y_{112}, y_{121}, y_{122}, \dots, y_{231}, y_{232})^t$, $\boldsymbol{\theta} = (\mu, \alpha_1, \dots, (\alpha\beta)_{23})^t$, and ϵ contains all residuals. The matrix $\mathbf{X}_F$ is then:

$$\left[\begin{array}{c|cc|ccc|cccccc}
1 & 1 & 0 & 1 & 0 & 0 & 1 & 0 & 0 & 0 & 0 & 0 \\
1 & 1 & 0 & 1 & 0 & 0 & 1 & 0 & 0 & 0 & 0 & 0 \\
1 & 1 & 0 & 0 & 1 & 0 & 0 & 1 & 0 & 0 & 0 & 0 \\
1 & 1 & 0 & 0 & 1 & 0 & 0 & 1 & 0 & 0 & 0 & 0 \\
1 & 1 & 0 & 0 & 0 & 1 & 0 & 0 & 1 & 0 & 0 & 0 \\
1 & 1 & 0 & 0 & 0 & 1 & 0 & 0 & 1 & 0 & 0 & 0 \\
1 & 0 & 1 & 1 & 0 & 0 & 0 & 0 & 0 & 1 & 0 & 0 \\
1 & 0 & 1 & 1 & 0 & 0 & 0 & 0 & 0 & 1 & 0 & 0 \\
1 & 0 & 1 & 0 & 1 & 0 & 0 & 0 & 0 & 0 & 1 & 0 \\
1 & 0 & 1 & 0 & 1 & 0 & 0 & 0 & 0 & 0 & 1 & 0 \\
1 & 0 & 1 & 0 & 0 & 1 & 0 & 0 & 0 & 0 & 0 & 1 \\
1 & 0 & 1 & 0 & 0 & 1 & 0 & 0 & 0 & 0 & 0 & 1
\end{array}\right] \tag{3.44}$$

which is highly structured. Several remarks are appropriate:

- The first column codes for the mean of the data.
- The pairs of identical rows in $\mathbf{X}_F$ refer to the two repeats in the ANOVA design.
- The second and third columns indicate the two levels of Factor A.
- The fourth, fifth, and sixth columns indicate the three levels of Factor B.
- The final columns indicate the levels of the interaction and can be obtained by element-wise products of the columns indicating levels of Factors A and B.

The matrix $\mathbf{X}_F$ has size (12×12); these numbers are accidentally the same for this example, but this is not general (for $N = 3$ this matrix would have size (18×12)). This matrix is also rank deficient, and it can be shown that the rank of $\mathbf{X}_F$ is six (see the last six columns of $\mathbf{X}_F$ which are linearly independent, and the first six columns can be written as combinations of those; it has also only six distinct rows). Hence, we are in the area of regression models not of full rank (Searle, 1971). The normal equations for this regression model are $(\mathbf{X}_F^t\mathbf{X}_F)\hat{\boldsymbol{\theta}} = \mathbf{X}_F^t\mathbf{y}$, and the matrix $(\mathbf{X}_F^t\mathbf{X}_F)(12 \times 12)$ has rank six and thus is rank-deficient. This means that the normal equations have an infinite number of solutions since the normal equations are a consistent set of equations (Searle, 1971). This matrix $\mathbf{X}_F^t\mathbf{X}_F$ is:

$$\left[\begin{array}{cccccccccccc}
12 & 6 & 6 & 4 & 4 & 4 & 2 & 2 & 2 & 2 & 2 & 2 \\
6 & 6 & 0 & 2 & 2 & 2 & 2 & 2 & 2 & 0 & 0 & 0 \\
6 & 0 & 6 & 2 & 2 & 2 & 0 & 0 & 0 & 2 & 2 & 2 \\
4 & 2 & 2 & 4 & 0 & 0 & 2 & 0 & 0 & 2 & 0 & 0 \\
4 & 2 & 2 & 0 & 4 & 0 & 0 & 2 & 0 & 0 & 2 & 0 \\
4 & 2 & 2 & 0 & 0 & 4 & 0 & 0 & 2 & 0 & 0 & 2 \\
2 & 2 & 0 & 2 & 0 & 0 & 2 & 0 & 0 & 0 & 0 & 0 \\
2 & 2 & 0 & 0 & 2 & 0 & 0 & 2 & 0 & 0 & 0 & 0 \\
2 & 2 & 0 & 0 & 0 & 2 & 0 & 0 & 2 & 0 & 0 & 0 \\
2 & 0 & 2 & 2 & 0 & 0 & 0 & 0 & 0 & 2 & 0 & 0 \\
2 & 0 & 2 & 0 & 2 & 0 & 0 & 0 & 0 & 0 & 2 & 0 \\
2 & 0 & 2 & 0 & 0 & 2 & 0 & 0 & 0 & 0 & 0 & 2
\end{array}\right] \tag{3.45}$$

which is highly structured without clear orthogonal parts. To solve the normal equations we have to use a generalized inverse.

Now $\mathbf{X}_F^t\mathbf{y}$ is $(y_{...}, y_{1..}, y_{2..}, y_{.1.}, y_{.2.}, y_{.3.}, y_{11.}, y_{12.}, y_{13.}, y_{21.}, y_{22.}, y_{23.})^t$, and there are several constraints active between the values of this $\mathbf{X}_F^t\mathbf{y}$ vector:

$$\begin{aligned}
y_{...} &= y_{1..} + y_{2..} \\
y_{...} &= y_{.1.} + y_{.2.} + y_{.3.} \\
y_{1..} &= y_{11.} + y_{12.} + y_{13.} \\
y_{2..} &= y_{21.} + y_{22.} + y_{23.} \\
y_{.1.} &= y_{11.} + y_{21.} \\
y_{.2.} &= y_{12.} + y_{22.} \\
y_{.3.} &= y_{13.} + y_{23.}
\end{aligned} \tag{3.46}$$

which can also concisely be written as:

$$\begin{bmatrix} 1 & -1 & -1 & 0 & 0 & 0 & 0 & 0 & 0 & 0 & 0 & 0 \\ 1 & 0 & 0 & -1 & -1 & -1 & 0 & 0 & 0 & 0 & 0 & 0 \\ 0 & 1 & 0 & 0 & 0 & 0 & -1 & -1 & -1 & 0 & 0 & 0 \\ 0 & 0 & 1 & 0 & 0 & 0 & 0 & 0 & 0 & -1 & -1 & -1 \\ 0 & 0 & 0 & 1 & 0 & 0 & -1 & 0 & 0 & -1 & 0 & 0 \\ 0 & 0 & 0 & 0 & 1 & 0 & 0 & -1 & 0 & 0 & -1 & 0 \\ 0 & 0 & 0 & 0 & 0 & 1 & 0 & 0 & -1 & 0 & 0 & -1 \end{bmatrix} \begin{bmatrix} y_{...} \\ y_{1..} \\ y_{2..} \\ y_{.1.} \\ y_{.2.} \\ y_{.3.} \\ y_{11.} \\ y_{12.} \\ y_{13.} \\ y_{21.} \\ y_{22.} \\ y_{23.} \end{bmatrix} = \mathbf{0} \tag{3.47}$$

where $\mathbf{0}$ is a vector of seven zeroes and since the matrix in this equation has rank six (rows $1-2+3+4-5-6-7=0$), there are six independent restrictions on the values in the vector $\mathbf{X}_F^t\mathbf{y}$. Thus, there are only six independent quantities in $\mathbf{X}_F^t\mathbf{y}$. This shows again that the essential rank of the system of normal equations is only six. One particular choice of the generalized inverse of $\mathbf{X}_F^t\mathbf{X}_F$ is (Schott, 2016):

$$\begin{bmatrix} 1 & 0 & 0 & 0 & 0 & 0 & 0 & 0 & 0 & 0 & 0 & 0 \\ 0 & 1 & -1 & 0 & 0 & 0 & 0 & 0 & 0 & 0 & 0 & 0 \\ 0 & -1 & 1 & 0 & 0 & 0 & 0 & 0 & 0 & 0 & 0 & 0 \\ 0 & 0 & 0 & 2 & -1 & -1 & 0 & 0 & 0 & 0 & 0 & 0 \\ 0 & 0 & 0 & -1 & 2 & -1 & 0 & 0 & 0 & 0 & 0 & 0 \\ 0 & 0 & 0 & -1 & -1 & 2 & 0 & 0 & 0 & 0 & 0 & 0 \\ 0 & 0 & 0 & 0 & 0 & 0 & 2 & -1 & -1 & -2 & 1 & 1 \\ 0 & 0 & 0 & 0 & 0 & 0 & -1 & 2 & -1 & 1 & -2 & 1 \\ 0 & 0 & 0 & 0 & 0 & 0 & -1 & -1 & 2 & 1 & 1 & -2 \\ 0 & 0 & 0 & 0 & 0 & 0 & -2 & 1 & 1 & 2 & -1 & -1 \\ 0 & 0 & 0 & 0 & 0 & 0 & 1 & -2 & 1 & -1 & 2 & -1 \\ 0 & 0 & 0 & 0 & 0 & 0 & 1 & 1 & -2 & -1 & -1 & 2 \end{bmatrix} \tag{3.48}$$

and this results in the estimates $(\mathbf{X}_F^t\mathbf{X}_F)^{-}\mathbf{X}_F^t\mathbf{y}$:

$$\hat{\theta} = \begin{bmatrix} \hat{\mu} \\ \hat{\alpha}_1 \\ \hat{\alpha}_2 \\ \hat{\beta}_1 \\ \hat{\beta}_2 \\ \hat{\beta}_3 \\ \widehat{(\alpha\beta)}_{11} \\ . \\ . \\ . \\ . \\ \widehat{(\alpha\beta)}_{23} \end{bmatrix} = \frac{1}{12}\begin{bmatrix} y_{...} \\ y_{1..} - y_{2..} \\ y_{2..} - y_{1..} \\ 2y_{.1.} - y_{.2.} - y_{.3.} \\ 2y_{.2.} - y_{.1.} - y_{.3.} \\ 2y_{.3.} - y_{.1.} - y_{.2.} \\ 2y_{11.} - 2y_{21.} - y_{12.} - y_{13.} + y_{22.} + y_{23.} \\ 2y_{12.} - 2y_{22.} - y_{11.} - y_{13.} + y_{21.} + y_{23.} \\ 2y_{13.} - 2y_{23.} - y_{11.} - y_{12.} + y_{21.} + y_{22.} \\ 2y_{21.} - 2y_{11.} - y_{22.} - y_{23.} + y_{12.} + y_{13.} \\ 2y_{22.} - 2y_{12.} - y_{21.} - y_{23.} + y_{11.} + y_{13.} \\ 2y_{23.} - 2y_{13.} - y_{21.} - y_{22.} + y_{11.} + y_{12.} \end{bmatrix} \tag{3.49}$$

which clearly shows that these estimates are dependent (see also Equation 3.46). It can be shown by using the relationships in Equation 3.46 that the estimates obtained in this way are $\hat{\mu} = \bar{y}_{...}$, $\hat{\alpha}_1 = \bar{y}_{1..} - \bar{y}_{...}$, $\hat{\alpha}_2 = \bar{y}_{2..} - \bar{y}_{...}$, $\hat{\beta}_1 = \bar{y}_{.1.} - \bar{y}_{...}$, etc. and writing out the equation $\hat{y} = \mathbf{X}\hat{\theta}$ then generates the decomposition as shown in Equation 3.9.

This generalized inverse comes down to applying the sum-to-zero constraints on the solution vector (Searle, 1971; Schott, 2016), and these are in this example:

$$\begin{aligned} \hat{\alpha}_1 + \hat{\alpha}_2 &= 0 \\ \hat{\beta}_1 + \hat{\beta}_2 + \hat{\beta}_3 &= 0 \\ (\widehat{\alpha\beta}_{11}) + (\widehat{\alpha\beta}_{21}) &= 0 \\ (\widehat{\alpha\beta}_{12}) + (\widehat{\alpha\beta}_{22}) &= 0 \\ (\widehat{\alpha\beta}_{13}) + (\widehat{\alpha\beta}_{23}) &= 0 \\ (\widehat{\alpha\beta}_{11}) + (\widehat{\alpha\beta}_{12}) + (\widehat{\alpha\beta}_{13}) &= 0 \\ (\widehat{\alpha\beta}_{21}) + (\widehat{\alpha\beta}_{22}) + (\widehat{\alpha\beta}_{23}) &= 0 \end{aligned} \tag{3.50}$$

which can be written concisely as

$$\begin{bmatrix} 1 & 1 & 0 & 0 & 0 & 0 & 0 & 0 & 0 & 0 & 0 \\ 0 & 0 & 1 & 1 & 1 & 0 & 0 & 0 & 0 & 0 & 0 \\ 0 & 0 & 0 & 0 & 0 & 1 & 0 & 0 & 1 & 0 & 0 \\ 0 & 0 & 0 & 0 & 0 & 0 & 1 & 0 & 0 & 1 & 0 \\ 0 & 0 & 0 & 0 & 0 & 0 & 0 & 1 & 0 & 0 & 1 \\ 0 & 0 & 0 & 0 & 0 & 1 & 1 & 1 & 0 & 0 & 0 \\ 0 & 0 & 0 & 0 & 0 & 0 & 0 & 0 & 1 & 1 & 1 \end{bmatrix} \begin{bmatrix} \hat{\alpha}_1 \\ \hat{\alpha}_2 \\ \hat{\beta}_1 \\ \hat{\beta}_2 \\ \hat{\beta}_3 \\ \widehat{(\alpha\beta)}_{11} \\ \widehat{(\alpha\beta)}_{12} \\ \widehat{(\alpha\beta)}_{13} \\ \widehat{(\alpha\beta)}_{21} \\ \widehat{(\alpha\beta)}_{22} \\ \widehat{(\alpha\beta)}_{23} \end{bmatrix} = \mathbf{0} \tag{3.51}$$

and the matrix in this equation has rank six since row 6+7 equals row 3+4+5. Hence, there are exactly six independent constraints as already encountered earlier.

Sum-coding was already explained earlier and works by skipping some of the estimates using the fact that there is redundancy in the parametrization. As stated earlier, there are six independent parameters to estimate, and since there are twelve in total, six have to be skipped. These are indicated with a "−1" in the design matrix. When skipping $\alpha_1, \beta_1, (\alpha\beta)_{11}, (\alpha\beta)_{12}, (\alpha\beta)_{13}, (\alpha\beta)_{21}$, this gives the matrix $\mathbf{X}_S$ as

$$\begin{bmatrix} 1 & -1 & -1 & -1 & 1 & 1 \\ 1 & -1 & -1 & -1 & 1 & 1 \\ 1 & -1 & 1 & 0 & -1 & 0 \\ 1 & -1 & 1 & 0 & -1 & 0 \\ 1 & -1 & 0 & 1 & 0 & -1 \\ 1 & -1 & 0 & 1 & 0 & -1 \\ 1 & 1 & -1 & -1 & -1 & -1 \\ 1 & 1 & -1 & -1 & -1 & -1 \\ 1 & 1 & 1 & 0 & 1 & 0 \\ 1 & 1 & 1 & 0 & 1 & 0 \\ 1 & 1 & 0 & 1 & 0 & 1 \\ 1 & 1 & 0 & 1 & 0 & 1 \end{bmatrix} \tag{3.52}$$

where the levels of Factor A are now indicated by −1 and 1 (second column); and for Factor B as −1, 0, or 1 (third and fourth columns) showing that α_1 and β_1 are skipped. This works in a similar way as explained in Elaboration 3.6 by inserting the constraints on the parameters, *e.g.*, by starting with the full-coding and inserting $\alpha_1 = -\alpha_2$ and $\beta_1 = -\beta_2 - \beta_3$ and likewise for the interactions. This then generates the design matrix $\mathbf{X}_S$. The interactions are represented in the last two columns and are again the element-wise multiplications of the second column with the third and fourth columns. Note that all columns sum to zero. This matrix $\mathbf{X}_S$ has size (12×6) and has rank six. The normal equations are in this parametrization again $(\mathbf{X}_S^t\mathbf{X}_S)\hat{\theta} = \mathbf{X}_S^t\mathbf{y}$ and now $\mathbf{X}_S^t\mathbf{X}_S$ is

$$\begin{bmatrix} 12 & 0 & 0 & 0 & 0 & 0 \\ 0 & 12 & 0 & 0 & 0 & 0 \\ 0 & 0 & 8 & 4 & 0 & 0 \\ 0 & 0 & 4 & 8 & 0 & 0 \\ 0 & 0 & 0 & 0 & 8 & 4 \\ 0 & 0 & 0 & 0 & 4 & 8 \end{bmatrix} \tag{3.53}$$

which is clearly invertible and shows orthogonality between the factor estimates. The resulting normal equations can be solved as $\hat{\theta} = (\mathbf{X}_S^t\mathbf{X}_S)^{-1}\mathbf{X}_S^t\mathbf{y}$ where now $\mathbf{X}_S^t\mathbf{y}$ is

$$\begin{bmatrix} y_{...} \\ y_{2..} - y_{1..} \\ -y_{11.} + y_{12.} - y_{21.} + y_{22.} \\ -y_{11.} + y_{13.} - y_{21.} + y_{23.} \\ y_{11.} - y_{12.} - y_{21.} + y_{22.} \\ y_{11.} - y_{13.} - y_{21.} + y_{23.} \end{bmatrix} \tag{3.54}$$

and the solution for the parameter vector $\hat{\theta}$ is now:

$$\hat{\theta} = \begin{bmatrix} \hat{\mu} \\ \hat{\alpha}_2 \\ \hat{\beta}_2 \\ \hat{\beta}_3 \\ \widehat{(\alpha\beta)}_{22} \\ \widehat{(\alpha\beta)}_{23} \end{bmatrix} = \frac{1}{12} \begin{bmatrix} y_{...} \\ y_{2..} - y_{1..} \\ 2y_{12.} + 2y_{22.} - y_{11.} - y_{13.} - y_{21.} - y_{23.} \\ 2y_{13.} + 2y_{23.} - y_{11.} - y_{12.} - y_{21.} - y_{22.} \\ 2y_{22.} - 2y_{12.} + y_{11.} + y_{13.} - y_{21.} - y_{23.} \\ 2y_{23.} - 2y_{13.} + y_{11.} + y_{12.} - y_{21.} - y_{22.} \end{bmatrix} \tag{3.55}$$

giving the solution $\hat{\mu} = \frac{1}{12}y_{...} = \bar{y}_{...}$, $\hat{\alpha}_2 = \frac{1}{12}(y_{2..} - y_{1..}) = \bar{y}_{2..} - \bar{y}_{...}$, $\hat{\beta}_2 = \frac{1}{12}(2y_{12.} + 2y_{22.} - y_{11.} - y_{13.} - y_{21.} - y_{23.}) = \bar{y}_{.2.} - \bar{y}_{...}$, and similarly for the other estimated parameters; where the latter two estimates are calculated using the rules in Equation 3.46. This gives the solution related to the sum-to-zero constraints which are thus built in implicitly. Calculating $\mathbf{X}_S\hat{\theta} = \mathbf{X}_S(\mathbf{X}_S^t\mathbf{X}_S)^{-}\mathbf{X}_S^t$ gives $\hat{\mathbf{y}}$ which generates the decomposition of Equation 3.9 showing (again) that now the sum-to-zero constraints are used implicitly.

For reference-coding a choice has to be made which effects to set to zero and thus serve as the reference. By setting to zero $\hat{\alpha}_1, \hat{\beta}_1, \widehat{(\alpha\beta)}_{11}, \widehat{(\alpha\beta)}_{12}, \widehat{(\alpha\beta)}_{13}, \widehat{(\alpha\beta)}_{21}$ this means that the columns representing these effects should be removed from the full coding matrix $\mathbf{X}_F$ (see Equation 3.44). Hence, the $\mathbf{X}_R$ for this reference-coding is:

$$\begin{bmatrix} 1 & 0 & 0 & 0 & 0 & 0 \\ 1 & 0 & 0 & 0 & 0 & 0 \\ 1 & 0 & 1 & 0 & 0 & 0 \\ 1 & 0 & 1 & 0 & 0 & 0 \\ 1 & 0 & 0 & 1 & 0 & 0 \\ 1 & 0 & 0 & 1 & 0 & 0 \\ 1 & 1 & 0 & 0 & 0 & 0 \\ 1 & 1 & 0 & 0 & 0 & 0 \\ 1 & 1 & 1 & 0 & 1 & 0 \\ 1 & 1 & 1 & 0 & 1 & 0 \\ 1 & 1 & 0 & 1 & 0 & 1 \\ 1 & 1 & 0 & 1 & 0 & 1 \end{bmatrix} \tag{3.56}$$

where now the factor estimates are expressed relative to the intercept which is an estimate for the observations at the first level of both factors.[2] The matrix $\mathbf{X}_R^t\mathbf{X}_R$ is now:

$$\begin{bmatrix} 12 & 6 & 4 & 4 & 2 & 2 \\ 6 & 6 & 2 & 2 & 2 & 2 \\ 4 & 2 & 4 & 0 & 2 & 0 \\ 4 & 2 & 0 & 4 & 0 & 2 \\ 2 & 2 & 2 & 0 & 2 & 0 \\ 2 & 2 & 0 & 2 & 0 & 2 \end{bmatrix} \tag{3.57}$$

2 Note that again the model parameters are different from the sum-coded ones, but we use the same symbols for simplicity.

which is full rank and can also be obtained by removing the rows and columns of $\mathbf{X}_F^t\mathbf{X}_F$ of Equation 3.45 related to the zeroed estimates. Note that there is no orthogonality between the columns. Because $\mathbf{X}_R^t\mathbf{X}_R$ is full rank, the normal equations can be solved. The $\mathbf{X}_R^t\mathbf{y}$ vector for this case is:

$$\begin{bmatrix} y_{...} \\ y_{2..} \\ y_{.2.} \\ y_{.3.} \\ y_{22.} \\ y_{23.} \end{bmatrix} \tag{3.58}$$

which shows only the "non-zeroed" factor levels. The factor estimates $\widehat{\theta}$ are:

$$\widehat{\theta} = \begin{bmatrix} \widehat{\mu} \\ \widehat{\alpha}_2 \\ \widehat{\beta}_2 \\ \widehat{\beta}_3 \\ \widehat{(\alpha\beta)}_{22} \\ \widehat{(\alpha\beta)}_{23} \end{bmatrix} = \frac{1}{2}\begin{bmatrix} y_{11.} \\ y_{21.} - y_{11.} \\ y_{12.} - y_{11.} \\ y_{13.} - y_{11.} \\ y_{11.} + y_{22.} - y_{12.} - y_{21.} \\ y_{11.} + y_{23.} - y_{13.} - y_{21.} \end{bmatrix} = \begin{bmatrix} \overline{y}_{11.} \\ \overline{y}_{21.} - \overline{y}_{11.} \\ \overline{y}_{12.} - \overline{y}_{11.} \\ \overline{y}_{13.} - \overline{y}_{11.} \\ \overline{y}_{11.} + \overline{y}_{22.} - \overline{y}_{12.} - \overline{y}_{21.} \\ \overline{y}_{11.} + \overline{y}_{23.} - \overline{y}_{13.} - \overline{y}_{21.} \end{bmatrix} \tag{3.59}$$

giving the factor estimates $\widehat{\mu} = \overline{y}_{11.}$, $\widehat{\alpha}_2 = \overline{y}_{21.} - \overline{y}_{11.}$, $\widehat{\beta}_2 = \overline{y}_{12.} - \overline{y}_{11.}$, $\widehat{\beta}_3 = \overline{y}_{13.} - \overline{y}_{11.}$ showing that now all estimates are expressed relative to $\widehat{\mu} = \overline{y}_{11.}$. The specific generalized inverse of size (12×12) for this situation is

$$\begin{bmatrix} (\mathbf{X}_R^t\mathbf{X}_R)^{-1} & \mathbf{0} \\ \mathbf{0} & \mathbf{0} \end{bmatrix} \tag{3.60}$$

where the matrices $\mathbf{0}$ have sizes to match the resulting (12×12) matrix (Searle, 1971). Using this generalized inverse, $\mathbf{X}_R\widehat{\theta} = \mathbf{X}_R(\mathbf{X}_R^t\mathbf{X}_R)^{-}\mathbf{X}_R^t$ gives again $\hat{\mathbf{y}}$.

3.4.1.3 Two-Way Nested Designs

The presented coding schemes can also be used for nested models (see Section 3.2.2.2). In Section 3.4.2 we will show how this works for the more general case of an unbalanced nested design.

3.4.2 Codings for Unbalanced Designs

For unbalanced designs, any coding will work and thus we can use full-coding, sum-coding, or reference-coding similar to in the balanced case. Simple ANOVA calculations do not work as explained in Elaboration 3.2. Elaboration 3.9 extends this to show how a regression model works in this case.

ELABORATION 3.9

Regression ANOVA retrieves the correct factor estimates (1)

The simple numerical example in Elaboration 3.2 is continued to show how proper factor estimates can be obtained using a regression approach. Recall that trial y_{12} is missing; hence, we have unbalanced data. We can use sum-coding in our regression model resulting in the design matrix:

$$\mathbf{X}_S = \begin{bmatrix} 1 & -1 \\ 1 & 1 \\ 1 & 1 \end{bmatrix} \tag{3.61}$$

and the parameter α_1 is skipped. Solving the normal equations gives estimates $\hat{\mu} = 2$ and $\hat{\alpha}_2 = -1$ which gives also $\hat{\alpha}_1 = 1$ because of the constraint. Thus, the original factor effect values are retrieved.

Also in the case of the two-way ANOVA model, regression models can retrieve the true factor effects, see Elaboration 3.10.

ELABORATION 3.10

Regression ANOVA retrieves the correct factor estimates (2)

The simple numerical example in Elaboration 3.3 is continued using a regression approach with the design matrix

$$\mathbf{X} = [\mathbf{X}_M|\mathbf{X}_A|\mathbf{X}_B|\mathbf{X}_{AB}] = \left[\begin{array}{cc:cc:cc} 1 & 1 & 1 & 0 & 1 & 0 \\ 1 & 1 & 1 & 0 & 1 & 0 \\ 1 & 1 & 0 & 1 & 0 & 1 \\ 1 & 1 & 0 & 1 & 0 & 1 \\ 1 & 1 & -1 & -1 & -1 & -1 \\ 1 & 1 & -1 & -1 & -1 & -1 \\ 1 & -1 & 1 & 0 & -1 & 0 \\ 1 & -1 & 1 & 0 & -1 & 0 \\ 1 & -1 & 0 & 1 & 0 & -1 \\ 1 & -1 & 0 & 1 & 0 & -1 \\ 1 & -1 & -1 & -1 & 1 & 1 \\ 1 & -1 & -1 & -1 & 1 & 1 \end{array}\right] \tag{3.62}$$

which is using the sum-coding. The matrix $\mathbf{X}^t\mathbf{X}$ is:

$$\left[\begin{array}{c:c:cc:cc} 12 & 0 & 0 & 0 & 0 & 0 \\ 0 & 12 & 0 & 0 & 0 & 0 \\ \hdashline 0 & 0 & 8 & 4 & 0 & 0 \\ 0 & 0 & 4 & 8 & 0 & 0 \\ \hdashline 0 & 0 & 0 & 0 & 8 & 4 \\ 0 & 0 & 0 & 0 & 4 & 8 \end{array}\right] \tag{3.63}$$

which shows that the submatrices in Equation 3.62 are orthogonal to each other. Both approaches (ANOVA calculations and regression estimates) give exactly the same result and retrieve the true factor estimates. Suppose now that trials y_{222} and y_{232} are missing; hence, we have unbalanced data. This means that the corresponding rows in $\mathbf{X}$ are missing, and this matrix loses its orthogonality properties. The matrix $\mathbf{X}^t\mathbf{X}$ for this case becomes:

$$\left[\begin{array}{c:c:cc:cc} 10 & 2 & 1 & 0 & 1 & 0 \\ 2 & 10 & -1 & 0 & -1 & 0 \\ \hdashline 1 & -1 & 7 & 3 & 7 & 3 \\ 0 & 0 & 3 & 6 & 3 & 6 \\ \hdashline 1 & -1 & 7 & 3 & 7 & 3 \\ 0 & 0 & 3 & 6 & 3 & 6 \end{array}\right] \tag{3.64}$$

which is obviously not of a simple form and shows that the submatrices $\mathbf{X}_M$, $\mathbf{X}_A$, $\mathbf{X}_B$, and $\mathbf{X}_{AB}$ are not orthogonal anymore.

The application of the regression-based solution using the proper $\mathbf{X}$ (with the two rows regarding the missing replicates deleted) retrieves exactly the true solution. Thus for unbalanced data, the regression approach is recommended, and this is what ASCA+ (see Section 4.3) uses.

As stated earlier, the codings as used in the balanced case can also be used in the unbalanced case, but some codings are more convenient than others. A particular new type of useful coding is the weighted-coding (Nieuwenhuis *et al.*, 2017; Te Grotenhuis *et al.*, 2017) which can also be used in (unbalanced) ASCA+ (Ali *et al.*, 2020). The basic idea in weighted-coding is that the effects are estimated relative to the sample mean. For a one-way ANOVA setup with levels $a = 1, \ldots, A$ and N_a replicates in group a, the overall sample mean is:

$$\bar{y}_{..} = \frac{1}{I}\sum_{a=1}^{A}\sum_{n}^{N_a} y_{an} \tag{3.65}$$

where $I = \sum_a N_a$ and this $\bar{y}_{..}$ is not the same as the average of the group means $\bar{y}_{a.}$ for the unbalanced case. In weighted-coding, there should also be a parameter skipped like in sum-coding. If we chose α_1 again, then the design matrix for this situation using weighted-coding is

$$\mathbf{X}_W = \begin{bmatrix} 1 & -\frac{N_2}{N_1} & \cdots & -\frac{N_A}{N_1} \\ \cdot & \cdot & \cdots & \cdot \\ \cdot & \cdot & \cdots & \cdot \\ 1 & -\frac{N_2}{N_1} & \cdots & -\frac{N_A}{N_1} \\ \hdashline 1 & 1\;0 & \cdots & 0 \\ \cdot & \cdot & \cdots & \cdot \\ \cdot & \cdot & \cdots & \cdot \\ 1 & 1\;0 & \cdots & 0 \\ \hdashline \cdot & \cdot & \cdots & \cdot \\ \cdot & \cdot & \cdots & \cdot \\ \cdot & \cdot & \cdots & \cdot \\ \hdashline 1 & 0 & \cdots & 1 \\ \cdot & \cdot & \cdots & \cdot \\ \cdot & \cdot & \cdots & \cdot \\ 1 & 0 & \cdots & 1 \end{bmatrix} \tag{3.66}$$

where the rows with dots mean repeats of the row above pertaining to the replicates in the same group. This coding is for the simple case of a one-way unbalanced ANOVA model, but this can be extended to more complex models with interactions (Nieuwenhuis *et al.*, 2017; Te Grotenhuis *et al.*, 2017). In particular, unlike reference-coding and sum-coding, the columns for the interaction terms are not simply obtained by multiplication of coded variables for the interacting factors. Instead, they are weighted as well, to take into account the possible unbalancedness. Below is an example of a weighted-coded design matrix for a two-factor design with Factor A at two levels and Factor B at three levels (with number of replicates indicated with N_a and N_b), where the skipped levels are the first for each factor ($a = 1$ and $b = 1$). Note that the columns (apart from the first one) add up to zero.

$$
\mathbf{X}_W = \left[\begin{array}{c:c:cc:cc}
1 & -\frac{N_{a=2}}{N_{a=1}} & -\frac{N_{b=2}}{N_{b=1}} & -\frac{N_{b=2}}{N_{b=1}} & \frac{N_{a=2\&b=2}}{N_{a=1\&b=1}} & \frac{N_{a=2\&b=3}}{N_{a=1\&b=1}} \\
\vdots & \vdots & \vdots & \vdots & \vdots & \vdots \\
1 & -\frac{N_{a=2}}{N_{a=1}} & -\frac{N_{b=2}}{N_{b=1}} & -\frac{N_{b=2}}{N_{b=1}} & \frac{N_{a=2\&b=2}}{N_{a=1\&b=1}} & \frac{N_{a=2\&b=3}}{N_{a=1\&b=1}} \\
\hdashline
1 & -\frac{N_{a=2}}{N_{a=1}} & 1 & 0 & -\frac{N_{a=2\&b=2}}{N_{a=1\&b=2}} & 0 \\
\vdots & \vdots & \vdots & \vdots & \vdots & \vdots \\
1 & -\frac{N_{a=2}}{N_{a=1}} & 1 & 0 & -\frac{N_{a=2\&b=2}}{N_{a=1\&b=2}} & 0 \\
\hdashline
1 & -\frac{N_{a=2}}{N_{a=1}} & 0 & 1 & 0 & -\frac{N_{a=2\&b=3}}{N_{a=1\&b=3}} \\
\vdots & \vdots & \vdots & \vdots & \vdots & \vdots \\
1 & -\frac{N_{a=2}}{N_{a=1}} & 0 & 1 & 0 & -\frac{N_{a=2\&b=3}}{N_{a=1\&b=3}} \\
\hdashline
1 & 1 & -\frac{N_{b=2}}{N_{b=1}} & -\frac{N_{b=2}}{N_{b=1}} & -\frac{N_{a=2\&b=2}}{N_{a=2\&b=1}} & -\frac{N_{a=2\&b=3}}{N_{a=2\&b=1}} \\
\vdots & \vdots & \vdots & \vdots & \vdots & \vdots \\
1 & 1 & -\frac{N_{b=2}}{N_{b=1}} & -\frac{N_{b=2}}{N_{b=1}} & -\frac{N_{a=2\&b=2}}{N_{a=2\&b=1}} & -\frac{N_{a=2\&b=3}}{N_{a=2\&b=1}} \\
\hdashline
1 & 1 & 1 & 0 & 1 & 0 \\
\vdots & \vdots & \vdots & \vdots & \vdots & \vdots \\
1 & 1 & 1 & 0 & 1 & 0 \\
\hdashline
1 & 1 & 0 & 1 & 0 & 1 \\
\vdots & \vdots & \vdots & \vdots & \vdots & \vdots \\
1 & 1 & 0 & 1 & 0 & 1
\end{array}\right] \tag{3.67}
$$

An illustration of the use of the weighted-coding scheme is discussed in Example 3.4.

Example 3.4: Weighted-coding for unbalanced ANOVA model with interactions

This example is drawn from Ali *et al.* (2020) and concerns a two-way ANOVA with interaction. There are two factors: Factor A at two levels and Factor B at three levels. All groups have one replicate except the last group ($a = 2, b = 3$) which has two replicates. Hence, in total, there are seven observations. The design matrix for this case using weighted-coding is:

$$
\mathbf{X}_W = \left[\begin{array}{c:c:cc:cc}
1 & 1 & 1 & 0 & 1 & 0 \\
1 & 1 & 0 & 1 & 0 & 1 \\
1 & 1 & -\frac{2}{3} & -\frac{2}{3} & -\frac{1}{1} & -\frac{1}{1} \\
1 & -\frac{3}{4} & 1 & 0 & -\frac{1}{1} & 0 \\
1 & -\frac{3}{4} & 0 & 1 & 0 & -\frac{1}{1} \\
1 & -\frac{3}{4} & -\frac{2}{3} & -\frac{2}{3} & \frac{1}{2} & \frac{1}{2} \\
1 & -\frac{3}{4} & -\frac{2}{3} & -\frac{2}{3} & \frac{1}{2} & \frac{1}{2}
\end{array}\right] \tag{3.68}
$$

where now the last levels of both factors are the omitted ones. This weighted-coding scheme is quite involved and is a function of the number of replicates in each group. Note that in this case, all columns in the matrix $\mathbf{X}_W$ (except the first one) add up to zero which would not have been the case if sum-coding was used. The cross-product matrix $\mathbf{X}_W^t\mathbf{X}_W$ is

$$\mathbf{X}_W^t\mathbf{X}_W = \left[\begin{array}{c|c|cc|cc} 7 & 0 & 0 & 0 & 0 & 0 \\ \hline 0 & 6 & \frac{7}{12} & \frac{7}{12} & 0 & 0 \\ \hline 0 & \frac{7}{12} & 6 & 4 & 0 & 0 \\ 0 & \frac{7}{12} & 4 & 6 & 0 & 0 \\ \hline 0 & 0 & 0 & 0 & \frac{7}{2} & \frac{3}{2} \\ 0 & 0 & 0 & 0 & \frac{3}{2} & \frac{7}{2} \end{array}\right] \tag{3.69}$$

which shows that indeed all other columns are orthogonal to the first column in $\mathbf{X}_W$. Moreover, the submatrix $\mathbf{X}_{\text{AB}}$ (last two columns of $\mathbf{X}_W$) is orthogonal to the other submatrices.

The best way to illustrate the different codings including the weighted-coding is by means of an extensive example, see Example 3.5.

Example 3.5: Different codings for a one-way unbalanced ANOVA model

In this example of an unbalanced one-way ANOVA, we will show how the different codings work. We will use a simple one-way ANOVA model with one factor at three levels and two replicates per cell for the first two levels and one replicate for the third level of the factor. Hence, the model is

$$y_{an} = \mu + \alpha_a + \epsilon_{an} \tag{3.70}$$

with Factor A ($a = 1, 2, 3$) and $\mathbf{y} = (y_{11}, y_{12}, y_{21}, y_{22}, y_{31})^t$. We start with the full-coding, thus

$$\mathbf{X}_F = \begin{bmatrix} 1 & 1 & 0 & 0 \\ 1 & 1 & 0 & 0 \\ 1 & 0 & 1 & 0 \\ 1 & 0 & 1 & 0 \\ 1 & 0 & 0 & 1 \end{bmatrix} \tag{3.71}$$

and this matrix is again rank deficient (the rank is three) since the last three columns add up to the first one. The resulting matrix $\mathbf{X}_F^t\mathbf{X}_F$ is

$$\mathbf{X}_F^t\mathbf{X}_F = \begin{bmatrix} 5 & 2 & 2 & 1 \\ 2 & 2 & 0 & 0 \\ 2 & 0 & 2 & 0 \\ 1 & 0 & 0 & 1 \end{bmatrix} \tag{3.72}$$

which has size (4×4) and thus is also rank deficient (rank is three as well). The other relevant matrix is $\mathbf{X}_F^t\mathbf{y}$ which is:

$$\mathbf{X}_F^t\mathbf{y} = \begin{bmatrix} 1 & 1 & 1 & 1 & 1 \\ 1 & 1 & 0 & 0 & 0 \\ 0 & 0 & 1 & 1 & 0 \\ 0 & 0 & 0 & 0 & 1 \end{bmatrix} \begin{bmatrix} y_{11} \\ y_{12} \\ y_{21} \\ y_{22} \\ y_{31} \end{bmatrix} = \begin{bmatrix} y_{..} \\ y_{1.} \\ y_{2.} \\ y_{3.} \end{bmatrix} \tag{3.73}$$

and thus the normal equations $(\mathbf{X}_F^t\mathbf{X}_F)\hat{\theta} = \mathbf{X}_F^t\mathbf{y}$ with $\hat{\theta} = (\hat{\mu}, \hat{\alpha}_1, \hat{\alpha}_2, \hat{\alpha}_3)^t$ are now

$$\begin{aligned} 5\hat{\mu} + 2\hat{\alpha}_1 + 2\hat{\alpha}_2 + \hat{\alpha}_3 &= y_{..} \\ 2\hat{\mu} + 2\hat{\alpha}_1 &= y_{1.} \\ 2\hat{\mu} + 2\hat{\alpha}_2 &= y_{2.} \\ \hat{\mu} + \hat{\alpha}_3 &= y_{3.} \end{aligned} \tag{3.74}$$

and using the constraint $\hat{\alpha}_1 + \hat{\alpha}_2 + \hat{\alpha}_3 = 0$ does not generate an easy solution.

The sum-coding goes as follows. Skipping the parameter $\hat{\alpha}_1$, the design matrix is:

$$\mathbf{X}_S = \begin{bmatrix} 1 & -1 & -1 \\ 1 & -1 & -1 \\ 1 & 1 & 0 \\ 1 & 1 & 0 \\ 1 & 0 & 1 \end{bmatrix} \tag{3.75}$$

and note that the columns do not add to zero anymore due to the unbalancedness. The resulting matrix $\mathbf{X}_S^t\mathbf{X}_S$ is

$$\mathbf{X}_S^t\mathbf{X}_S = \begin{bmatrix} 5 & 0 & -1 \\ 0 & 4 & 2 \\ -1 & 2 & 3 \end{bmatrix} \tag{3.76}$$

which has size (3×3) and has full rank. The normal equations for this system are

$$\begin{aligned} 5\hat{\mu} - \hat{\alpha}_3 &= y_{..} \\ 4\hat{\alpha}_2 + 2\hat{\alpha}_3 &= y_{2.} - y_{1.} \\ -\hat{\mu} + 2\hat{\alpha}_2 + 3\hat{\alpha}_3 &= y_{3.} - y_{1.} \end{aligned} \tag{3.77}$$

which is solvable but has no trivial solution.

The reference-coding goes as follows using $\hat{\alpha}_1 = 0$ as reference:

$$\mathbf{X}_R = \begin{bmatrix} 1 & 0 & 0 \\ 1 & 0 & 0 \\ 1 & 1 & 0 \\ 1 & 1 & 0 \\ 1 & 0 & 1 \end{bmatrix} \tag{3.78}$$

and the resulting matrix $\mathbf{X}_R^t\mathbf{X}_R$ is

$$\mathbf{X}_R^t\mathbf{X}_R = \begin{bmatrix} 5 & 2 & 1 \\ 2 & 2 & 0 \\ 1 & 0 & 1 \end{bmatrix} \tag{3.79}$$

which has size (3×3) and is full rank. The normal equations for this system are

$$\begin{aligned} 5\hat{\mu} + 2\hat{\alpha}_2 + \hat{\alpha}_3 &= y_{..} \\ 2\hat{\mu} + 2\hat{\alpha}_2 &= y_{2.} \\ \hat{\mu} + \hat{\alpha}_3 &= y_{3.} \end{aligned} \tag{3.80}$$

which can be solved and has the solution:

$$\begin{aligned}\hat{\mu} &= \frac{1}{2}(y_{..} - y_{2.} - y_{3.})\\ \hat{\alpha}_1 &= 0\\ \hat{\alpha}_2 &= y_{2.} + \frac{1}{2}y_{3.} - \frac{1}{2}y_{..}\\ \hat{\alpha}_3 &= \frac{3}{2}y_{3.} + \frac{1}{2}y_{2.} - \frac{1}{2}y_{..}\end{aligned} \tag{3.81}$$

which is not as easy as in the balanced case. Note that for simplicity we use again the same symbols $\hat{\mu}$ and $\hat{\alpha}_a$, but the interpretation of these estimates differs between the codings.

The weighted-coding with the first level as reference has the design matrix:

$$\mathbf{X}_W = \begin{bmatrix} 1 & -1 & -\frac{1}{2} \\ 1 & -1 & -\frac{1}{2} \\ 1 & 1 & 0 \\ 1 & 1 & 0 \\ 1 & 0 & 1 \end{bmatrix} \tag{3.82}$$

and note that the columns of this matrix add up to zero (except for the first column). The resulting matrix $\mathbf{X}_W^t\mathbf{X}_W$ is

$$\mathbf{X}_W^t\mathbf{X}_W = \begin{bmatrix} 5 & 0 & 0 \\ 0 & 4 & 1 \\ 0 & 1 & \frac{3}{2} \end{bmatrix} \tag{3.83}$$

which has size (3×3) and is full rank. The normal equations for this system are

$$\begin{aligned} 5\hat{\mu} &= y_{..}\\ 4\hat{\alpha}_2 + \hat{\alpha}_3 &= y_{2.} - y_{1.}\\ \hat{\alpha}_2 + \frac{3}{2}\hat{\alpha}_3 &= y_{3.} - \frac{1}{2}y_{1.}\end{aligned} \tag{3.84}$$

which can be solved rather simply and has the solution:

$$\begin{aligned}\hat{\mu} &= \bar{y}_{..}\\ \hat{\alpha}_2 &= \bar{y}_{2.} - \bar{y}_{..}\\ \hat{\alpha}_3 &= \bar{y}_{3.} - \bar{y}_{..}\end{aligned} \tag{3.85}$$

which is a nice generalization of the balanced case. For estimating α_1, an analysis can be performed while skipping either α_2 or α_3 (Nieuwenhuis *et al.*, 2017; Te Grotenhuis *et al.*, 2017).

Finally, using the means model gives the design matrix

$$\mathbf{X}_{MM} = \begin{bmatrix} 1 & 0 & 0 \\ 1 & 0 & 0 \\ 0 & 1 & 0 \\ 0 & 1 & 0 \\ 0 & 0 & 1 \end{bmatrix} \tag{3.86}$$

and this gives the normal equations

$$2\hat{\alpha}_1 = y_{1.} \qquad (3.87)$$
$$2\hat{\alpha}_2 = y_{2.}$$
$$\hat{\alpha}_3 = y_{31}$$

with solution $\hat{\alpha}_1 = \bar{y}_{1.}$, $\hat{\alpha}_2 = \bar{y}_{2.}$, and $\hat{\alpha}_3 = y_{31}$.

For nested models the coding goes as follows. We will use an example of coding a nested ANOVA model where the Factor A has two levels and the Factor B is nested in Factor A; where for level 1 of A the Factor B has two levels and for level 2 of Factor A, the Factor B has three levels. Hence, this is an example of an unbalanced nested model. We will assume that the number of replicates per cell is $N = 2$. Using full coding the model is then:

$$\begin{bmatrix} y_{111} \\ y_{112} \\ y_{121} \\ y_{122} \\ y_{211} \\ y_{212} \\ y_{221} \\ y_{222} \\ y_{231} \\ y_{232} \end{bmatrix} = \begin{bmatrix} 1 & 1 & 0 & 1 & 0 & 0 & 0 & 0 \\ 1 & 1 & 0 & 1 & 0 & 0 & 0 & 0 \\ 1 & 1 & 0 & 0 & 1 & 0 & 0 & 0 \\ 1 & 1 & 0 & 0 & 1 & 0 & 0 & 0 \\ 1 & 0 & 1 & 0 & 0 & 1 & 0 & 0 \\ 1 & 0 & 1 & 0 & 0 & 1 & 0 & 0 \\ 1 & 0 & 1 & 0 & 0 & 0 & 1 & 0 \\ 1 & 0 & 1 & 0 & 0 & 0 & 1 & 0 \\ 1 & 0 & 1 & 0 & 0 & 0 & 0 & 1 \\ 1 & 0 & 1 & 0 & 0 & 0 & 0 & 1 \end{bmatrix} \begin{bmatrix} \mu \\ \alpha_1 \\ \alpha_2 \\ \beta_{11} \\ \beta_{12} \\ \beta_{21} \\ \beta_{22} \\ \beta_{32} \end{bmatrix} + \epsilon \qquad (3.88)$$

where the design matrix $\mathbf{X}_F(10 \times 8)$ has rank five (the last five columns are independent and the first three can be written as combinations of those last five). Hence, we have to put restrictions on the 8 parameters, and the easiest is to impose the constraints $\hat{\mu} = \hat{\alpha}_1 = \hat{\alpha}_2 = 0$ which is reference-coding. This comes down to use the generalized inverse:

$$(\mathbf{X}^t\mathbf{X})^- = \begin{bmatrix} \mathbf{0} & \mathbf{0} \\ \mathbf{0} & \mathbf{D}(1/2) \end{bmatrix} \qquad (3.89)$$

where $\mathbf{0}$ are matrices conformable to $(\mathbf{X}^t\mathbf{X})^-$ and $\mathbf{D}(1/2)(5 \times 5)$ is a diagonal matrix with the values $1/2$ on its diagonal (Searle, 1971). This generates the following solution for the vector of parameters: $\hat{\theta} = (0, 0, 0, \bar{y}_{11.}, \bar{y}_{12.}, \bar{y}_{21.}, \bar{y}_{22.}, \bar{y}_{23.})$ and thus gives the decomposition $y_{abn} = \bar{y}_{ab.} + e_{abn} = \bar{y}_{ab.} + (y_{abn} - \bar{y}_{ab.})$.

Sum-coding can also be used for nested models, and we will continue the example of coding the nested ANOVA model as above. Using sum-coding, the design matrix $\mathbf{X}_S$ is then:

$$\mathbf{X}_S = \begin{bmatrix} 1 & 1 & 1 & 0 & 0 \\ 1 & 1 & 1 & 0 & 0 \\ 1 & 1 & -1 & 0 & 0 \\ 1 & 1 & -1 & 0 & 0 \\ 1 & -1 & 0 & 1 & 0 \\ 1 & -1 & 0 & 1 & 0 \\ 1 & -1 & 0 & 0 & 1 \\ 1 & -1 & 0 & 0 & 1 \\ 1 & -1 & 0 & -1 & -1 \\ 1 & -1 & 0 & -1 & -1 \end{bmatrix} \qquad (3.90)$$

where the first column pertains to the mean; the second column shows that the estimate for α_2 is skipped; the third column shows that the estimate for $\beta_{2(1)}$ is skipped; and finally the last two columns show that the estimate for $\beta_{3(2)}$ is skipped.

There are relationships between the different codings and also practical aspects related to their use. This is explained in Sections 3.A.2 and 3.A.3.

3.5 Advanced Models

3.5.1 Variance Component Models

In the introduction (Section 3.1), the difference between fixed and random effects was already explained. Simplifying the example of a variance component model in the introduction, consider levels of a hormone in a group of individuals (Factor A) measured N times using the same instrument. A model for such data could be:

$$y_{an} = \mu + \alpha_a + \epsilon_{an} \tag{3.91}$$

where a indicates individual a ($a = 1, \ldots, A$) and the same number of observations N ($n = 1, \ldots, N$) per individual is used for simplicity (*i.e.*, a balanced design). The interest is now in the interindividual variability of such hormone levels. Note that this resembles the one-way ANOVA model of Equation 3.2, but there are crucial differences in assumptions. The assumptions for model 3.91 are

1) $\epsilon_{an} \sim N(0, \sigma_e^2)$
2) $\alpha_a \sim N(0, \sigma_a^2)$
3) $cov(\alpha_a, \epsilon_{an}) = 0; \forall a, n$
4) $cov(\alpha_a, \alpha_{a'}) = 0; \forall a, a'$

where a and a' are two different individuals and the effects α_a are considered random. Assumption 1 is the same as in the fixed effect model; assumption 2 defines the properties of the random variable α_a; assumption 3 shows that the random components α_a and ϵ_{an} are assumed to be independent; and assumption 4 asserts that the individuals are independent drawings. The ANOVA table for this model is shown in Table 3.7.

This table has some resemblance with the one of one-way fixed effect models but with some clear differences. The term EMS stands for Expected Mean Squares and is derived using model 3.91 and the assumptions mentioned above (Searle *et al.*, 1992). The concept of EMS is explained in a bit more detail in Elaboration 3.11 and Section 3.7.2.

Table 3.7 The ANOVA table for a one-way variance component model for the balanced case.

Source	SS	df	EMS	MS	F
Factor A	SS_A	$A-1$	$\sigma_\epsilon^2 + N\sigma_\alpha^2$	$MS_A = \hat{\sigma}_\epsilon^2 + N\hat{\sigma}_\alpha^2 = SS_A/(A-1)$	MS_A/MS_E
Error	SS_E	$A(N-1)$	σ_ϵ^2	$MS_E = \hat{\sigma}_\epsilon^2 = SS_E/(A(N-1))$	
Total	SS_T	$I-1$			

ELABORATION 3.11

Expected Mean Squares (EMS)

The concept of EMS is useful in the context of ANOVA models to test the significance of the effect of factors/interactions and to calculate variance components. As the name suggests, EMS is the expected value of the MS of any term resulting from an ANOVA decomposition. EMS plays an important role in the context of ANOVA models, since its calculation may guide the experimenter in:

- choosing which MS should be compared (*e.g.*, in an F-test) to test the significance of a factor or interaction;
- defining how to linearly combine individual MS to calculate variance components.

If the ANOVA model involves K terms, EMS has the general form:

$$EMS = c_1\sigma_1^2 + c_2\sigma_2^2 + \cdots + c_k\sigma_k^2 + \cdots + c_K\sigma_K^2 \tag{3.92}$$

where σ_k^2 are the variance components associated with the different terms resulting from the ANOVA partition, and the associated c_k are coefficients which have to be determined and that can also be zero if a particular term does not take part in the EMS. More specifically, the values of c_k depend, *a.o.*, on the statistical design, on whether the associated term is fixed or random, on whether the design is balanced or unbalanced, and on the number of levels. Note that for fixed factors or interactions, the variance σ_k^2 is replaced by the variance-like quadratic term τ_k^2 (see Example 3.9).

While the calculation of the EMS for balanced designs is relatively straightforward (see, *e.g.*, Cornfield and Tukey, 1956) many algorithms have been proposed for the calculation of EMS for unbalanced design, and the interested reader is referred to Snee (1974) for a more detailed overview.

The MS in Table 3.7 is estimated by using the variance estimates $\hat{\sigma}_\epsilon^2$ and $\hat{\sigma}_\alpha^2$ where the latter can be expressed explicitly as

$$\hat{\sigma}_\alpha^2 = \left(\frac{SS_A}{A-1} - \hat{\sigma}_\epsilon^2\right)/N = \frac{MS_A - MS_E}{N} \tag{3.93}$$

and the former ($\hat{\sigma}_\epsilon^2$) is given in Table 3.7. The hypothesis that can be tested now is in terms of a variance:

$$\begin{aligned} &H_0 : \sigma_\alpha^2 = 0 \\ &H_1 : \sigma_\alpha^2 > 0 \\ &F = MS_A/MS_E; df_1 = (A-1), df_2 = A(N-1) \end{aligned} \tag{3.94}$$

and interpretation of a significant result should thus also be in terms of a variance. This way of estimating variances is called the ANOVA way. There are alternatives such as maximum-likelihood (Searle *et al.*, 1992). Note that treating an effect as fixed or random has repercussions for the testing and thus the significance of such effects (Næs and Langsrud, 1998).

It is also possible to obtain predictors for the individual contributions $\hat{\alpha}_a$:

$$\hat{\alpha}_a = (\bar{y}_{a.} - \bar{y}_{..}) - \frac{\hat{\sigma}_\epsilon^2/N}{\hat{\sigma}_\alpha^2 + \sigma_\epsilon^2/N}(\bar{y}_{a.} - \bar{y}_{..}) \tag{3.95}$$

which is an interesting equation (Searle *et al.*, 1992). For large values of N this predictor tends to be equal to $(\bar{y}_{a.} - \bar{y}_{..})$ which is the observed deviation in the sample. For small values of N, this predictor is shrunken toward zero. Hence, for small values of N we are less certain about the effect for individual a. Such predictors are called BLUPs (Best Linear Unbiased Predictors (Searle *et al.*, 1992)) and can be used to study individual differences, which will become apparent in Chapter 5.

3.5.2 Linear Mixed Models

3.5.2.1 General Idea

As explained in Section 3.1, Linear Mixed Models (LMM) are linear models with both fixed and random effects. A factor effect is considered fixed if there is specific interest in the chosen levels and in the differences between these levels. A factor effect is considered random if the levels are not of interest, but they provide a known cause for an amount of variation in the response. Examples of Section 3.1 are different batches, different individuals, or between-day variations. In most statistical designs, the treatment factors are considered as fixed, while the way in which the treatments are applied to the experimental units, using blocks, batches, and their randomization structure is defined in random factors.

The simplest LMM can be formulated as having two factors: one fixed and one random:

$$y_{abn} = \mu + \alpha_a + \beta_b + \epsilon_{abn} \tag{3.96}$$

where now – for instance – the Factor A is fixed and represents A treatment levels, while the Factor B is random and represents B blocks in which n replicates were measured; μ represents the overall mean, α_a represents the effect of treatment level a, while β_b represents the random block effect. LMMs encompass a wide range of models, *e.g.*, the RM-ANOVA models (see Equation 3.109). Estimation of LMMs usually proceeds by ML or REstricted Maximum Likelihood (REML) methods (Searle *et al.*, 1992) (see Elaboration 2.4). In Section 3.5.2.5, we will discuss a much-used special case of LMMs, namely those that model time-resolved experiments, so-called longitudinal LMMs.

3.5.2.2 Estimation of Model Parameters

In LMMs, it is common to write the fixed and random parts of the design in two separate design matrices.

$$\mathbf{y} = \mathbf{X}\boldsymbol{\theta} + \mathbf{Z}\boldsymbol{\gamma} + \boldsymbol{\epsilon} \tag{3.97}$$

Here $\mathbf{y}$ represents the vector with all response values, $\mathbf{X}$ is the design matrix for the fixed factors, and $\boldsymbol{\theta}$ contains the estimated effects for each column in $\mathbf{X}$; $\mathbf{Z}$ represents the design matrix for the random factors, while $\boldsymbol{\gamma}$ contains the effects for each of the columns in $\mathbf{Z}$. The estimation of the fixed and random model parameters in a linear mixed model such as Equation 3.97 is usually performed in four steps. The first step is to estimate the variance parameters of $\boldsymbol{\gamma}$ and $\boldsymbol{\epsilon}$ using REML. The random model parameters are distributed as:

$$\begin{pmatrix} \boldsymbol{\gamma} \\ \boldsymbol{\epsilon} \end{pmatrix} \sim N\left(\begin{pmatrix} 0 \\ 0 \end{pmatrix}, \begin{pmatrix} \boldsymbol{\Phi} & 0 \\ 0 & \boldsymbol{\Sigma} \end{pmatrix}\right) \tag{3.98}$$

Both the random effect parameters $\boldsymbol{\gamma}$ and the residual variation $\boldsymbol{\epsilon}$ are normally distributed with mean 0 and covariance $\boldsymbol{\Phi}$ and $\boldsymbol{\Sigma}$. This estimation is performed using an iterative procedure as usually analytical solutions for the estimation do not exist. After convergence of the procedure, the estimated covariance matrix $\hat{\mathbf{V}} = \mathbf{Z}\hat{\boldsymbol{\Phi}}\mathbf{Z}^t + \hat{\boldsymbol{\Sigma}}$ contains the estimates for the covariances $\boldsymbol{\Phi}$ due to the random effects defined in $\mathbf{Z}$ and for the covariance $\boldsymbol{\Sigma}$ due to residual variation.

In the second step the model parameters of the fixed effect θ are estimated using weighted least squares, with the weights $\hat{\mathbf{V}}^{-1}$.

$$\hat{\theta} = (\mathbf{X}^t\hat{\mathbf{V}}^{-1}\mathbf{X})^{-1}\mathbf{X}^t\hat{\mathbf{V}}^{-1}\mathbf{y} \tag{3.99}$$

The $\hat{\theta}$ parameters are indicated as BLUE (see Section 2.2.1.1) of the fixed effects.

The third step is the prediction of the values for the random effect values γ for each column of $\mathbf{Z}$. These values are predicted as:

$$\hat{\gamma} = \hat{\mathbf{\Phi}}\mathbf{Z}^t\hat{\mathbf{V}}^{-1}(\mathbf{y} - \mathbf{X}\hat{\theta}) \tag{3.100}$$

The parameters $\hat{\gamma}$ are usually indicated as the Best Linear Unbiased Predictors (BLUP). These values are shrunken predictors, which means that their values are smaller (closer to 0) than if they would have been estimated as a parameter of a fixed effect. These BLUPs are exactly the same as the earlier mentioned predictors for the individual contributions in variance component models of Equation 3.95; see Elaboration 3.12. Finally, in the last (fourth) step the residuals are calculated as the remaining variation: $\mathbf{e} = \mathbf{y} - \mathbf{X}\hat{\theta} - \mathbf{Z}\hat{\gamma}$.

ELABORATION 3.12

More on the BLUPs

Equations 3.100 and 3.95 are essentially equal. This will be shown for a variance component model with one random Factor A at three levels and two replicates per cell. The model in that case reads

$$y_{an} = \mu + \alpha_a + \epsilon_{an} \tag{3.101}$$

and the relevant matrices in Equation 3.97 are $\mathbf{X} = \mathbf{1}_6$ and

$$\mathbf{Z} = \begin{bmatrix} 1 & 0 & 0 \\ 1 & 0 & 0 \\ 0 & 1 & 0 \\ 0 & 1 & 0 \\ 0 & 0 & 1 \\ 0 & 0 & 1 \end{bmatrix} \tag{3.102}$$

Moreover, $\hat{\mathbf{\Phi}} = \hat{\sigma}_\alpha^2 I_3$ and $\hat{\mathbf{\Sigma}} = \hat{\sigma}_\epsilon^2 I_6$. Combining these in the matrix $\hat{\mathbf{V}}$ gives $\hat{\mathbf{V}} = \hat{\sigma}_\alpha^2\mathbf{Z}\mathbf{Z}^t + \hat{\sigma}_\epsilon^2 I_6$. This matrix has a special structure:

$$\hat{\mathbf{V}} = \left[\begin{array}{cc:cc:cc} \hat{\sigma}_\alpha^2 + \hat{\sigma}_\epsilon^2 & \hat{\sigma}_\alpha^2 & 0 & 0 & 0 & 0 \\ \hat{\sigma}_\alpha^2 & \hat{\sigma}_\alpha^2 + \hat{\sigma}_\epsilon^2 & 0 & 0 & 0 & 0 \\ \hdashline 0 & 0 & \hat{\sigma}_\alpha^2 + \hat{\sigma}_\epsilon^2 & \hat{\sigma}_\alpha^2 & 0 & 0 \\ 0 & 0 & \hat{\sigma}_\alpha^2 & \hat{\sigma}_\alpha^2 + \hat{\sigma}_\epsilon^2 & 0 & 0 \\ \hdashline 0 & 0 & 0 & 0 & \hat{\sigma}_\alpha^2 + \hat{\sigma}_\epsilon^2 & \hat{\sigma}_\alpha^2 \\ 0 & 0 & 0 & 0 & \hat{\sigma}_\alpha^2 & \hat{\sigma}_\alpha^2 + \hat{\sigma}_\epsilon^2 \end{array}\right] \tag{3.103}$$

which can also concisely be written as

$$\hat{\mathbf{V}} = \left[\begin{array}{c:c:c} \mathbf{Q} & \mathbf{0} & \mathbf{0} \\ \hdashline \mathbf{0} & \mathbf{Q} & \mathbf{0} \\ \hdashline \mathbf{0} & \mathbf{0} & \mathbf{Q} \end{array}\right] \tag{3.104}$$

where $\mathbf{Q}$ is implicitly defined. The inverse of such a matrix is simply

$$\hat{\mathbf{V}}^{-1} = \left[\begin{array}{c:c:c} \mathbf{Q}^{-1} & \mathbf{0} & \mathbf{0} \\ \hdashline \mathbf{0} & \mathbf{Q}^{-1} & \mathbf{0} \\ \hdashline \mathbf{0} & \mathbf{0} & \mathbf{Q}^{-1} \end{array}\right] \tag{3.105}$$

which follows from the special structure of $\mathbf{V}$ (Schott, 2016). The inverse of the matrix $\mathbf{Q}$ is

$$\hat{\mathbf{Q}}^{-1} = \frac{1}{\hat{\sigma}_\epsilon^2(\hat{\sigma}_\epsilon^2 + 2\hat{\sigma}_\alpha^2)} \begin{bmatrix} \hat{\sigma}_\epsilon^2 + 2\hat{\sigma}_\alpha^2 & -\hat{\sigma}_\alpha^2 \\ -\hat{\sigma}_\alpha^2 & \hat{\sigma}_\epsilon^2 + 2\hat{\sigma}_\alpha^2 \end{bmatrix} \tag{3.106}$$

Substituting this inverse in Equation 3.100 for the ath component of $\hat{\gamma}$ gives then (after some rearrangements):

$$\hat{\alpha}_a = (\bar{y}_{a.} - \bar{y}_{..}) - \frac{\hat{\sigma}_\epsilon^2/2}{\hat{\sigma}_\alpha^2 + \sigma_\epsilon^2/2}(\bar{y}_{a.} - \bar{y}_{..}) \tag{3.107}$$

which for the general case of N replicates gives exactly Equation 3.95.

3.5.2.3 Repeated Measures ANOVA

A special class of linear mixed models is obtained for designs in which repeated measurements are performed on the same subject. A prime example is when measurements are performed on the same individual at several points in time. Figure 3.11 shows data from a (fictive) experiment. The separate measurements in this fictive experiment are indicated by a star and are measurements for each individual per time point (for convenience this is only shown for one individual). Ignoring the fact that there is repeated sampling, the linear model may look like

$$y_{it} = \mu + \alpha_t + \epsilon_{it} \tag{3.108}$$

where i indicates the individual or subject and t the time point. The fitted values of the fixed effect α are indicated by the solid line which for simplicity is assumed to be linear. The interest may then be in the time behavior of the variable in a population. The residuals are then the

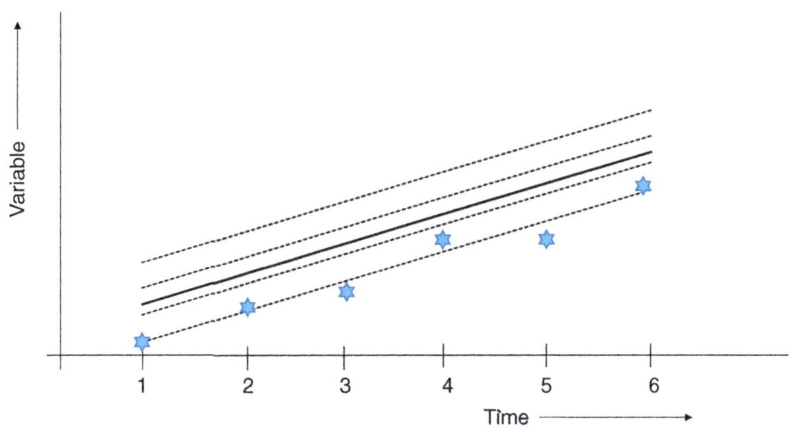

Figure 3.11 A variable is repeatedly measured for four individuals at six time points (for visual convenience represented by lines). The solid line is the average across all individuals; the dotted line is the best linear fit per individual and the stars are the individual measurements (shown for only one individual to not clog the picture).

deviations of the stars from this line, and these are clearly not independently distributed: they are all below the drawn line. The reason is that this specific individual has lower values than the average of all individuals and remains to do so in time. The dotted lines represent the fitted values per individual, and the stars are randomly distributed around these lines. Hence, a better model may be:

$$y_{it} = \mu + \alpha_t + \gamma_i + \epsilon_{it} \tag{3.109}$$

where γ_i represents the individual-specific intercept. Now the ϵ_{it} can be assumed to be *iid*. It is customary to assume that the γ_i components are random since the specific individuals are usually not of interest, but these are assumed to be drawn from a population.

The model of Equation 3.109 is called a Repeated Measures model, and the resulting analysis is then called Repeated-Measures ANOVA (RM-ANOVA). If the fitted lines per individual also differ with respect to the slope instead of only the intercept, then this difference can also be accounted for by including extra random terms.

3.5.2.4 Cross-over Designs and Models

In a cross-over design, the subjects undergo all the treatments ($a = 1, \ldots, A$) in a random order. Typically, this is done with a wash-out period in between the treatments to avoid carry-over effects. The simplest model for such a design could be

$$y_{ai} = \mu + \alpha_a + \beta_i + \epsilon_{ai} \tag{3.110}$$

which shows that the subjects are not nested in the treatments since each subject undergoes all treatments. The subjects are treated as a random effect, and the main interest is in estimating the treatment effects.

3.5.2.5 Longitudinal LMMs

Categorical Time

Longitudinal LMMs can be described using time as a categorical variable or as a quantitative variable using a dynamic model of time. We will start treating time as a categorical factor where the natural order of the different time levels is not explicitly taken into account. Quantitative time models will be discussed in a next subsection.

We will illustrate the categorical longitudinal LMM with a study where $I(i = 1, \ldots, I)$ individuals take part in a study with $A = 3$ different treatments. Each individual only receives a single treatment. After the treatment has been provided, measurements are obtained at $T = 4$ different time points for each individual. The LMM for y_{ait} is:

$$y_{ait} = \mu + \gamma_i + \alpha_a + \beta_t + (\alpha\beta)_{at} + \epsilon_{ait} \tag{3.111}$$

where y_{ait} is the measurement for individual i after treatment a at time t. This model is sometimes called the unconstrained model as no constraints with respect to group levels at baseline are applied. In this example, a mixed model will be used as the measurements of the same individual are not independent. The dependence between the samples of the same individual is defined by the random intercept γ_i. All values of individual i have the same intercept, which is different from the intercepts of all other individuals. Furthermore, the individual itself is not of interest in the model but serves as a manner to study the variation of the population.

When using LMMs, the different levels of all factors are summarized in a design matrix. As an example, we will show a repeated measures model with three treatments ($a = 0, 1, 2$) and four time points ($t = 0, 1, 2, 3$). Each individual only receives a single treatment. All individuals are assumed to have their own intercept which then serves as a random factor. Hence, this design has two fixed factors (*Treatment* and *Time*) and one random factor (*Individual*) where we will

Table 3.8 Reference-coded design matrix for categorical Factor *Time*.

Time X_T	T_1	T_2	T_3
$t = 0$	0	0	0
$t = 1$	1	0	0
$t = 2$	0	1	0
$t = 3$	0	0	1

assume *Time* as a categorical factor for the moment. Upon using reference-coding, the Factor *Time* can be encoded as in Table 3.8 which will generate the part $\mathbf{X}_T$ of the design matrix. For both the *Time* and the *Treatment* factor, the 0 level is considered the reference level.

The Factor *Treatment* (using also reference-coding) can be encoded as in Table 3.9 which will generate the part $\mathbf{X}_A$ of the design matrix.

Table 3.9 Reference-coded design matrix for categorical Factor *Treatment*.

Treatment X_A	A_1	A_2
$a = 0$	0	0
$a = 1$	1	0
$a = 2$	0	1

The interactions of this design are then encoded as shown in Table 3.10 which will generate the part $\mathbf{X}_{AT}$ of the design matrix.

Table 3.10 Design matrix for *Treatment:Time* interaction.

Treatment:Time X_{AT}	T_1A_1	T_2A_1	T_3A_1	T_1A_2	T_2A_2	T_3A_2
$t = 0, a = 0$	0	0	0	0	0	0
$t = 1, a = 0$	0	0	0	0	0	0
$t = 2, a = 0$	0	0	0	0	0	0
$t = 3, a = 0$	0	0	0	0	0	0
$t = 0, a = 1$	0	0	0	0	0	0
$t = 1, a = 1$	1	0	0	0	0	0
$t = 2, a = 1$	0	1	0	0	0	0
$t = 3, a = 1$	0	0	1	0	0	0
$t = 0, a = 2$	0	0	0	0	0	0
$t = 1, a = 2$	0	0	0	1	0	0
$t = 2, a = 2$	0	0	0	0	1	0
$t = 3, a = 2$	0	0	0	0	0	1

Finally, Table 3.11 shows the design matrix for the random individual factor.

Table 3.11 Design matrix for random effect matrix for the Factor *Individual*.

Z	Z_1	Z_2	Z_3	$Z_{...}$	Z_I
$i = 1,$	1	0	0	0	0
$i = 2,$	0	1	0	0	0
$i = 3,$	0	0	1	0	0
$i = ...,$	0	0	0	1	0
$i = I,$	0	0	0	0	1

Note that $\mathbf{X}_T$, $\mathbf{X}_A$, $\mathbf{X}_{AT}$, and $\mathbf{Z}$ all have (IT) rows, and only their different rows are shown in the tables. Using the *Time* and *Treatment*, design schemes the model for y_{ait} can be written as

$$y_{ait} = (\theta_0 + \gamma_{0i}) + \theta_1 T_{1t} + \theta_2 T_{2t} + \theta_3 T_{3t} + \theta_4 A_{1a} + \theta_5 A_{2a} + \theta_6 T_{1t} A_{1a} + \theta_7 T_{2t} A_{1a} + \theta_8 T_{3t} A_{1a} + \theta_9 T_{1t} A_{2a} + \theta_{10} T_{2t} A_{2a} + \theta_{11} T_{3t} A_{2a} + e_{ait} \tag{3.112}$$

The intercept for response y_{ait}, $(\theta_0 + \gamma_{0i})$ is individual specific due to the random intercept coefficient γ_{0i}. Collecting all indicator variables for the fixed effects (*Time, Treatment* and their interactions) in matrix $\mathbf{X}$, and the random component in matrix $\mathbf{Z}$ leads to the following equation for response $\mathbf{y}$:

$$\mathbf{y} = \mathbf{X}\boldsymbol{\theta} + \mathbf{Z}\boldsymbol{\gamma} + \boldsymbol{\epsilon} \tag{3.113}$$

where $\mathbf{y}$ is a $(IT \times 1)$ response vector, $\mathbf{X}$ is the $(IT \times 12)$ design matrix for all fixed effects starting with a column of ones (for the intercept), three columns for the *Time* indicator matrix $\mathbf{X}_T$, two columns for the *Treatment* indicator matrix $\mathbf{X}_A$ and 6 interaction columns for $\mathbf{X}_{AT}$; $\boldsymbol{\beta}$ is the vector of 12 parameters for the contrasts (columns) defined in $\mathbf{X}$; $\mathbf{Z}$ is the $(IT \times I)$ indicator matrix for the random component, in this case the random intercept for each of the I individuals, and $\boldsymbol{\gamma}$ contains the random intercept coefficients for all individuals in response $\mathbf{y}$.

Longitudinal LMMs with a Baseline Measurement

In many clinical trials, where the effect of different treatments is studied over time, a baseline measurement, obtained before the treatment is started, is included to relate the change due to the treatment to the baseline level of the individual. The baseline measurement provides additional information that is different for each individual. There are different options on how to deal with a baseline measurement, and a clear overview is provided in Twisk *et al.* (2018). Both the "Analysis of Changes" where the baseline measurement is first subtracted from all other time points of the measurements of an individual, and "longitudinal Analysis of Covariance (ANCOVA)" require the presence of the baseline measurement, and individuals of which the baseline measurement is missing cannot be used in the analysis. The *constrained* longitudinal data analysis, in which the main treatment effect is removed from the model, does not suffer from this drawback and will be discussed here.

In the next subsection, the example of Section 3.5.2.5 is continued assuming time point $t = 0$ as the baseline measurement. At baseline, the treatment cannot have an effect as the treatment has not been given yet. A randomized controlled trial (RCT) assumes that no differences exist between the groups because of the random selection of individuals in the treatment groups, and thus, the group average of all treatment groups at baseline should be equal. This assumption is forced into the model by removing the main treatment effect (α_a) and only estimating the treatment interaction with time. In that case, the unconstrained model in Equation 3.111

transforms into

$$y_{ait} = \mu + \gamma_i + \beta_t + (\alpha\beta)_{at} + \epsilon_{ait} \tag{3.114}$$

and due to the removal of the main treatment effect, the interaction effect alone describes the effect of treatment for each time point. By removing the main treatment effect, the estimates are adjusted for baseline level (Twisk *et al.*, 2018).

Longitudinal LMMs with Quantitative Time

If the Factor *Time* is considered as a categorical variable, then no relationship between the time points is modeled. The model has no information that level $t = 0$ is the first time point and about the order of the time points. Using a quantitative time model allows for measurements at different time points and for a different number of time points for the individuals. Such a model makes it possible to interpolate responses at times without a measurement, to be able to compare different individuals that have measurements at different time points. If the number of time points increases, it can be also favorable to use a quantitative time model as the number of parameters that describe the time profile can be less with a quantitative time model.

One approach to include time as a quantitative covariate[3] to model time profiles is by using time polynomials:

$$\begin{aligned} y_{ait} = {} & (\theta_0 + \gamma_{0i}) + \theta_1 T_{it} + \theta_2 T_{it}^2 + \theta_3 A_{1a} + \theta_4 A_{2a} + \theta_5 T_{it} A_{1a} \\ & + \theta_6 T_{it}^2 A_{1a} + \theta_7 T_{it} A_{2a} + \theta_8 T_{it}^2 A_{2a} + \epsilon_{ait} \end{aligned} \tag{3.115}$$

and now instead of the categorical Factor *Time* as was used in Equation 3.111, the actual time value and its squared value are used in the model as main effects and also in interaction with the *Treatment*. Equation 3.115 represents an unconstrained model, as the main *Treatment* effects $\theta_3 A_{1a}$ and $\theta_4 A_{2a}$ are present. In the equation, T_{it} represents the actual time of the tth timepoint for individual i, as each individual can have their own time points. In Equation 3.115, the only random effect taken into account is for the intercept, as each subject is allowed to start at a different value at time 0. As discussed in Section 3.5.2.5, this model is known as the random intercept model.

Equation 3.116 shows the model with random factors for the intercept (γ_{0i}), linear slope (γ_{1i}), and quadratic slope (γ_{2i}), and all interactions between time and treatment ($\gamma_{5i}, \dots, \gamma_{8i}$):

$$\begin{aligned} y_{ait} = {} & (\theta_0 + \gamma_{0i}) + (\theta_1 + \gamma_{1i}) T_{it} + (\theta_2 + \gamma_{2i}) T_{it}^2 + \theta_3 A_{1a} + \theta_4 A_{2a} \\ & + (\theta_5 + \gamma_{5i}) T_{it} A_{1a} + (\theta_6 + \gamma_{6i}) T_{it}^2 A_{1a} + (\theta_7 + \gamma_{7i}) T_{it} A_{2a} \\ & + (\theta_8 + \gamma_{8i}) T_{it}^2 A_{2a} + \epsilon_{ait} \end{aligned} \tag{3.116}$$

Such a model might be too flexible, and some of the random effects parameters could be removed from the model if necessary. The model can also be written per individual in matrix form as

$$\mathbf{y}_i = \mathbf{X}_i \boldsymbol{\theta} + \mathbf{Z}_i \boldsymbol{\gamma}_i + \boldsymbol{\epsilon}_i \tag{3.117}$$

Now $\mathbf{y}_i$ is a vector with responses of size T_i and $\mathbf{X}_i(T_i \times (K_F + 1))$ contains the settings of the K_F fixed effect factors and the intercept for each subject, while its coefficients $\boldsymbol{\theta}((K_F + 1) \times 1)$ are the same for all subjects. The matrix $\mathbf{Z}_i(T_i \times (K_R + 1))$ contains the settings for the K_R random effect factors and $\boldsymbol{\gamma}_i((K_R + 1) \times 1)$ contains the subject specific random effects for variable $\mathbf{y}_i$. The vector T_i represents the total number of time points for individual i.

3 The term factor is usually reserved for categorical variables.

Table 3.12 Design matrix for quantitative time factor and its interactions for **fixed effects**.

$X_i(a=1)$	Intercept	A_1	A_2	T	T^2	TA_1	TA_2	T^2A_1	T^2A_2
t_0	1	1	0	1.5	1.5^2	1.5	0	1.5^2	0
t_1	1	1	0	2.8	2.8^2	2.8	0	2.8^2	0
t_2	1	1	0	4.3	4.3^2	4.3	0	4.3^2	0
t_3	1	1	0	6.7	6.7^2	6.7	0	6.7^2	0
$t_{\dots}$	1	1	0	…	$\dots^2$	…	0	$\dots^2$	0

Table 3.12 shows the $\mathbf{X}_i$ values for individual i for time points (1.5, 2.8, 4.3, 6.7, …). The first column is for the intercept, columns 2 and 3 indicate which of the treatments individual i has obtained (in this case treatment $a = 1$). Columns 4 and 5 represent the actual time of the measurement and its square and the final columns are for the interaction of the treatment columns A_1 and A_2 and the time and squared time values. Likewise, Table 3.13 shows the $\mathbf{X}_i$ values for individual i for time points (1.5, 2.8, 4.3, 6.7, …) for the random effects.

Extending Equation 3.117 for all individuals ($i = 1, \dots, I$),

$$\mathbf{y} = \mathbf{X}\theta + \mathbf{Z}\gamma + \epsilon \tag{3.118}$$

$\mathbf{y}$ is obtained by stacking all $\mathbf{y}_i$ vectors below each other; $\mathbf{X}$ is formed by stacking all $\mathbf{X}_i$ matrices below each other, similar to $\mathbf{y}$; $\mathbf{X} = \left(\mathbf{X}_1^t|\mathbf{X}_2^t|..|\mathbf{X}_i^t|..|\mathbf{X}_I^t\right)^t$. The size of $\mathbf{X}$ is $(\sum_{i=1}^{I} T_i \times (K_F + 1))$ and θ is the same as in Equation 3.117. The matrix $\mathbf{Z}$ will be formed by placing all $\mathbf{Z}_i$ matrices in the diagonal of a huge matrix.

$$\mathbf{Z} = \begin{pmatrix} \mathbf{Z}_1 & 0 & 0 & 0 & 0 & 0 \\ 0 & \mathbf{Z}_2 & 0 & 0 & 0 & 0 \\ 0 & 0 & . & 0 & 0 & 0 \\ 0 & 0 & 0 & \mathbf{Z}_i & 0 & 0 \\ 0 & 0 & 0 & 0 & . & 0 \\ 0 & 0 & 0 & 0 & 0 & \mathbf{Z}_I \end{pmatrix} \tag{3.119}$$

and its size is $(\sum_{i=1}^{I} T_i \times IK_R)$; γ is obtained by stacking all γ_i below each other and thus its size becomes $IK_R \times 1$.

An extensive example of the use of these different kinds of longitudinal LMMs is given in Example 3.6.

Table 3.13 Design matrix for quantitative time factor and its interactions for **random effects**.

$Z_i(a=1)$	Intercept	T	T^2	TA_1	TA_2	T^2A_1	T^2A_2
t_0	1	1.5	1.5^2	1.5	0	1.5^2	0
t_1	1	2.8	2.8^2	2.8	0	2.8^2	0
t_2	1	4.3	4.3^2	4.3	0	4.3^2	0
t_3	1	6.7	6.7^2	6.7	0	6.7^2	0
$t_{\dots}$	1	…	$\dots^2$	…	0	$\dots^2$	0

Example 3.6: Vaccine responses

Vaccines are often administered together with an adjuvant. In a clinical study, subjects were administered the same vaccine with different adjuvants (Hasdemir *et al.*, 2020). There were five groups of subjects, where the same adjuvant was given within the same group. There were two vaccinations and the antibody (Ab) counts were measured in the blood of the subjects at several days. An example of representative profiles is given in Figure 3.12.

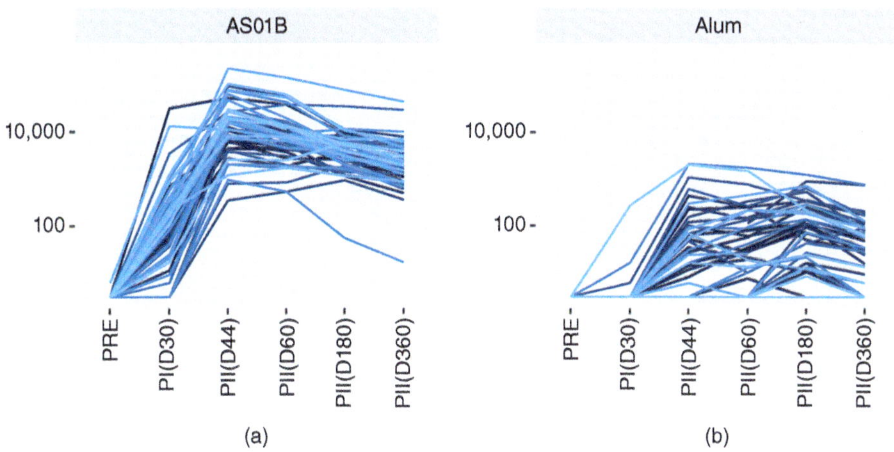

Figure 3.12 Ab profiles of different subjects. PRE is pre-dose, PI(D30) is at 30 days (and right after the first vaccination), PII(D44) is at 44 days (and right after the second vaccination), D60, D180, and D360 refer to the days after the start of the study. Each line in the plot is one subject ((a) adjuvant AS01B; (b) adjuvant Alum). Source: Hasdemir *et al.* (2020)/Springer Nature/CC BY 4.0.

This figure shows two issues: (i) the responses between the groups are different and (ii) the inter-individual variation (within a group) can be large. Since modeling all data simultaneously would require indicator variables for distinguishing the different adjuvant groups including many interaction terms to account for differences between the fixed and random effect, it was decided to build models per group. Such a model should accommodate: (i) a fixed time effect describing the overall behavior of the group, (ii) random effects for describing the inter-individual variation, and (iii) a way to model the two interventions (vaccination I and II). One choice to make was to model the time as a continuous or categorical factor. Many models were considered and tested, such as piecewise polynomials of different degrees (with fixed and variable knot locations), cubic polynomial models (continuous-time models), and categorical-time models with different types of random intercepts. Using established model selection criteria such as the AIC (the AIC estimates the generalization error of a model) was not straightforward and did not necessarily result in a good model; they tended to select the most complex model which was difficult to interpret. Finally, it was decided to use a categorical time model with a selection of random intercepts:

$$y_{ai} = \theta_0 D_0 + (\theta_1 + \gamma_{1i})D_1 + (\theta_2 + \gamma_{2i})D_2 + (\theta_3 + \gamma_{2i})D_3 + (\theta_4 + \gamma_{3i})D_4 + (\theta_5 + \gamma_{3i})D_5 + \epsilon_{ai} \tag{3.120}$$

where θ_a are the fixed effects; γ_{ai} the random effects (i is the index for individual); and D_a the indicators for the time points. Time point zero corresponds to the PRE dose and note that time-points 2 and 3 share the same random effect for the individual, as do time-points 4 and 5. This final linear mixed model represented a good compromise between model complexity and interpretability. Some model fits are shown in Figure 3.13. The fit for individual 315 is too good, but the other individuals show reasonable fits.

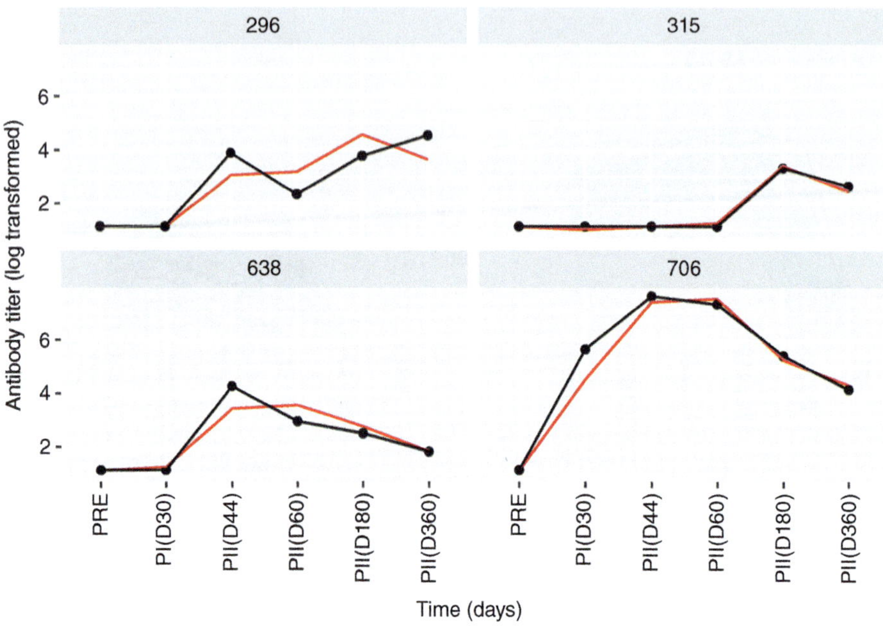

Figure 3.13 Fit of the categorical model of Equation 3.120 for the Alum group. Four panels show the fit for four individuals in that group. The fitted values (red) are the addition of the fixed and random effects (BLUP) for that particular individual (measurements in black). Source: Hasdemir *et al.* (2020)/Springer Nature/CC BY 4.0.

This example shows that modeling longitudinal data is not straightforward and requires considering many steps and approaches. This becomes even more challenging for multivariate longitudinal data.

Polynomial time models have the tendency to oscillate between data points, and especially at the boundaries of time models. This makes interpolation and small extrapolations problematic. A different approach is to use splines. Splines are constructs of multiple piecewise polynomial functions that usually are better at interpolation. Natural cubic splines are functions that are much stiffer than normal polynomials and have the property that the second-order derivative of the function (also called its bending moments) is made 0 at the boundaries (first and last time point) (Hastie *et al.*, 2009). It is possible to calculate basis functions $\mathbf{S}$ such that a longitudinal response $\mathbf{y}$ can be modeled by the basis functions: $\mathbf{y} = \mathbf{Sc} + \mathbf{e}$ where $\mathbf{c}$ represents the coefficients for each of the basis functions.

3.6 Hasse Diagrams

Hasse diagrams (Lohr, 1995; Oehlert, 2000) are a powerful tool for visualizing the structure of a statistical design and can provide a straightforward way of finding properties such as the number of degrees of freedom, the experimental units, and the correct denominators for the F-test; consequently, they also represent a valid support for properly conducting permutation tests, as a clear indication of the exchangeable units to be permuted or of how to restrict the permutations can be obtained from the diagrams (which will be discussed in Section 3.7.3).

There is a direct correspondence between the linear (mixed) models and the diagrams themselves which can be exploited for model definition and calculation. Hasse diagrams can be used both for orthogonal and non-orthogonal designs and allow for easily finding the sum of squares, expected mean squares, and the appropriate hypothesis test. More specifically, they allow for identifying the structure in the experimental units and the treatment design, providing insights into the fixed and random effects to be included.

Operationally, they show the relationships among the factors in ANOVA-type designs using nesting as the partial ordering criterion, and they highlight the structure in the experimental design induced by blocking and randomization. In a Hasse diagram, each vertex/node is a factor reflecting both the treatment or the experimental design, while edges represent nesting relationships, including those determined by each level of randomization, and any subsampling within experimental units; in particular, the top node M is always the coarsest summary of the data, usually the grand mean; every other design term is nested in M. On the other hand, the bottom vertex is the finest, *e.g.*, the residuals E. If a generic design term F_2 is nested in another design term F_1 it is drawn below it and the two nodes are connected with an edge, while if they are crossed they are represented next to one another and there is no edge connecting them. In the latter case, the two terms will give rise to an interaction that is nested in both.

3.6.1 Building a Hasse Diagram

The construction of any Hasse diagram, irrespective of whether the underlying design is orthogonal or not, is based on four steps that will be described in detail below and illustrated by means of a worked-out example.

1) As anticipated, the coarsest partition (*i.e.*, the grand mean M) should be positioned on the top, while the finest (the experimental or observational unit, sometimes also indicated as residual E) should be placed on the bottom; all the other design terms should be written between these two items.
2) If a Factor `B` is nested in Factor `A`, `B` is placed below `A` and the two terms are connected with an edge. If `A` and `B` are crossed, they are placed next to one another and each is connected to the next factor that it is nested in, which in this case, could be the interaction `AB`. All factors are nested in the top node M. Accordingly when dealing with a fully crossed design, all the main effects will be placed in the first level under M, all the binary interactions in the level immediately below, and so on until the higher-order interaction, which will be placed just above E.
3) Random terms should be identified (they are usually highlighted in the diagram by being enclosed in parentheses). The residual term E is always random. Additionally, any term placed between two random terms along a path is also random.
4) For each term in the diagram, the number of levels should be written as superscript.
5) For each term, the number of degrees of freedom should be indicated as a subscript, proceeding as follows. The mean M has one degree of freedom; for any other term f, the number

of degrees of freedom is calculated as:

$$df_f = levels_f - \sum_{termsabove_f} df_{termsabove_f} \tag{3.121}$$

where $levels_f$ is the superscript of the term while $df_{termsabove_f}$ are the subscripts of all the design terms the factor is nested in.

To illustrate the procedure for building a Hasse diagram and extracting relevant information from it, Example 3.7 provides for a crossed two-factor design.

Example 3.7: Hasse diagram for a crossed two-factor design

Consider the case of a balanced two-factor crossed design, three different drugs (Factor A, fixed) at two different doses (Factor B, fixed) are administered to 30 individuals (experimental units), so that each combination of treatment factor levels has 5 replicates. In particular, to emphasize the two components of each statistical design, we will illustrate how to draw separate Hasse diagrams for the treatment and experimental (unit) design structures, and how to combine them into the overall diagram by considering the randomization.

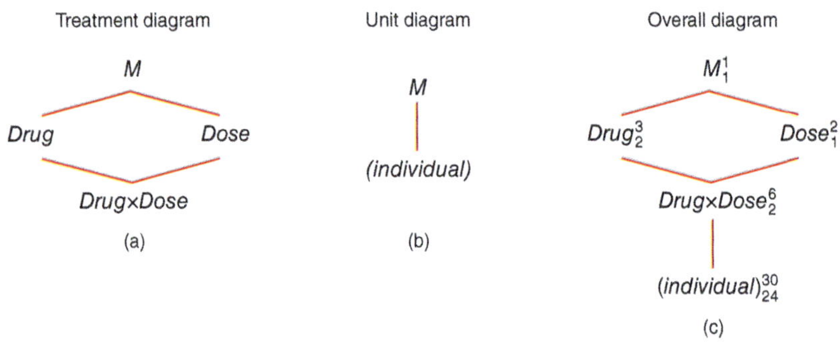

Figure 3.14 Hasse diagrams for the crossed two-factor design: (a) diagram for the treatment design; (b) diagram for the experimental design; (c) overall diagram for the statistical design, resulting from the combination of (a) and (b).

Based on the general pipeline described above, for the treatment design (panel (a) of Figure 3.14), the construction of the corresponding Hasse diagram should start from placing the grand mean *M* as the top node, followed by the two factors *Drug* and *Dose* as crossed terms in the layer immediately below. Each of them is nested in *M* and, therefore, the corresponding vertex is connected to it by an edge, but they are not connected to one another. Their binary interaction *Drug:Dose* is drawn in the successive layer, connected by an edge to each of them since it is nested in both. As for the experimental design (panel (b) of Figure 3.14), it contains only the Factor *Individual*, which is the experimental unit, nested in *M*. Then, to construct the overall diagram for the statistical design, we proceed as follows: first, we merge the two top nodes *M* of the treatment and unit diagrams, into a single vertex. Successively, an edge is drawn from each treatment factor to its experimental unit factor: since *Drug* and *Dose* were completely randomized on *Individual*, we should draw an edge connecting *Drug* and *Individual*, another connecting *Dose* and *Individual* and a third connecting

Drug:Dose and *Individual*. However, since the nesting relations between *Drug* and *Individual* or *Dose* and *Individual* are already implied by the nesting path through *Drug:Dose*, they can be removed to make the resulting diagram clearer. The result is shown in panel (c) of Figure 3.14.

The superscripts are the number of factor levels: 1 for *M*, 3 for *Drug*, 2 for *Dose*, and 30 for *Individual*. Accordingly, the degrees of freedom for *Drug* are $3 - 1 = 2$, those for *Dose* are $2 - 1 = 1$ and, consequently, the ones for their binary interaction are $6 - 2 - 1 - 1 = 2$. The degrees of freedom for *Individual* are obtained by subtracting from its number of levels (30) the degrees of freedom all the factors it is nested in, *i.e.*, *Drug*, *Dose*, *Drug*×*Dose*, and *M*, and they are equal to $30 - 2 - 1 - 2 - 1 = 24$.

3.7 Validation

3.7.1 Classical Tests Revisited

Evaluation of linear mixed models is not straightforward. These models have two parts, the fixed part containing information on the effect of treatments and the random part, containing information on the randomization and blocking structure used to perform the experiment. In the general case, these two parts of the model are not orthogonal and the design may be unbalanced. For the random part, REML is used to estimate the variance components for each of the random factors. These variance components are then used in a weighted least squares approach to estimate the model parameters of the fixed part of the model.

To evaluate the fixed part of the model, we must realize that the parameters of the fixed part are not independent of each other due to the nonorthogonality of the design. Therefore, the increase in the performance of the model when adding the parameters and the decrease in performance when removing the parameter must be considered.

The general way of evaluating the significance of a fixed model parameter is the Wald F-statistic, which equals the square of the ratio of the estimated effect divided by its estimated standard error (Welham *et al.*, 2014). However, due to the complex structure of the design, the number of degrees of freedom used to estimate the standard error is unclear, and therefore the Wald statistic value is not easily related to a p-value. A much-used approximation to obtain p-values assuming that the Wald F-statistic is normally distributed, therefore, leads to over-optimistic results (*i.e.*, many false positives) (Welham *et al.*, 2014).

Another approach commonly used for evaluating the significance of model parameters of fixed effect in LMMs is the likelihood ratio tests (Martin and Govaerts, 2020). The Log-Likelihood Ratio (LLR) and the Restricted LLR (RLLR) tests can be used to assess the significance of either one or several fixed effects or one or several variance components by comparing two nested models. For the evaluation of fixed effect parameters, the LLR should be used, as this test uses ML likelihood values, while for random parameters the RLLR should be used which uses REML likelihood values. The LLR and RLLR tests are defined as $2[log(L_{H1}) - log(L_{H0})]$, in which L_{H1} represents the likelihood of the model with the parameter that is tested and L_{H0} represents the likelihood of the model without that parameter. For fixed model parameters the LLR test follows a χ^2_k distribution, with k being the number of parameters tested. However, this approach is known to provide many false positives. An alternative approach is to use bootstrap resampling in which the LLR value is compared to its bootstrap distribution. This approach is provided in more detail in Section 5.3.3.1.

For parameters in the random part of the mixed model the RLLR test can also be used using the bootstrap procedure. Possible parameters to test are, *e.g.*, whether a random slope parameter in a model should be used or not. An advantage of the LLR and RLLR tests is that it can be used even for complex random effect structures, and no assumptions of degrees of freedom have to be made.

3.7.2 Expected Mean Squares from Hasse Diagrams

Calculation of EMS plays a fundamental role in defining what MS should be used as denominator for significance testing when performing an F-test. Indeed, for a generic model term f (factor or interaction), the F_f-ratio is defined as:

$$F_f = \frac{MS_{numerator}}{MS_{denominator}} \tag{3.122}$$

where $MS_{numerator}$ is the observed mean square for the factor/interaction of interest, while $MS_{denominator}$ is the error mean square, which can be defined as the MS whose expected value (EMS) would be the same as the one of $MS_{numerator}$ in case that the effect of the factor is absent (*i.e.*, if the null hypothesis is true).

Hasse diagrams constitute a valid and straightforward tool to be used for the calculation of the expected mean squares and to find the appropriate test statistics. However, in this case, the calculation depends on the specific formulation of the linear mixed model which, in turn, reflects the covariance structure of the data. In any case, the starting point is the "standard" Hasse diagram with levels as superscripts and degrees of freedom as subscripts of the design terms, like the one depicted in Figure 3.15 in Example 3.8. First of all, it is worth mentioning that, for a crossed two-factor design, where Factor A is fixed and Factor B is random, the corresponding linear mixed model can be formulated as:

$$y_{abn} = \mu + \alpha_a + \beta_b + (\alpha\beta)_{ab} + \epsilon_{abn} \tag{3.123}$$

Depending on the assumptions on the mixed interaction term $(\alpha\beta)_{ab}$, the variance components associated to the model in Equation 3.123 can be calculated in either a constrained (restricted LMM) or unconstrained (unrestricted LMM) way. In particular, in an unrestricted model the mixed interaction term is treated as a completely random factor, whereas in a restricted one it is constrained to satisfy:

$$\sum_{\alpha=1}^{A} (\alpha\beta)_{ab} = 0 \quad \forall b = 1 : B \tag{3.124}$$

(see, *e.g.*, Schwarz (1993) for more details).

Calculation of the expected mean squares from the Hasse diagram is accomplished by starting from the terms at the bottom of the graph and proceeding upwards to the top. For any generic term f (factor or interaction), the procedure involves the following steps:

1) The first term contributing to the EMS of f is given by:

$$\frac{total\ number\ of\ observations}{superscript\ of\ f} \times \sigma_f^2 \tag{3.125}$$

if f is a random term, or

$$\frac{total\ number\ of\ observations}{superscript\ of\ f} \times \tau_f^2 \tag{3.126}$$

if f is fixed.

2) For every random term R below f, add the following contribution:

$$\frac{total\ number\ of\ observations}{superscript\ of\ R} \times \sigma_R^2 \tag{3.127}$$

3) When the variance components of the mixed model are estimated in a restricted way, (step 2) is limited only to the random terms R that contain the same fixed factors as f.

How to practically perform these calculations will be shown in Example 3.8.

Example 3.8: Calculating expected mean squares from a Hasse diagram

The calculation of EMSs for any ANOVA model involves understanding how the variance components are distributed across the different factors, including both fixed and random effects. Casella (2008) provides an example of a randomized complete block design, where four *Varieties* ($v = 1, \dots, V$) of alfalfa are grown. The experiments are performed within four blocks, ($b = 1, \dots, B$) where in each block three plants ($n = 1, \dots, N$) of each variety were grown.

As discussed above, calculations are carried out from bottom to top, so the first term to be estimated is the EMS for the *Within* variation. According to the rules presented above, since the term is a random term with no further term below, its EMS is:

$$EMS_{Within} = \frac{48}{48}\sigma^2_{Within} = \sigma^2_{Within} \tag{3.128}$$

In the case of the *Varieties:Blocks* interaction, random as well, step 1 of the procedure indicates that the first term should be $\frac{48}{16}\sigma^2_{AB}$. Since the only random term below *Varieties:Blocks* is *Within*, the only other contribution to the EMS is $\frac{48}{48}\sigma^2_{Within}$:

$$EMS_{Varieties:Blocks} = \frac{48}{16}\sigma^2_{Varieties:Blocks} + \frac{48}{48}\sigma^2_{Within} = 3\sigma^2_{Varieties:Blocks} + \sigma^2_{Within} \tag{3.129}$$

Both for Factor *Varieties* and *Blocks*, the random terms in the layers below are *Varieties:Blocks* and *Within*. However, while in the case of Factor *Varieties* all the terms contribute to the EMS since the random term *Varieties:Blocks* contains the same fixed factor as *Varieties*, the same does not hold for Factor *Blocks*, for which the mixed interaction does not contribute to the EMS.

$$\begin{aligned} EMS_{Varieties} &= \frac{48}{4}\tau^2_{Varieties} + \frac{48}{16}\sigma^2_{Varieties:Blocks} + \frac{48}{48}\sigma^2_{Within} \\ &= 12\tau^2_{Varieties} + 3\sigma^2_{Varieties:Blocks} + \sigma^2_{Within} \end{aligned} \tag{3.130}$$

$$EMS_{Blocks} = \frac{48}{4}\sigma^2_{Blocks} + \frac{48}{48}\sigma^2_{Within} = 12\sigma^2_{Blocks} + \sigma^2_{Within} \tag{3.131}$$

The resulting Hasse diagram is displayed in Figure 3.15.

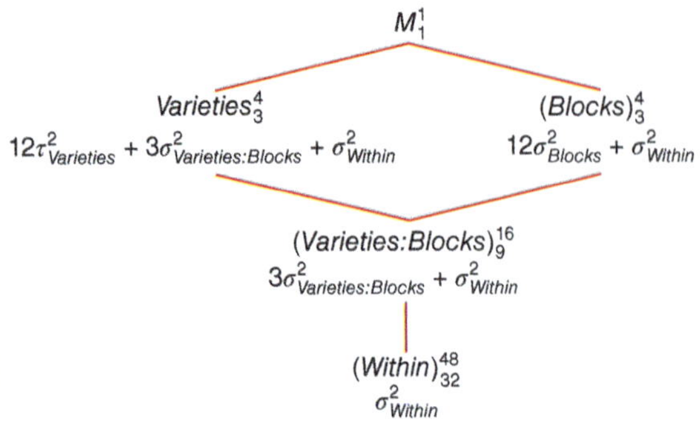

Figure 3.15 Hasse diagram for the calculation of the expected mean squares of a crossed two-factor design under the unrestricted linear mixed model.

These calculations also allow us to derive rules about which mean square to use as the denominator in the F-test for checking the significance of the effect of a factor or interaction. The identification of the denominator mean square in the F-ratio also plays a fundamental role in permutation testing (as will be discussed in Section 3.7.3), as it is related to the concept of *exchangeable units*, *i.e.*, the elements of a design which are exchangeable under the null hypothesis. Indeed, building the F-ratio allows identifying the contributions to the variability of the term being tested, through the comparison of the expected mean squares of individual terms. Accordingly, if the numerator's expected mean square includes components of variation that are not being tested, they must be present as well in the expected mean square of the denominator. Thus, in any ANOVA model, the denominator mean square identifies the components that contribute to the variability of the term being tested and, as a consequence, the appropriate exchangeable units for a permutation test. The denominator can also be identified in the Hasse diagram: it is the mean square of the nearest random term below the factor/interaction to be tested.

3.7.3 Permutation Tests

Statistical significance for factors in ANOVA or model terms in a linear model is usually obtained from an F-test. The F-distribution, however, relies on assumptions, *e.g.*, the normal distribution of the errors. When the response data does not provide normally distributed residuals, and thus the F-distribution cannot be used for validation, one of the alternatives is to use permutations for testing the significance of the factors.

Although for ANOVA where univariate responses are modeled, F-tests are often sufficient, this situation becomes different for multivariate responses where the significance of effect on a multivariate response has to be assessed. Furthermore, when distances between samples based on their multivariate response are used to study treatment effects, permutation tests become more attractive. Permutation tests for HD-ANOVA applications are discussed in Sections 4.1.3, 5.3.2.2, and 6.2.4. However, we will already explain the functioning of permutation tests in this chapter as the explanation will use a single response example of an experiment with multiple levels.

In Section 2.2.3.2 the main idea of permutation tests was introduced using the example of a two-sample t-test. The goal of the permutation test is to create an alternative H_0 distribution by reshuffling the response values in such a way that it mimics the situation in which the factor that is tested would have had no effect on the response (Fisher, 1935). Similar to the t-test, the alternative distribution to find the proper type-1 error cut-off value, will be created by reshuffling samples under the assumption that H_0 is true. The actual value of the test can then be compared to this alternative distribution to provide a p-value.

The reference H_0 distribution of a permutation test is constructed by calculating all (or a large random subset) of permutations of the observations into different groups. The F-value of the original (non-permuted) observations can then be compared to the reference distribution to decide on the fate of the H_0. A p-value can be calculated from the permutation results using

$$p_{perm} = \frac{(\#F^p \geq F_{obs}) + 1}{P + 1} \tag{3.132}$$

where $\#F^p \geq F_{obs}$ represents the number of permutations with an equal or higher value than F_{obs}, the F-value of the non-permuted data, and P represents the number of permutations applied. Note that both the numerator and denominator of Equation 3.132 include +1 to represent that the F-value of the non-permuted data is also included in the distribution of permuted F-values. This prevents p_{perm} to become 0.

There are exact permutation tests and approximate permutation tests. For an exact test, the probability of rejecting the null hypothesis when it is true is exactly the level of significance α chosen for the test *a priori* (Anderson, 2001). For an approximate test, the chosen significance level is asymptotically approximated when using a sufficient number of replicates. Approximate tests often first subtract the estimated effect of another factor and then reshuffle the residuals to test the factor of interest.

The appropriate permutation procedure for the test of interest needs input on the following questions:

- Should I use an exact test or an approximate test?
- Should raw response values or residuals be permuted?
- Should separate observations or groups of observations be permuted?
- Should permutation be free or restricted within levels of other factors?
- Which test statistic should be used?

To start with the first question, exact tests are usually easier than approximate tests. They do not suffer from inflated type-1 errors. Sometimes, the power of exact tests can be low when a sufficient number of permutations cannot be performed, for example when the number of replicates or units of replicates is low. Thus, an approximate test can be more powerful than an exact test and can be preferred. For some tests, such as the interaction in a crossed two-factor design, no exact test exists and only an approximate test can be used. In the remainder, we will explain how an exact test and an approximate test can be created in general.

3.7.3.1 Exact Tests

An exact permutation test in an ANOVA model is achieved by permuting exchangeable units, identified by the denominator mean square of the F-ratio of the corresponding null hypothesis and restricting these permutations to occur within the levels of terms that are above or on the same level in the Hasse diagram. This is a free translation of the guideline posted by Anderson and Ter Braak (2003). Using the Hasse diagram, the factors of the ANOVA model can be ordered according to their hierarchy with the intercept on top and main effects and interaction below, while the residuals are on the bottom, as was explained in Section 3.6.

Example 3.8 shows, for the alfalfa experiment, how expected mean squares for each of the factors in the model can be calculated. The test of differences between the 4 *Varieties* has the F-test $F_{Varieties} = \frac{MS_{Variety}}{MS_{Variety:Blocks}}$ where *Varieties* is the fixed factor and the *Blocks* is the random factor. The interaction *Variety:Blocks* is in the denominator of the F-ratio. In that case, the exchangeable units are the 4×4 groups of 3 measurements from each *Variety* identified by that interaction (*Variety:Blocks*). The permutation of units has to be restricted within the same level of the Block factor. In this case only the $MS_{Varieties}$ and the $MS_{Variety:Blocks}$ change while the MS_{Blocks} and MS_{Within} stay equal.

The alfalfa example discussed in Example 3.8 will be continued in Example 3.9 to show how the exchangeable units for permutation testing can be obtained from EMSs.

Example 3.9: Use of EMS to defined exchangeable units in alfalfa example

The calculation of EMSs for any ANOVA model involves understanding how the variance components are distributed across the different factors, including both fixed and random effects. The alfalfa experiment, already discussed in Example 3.8, provides a randomized complete block design, where $v = 1, \ldots, 4$ alfalfa *Varieties* are grown in $b = 1 \ldots, 4$ *Blocks*, where in each block $n = 1, \ldots, 3$ plants of each variety were grown.

Varieties is the fixed factor and *Blocks* is the random factor. The interest is in the difference between the *Variety* levels, and these differences can only be analyzed within each block. The three replicate plants of each *Variety* in each block are an example of subsampling of observational units, and their variation is called *Within* variation. The corresponding ANOVA mixed model is:

$$y_{vbn} = \mu + Varieties_v + Blocks_b + (Varieties{:}Blocks)_{vb} + \epsilon_{vbn} \tag{3.133}$$

This is similar to Equation 3.8; however, here the Factor *Blocks* is random. Accordingly, the EMSs for each of the terms in the ANOVA model are shown in Table 3.14.

Table 3.14 Results of EMS calculation for a two-factor LMM.

	EMS
Varieties	$\sigma_\epsilon^2 + N\sigma^2_{Varieties:Blocks} + BN\tau^2_{Varieties}$
Blocks	$\sigma_\epsilon^2 + VN\sigma^2_{Blocks}$
Varieties:Blocks	$\sigma_\epsilon^2 + N\sigma^2_{Varieties:Blocks}$
Within	σ_ϵ^2

The variance contribution for the fixed Factor *Varieties* is given by:

$$\tau^2_{Varieties} = \frac{\sum_{v=1}^{V} Varieties_v}{V-1} \tag{3.134}$$

Accordingly, considering the design in Example 3.9, to identify which MS should be put as the denominator for an F-test with $MS_{Varieties}$ as numerator, we should examine the EMS. Under the assumption that the Factor *Varieties* has no effect, its corresponding EMS would be:

$$E(MS_{Varieties} | \tau^2_{Varieties} = 0) = \sigma_\epsilon^2 + N\sigma^2_{Varieties:Blocks} \tag{3.135}$$

Since $E(MS_{Varieties} | \tau^2_{Varieties} = 0)$ is equal to the EMS of the interaction *Varieties:Blocks*, this indicates that $MS_{Varieties:Blocks}$ should be used as the denominator in the F-test for assessing the significance of Factor *Varieties*.

The exchangeable units for permutation thus are the groups of 3 plants, under the H_0 hypothesis of no difference between the *Varieties*, they can be permuted within each block (see Figure 3.16).

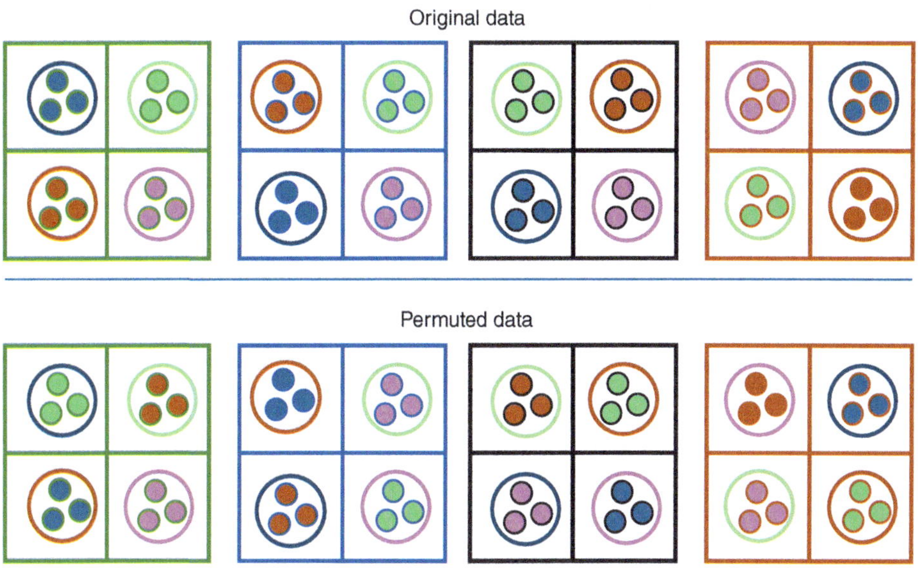

Figure 3.16 Randomized complete block design of the Alfalfa experiment with four *Blocks*, where in each block three plants of four *Varieties* are grown. Top row: Original data; Bottom row: Permuted data. The blocks are represented by the 4 squares and within each block are 4 *Varieties* of alfalfa indicated by the colored circle. Within each circle are the three plants of corresponding variety. The block in which the plant is present is indicated by the colored circle for each plant. In the permuted data, the groups of three plants, of the same variety, together change position to a randomly chosen new variety inside their own block. The circles representing *Varieties* do not change position.

3.7.3.2 Approximate Tests

The exact permutation test requires that only the exchangeable units under the H_0 are reshuffled within the levels of factors that are at equal or higher positions in the Hasse diagram to make sure that only the sources of variation indicated by the F-ratio for the test are varied. For some tests, this is not possible.

An example is the interaction term of a design with two (or more) Factors A and B. The model for the fully crossed design can be written as in Equation 2.73. The Hasse diagram for this model was introduced in Example 3.7.

$$y_{abn} = \mu + \alpha_a + \beta_b + (\alpha\beta)_{ab} + \epsilon_{abn} \tag{3.136}$$

The F-test for the interaction $F_{\text{AB}} = \frac{MS_{\text{AB}}}{MS_E}$ has the residual mean square MS_E as its denominator. The MS_E indicates that the independent observations are the exchangeable units that should be permuted within the levels of both factors that are above the interaction in the Hasse diagram. The only possible way to reshuffle observations within the levels of both factors is to keep them in the same AB group, which would not change the estimate of the interaction. Another approach to correct for the effect of Factors A and B is by correcting the response for the effect of these factors and using the residuals of the reduced model, that is, from the model

$$\hat{r}_{abn} = y_{abn} - \hat{\mu} - \hat{\alpha}_a - \hat{\beta}_b = \widehat{(\alpha\beta)}_{ab} + \hat{\epsilon}_{abn} \tag{3.137}$$

As $\hat{\mu}$, $\hat{\alpha}_a$, and $\hat{\beta}_b$ are approximated from the provided data, this is an approximate test.

If H_0 that $(\alpha\beta)_{ab} = 0$ is true, the estimated variability due to the interaction and due to the residuals is comparable, and the F-value $F_{\mathrm{AB}} = \frac{MS_{\mathrm{AB}}}{MS_E}$ is close to 1. The residuals $\hat{r}_{abn}$ are exchangeable over the samples, and an unrestricted permutation of the residuals can be applied to calculate the permutation distribution for H_0. The F-ratio of the non-permuted situation compares $\hat{MS}_{(\mathrm{AB})}$ with $\hat{MS}_E$. After permutation of the residuals, $\widehat{(\alpha\beta)}_{ab}$ and $\hat{\epsilon}_{abn}$ are re-estimated from permuted r^p_{abn}, and F^p is calculated in the same manner.

If the interaction is found not significant the model simplifies to

$$y_{abn} = \mu + \alpha_a + \beta_b + \epsilon_{abn} \tag{3.138}$$

For the main Factors A and B the significance can be obtained by restricted permutation in which permutation is only allowed within levels of the other factor, leading to an exact test. Another approach is again the approximate test where the effect of the other factor is first removed and the residuals are being permuted. For Factor A this would lead to residuals: $\hat{r}_{abn} = y_{abn} - \hat{\mu} - \hat{\beta}_b = \hat{\alpha}_a + \hat{\epsilon}_{abn}$. For the H_0 that there is no difference between the levels of Factor A, the residuals are exchangeable over all these levels and can thus be permuted without restriction. The exact test is often preferable with respect to power and type 1 error inflation, but in case of a limited number of possible permutations, the approximate test can also be used.

3.8 Miscellaneous Models

3.8.1 Multivariate Analysis of Variance

3.8.1.1 Traditional Multivariate Analysis of Variance

With the availability of multiple responses as a function of factors or grouping, the ANOVA methods thus far need to be extended to the multivariate domain. The standard technique for doing so is called multivariate analysis of variance or MANOVA (Mardia *et al.*, 1979; Stahle and Wold, 1990; Krzanowski, 2000).

Suppose that J responses are measured in a designed experiment. For the moment, we start with a balanced one-way setup, so there is one underlying experimental Factor A. Instead of a univariate response y_{an} we have now a multivariate response $\mathbf{y}_{an}(J \times 1)$. The one-way MANOVA model for this response vector is:

$$\mathbf{y}_{an} = \boldsymbol{\mu} + \boldsymbol{\alpha}_a + \boldsymbol{\epsilon}_{an} \tag{3.139}$$

where all vectors on the RHS are obvious generalizations of their univariate counterparts and of size $(J \times 1)$. The usual assumption for $\boldsymbol{\epsilon}_{an}$ is that these are multivariate normal distributed with mean $\mathbf{0}$ and $(J \times J)$ covariance matrix $\boldsymbol{\Sigma}$. Three crucial matrices can now be defined as generalizations of the between- and within-group variation:

$$\begin{aligned}
\mathbf{W} &= \sum_a \sum_n (\mathbf{y}_{an} - \bar{\mathbf{y}}_{a.})(\mathbf{y}_{an} - \bar{\mathbf{y}}_{a.})^t \\
\mathbf{B} &= \sum_a N(\bar{\mathbf{y}}_{a.} - \bar{\mathbf{y}}_{..})(\bar{\mathbf{y}}_{a.} - \bar{\mathbf{y}}_{..})^t \\
\mathbf{T} &= \sum_a \sum_n (\mathbf{y}_{an} - \bar{\mathbf{y}}_{..})(\mathbf{y}_{an} - \bar{\mathbf{y}}_{..})^t
\end{aligned} \tag{3.140}$$

where we used again the "dot" notation. It now holds that $\mathbf{T} = \mathbf{W} + \mathbf{B}$ which is the central decomposition in MANOVA.

The null hypothesis to test is $H_0 : \boldsymbol{\alpha}_a = 0; \forall a$ and there are different statistics for testing this null hypothesis (Stahle and Wold, 1990; Krzanowski, 2000). All these test statistics depend on the "size" of $\mathbf{B}$ relative to $\mathbf{W}$ since they use the matrix $\mathbf{W}^{-1}\mathbf{B}$ that is the natural extension of the univariate F-ratio. This already points to a problem with MANOVA in the high-dimensional context: the matrix $\mathbf{W}$ must be non-singular and well-conditioned. The matrix $\mathbf{W}$ has size $(J \times J)$ and it holds that $rank(\mathbf{W}) \leq min(A(N-1), J)$. Hence, if the number of variables is high relative to the sample size, MANOVA fails (see Elaboration 1.2). One way to circumvent this problem is by using techniques to regularize the matrix $\mathbf{W}$ which will be discussed in Chapter 5.

The MANOVA approach can readily be extended to more complicated designs such as two-way ANOVA with interaction. Then, by calling $\mathbf{y}_{abn}(J \times 1)$ the multivariate response vector collected on the nth replicate when Factor A is at level a and Factor B is at level b, the MANOVA model is:

$$\mathbf{y}_{abn} = \boldsymbol{\mu} + \boldsymbol{\alpha}_a + \boldsymbol{\beta}_b + (\boldsymbol{\alpha\beta})_{ab} + \boldsymbol{\epsilon}_{abn} \tag{3.141}$$

Analogously to what is shown in Equation 3.140, the total variation $\mathbf{T}$ can be split into the contribution of the between-group matrices corresponding to the individual factors and the binary interaction ($\mathbf{B}_{\mathrm{A}}$, $\mathbf{B}_{\mathrm{B}}$, and $\mathbf{B}_{\mathrm{AB}}$, respectively) and that of the within-group scatter ($\mathbf{W}$):

$$\mathbf{T} = \mathbf{B}_{\mathrm{A}} + \mathbf{B}_{\mathrm{B}} + \mathbf{B}_{\mathrm{AB}} + \mathbf{W} \tag{3.142}$$

where:

$$\mathbf{W} = \sum_a \sum_b \sum_n (\mathbf{y}_{abn} - \bar{\mathbf{y}}_{ab.})(\mathbf{y}_{abn} - \bar{\mathbf{y}}_{ab.})^t \tag{3.143}$$

$$\mathbf{B}_{\mathrm{A}} = \sum_a BN(\bar{\mathbf{y}}_{a..} - \bar{\mathbf{y}}_{...})(\bar{\mathbf{y}}_{a..} - \bar{\mathbf{y}}_{...})^t$$

$$\mathbf{B}_{\mathrm{B}} = \sum_b AN(\bar{\mathbf{y}}_{.b.} - \bar{\mathbf{y}}_{...})(\bar{\mathbf{y}}_{.b.} - \bar{\mathbf{y}}_{...})^t$$

$$\mathbf{B}_{\mathrm{AB}} = \sum_a \sum_b N(\bar{\mathbf{y}}_{ab.} - \bar{\mathbf{y}}_{...})(\bar{\mathbf{y}}_{ab.} - \bar{\mathbf{y}}_{...})^t$$

$$\mathbf{T} = \sum_a \sum_b \sum_n (\mathbf{y}_{abn} - \bar{\mathbf{y}}_{...})(\mathbf{y}_{abn} - \bar{\mathbf{y}}_{...})^t$$

To test the significance of each factor or of the interaction, the corresponding between-groups matrix ($\mathbf{B}_f$, with f = A, B, AB) is compared to the within-group scatter $\mathbf{W}$, through one of the statistics, described in Section 3.8.1.2. However, also in these cases, the problem related to the matrix $\mathbf{W}$ being ill-conditioned remains.

3.8.1.2 Significance Testing in MANOVA

Analogously to what happens in ANOVA, in MANOVA, for each factor or interaction, testing is based on a comparison between the between- and the within-level dispersion, which is accomplished by calculating statistical variables which are functions of the eigenvalues of the matrix

$$\mathbf{J}_f = \mathbf{B}_f \mathbf{W}^{-1} \qquad f = \mathrm{A}, \mathrm{B}, \mathrm{AB}, \ldots \tag{3.144}$$

The four statistics that are used to test for the significance of the multivariate effect in MANOVA are: Wilks' lambda Λ_f, the Hotelling–Lawley trace H_f, the Pillai–bartlett trace P_f and Roy's maximum root ϱ_f.

Wilks' lambda is defined as:

$$\Lambda_f = \sum_{r=1}^{R} \frac{1}{1+\lambda_r^{(f)}} = \frac{1}{|\mathbf{I}+\mathbf{J}_f|} = \frac{|\mathbf{W}|}{|\mathbf{W}+\mathbf{B}_f|} \tag{3.145}$$

where $\lambda_r^{(f)}$ and R are the rth eigenvalue and the rank of $\mathbf{J}_f$, respectively, and $\mathbf{I}$ is the identity matrix. Due to its definition, the larger the significance of the effect of the design term, the smaller the value of Λ_f.

The Pillai-Bartlett trace is defined as:

$$P_f = \sum_{r=1}^{R} \frac{\lambda_r^{(f)}}{1+\lambda_r^{(f)}} = tr\left(\mathbf{J}_f(\mathbf{I}+\mathbf{J}_f)^{-1}\right) = tr\left(\mathbf{B}_f(\mathbf{W}+\mathbf{B}_f)^{-1}\right) \tag{3.146}$$

where once again $\lambda_r^{(f)}$ denotes the rth eigenvalue of $\mathbf{J}_f$. Contrarily than for Wilks' lambda, in this case, a significant effect corresponds to a large value of the test statistics.

The Hotelling–Lawley trace is defined as:

$$H_f = \sum_{r=1}^{R} \lambda_r^{(f)} = tr(\mathbf{J}_f) \tag{3.147}$$

where $\lambda_r^{(f)}$ has the same meaning as above. Also in this case, a significant effect corresponds to a large value of the test statistics.

Lastly, Roy's largest characteristic root is defined as:

$$\varrho_f = \frac{\lambda_1^{(f)}}{1+\lambda_1^{(f)}} \tag{3.148}$$

where $\lambda_1^{(f)}$ denotes the largest eigenvalue of $\mathbf{J}_f$. Again a significant effect corresponds to a large value of the test statistics.

For all of these four test statistics, there are mathematical transformations that lead to null distributions that are approximately (or, in some cases, exactly) F-distributed. On the other hand, it should be highlighted that they do not necessarily agree, as their statistical power is different (Olson, 1976; Stevens, 1979). Indeed, when differences among levels are concentrated mostly along a single direction in multivariate space, the statistical power follows the order: $\varrho_f > H_f > \Lambda_f > P_f$, whereas, if the differences are more spreadly diffused, the order is reversed ($P_f > \Lambda_f > H_f > \varrho_f$).

3.8.1.3 Regression Formulation of MANOVA

MANOVA models can also be written in a regression context. Collecting all response variables in the matrix $\mathbf{Y}(I \times J)$ gives the model:

$$\mathbf{Y} = \mathbf{X\Theta} + \mathbf{Y}_{\mathtt{Ind}} = \hat{\mathbf{Y}} + \mathbf{Y}_{\mathtt{Ind}} \tag{3.149}$$

where $\mathbf{X}$ is the design matrix; $\mathbf{\Theta}$ are the parameters; and $\mathbf{Y}_{\mathtt{Ind}}$ is the matrix of residuals. This boils down to a multivariate multiple regression problem.

In the case of a two-way MANOVA model with interaction, as the one in Equation 3.141, the response matrix $\mathbf{Y}$ is made as follows:

$$\mathbf{Y} = \begin{bmatrix} \mathbf{y}_{111}^t \\ \vdots \\ \mathbf{y}_{abn}^t \\ \vdots \\ \mathbf{y}_{ABN}^t \end{bmatrix} \tag{3.150}$$

and $\mathbf{X}$ is the coded design matrix (usually, sum-coded) which, for a two-way crossed design is partitioned as:

$$\mathbf{X} = \left[\mathbf{1}_I \,\middle|\, \mathbf{X}_{\mathtt{A}} \,\middle|\, \mathbf{X}_{\mathtt{B}} \,\middle|\, \mathbf{X}_{\mathtt{AB}} \right] \tag{3.151}$$

Accordingly, the model parameter matrix in Equation 3.149 corresponds to:

$$\mathbf{\Theta} = \begin{bmatrix} \boldsymbol{\mu}^t \\ \boldsymbol{\alpha}_1^t \\ \vdots \\ \boldsymbol{\alpha}_{A-1}^t \\ \boldsymbol{\beta}_1^t \\ \vdots \\ \boldsymbol{\beta}_{B-1}^t \\ (\boldsymbol{\alpha\beta})_{11}^t \\ \vdots \\ (\boldsymbol{\alpha\beta})_{(A-1)(B-1)}^t \end{bmatrix} \tag{3.152}$$

Based on the regression formulation, it is also possible to calculate straightforwardly the between-groups and the within-group matrices:

$$\begin{aligned} \mathbf{B}_{\mathtt{A}} &= \mathbf{Y}^t\mathbf{X}_{\mathtt{A}}(\mathbf{X}_{\mathtt{A}}^t\mathbf{X}_{\mathtt{A}})^{-1}\mathbf{X}_{\mathtt{A}}^t\mathbf{Y} \\ \mathbf{B}_{\mathtt{B}} &= \mathbf{Y}^t\mathbf{X}_{\mathtt{B}}(\mathbf{X}_{\mathtt{B}}^t\mathbf{X}_{\mathtt{B}})^{-1}\mathbf{X}_{\mathtt{B}}^t\mathbf{Y} \\ \mathbf{B}_{\mathtt{AB}} &= \mathbf{Y}^t\mathbf{X}_{\mathtt{AB}}(\mathbf{X}_{\mathtt{AB}}^t\mathbf{X}_{\mathtt{AB}})^{-1}\mathbf{X}_{\mathtt{AB}}^t\mathbf{Y} \\ \mathbf{W} &= \mathbf{Y}_{\mathtt{Ind}}^t\mathbf{Y}_{\mathtt{Ind}} = \mathbf{Y}^t(\mathbf{I} - \mathbf{X}(\mathbf{X}^t\mathbf{X})^{-1}\mathbf{X}^t)\mathbf{Y} \end{aligned} \tag{3.153}$$

Each of the matrices $\mathbf{B}_f$ ($f = \mathtt{A}, \mathtt{B}, \mathtt{AB}$) accounts for the scatter of the mean profiles associated with the levels of the corresponding factor or interaction. The dispersion induced by the factor (or interaction) is compared to the variability associated with the residuals which are expressed as within levels scatter matrix $\mathbf{W}$.

3.8.2 Multivariate LMMs

There exist also generalizations of LMMs to the multivariate case (Sammel *et al.*, 1999; Zhou and Stephens, 2014). These are, however, not trivial. They are either used for multivariate data with a limited set of features (in the range of 5–10)(Zhou and Stephens, 2014) or use factor analysis as a dimension reduction step taking one latent variable (Sammel *et al.*, 1999). These models are not easy to use or interpret. The factor analysis-based models show some resemblance with LiMM-PCA and PC-ANOVA (see Chapter 5).

3.A Appendix

3.A.1 Proof

In this section, we will prove Equation 3.16. We start with the decomposition of Equation 3.15, and then after taking squares on both sides of this equation we focus on the cross-term

$$\sum_{a=1}^{A}\sum_{b=1}^{B}\sum_{n=1}^{N}(\bar{y}_{a..} - \bar{y}_{...})(\bar{y}_{.b.} - \bar{y}_{...}) \tag{3.A.1}$$

and expanding this into

$$\sum_{a=1}^{A}\sum_{b=1}^{B}\sum_{n=1}^{N}(\bar{y}_{a..}\bar{y}_{.b.}-\bar{y}_{...}\bar{y}_{a..}-\bar{y}_{...}\bar{y}_{.b.}+(\bar{y}_{...})^2) \tag{3.A.2}$$

Upon introducing the notation $N_{.b}=\sum_a N_{ab}$, $N_{a.}=\sum_b N_{ab}$, and $N_{..}=\sum_a\sum_b N_{ab}=I$ it holds that (after some simplifications) Equation 3.A.2 can be written as:

$$\sum_{a=1}^{A}\sum_{b=1}^{B}N_{ab}\bar{y}_{a..}\bar{y}_{.b.}-\bar{y}_{...}\sum_{a=1}^{A}\bar{y}_{a..}N_{a.}-\bar{y}_{...}\sum_{b=1}^{B}\bar{y}_{.b.}N_{.b}+I(\bar{y}_{...})^2) \tag{3.A.3}$$

and since

$$\bar{y}_{...}=\frac{\sum_{b=1}^{B}\bar{y}_{.b.}}{N_{..}}=\frac{\sum_{a=1}^{A}\bar{y}_{.a.}}{N_{..}} \tag{3.A.4}$$

Equation 3.A.3 becomes

$$\sum_{a=1}^{A}\sum_{b=1}^{B}N_{ab}\bar{y}_{a..}\bar{y}_{.b.}-(\bar{y}_{...})^2N_{..}-(\bar{y}_{...})^2N_{..}+I(\bar{y}_{...})^2) \tag{3.A.5}$$

which gives

$$\sum_{a=1}^{A}\sum_{b=1}^{B}N_{ab}\bar{y}_{a..}\bar{y}_{.b.}-I(\bar{y}_{...})^2) \tag{3.A.6}$$

which is not equal to zero. If $N_{ab}=N$ then:

$$\begin{aligned}\sum_{a=1}^{A}\sum_{b=1}^{B}N\bar{y}_{a..}\bar{y}_{.b.}&=N\sum_{a=1}^{A}\sum_{b=1}^{B}\bar{y}_{a..}\bar{y}_{.b.}=N\sum_{a=1}^{A}\bar{y}_{a..}\sum_{b=1}^{B}\bar{y}_{.b.}\\&=\sum_{a=1}^{A}B\bar{y}_{...}\bar{y}_{a..}=NB\bar{y}_{...}\sum_{a=1}^{A}\bar{y}_{a..}=ABN(\bar{y}_{...})^2=I(\bar{y}_{...})^2\end{aligned} \tag{3.A.7}$$

and thus in that case Equation 3.A.6 generates a zero as expected for the balanced design.

3.A.2 Relationships Between Codings

The codings as discussed in Section 3.4 are just four examples of possible codings. There is an intimate relationship between a coding, restrictions on the estimated parameters which are needed to solve the normal equations, and the type of generalized inverse used to obtain a solution. In terms of estimated y-values all these codings give exactly the same result which is shown in Elaboration 2.3.

The selection of the type of coding depends mostly on the interpretation that should be given to the estimated factor effects. The type of coding defines what kind of contrasts we are interested in; in reference-coding the interest is in contrasts such as $\alpha_a-\alpha_1$; $a=2,\dots,A$ if the first level is taken as the reference level. Similar reasonings can be made for more complicated models (Searle, 1971).

Sum-coding is the type of coding used in ASCA+ and reference-coding is used in PRC and SMART (see Chapter 4).

Estimated factor effects obtained via different codings are transferable. For sum- and reference-coding this can be seen as follows. Suppose that the sum-coding design matrix is called $\mathbf{X}_S$ which has full column-rank; likewise, define $\mathbf{X}_R$ which also has full column-rank. Then it holds that:

$$\hat{\mathbf{y}} = \mathbf{X}_S\hat{\boldsymbol{\theta}}_S = \mathbf{X}_R\hat{\boldsymbol{\theta}}_R \tag{3.A.8}$$

and using the fact that both $\mathbf{X}_S$ and $\mathbf{X}_R$ have full column-rank, we can conclude that

$$\begin{aligned}\hat{\boldsymbol{\theta}}_R &= \mathbf{X}_R^{+}\mathbf{X}_S\hat{\boldsymbol{\theta}}_S \\ \hat{\boldsymbol{\theta}}_S &= \mathbf{X}_S^{+}\mathbf{X}_R\hat{\boldsymbol{\theta}}_R\end{aligned} \tag{3.A.9}$$

where $\hat{\boldsymbol{\theta}}_R$ and $\hat{\boldsymbol{\theta}}_S$ are the main effect and interaction estimates obtained via reference or sum-coding, respectively, and the superscript $\mathbf{X}^{+}$ indicates the Moore–Penrose inverse (see Section 2.1.3). This also holds for the multivariate case with factor estimates $\hat{\boldsymbol{\Theta}}_R$ and $\hat{\boldsymbol{\Theta}}_S$ since these estimates can be regarded as having been obtained through multiple (univariate) regressions. Note that this conversion is less easy for the full-coding case, since the full-coding design matrix $\mathbf{X}_F$ does not have full column-rank.

Having shown different coding schemes including the weighted-coding, it may be worthwhile to connect all these codings and see visually how they are related. A nice illustration is given in Nieuwenhuis *et al.* (2017) which is adapted and shown in Figure 3.A.1. The first thing to notice in Figure 3.A.1 is that the grand mean is not the same as the weighted mean. The grand mean is simply the average of the group means, and the weighted mean is the overall sample mean, as already explained in Section 3.4.2. The size of the bubbles in the plot is representative of the number of replicates in this cell (*e.g.*, for level 2 of the factor, many replicates are available). Hence, this design is unbalanced and because of the fact that the second group is substantially larger than the others, the weighted mean is drawn toward that group. The sum-coding (dashed arrow) measures the distance of a group-mean relative to the grand mean. The reference-coding selects one of the groups as a reference group which in this case is

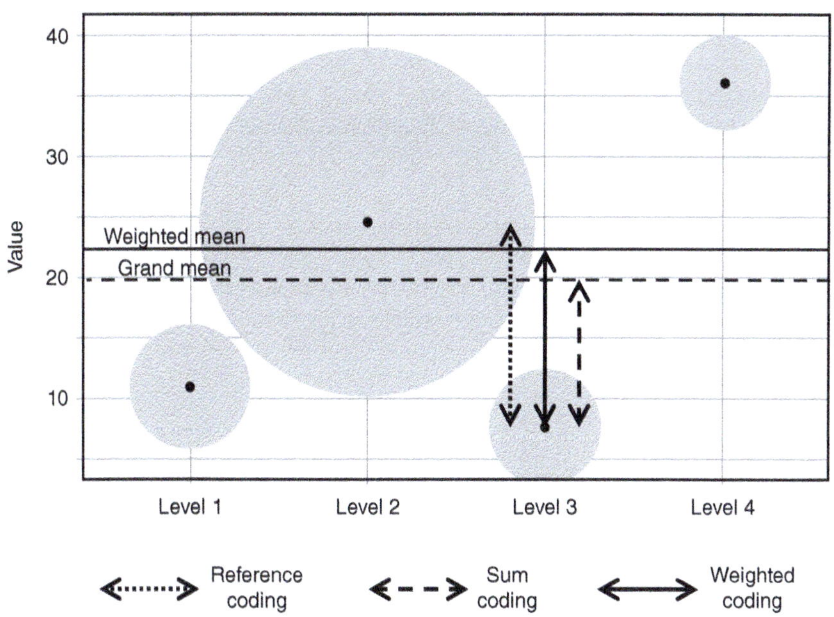

Figure 3.A.1 An unbalanced one-way ANOVA with four levels. The bubble size refers to the number of replicates in each group and the arrows show various coding schemes for level 3. For more explanation, see the text.

Table 3.A.1 Summary of the centering properties of the different coding systems.

	Balanced		Unbalanced	
	Main effects	Interactions	Main effects	Interactions
Sum-coding	Centered	Centered	No	No
Reference-coding	No	No	No	No
Weighted-coding	Centered	Centered	Centered	Centered

group two, and then measures effects relative to this group (dotted arrow). Weighted-coding, finally, measures effects relative to the weighted mean (drawn arrow). From this figure it is also clear that all effects are "transferable": solutions can be transferred into one other. This is a consequence of the over-parametrization of ANOVA models.

All the coding schemes end up in a design matrix $\mathbf{X}$ that can be partitioned as $\mathbf{X} = [\mathbf{X}_M | \mathbf{X}_A | \mathbf{X}_B | \mathbf{X}_{AB}]$ for the case of a two-factor crossed design (and analogously for other designs). It is then interesting to investigate what the centering properties are of the different sub-design matrices. Those properties are summarized in Table 3.A.1.

These centering properties may have repercussions for the HD-ANOVA generalizations as discussed in Chapter 4–6.

3.A.3 Practical Aspects of Codings

Although the different factor codings generate the same fitted response values, they may give different auxiliary results, especially in unbalanced data. This is not always trivial and selecting the type of coding scheme thus needs careful thinking. It has also consequences in conjunction with sums-of-squares Types I, II, and III, and this is not easy to see from the formulas. Therefore, we simulate a dataset where we have exact control of the variation and corresponding significances. The simulation contains 120 samples, three effects with all possible interactions, and has 5 replicates for each factor combination, see Table 3.A.2. A small amount of noise was added to generate realistic data.

Table 3.A.2 Simulation setup for data used in three-factor fixed effect ANOVA.

Source	Levels	df	*p*-value
f1	3	2	0.001
f2	4	3	0.02
f3	2	1	0.0005
f1:f2	12	6	0.03
f1:f3	6	2	0.0485
f2:f3	8	3	0.1
f1:f2:f3	24	6	0.78
Errors	120	96	

Table 3.A.3 Cell counts for unbalanced data, *i.e.*, the number of samples for each factor level combination after unbalancing the simulated data.

		f3 = 1				f3 = 2			
f2		1	2	3	4	1	2	3	4
f1	1	1	4	4	5	4	4	5	5
	2	5	5	5	5	5	5	5	5
	3	5	5	4	4	5	4	4	4

After the data have been simulated, some samples are removed to create an unbalanced data set. Single samples are removed randomly, and all but one samples are removed from the factor combination $f1 = 1, f2 = 1, f3 = 1$ to force severe imbalance. Table 3.A.3 shows the counts of each factor level combination after unbalancing.

For each of the nine combinations of sum-coding, reference-coding, and weighted-coding combined with sums-of-squares Type (I, II, III), full factorial fixed effect ANOVAs were fitted. This was done first for the balanced data, shown in Table 3.A.4. In the case of reference-coding, there is an extra decision to make regarding which level is chosen as the reference level; this is indicated in the table.

For the balanced case, the sum-coding and weighted coding give equivalent results reflecting the original p-values. This is not the case for reference-coding with SS-Type III: the results are quite different and also depend on the chosen reference level. This will be explained in the sequel.

For the unbalanced case, the results are shown in Table 3.A.5.

For the SS-Type I and II the results for all codings are equivalent. Yet, the p-values differ a bit from the simulated ones, and upon taking a cut-off value of $\alpha = 0.05$ the conclusions regarding Factor *f2* and the interaction *f1:f2* will be different. This also shows that hard cut-off values can be problematic. The SS-Type III reference-coding results are very different, both compared to the other coding schemes as well as regarding which level is chosen as the reference level.

Table 3.A.4 *P*-values for fixed effect ANOVAs using combinations of types of sums-of-squares and factor codings in the *balanced* case.

	SS-Type I			SS-Type II			SS-Type III			
Factor	sum	ref.	wei.	sum	ref.	wei.	sum	ref.1	ref.end	wei.
f1	0.001	0.001	0.001	0.001	0.001	0.001	0.001	0.009	0.937	0.001
f2	0.020	0.020	0.020	0.020	0.020	0.020	0.020	0.151	0.014	0.020
f3	0.000	0.000	0.000	0.000	0.000	0.000	0.000	0.218	0.938	0.000
f1:f2	0.030	0.030	0.030	0.030	0.030	0.030	0.030	0.278	0.126	0.030
f1:f3	0.049	0.049	0.049	0.049	0.049	0.049	0.049	0.069	0.926	0.049
f2:f3	0.100	0.100	0.100	0.100	0.100	0.100	0.100	0.744	0.073	0.100
f1:f2:f3	0.780	0.780	0.780	0.780	0.780	0.780	0.780	0.780	0.780	0.780

A value of 0.000, means < 0.001. "ref.1" and "ref.end" refer to using the first and last factor levels as the reference level.

Table 3.A.5 *P*-values for fixed effect ANOVAs using combinations of types of sums-of-squares and factor coding in the *unbalanced* case.

	SS-Type I			SS-Type II			SS-Type III			
Factor	sum	ref.	wei.	sum	ref.	wei.	sum	ref.1	ref.end	wei.
f1	0.018	0.018	0.018	0.040	0.040	0.040	0.035	0.774	0.946	0.037
f2	0.066	0.066	0.066	0.070	0.070	0.070	0.058	0.467	0.035	0.071
f3	0.000	0.000	0.000	0.000	0.000	0.000	0.000	0.094	0.488	0.000
f1:f2	0.058	0.058	0.058	0.095	0.095	0.095	0.120	0.794	0.110	0.095
f1:f3	0.049	0.049	0.049	0.068	0.068	0.068	0.072	0.090	0.748	0.068
f2:f3	0.130	0.130	0.130	0.130	0.130	0.130	0.122	0.559	0.170	0.130
f1:f2:f3	0.851	0.851	0.851	0.851	0.851	0.851	0.851	0.851	0.851	0.851

A value of 0.000, means < 0.001. "ref.1" and "ref.end" refer to using the first and last factor levels as the reference level.

To shed some light on the background of these results, it is worthwhile to study the properties of the design matrix $\mathbf{X}$ which results from the different types of codings. This is visualized in Figure 3.A.2. Here the cross-products of the coded factors are plotted with black where there is non-orthogonality, and a green overlay represents the balanced sum-coding case as a reference.

In the balanced case, sum-coding and weighted coding are perfectly orthogonal across factors but are not orthogonal within factors. Reference-coding has a pattern of overlap between all factors and interactions, but these are internally orthogonal. The latter pattern stays consistent in the unbalanced case. For sum-coding, the imbalance leads to non-orthogonality scatter throughout all factor combinations. Finally, for the weighted coding used on unbalanced data, there are different but logical patterns. All factors and interactions are orthogonal to their higher-order interactions, *e.g.*, *f1* is orthogonal to *f1:f2*, *f1:f3*, and *f1:f2:f3*, but not to *f2:f3*.

The different (non-)orthogonality properties, indeed, help to understand what is happening in terms of types of sum-of-squares. The results for the significance of Factor *f1* in the unbalanced reference-coding SS-Type III can be understood as follows. The SS-Type III works by first using all factors except *f1* (the reduced model) and subsequently fitting a full model with all factors. In the reduced model almost all variation is 'eaten away' by the other factors which are not orthogonal to *f1*. Hence, little variation to explain is left over for Factor *f1* resulting in a very high p-value.

There is a debate in the literature about using which type of sum-of-squares and some authors recommend SS-type II for unbalanced data (Langsrud, 2003). It is clear from this small example that also the type of coding matters. Especially reference-coding can give unexpected results and should be handled with care. The combination of reference-coding and SS-type III seems problematic. Note that in all cases the fitted values for the models are the same across all coding schemes.

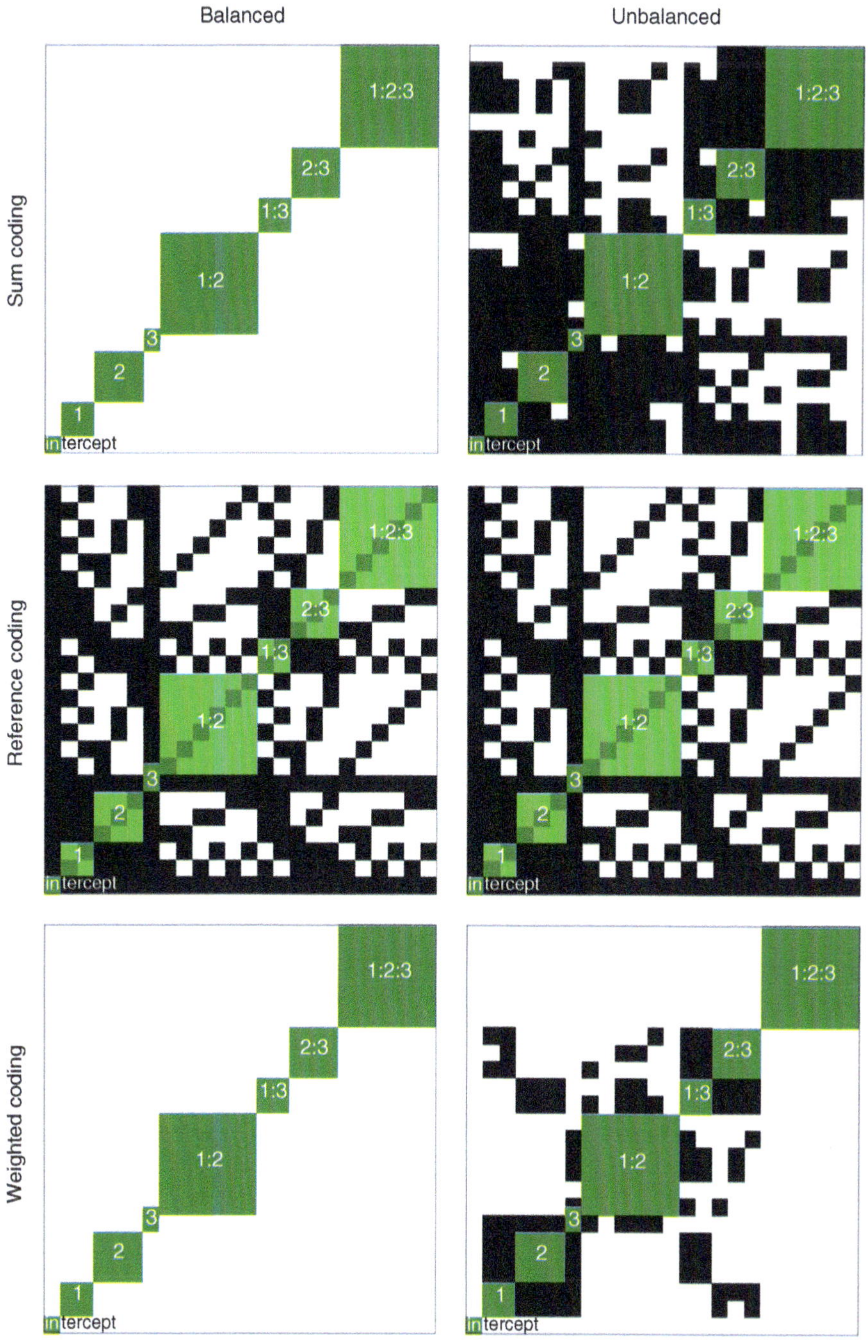

Figure 3.A.2 Cross-products of the different factor codings (*i.e.*, the $\mathbf{X}^t\mathbf{X}$ matrices) for three-factor data. Colors are white for zero cross-products and black for non-zero cross-products, due to lack of orthogonality. Green is overlayed on the block-diagonal corresponding to the balanced sum-coding case.

4

ASCA and Related Methods

In this chapter, we will present methods that are all related to the original formulation of ANOVA-simultaneous component analysis (ASCA) (Smilde *et al.*, 2005). All these methods can be written as a restricted multivariate multiple regression problem using certain types of coding schemes. We will explain for each method how this actually works, and this gives a generic framework for these methods. Alternative methods that do not fit into this scheme are presented in Chapter 5.

We will reserve the term **ASCA** for those models of which the parameters can be estimated in their simplest form, *i.e.*, assuming balanced data. This means that per variable an analysis of variance (ANOVA) model is calculated where the estimated parameters are obtained by simple averages among groups which is the ANOVA way of estimating those parameters (see Chapter 3). This works fine for the simplest balanced cases, but for more complicated cases we have to use the regression framework. Then the subsequent methods will be called differently, such as ASCA+ and PRC.

4.1 ASCA

4.1.1 Basic Idea of ASCA

The ASCA method will be explained by using an example of an experiment in which J responses y_{abnj} are measured as a function of a treatment design consisting of two crossed. Factors A and B with treatment levels $a = 1, \dots, A$ and $b = 1, \dots, B$, respectively, and $n = 1, \dots, N$ repeats per treatment group. Hence, we assume balanced data. The general analysis of variance (ANOVA) model for response j is then:

$$y_{abnj} = \mu_j + \alpha_{aj} + \beta_{bj} + (\alpha\beta)_{abj} + \epsilon_{abnj} \tag{4.1}$$

and under the sum-to-zero constraints (see Chapter 3) the ANOVA decomposition for the jth response in terms of the estimated factor effects is then:

$$\begin{aligned} y_{abnj} &= \overline{y}_{...j} + (\overline{y}_{a..j} - \overline{y}_{...j}) + (\overline{y}_{.b.j} - \overline{y}_{...j}) \\ &\quad + (\overline{y}_{ab.j} - \overline{y}_{a..j} - \overline{y}_{.b.j} + \overline{y}_{...j}) + (y_{abnj} - \overline{y}_{ab.j}) \end{aligned} \tag{4.2}$$

in which the notation regarding overlines and dots is explained in Section 1.7.2. The first term on the RHS of the equation is the general mean; the second term is the estimated main effect of Factor A at level a; the third term is the estimated main effect of Factor B at level b; the fourth term is the estimated interaction between Factors A and B at levels a and b; and the last term is the unexplained or individual variation. This is a repeat of the ANOVA model as discussed

Analysis of Variance for High-Dimensional Data: Applications in Life, Food and Chemical Sciences,
First Edition. Age K. Smilde, Federico Marini, Johan A. Westerhuis, and Kristian H. Liland.

in Chapter 3 with an extra index indicating the response variable j. Due to the simplicity of the design, all the factor effects can be estimated by calculating means, as shown in Equation 4.2.

The terms of Equation 4.2 can now be collected for all J responses into matrices $\mathbf{Y}$, $\overline{\mathbf{Y}}_{\mathtt{M}}$, $\overline{\mathbf{Y}}_{\mathtt{A}}$, $\overline{\mathbf{Y}}_{\mathtt{B}}$, $\overline{\mathbf{Y}}_{\mathtt{AB}}$, and $\mathbf{Y}_{\mathtt{Ind}}$ which are all of size $(ABN \times J)$. We will denote ABN by the index I from now on. This results then in the equation:

$$\mathbf{Y} = \overline{\mathbf{Y}}_{\mathtt{M}} + \overline{\mathbf{Y}}_{\mathtt{A}} + \overline{\mathbf{Y}}_{\mathtt{B}} + \overline{\mathbf{Y}}_{\mathtt{AB}} + \mathbf{Y}_{\mathtt{Ind}} \tag{4.3}$$

where the term $\mathbf{Y}_{\mathtt{Ind}}$ collects the individual contributions $(y_{abnj} - \overline{y}_{ab.j})$ also sometimes called residuals. This is the basic decomposition in the ASCA approach. As an example, the matrix $\overline{\mathbf{Y}}_{\mathtt{A}}$ is

$$\overline{\mathbf{Y}}_{\mathtt{A}} = \begin{bmatrix} \overline{\mathbf{y}}_{1..1} & \cdots & \overline{\mathbf{y}}_{1..J} \\ \vdots & \ddots & \vdots \\ \overline{\mathbf{y}}_{A..1} & \cdots & \overline{\mathbf{y}}_{A..J} \end{bmatrix} \tag{4.4}$$

where $\overline{\mathbf{y}}_{a..j}(BN \times 1)$ is the vector containing all estimates of Factor A at level a for response j. Hence, it is assumed that the indices n and b are running within this vector (but all values are the same within this vector). Clearly this matrix $\overline{\mathbf{Y}}_{\mathtt{A}}$ is highly structured. The matrix $\overline{\mathbf{Y}}_{\mathtt{M}}$ equals $[\overline{\mathbf{y}}_{...1}| \cdots |\overline{\mathbf{y}}_{...J}]$ where each vector $\overline{\mathbf{y}}_{...j}(I \times 1)$ contains the means of response j across all samples. The matrix $\overline{\mathbf{Y}}_{\mathtt{M}}$ has rank one and can be written as $\mathbf{1}_I\overline{\mathbf{y}}_M^t$ in which $\overline{\mathbf{y}}_M(J \times 1)$ contains the mean per response across all samples. The other matrices on the RHS of Equation 4.3 are also highly structured. A pictorial overview of these matrices is given in Example 4.1.

Example 4.1: Schematic of the ASCA model

In this example, we assume that we have a treatment design (a randomized controlled design) with a Factor A (*Time*) at four levels, a Factor B (*Treatment*) at three levels, and two replicates per group. For simplicity, we do not assume an interaction. Multiple responses are measured, and the ANOVA decompositions for the two metabolites tyrosine and lysine are shown in Figure 4.1

Sample	Time	Treatment	Tyrosine	Lysine		Time	Time		Treatment	Treatment			
1	1	1	−0.3966	−0.2017		−0.0753	0.1822		−0.4702	−0.5567		0.1489	0.1728
2	1	1	−0.3879	−0.4718		−0.0753	0.1822		−0.4702	−0.5567		0.1576	−0.0973
3	1	2	−0.2193	0.3284		−0.0753	0.1822		−0.0559	−0.088		−0.0881	0.2342
4	1	2	−0.0334	0.1096		−0.0753	0.1822		−0.0559	−0.088		0.0978	0.0154
5	1	3	0.1903	0.4425		−0.0753	0.1822		0.526	0.6447		−0.2604	−0.3844
6	1	3	0.3952	0.8862		−0.0753	0.1822		0.526	0.6447		−0.0555	0.0593
7	2	1	−0.4892	−0.7128		0.1923	0.062		−0.4702	−0.5567		−0.2113	−0.2181
8	2	1	−0.3101	−0.3931		0.1923	0.062		−0.4702	−0.5567		−0.0322	0.1016
9	2	2	−0.0788	0.3619		0.1923	0.062		−0.0559	−0.088		−0.2152	0.3879
10	2	2	0.3941	0.3118		0.1923	0.062		−0.0559	−0.088		0.2577	0.3378
11	2	3	−0.1609	0.2076		0.1923	0.062		0.526	0.6447		−0.8792	−0.4991
12	2	3	1.7989	0.5966	=	0.1923	0.062	+	0.526	0.6447	+	1.0806	−0.1101
13	3	1	−0.4608	−0.6939		−0.2169	−0.1015		−0.4702	−0.5567		0.2263	−0.0357
14	3	1	−0.6839	−0.7274		−0.2169	−0.1015		−0.4702	−0.5567		0.0032	−0.0692
15	3	2	−0.0912	−0.516		−0.2169	−0.1015		−0.0559	−0.088		0.1816	−0.3265
16	3	2	−0.1951	−0.4889		−0.2169	−0.1015		−0.0559	−0.088		0.0777	−0.2994
17	3	3	0.1603	1.0772		−0.2169	−0.1015		0.526	0.6447		−0.1488	0.534
18	3	3	−0.031	0.7402		−0.2169	−0.1015		0.526	0.6447		−0.3401	0.197
19	4	1	−0.5412	−0.7148		0.0999	−0.1427		−0.4702	−0.5567		−0.1709	−0.0154
20	4	1	−0.4916	−0.5382		0.0999	−0.1427		−0.4702	−0.5567		−0.1213	0.1612
21	4	2	0.0237	−0.3661		0.0999	−0.1427		−0.0559	−0.088		−0.0203	−0.1354
22	4	2	−0.247	−0.4446		0.0999	−0.1427		−0.0559	−0.088		−0.291	−0.2139
23	4	3	1.7142	0.1468		0.0999	−0.1427		0.526	0.6447		1.0883	−0.3552
24	4	3	0.1413	1.0603		0.0999	−0.1427		0.526	0.6447		−0.4846	0.5583

Figure 4.1 Pictorial presentation of a two-way ASCA model without interaction for two metabolites. Source: Reproduced from Smilde *et al.* (2022)/with permission of John Wiley & Sons.

where the design is indicated with the columns Time and Treatment. Note that these are not the coding variables for the design but simply the levels of the design. The data matrix $\mathbf{Y}$ is assumed to be centered and then the first matrix on the RHS represents $\overline{\mathbf{Y}}_{\mathtt{A}}$. The second matrix is then $\overline{\mathbf{Y}}_{\mathtt{B}}$ and the last part on the RHS are the individual contributions (or residuals) $\mathbf{Y}_{\mathtt{Ind}}$.

The factor effect estimates matrices are seen to be highly structured. This can easily be extended to more than two metabolites.

The next step in the ASCA model is to perform a principal component analysis (PCA) on the matrices of estimated effects, which thus results in the basic ASCA equation:

$$\mathbf{Y} - \overline{\mathbf{Y}}_{\mathtt{M}} = \mathbf{T}_{\mathtt{A}}\mathbf{P}_{\mathtt{A}}^{t} + \mathbf{T}_{\mathtt{B}}\mathbf{P}_{\mathtt{B}}^{t} + \mathbf{T}_{\mathtt{AB}}\mathbf{P}_{\mathtt{AB}}^{t} + \mathbf{T}_{\mathtt{Ind}}\mathbf{P}_{\mathtt{Ind}}^{t} + \mathbf{E}_{ASCA} \tag{4.5}$$

in which the matrices $\mathbf{T}$ and $\mathbf{P}$ are scores and loadings, respectively, and the term $\mathbf{E}_{ASCA}$ contains all residuals of the (truncated) PCA models of the effect matrices, *i.e.*, $\mathbf{E}_{ASCA} = \mathbf{E}_{\mathtt{A}} + \mathbf{E}_{\mathtt{B}} + \mathbf{E}_{\mathtt{AB}} + \mathbf{E}_{\mathtt{Ind}}$. The scores and loadings per effect matrix can be plotted and studied, which will be shown in many examples later in this book. It is not necessary to have the same number of components in the PCA models of the separate effect matrices; this depends on the complexity of these matrices. The upper bound of the number of components for each effect matrix is its rank which, for the main effects, is usually the number of levels minus one, see Section 4.1.2. Since a PCA model per effect matrix is performed across all blocks of the structured effect matrix (see Equation 4.4) this is in fact a simultaneous component analysis (SCA) model (Smilde *et al.*, 2022), hence the name ASCA (ANOVA–Simultaneous Component Analysis). The score matrices of the PCA models for the factor effect estimates are also highly structured because of the structure of the matrices in Equation 4.3. For the Factor $\mathtt{A}$, *e.g.*, such a score matrix will only have A distinct points centered around zero as can be inferred from the structure of $\overline{\mathbf{Y}}_{\mathtt{A}}$ (see Equation 4.4).

Note the subtle interpretation of the PCA models of the separate effect matrices. Such a PCA does not model correlations but similarities of effects across the responses (see Section 2.3.2). If all systematic variation is described by the factors, then $\mathbf{Y}_{\mathtt{Ind}}$ contains the within-group deviations for all variables which can also be studied with a PCA albeit with a different interpretation (see Figure 2.12). The ASCA algorithm as expressed in ANOVA models as shown above is summarized in Algorithm 4.1.

Algorithm 4.1

ASCA IN ANOVA TERMS

The ASCA model as expressed in ANOVA terms can be summarized in the following steps:

1) Perform individual ANOVA models per response variable *j*.
2) Collect all factor estimates of these ANOVA models and build the factor estimate matrices $\overline{\mathbf{Y}}_{\mathtt{A}}$, $\overline{\mathbf{Y}}_{\mathtt{B}}$, $\overline{\mathbf{Y}}_{\mathtt{AB}}$ and $\mathbf{Y}_{\mathtt{Ind}}$.
3) Perform PCAs on these matrices; plot and interpret the results.

Sometimes it is convenient for reasons of interpretability to combine effects in ASCA. That means that a PCA is then performed on $\overline{\mathbf{Y}}_{\mathtt{B}} + \overline{\mathbf{Y}}_{\mathtt{AB}}$ instead of on the individual matrices $\overline{\mathbf{Y}}_{\mathtt{B}}$ and $\overline{\mathbf{Y}}_{\mathtt{AB}}$. This is shown in Example 4.2.

Example 4.2: Example of ASCA

This example expands on the plant science study of *Arabidopsis* already used in Section 3.1. Changes in metabolism related to growth under different light (and temperature) conditions are recorded at several time points. *Time* and the combined *Light* + *Time:Light* interaction are included in the ASCA analysis. The combined effect, whose loadings and scores are shown in Figure 4.2, explains 25.18% of the total variation in the data (see later for how this is calculated). ASCA scores start at zero effect at time point zero and show the *Time:Light* effect as time

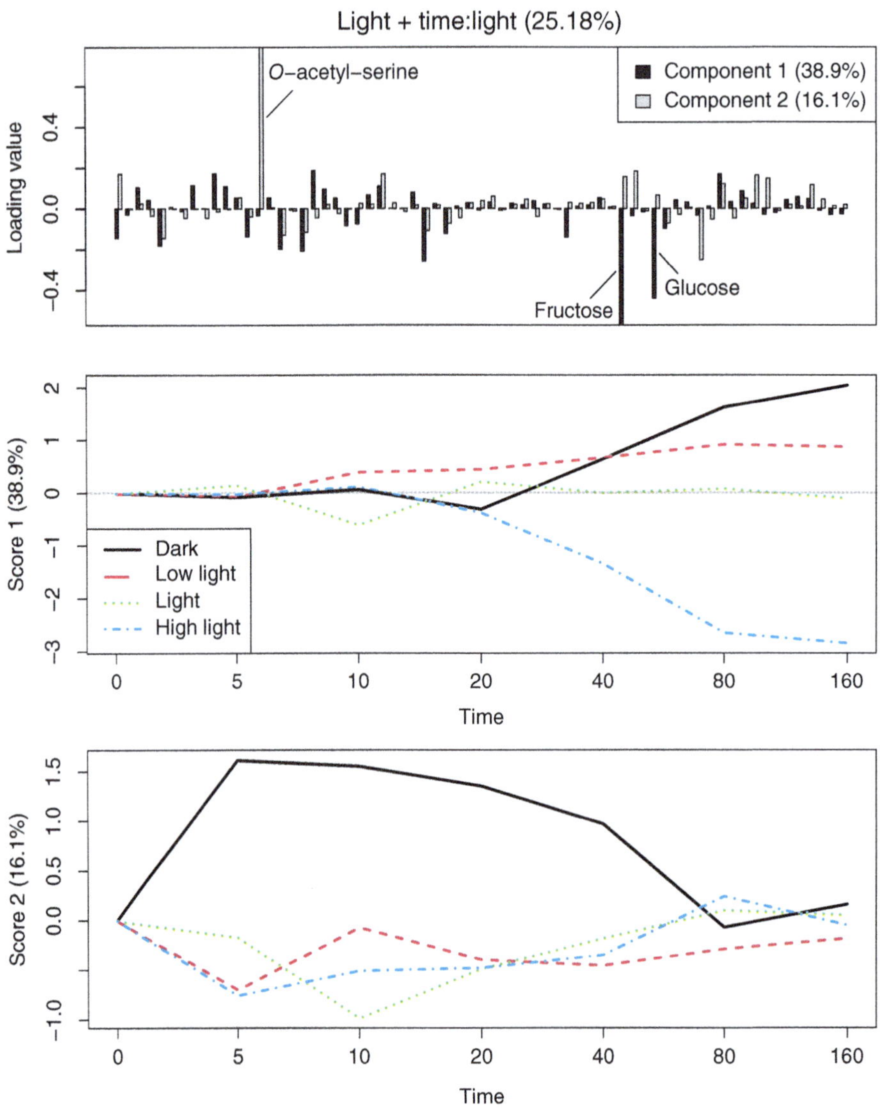

Figure 4.2 ASCA analysis of metabolomics data. The scores and loadings of PC1 and PC2 of the *Light + Time:Light* combination are provided. Source: Adapted from Caldana *et al.* (2011).

progresses with the factor levels *Dark* on top, *Low light* and *Light* in the middle, and *High light* clearly separated below. The explained variance of the first component of ASCA (38.9%) is mainly due to glucose and fructose.

The second component of the ASCA model with 16.1% explained variance clearly shows the deviating behavior at the *Dark Light* level which is mainly due to *O*-acetyl-serine. In total, 38.9% + 16.1% = 55% of the combined *Light* + *Time:Light* variation is explained by the two plotted components.

4.1.2 Properties of ASCA

The ASCA model has several properties of which the most important one is that the total data variation can be separated into the contribution of each effect and their interactions:

$$||\mathbf{Y}||^2 = ||\overline{\mathbf{Y}}_{\mathrm{M}}||^2 + ||\overline{\mathbf{Y}}_{\mathrm{A}}||^2 + ||\overline{\mathbf{Y}}_{\mathrm{B}}||^2 + ||\overline{\mathbf{Y}}_{\mathrm{AB}}||^2 + ||\mathbf{Y}_{\mathtt{Ind}}||^2 \tag{4.6}$$

which rests on the property that the matrices in Equation 4.3 are orthogonal to each other (Jansen *et al.*, 2005) and holds for the balanced design case (for a simple proof, see Section 4.7). This implies that Equation 4.3 can also be understood in terms of direct sums of subspaces, see Elaboration 4.1 and Chapter 2. Note that Equation 4.6 is in terms of sums-of-squares and not mean sums-of-squares. For the latter, we would need degrees of freedom that are not trivial and not yet known for this situation.

ELABORATION 4.1

ASCA in terms of a Direct Sum of subspaces

Combining Equation 4.3 and the fact that the different matrices on the RHS of that equation are orthogonal, means that we can actually also write the same equation in the context of direct sums:

$$R(\mathbf{Y}) = R(\overline{\mathbf{Y}}_{\mathrm{M}}) \oplus R(\overline{\mathbf{Y}}_{\mathrm{A}}) \oplus R(\overline{\mathbf{Y}}_{\mathrm{B}}) \oplus R(\overline{\mathbf{Y}}_{\mathrm{AB}}) \oplus R(\mathbf{Y}_{\mathtt{Ind}}) \tag{4.7}$$

which shows that the ASCA (and ANOVA) decomposition can also be understood in terms of separating subspaces.

Equation 4.6 is the basis for a convenient diagnostic: the explained sum of squares of a factor f or an interaction. For an arbitrary factor f (or interaction or the term $\mathbf{Y}_{\mathtt{Ind}}$) this can be quantified as:

$$Var_f = \frac{||\overline{\mathbf{Y}}_f||^2}{||\mathbf{Y} - \overline{\mathbf{Y}}_M||^2} \times 100\% \tag{4.8}$$

which is the amount of explained variation of factor f in percentage (see Equation 4.6). The sum of these numbers across all factors and interactions equals 100% due to the breakdown as shown in Equation 4.6.

Another property that is of interest is the rank of the factor estimate matrices. Usually, the number of responses is much higher than the number of factor levels, and thus these numbers of factor levels are limiting the ranks of the factor estimate matrices. From Equation 4.4 which displays the factor estimate matrix $\overline{\mathbf{Y}}_{\mathrm{A}}$, it is clear that this matrix is rank-deficient. It has A distinct rows and J columns. There is a linear dependency betw een the rows since each column in this matrix adds up to zero because it is orthogonal to $\overline{\mathbf{Y}}_{\mathrm{M}}$. Hence, upon assuming that $A < J$ the maximum rank of $\overline{\mathbf{Y}}_{\mathrm{A}}$ is $(A-1)$. Likewise, the maximum rank of $\overline{\mathbf{Y}}_{\mathrm{B}}$ is $(B-1)$ and for $\overline{\mathbf{Y}}_{\mathrm{AB}}$ this is $(A-1)(B-1) = (AB - A - B + 1)$. Thus, for the systematic part of $\mathbf{Y}$ (*i.e.*, $\hat{\mathbf{Y}}$) the total maximum rank is $(1 + A - 1 + B - 1 + AB - A - B + 1) = AB$ using the fact that $rank(\overline{\mathbf{Y}}_{\mathrm{M}}) = 1$ and all matrices are mutually orthogonal. These maximum rank numbers also give an upper limit of the number of PCs to extract from each factor estimate matrix. The minimum rank of the matrix $\mathbf{Y}_{\mathtt{Ind}}$ is thus $rank(\mathbf{Y}) - AB$. If a factor only has two levels, then this represents a special case, see Elaboration 4.2.

ELABORATION 4.2

ASCA factors at two levels

When a factor has only two levels, then the maximum rank of the corresponding factor estimate matrix is only 1. Thus a PCA of such a factor estimate matrix has only one score vector which can be displayed as two points on a line. This is a rather boring plot, but an alternative is presented in Chapter 1 as a spider plot (Example 1.1) and explained more in Chapter 8.

4.1.3 Permutation Tests for ASCA

ASCA as described so far is mainly a technique to visualize high-dimensional designed data. It would be worthwhile to also have some machinery to assess the significance of treatment factors. Unfortunately, the ANOVA machinery for testing significance is not easily generalized to the ASCA context since a full account of, *e.g.*, degrees of freedom is very difficult to give. One of the possibilities is to use resampling techniques of which permutation testing is very promising. The idea of permutation testing in ANOVA was already explained in Section 3.7.3.

Extending permutation testing to the high-dimensional (HD)-ANOVA case is not straightforward due to correlations between the variables in the response matrix $\mathbf{Y}$. Instead of aiming for significance of a certain factor for a single variable, the goal is to find significance of that factor for the whole effect matrix.

Vis *et al.* (2007) were the first to introduce permutation tests for ASCA. The same permutation guidelines as for ANOVA that were described in Section 3.7.3 also apply to ASCA. Both exact and approximate permutation tests can be performed. Exchangeable units are again defined by the denominator mean square F-ratio for the factor to be tested. For approximate tests, residuals can be permuted; *e.g.*, for testing the interaction AB in a two-factor design with fixed factors, $\mathbf{R}_{\mathrm{AB}} = \mathbf{Y} - \overline{\mathbf{Y}}_{\mathrm{M}} - \overline{\mathbf{Y}}_{\mathrm{A}} - \overline{\mathbf{Y}}_{\mathrm{B}}$, and this residual matrix $\mathbf{R}_{\mathrm{AB}}$ can be permuted for the approximate permutation test.

The difference with the permutations for a single response is the selection of the test statistic. For ANOVA the F-ratio of the effect being tested was used: $F_{\mathrm{A}} = \frac{MS_{\mathrm{A}}}{MS_{Ind}}$. As the F-ratio is compared to other F-ratios of permuted data sets, the degrees of freedom have no effect and $F_{\mathrm{A}} = \frac{SS_{\mathrm{A}}}{SS_{Ind}}$ will work as well. For exact tests, permutations are done in such a way that all other effects stay constant and only SS_{A} and SS_{Ind} will change. Thus $SS_{\mathrm{A}} + SS_{Ind}$ is also constant. Therefore, SS_{A} can equally be used as a test statistic for the permutations. Example 4.3 shows the use of such a permutation test. Instead of using the sum-of-squares as a test statistic it is also possible to use another test statistic. One possibility is to use the scores of the PCs of the factor estimate matrices and calculate the sum-of-squares of those in the permutations (Langsrud, 2002; Zwanenburg *et al.*, 2011). In that case, the sum-of-squares of the effect matrix is first divided into a systematic part (the PC's) and a non-systematic part (the residuals), and this division is expected to improve the power of the test.

For more complicated experimental designs it is not always clear which of the tests has the most power. This may also depend on the amount of correlation among the responses and in particular on the (differences in) within-group correlation.[1] For the test of certain factors, both exact and approximate permutation tests exist, and different types of approximate tests can be applied. Camacho *et al.* (2023) use power simulations to find out which permutation tests combined with which test statistics are most powerful to detect small effects. For unbalanced designs, permutations are more complicated due to the correlated effect matrices that appear because of the unbalancedness. For univariate data, (Kherad-Pajouh and Renaud, 2010) discussed an approach based on Type III sum-of-squares. However, the exchangeability of experimental units has to be taken into account by selecting appropriate sets of permutations. How to extend this to the high-dimensional case is not yet completely clear. A new approach by Rasmussen *et al.* (2024) to define allowable permutations for high-dimensional data seems promising, but rigorous testing for sensitivity is still needed.

Example 4.3: Permutation tests in ASCA

For this example, we will look at validation of an ASCA model with two factors and their interaction. The full ASCA model decomposes $\mathbf{Y}$ into its contributions as described in Equation 4.3. First, the interaction will be tested using an approximate test in which the contributions of the main factors will be removed from $\mathbf{Y}$.

$$\mathbf{R}_{\mathrm{AB}} = \mathbf{Y} - \overline{\mathbf{Y}}_{\mathrm{M}} - \overline{\mathbf{Y}}_{\mathrm{A}} - \overline{\mathbf{Y}}_{\mathrm{B}} \tag{4.9}$$

Now $\mathbf{R}_{\mathrm{AB}}$ contains variation due to the interaction $\overline{\mathbf{Y}}_{\mathrm{AB}}$ and the individual source $\mathbf{Y}_{\mathtt{Ind}}$. The ASCA model thus becomes $\mathbf{R}_{\mathrm{AB}} = \overline{\mathbf{Y}}_{\mathrm{AB}} + \mathbf{Y}_{\mathtt{Ind}}$ with test statistic the F-ratio $F_{\mathrm{AB}} = \frac{SS_{\mathrm{AB}}}{SS_{\mathtt{Ind}}}$. Unrestricted permutations can be applied to $\mathbf{R}_{\mathrm{AB}}$ as no other factor is present anymore. Then, SS^p_{AB} and $SS^p_{\mathtt{Ind}}$ can be calculated, and the F-ratio of the permuted $\mathbf{R}^p_{\mathrm{AB}}$ equals $F^p_{\mathrm{AB}} = \frac{SS^p_{\mathrm{AB}}}{SS^p_{\mathtt{Ind}}}$. The p-value equals the probability that $F^p_{\mathrm{AB}} \geq F_{\mathrm{AB}}$.

If the interaction is not found significant, the model of Equation 4.3 simplifies to

$$\mathbf{Y} = \overline{\mathbf{Y}}_{\mathrm{M}} + \overline{\mathbf{Y}}_{\mathrm{A}} + \overline{\mathbf{Y}}_{\mathrm{B}} + \mathbf{Y}_{\mathtt{Ind}} \tag{4.10}$$

For each main Factors A and B an exact test can be applied. For the test of A the F-ratio $F_{\mathrm{A}} = \frac{SS_{\mathrm{A}}}{SS_{\mathtt{Ind}}}$ applies. The denominator term dictates that the individual observations (which are rows in $\mathbf{Y}$) are exchangeable and can be permuted, restricted within the levels of Factor B. Then, SS^p_{A} and $SS^p_{\mathtt{Ind}}$ can be calculated, and the statistic $F^p_{\mathrm{A}} = \frac{SS^p_{\mathrm{A}}}{SS^p_{\mathtt{Ind}}}$ can be obtained. The p-value can be obtained as the fraction $F^p_{\mathrm{A}} \geq F_{\mathrm{AB}}$.

4.1.4 Back-Projection

In the ASCA model, the score plots of the factor estimates do not reveal the individual variability anymore. This is a drawback since such variability may be useful for visualizing and assessing the significance of the obtained factor estimates. A simple method to visualize such variability is by a method called back-projection (Zwanenburg *et al.*, 2011). For the Factor A this goes as follows. After having obtained the matrices $\overline{\mathbf{Y}}_{\mathrm{A}}$ and $\mathbf{Y}_{\mathtt{Ind}}$ a PCA is performed on $\overline{\mathbf{Y}}_{\mathrm{A}}$ resulting in $\mathbf{T}_{\mathrm{A}}(I \times R_{\mathrm{A}})$ and $\mathbf{P}_{\mathrm{A}}(J \times R_{\mathrm{A}})$ where R_{A} is the number of PCs. As mentioned in

1 We thank Jasper Engel for pointing that out.

Section 4.1.1 the score plot of matrix $\mathbf{T}_{\mathtt{A}}(I \times R_{\mathtt{A}})$ is highly structured and only shows A points. The back-projection is now performed by calculating

$$\mathbf{T}_{\mathtt{A},bp} = (\overline{\mathbf{Y}}_{\mathtt{A}} + \mathbf{Y}_{\mathtt{Ind}})\mathbf{P}_{\mathtt{A}} \tag{4.11}$$

where $\mathbf{T}_{\mathtt{A},bp}$ stands for back-projected scores and $\mathbf{P}_{\mathtt{A}}^t\mathbf{P}_{\mathtt{A}} = \mathbf{I}$. These scores show the individual variation due to the addition of $\mathbf{Y}_{\mathtt{Ind}}$, and confidence ellipsoids can be calculated around such scores and shown in the plot (see Example 4.8).

4.1.5 Scaling in ASCA

Centering is rather straightforward in ASCA modeling since it is part of the model specification and generalizes from (univariate) ANOVA models to HD-ANOVA models. In component analysis of a matrix $\mathbf{Y}(I \times J)$ (*e.g.*, in PCA), it is often the practice to use per variable j the observed standard deviation (s_j) and scale the matrix $\mathbf{Y}$ accordingly. Scaling in ASCA, however, is certainly not trivial since many options are available, and the results of subsequent analyses may depend on the type of scaling (Timmerman *et al.*, 2015). The basic idea is to use a type of scaling that does not affect the treatment information and reduces the nuisance variation.

We will discuss the different options for scaling based on an ASCA model with two factors: a factor *Treatment* (Factor A) and a factor *Time* (Factor B) based on the results in Timmerman *et al.* (2015). For simplicity, we assume balanced data. The ASCA decomposition is then:

$$\mathbf{Y} = \overline{\mathbf{Y}}_M + \overline{\mathbf{Y}}_{\mathtt{A}} + \overline{\mathbf{Y}}_{\mathtt{B}} + \overline{\mathbf{Y}}_{\mathtt{AB}} + \mathbf{Y}_{\mathtt{Ind}} \tag{4.12}$$

and we will consider PCA models of $(\overline{\mathbf{Y}}_{\mathtt{A}} + \overline{\mathbf{Y}}_{\mathtt{AB}})$ and $\mathbf{Y}_{\mathtt{Ind}}$ where the first is the main effect of *Treatment* and its differential effects across *Time* (*Treatment* + *Time:Treatment*; called between-effect) and the second describes the natural variation across individuals (called within-effect). We will discuss scaling for the two types of effects separately. For the between-effect, the options are:

No scaling: This is always an option but not necessarily the best one.

Autoscaling: This is probably the one that is used in the default mode. This scaling pertains to the centered matrix $\mathbf{Y}$ (thus of $\mathbf{Y} - \overline{\mathbf{Y}}_M$).

Reference residual scaling: If the design contains a reference group (control group, say level $a = 1$ of Factor A) then the residuals of this group ($\mathbf{Y}_{\mathtt{Ind(ref)}}$, *i.e.*, the part of $\mathbf{Y}_{\mathtt{Ind}}$ pertaining to the reference group $a = 1$) can be used in the scaling. Specifically, if one of the groups is a reference group, then we can use the standard deviations of the residuals in $\mathbf{Y}_{\mathtt{Ind(ref)}}$. A similar type of scaling can be used if there is a pre-dose group (say, for level $b = 1$ of Factor B).

Reference group scaling: If there are variables with a substantial time effect in the reference group (to be seen in $\mathbf{Y}$ at $a = 1$ across all time points), then this may be taken into account by using the standard deviations of the columns of the reference group (across time) which can be derived from $\mathbf{Y}$. This implies that variables with substantive variation in the reference group will be down-weighted. Note that these are not the residuals; hence, this differs from reference residual scaling.

All these scalings operate on the columns of the data matrix $\mathbf{Y}$ albeit in different ways. Note that there is also an option to scale using the within-group covariance matrices (Engel *et al.*, 2015), but this was not explored further.

For the within-effect, there are also different possibilities for scaling:

No scaling: This is always an option but not necessarily the best one.

Autoscaling: This is probably the one that is used in the default mode and is the same as mentioned above for the between-effect. It uses the standard deviations across all samples in $\mathbf{Y}$.

Reference residual scaling: If the design contains a reference group (control group) then this can be used in the scaling. Specifically, if one of the groups is a reference group, then we can use the standard deviations of the columns of $\mathbf{Y}_{\texttt{Ind(ref)}}$, *i.e.*, the part of $\mathbf{Y}_{\texttt{Ind}}$ pertaining to the reference group.

Note that if the residual variation differs a lot between cells then a preprocessing transformation step may be included to homogenize such variation.

A simulation study was performed to study all these types of scalings. The experimental design pertains to three treatments (Factor $\texttt{A}$ at three levels) including a reference condition. Per condition, 20 measurements on subjects were performed at $B = 10$ non-equidistant time points reflecting $t = 0.0001, \ldots, 34$ h after intake of a polyphenol capsule (placebo and two different types of capsules) and $J = 7$ variables were measured. The raw data is visualized in Figure 4.3. In these simulations, variables 1–4 have an early treatment effect peak with different time trends and amplitudes of the effects. Variables 5 and 6 show a medium-to-late treatment effect. Finally, variable 7 shows no effect.

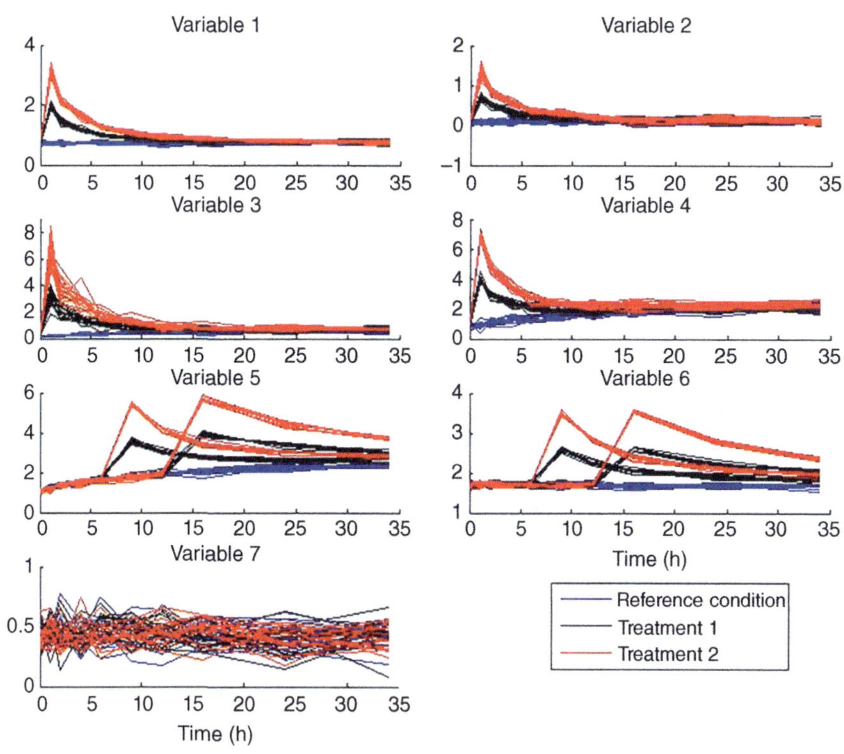

Figure 4.3 Simulated variables for the ASCA model showing different time and peaking trends. Source: Timmerman *et al.* (2015)/Springer Nature/CC BY 4.0.

After applying the different types of between-effect scaling, the results are as shown in Figure 4.4. For visualization purposes, the ASCA scores are expressed relative to the reference group as a postprocessing step (see also Smilde *et al.* (2012), *i.e.*). To be precise, the time profile of the reference group is put to zero. This figure shows the effect of these different types of scaling on the first between-effect component. The time-resolved scores on these first components are highly similar for all types of scaling (left panel); thus, their loadings (right panel) can be compared. The no-scaling situation reflects the relative effect sizes of variables 3 and 4. The reference residual scaling reflects the differences between effect sizes relative to the residual variances, and the order of importance is variable 3, 1, 4, and 2 which can be understood from the raw data. Comparing the reference group scaling results with the reference residual scaling shows a down-weighting of variables 3–5 because these variables have a time trend. Autoscaling, finally, expresses the treatment effect sizes relative to the overall variance of each variable, which is about equal for variables 1–4. This shows that autoscaling fails to reflect the relative sizes of the effect.

The results for the within-effects are shown in Figure 4.5 for the first component. Autoscaling shows a high importance for variable 7 which is the noisy variable and thus not interesting. The other two types of scaling show the two sub-types of responders in variables 5 and 6. Finally, it is highly recommended to make plots both of the raw data and of the pre-processed data to investigate the appropriateness of the scaling. An extensive example using a real-life data set is shown in Example 4.4.

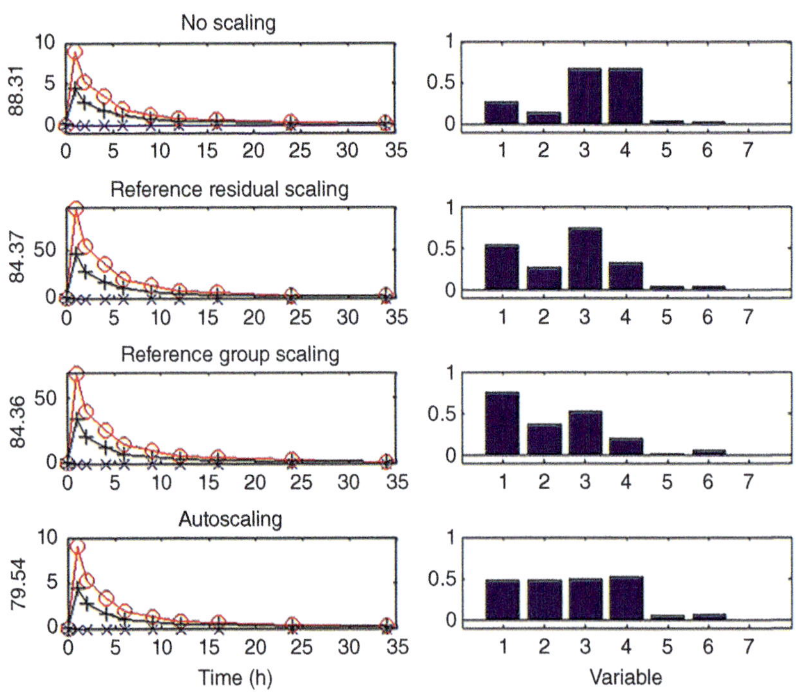

Figure 4.4 Results of different types of scaling of the ASCA between-effects. Left panel: time-resolved scores on the first between component (with their explained variance in percentages along the y-axis). Right panel: their corresponding loadings. For more explanation, see the text. Source: Timmerman *et al.* (2015)/Springer Nature/CC BY 4.0.

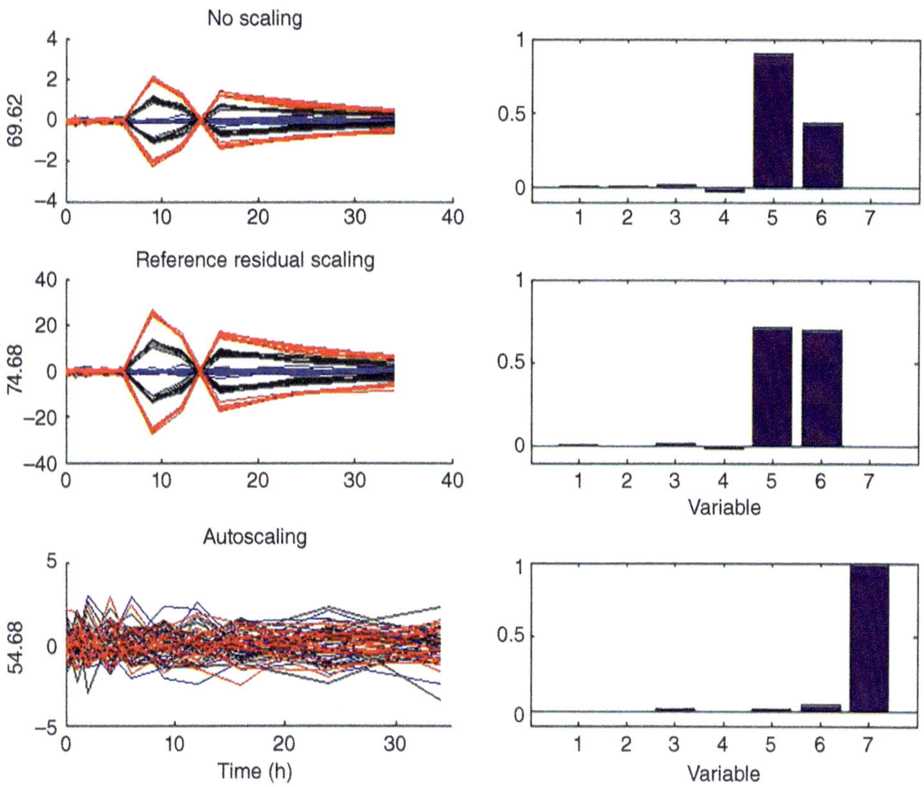

Figure 4.5 Results of different types of scaling of the ASCA within-effects. Left panel: time-resolved scores on the first between component (with their explained variances along the y-axis). Right panel: their corresponding loadings. For more explanation, see the text. Source: Timmerman *et al.* (2015)/Springer Nature/CC BY 4.0.

Example 4.4: Scaling in an ASCA model of nutrikinetics

The example is from the field of nutrikinetics and concerns measurements to study the bioavailability of polyphenols in a group of 20 male subjects (van Velzen *et al.*, 2014). They underwent a treatment with a tea extract, a wine extract, or a placebo extract in a cross-over design. Blood samples were collected just before (0 h) and at 1, 2, 6, 9, 12, 24, and 36 hours after the intake. From the data, we use nine measured metabolites for illustrative purposes. For details about these measurements and their quantification, we refer to van Velzen *et al.* (2014). The raw data are shown in Figure 4.6 and show (i) large individual differences, (ii) large-scale differences between the concentrations of the metabolites (see the y-scale in the plots), and (iii) variables 8 and 9 have their peaks at a later point in time; these are gut-mediated metabolites which are products of the gut-microbiome.

The model we use for this data is the same as for the simulations, see Equation 4.12 where we again combine *Treatment* and *Treatment:Time* interaction as the between-variation and the

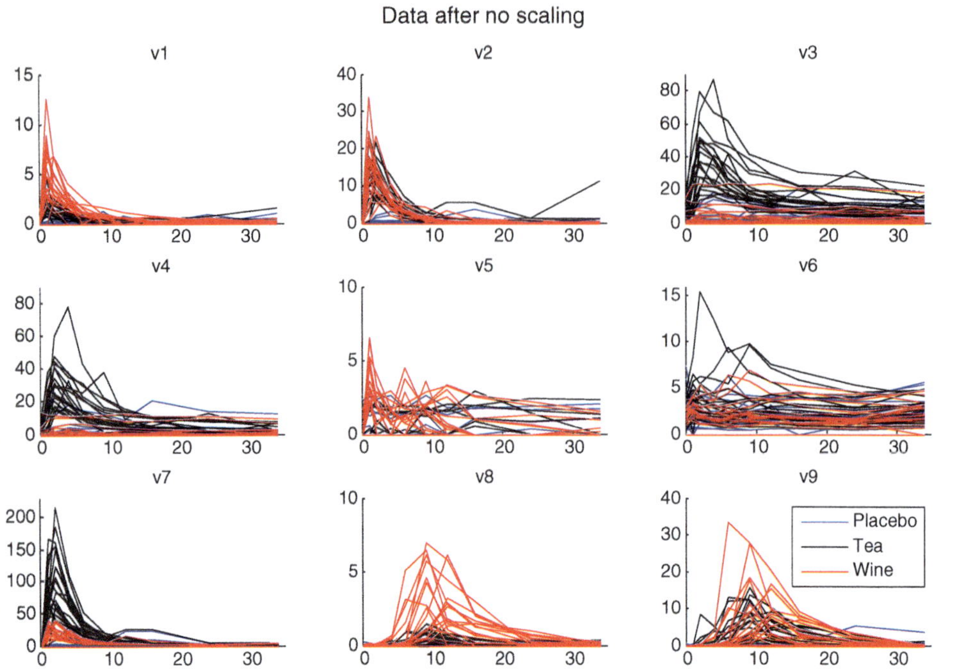

Figure 4.6 The profiles of the concentrations of the blood metabolites (v1–v9) of the nutrikinetics study. Source: van Velzen *et al.* (2014) and Timmerman *et al.* (2015)/Springer Nature/CC BY 4.0.

term $\mathbf{Y}_{\mathtt{Ind}}$ as the within-variation. Again, we chose to postprocess the ASCA scores relative to the reference group as in the simulations. To illustrate the scaling steps in ASCA for the between-variation, the data were subjected to four types of scaling: no scaling, reference residual scaling (using the placebo group residuals), reference group scaling (using the measurements from the placebo group), and autoscaling. After reference residual scaling the large-scale differences between the metabolites are changed; now metabolites v3 and v7 (with the largest variation in the raw data) are at the 3rd and 6th positions, respectively. This shows the effect of scaling of the raw data. Figure 4.7 shows the result of the different types of scaling on the first between-component.

The scores on the first component for the no-scaling and autoscaling situation are very similar and focus on the early responses, albeit with different emphasis on variables 3, 4, and 7; the last variable has the largest variation and thus dominates the no-scaling case (see also the high amount of explained variation of this component in the between-effect model). The scores for the reference residual scaling and the reference group scaling are very similar which is expected given the raw profiles. Also, the associated loadings are very similar. The two groups of scaling – no-scaling and autoscaling on the one hand and reference residual scaling and reference groups scaling on the other hand – show different aspects of the data. Clearly, there is a need for a second component and this is shown in Figure 4.8.

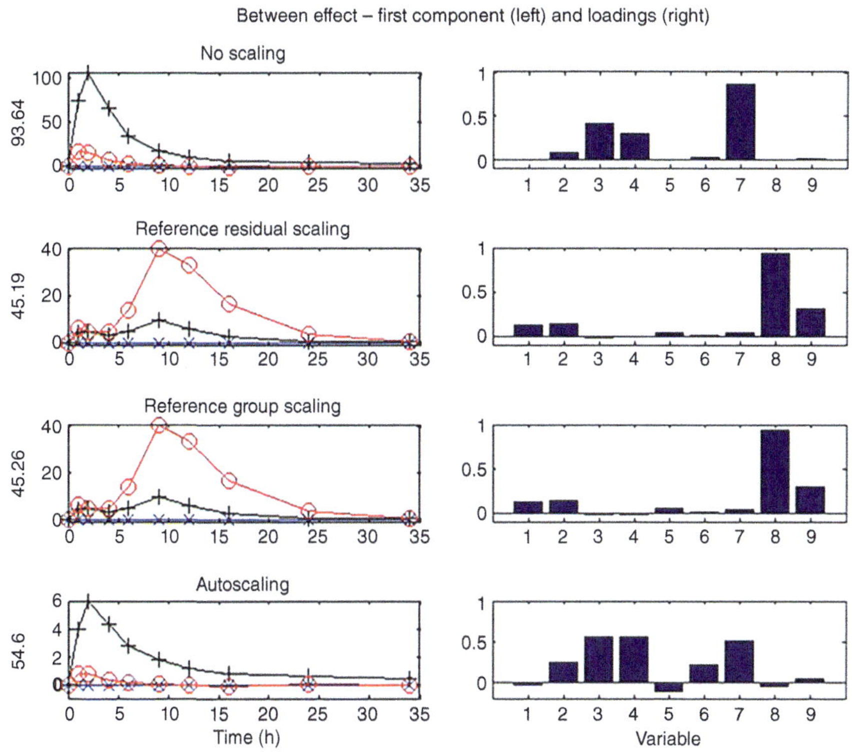

Figure 4.7 The between-effect results for the first component. Left panels: scores as a function of time (x-axis) with percentages of explained variation as numbers on the y-axis; right panel: the corresponding loadings. Source: Timmerman *et al.* (2015)/Springer Nature/CC BY 4.0.

In the scores of the second component, the variables with the early peaks still dominate the no-scaling situation. The variables 3, 4, and 7 are so dominant in terms of variation that they dominate the first two components of the ASCA between-effect model. With reference residual scaling and reference group scaling the variables with the early peaks become visible now. To some extent, this also happens with autoscaling. All in all, for the between-effect the preferred scaling is reference residual scaling or reference group scaling since these clearly show the characteristic features in the data.

It is also interesting to investigate the within model to study individual differences. The results are shown in Figure 4.9. In this figure, the results are shown for no-scaling, reference residual scaling and autoscaling. In the no-scaling case, again variable 7 dominates because of its high variance and its large individual variation. After both autoscaling and reference residual scaling, the first within-component describes the individual differences for variables 8 and 9.

This extensive example shows that scaling in ASCA really makes a difference and should be considered carefully. Making plots of data – before and after the scaling operation – facilitates making proper decisions.

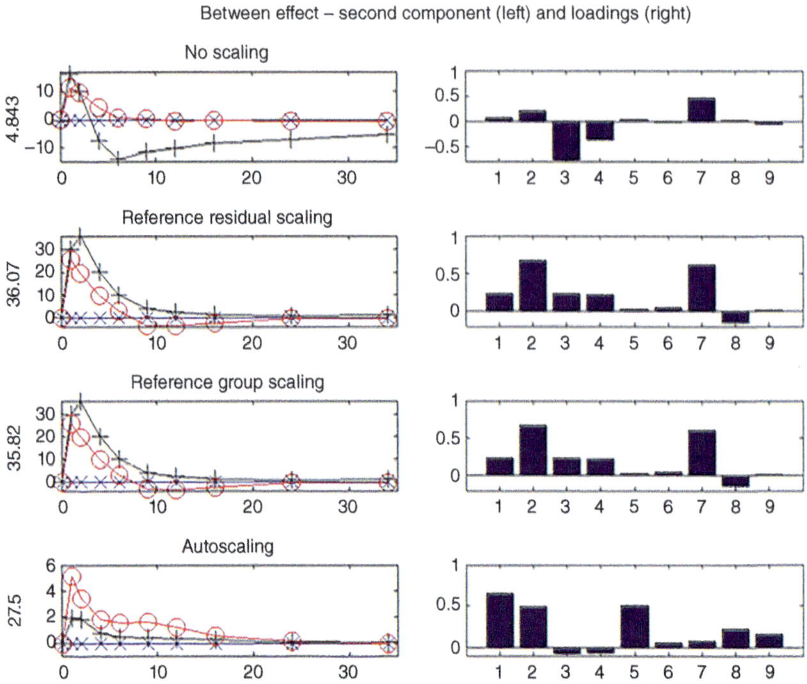

Figure 4.8 The between-effect results for the second component. Left panels: scores as a function of time (x-axis) with percentages of explained variation as numbers on the y-axis; right panel: the corresponding loadings. Source: Timmerman *et al.* (2015)/Springer Nature/CC BY 4.0.

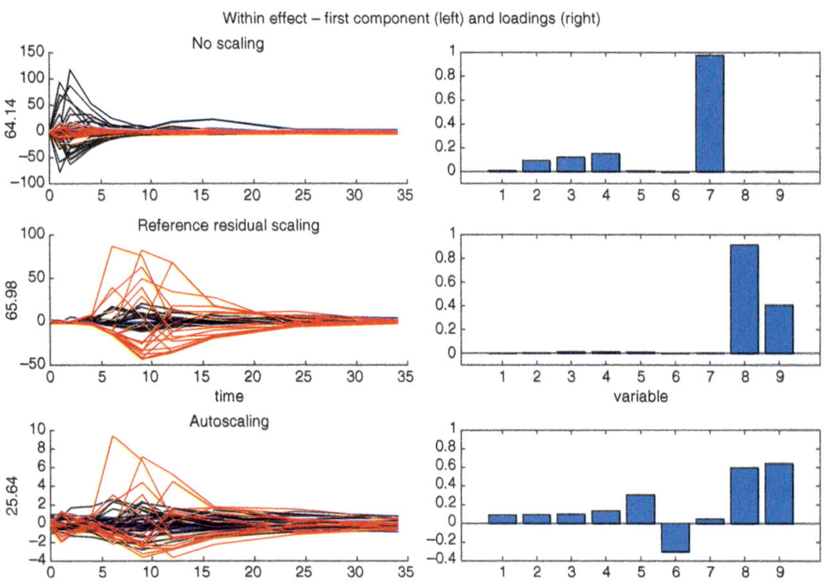

Figure 4.9 The within-effect results for the first component. Left panels: scores as a function of time (x-axis) with percentages of explained variation as numbers on the y-axis; right panel: the corresponding loadings. Source: Timmerman *et al.* (2015)/Springer Nature/CC BY 4.0.

4.1.6 Group-wise ASCA

A drawback of ASCA – which is actually also a drawback of many other methods such as PCA and partial least squares (PLS) – is that the PCA step in ASCA does not provide sparse loadings. This may hamper interpretation as the effect of a certain factor is "smeared out" across many variables in the principal components. This may of course be a reality in certain substantive domains but is oftentimes more an effect of the construct of a linear combination. In Chapter 2 we already addressed this issue to some extent (see Section 2.2.2.4), and we will discuss this issue now in the context of ASCA, specifically regarding the method group-wise ASCA or GASCA for short (Saccenti *et al.*, 2018). GASCA is based on the idea of group-wise PCA, see Algorithm 4.2.

Algorithm 4.2

GROUP-WISE PCA

The group-wise PCA of $\mathbf{Y}(I \times J)$ proceeds as follows:

1) Calculate a summary of the associations between the variables, *e.g.*, in the form of $\mathbf{Y}^t\mathbf{Y}$, where $\mathbf{Y}$ is centered and possibly scaled.
2) Use a clustering method to find K groups of related variables in this summary, *e.g.*, those that within a group have a correlation exceeding a preset threshold.
3) Select for the first loading the group of variables k among the K preselected ones that explain the highest variation in the data set $\mathbf{Y}$.
4) Deflate the data set $\mathbf{Y}$ using the variables in the selected set k.
5) Consider the next principal component using the left-over groups.
6) Stop when a sufficient number of components is available.
7) Plot and interpret the results.

This is the general scheme; for more details we refer to the paper of Camacho *et al.* (2017).

GASCA is basically a group-wise PCA on the factor effect matrices and shown in Algorithm 4.3.

Algorithm 4.3

GROUP-WISE ASCA

The group-wise ASCA of $\mathbf{Y}(I \times J)$ proceeds as follows:

1) Perform an ASCA analysis on $\mathbf{Y}$ generating effect matrices $\overline{\mathbf{Y}}_f$, where the f refers to a main effect, an interaction or the individual contributions.
2) Perform a test to check whether a factor effect matrix has a significant contribution, *e.g.*, by using a permutation test.
3) Summarize the effect matrices $\mathbf{Y}_f$ in suitable association matrices, *e.g.*, in the form of matrices $\overline{\mathbf{Y}}_f^t\overline{\mathbf{Y}}_f$.
4) Use the group-wise PCA method for each effect matrix summary.
5) Plot and interpret the results.

Note that – strictly speaking – the matrices $\overline{\mathbf{Y}}_f^t\overline{\mathbf{Y}}_f$ do not contain correlations but merely associated changes in variables affected by an experimental factor. When there are only a few levels of a factor then such associations should be handled with care. For instance, if there are only two levels for a factor then this will result in correlations of 1 or −1 which is clearly a degenerate case.

Potentially, the GASCA method can give "cleaner" results than using (ordinary) ASCA (Saccenti *et al.*, 2018).

4.1.7 Variable-Selection ASCA

Another way of dealing with the disadvantage of ASCA to generate loadings of the factor estimate matrices that have contributions from all variables is tackled by the Variable-selection ASCA (VASCA) approach (Camacho *et al.*, 2023). The basic idea is to perform a forward variable selection on each factor estimate (or interaction estimate) matrix. An outline of this method is given in Algorithm 4.4.

Algorithm 4.4

VARIABLE-SELECTION ASCA

An outline of the VASCA method is as follows:

1) Perform a regular ASCA model on the data.
2) Select a test statistic S to be used in testing the significance of a factor or interaction effect matrix.
3) Calculate this test statistic S for the effect matrix of interest per variable j and sort the variables in decreasing ranking of these values S_j.
4) Perform a permutation test on this effect matrix with an increasing number of variables sequentially chosen according to their ranking in the previous step.
5) Select the maximum number of variables still being significant.

This algorithm generates a per factor/interaction estimate matrix and a list of variables that are important. There are many details on how to perform the permutations, see Camacho *et al.* (2023).

Several remarks regarding the VASCA method are appropriate:

- The way permutations are performed is along the lines as explained in Section 4.1.3.
- There are several options for selecting the statistic S (and S_j).
- It may be that within the group of selected variables, there are still some unimportant ones. This is the general drawback of a forward selection procedure, and the authors recommend a bootstrap method to diagnose this problem.
- The selected variables are in general not the same for each factor/interaction estimate matrix.

4.1.8 REP-ASCA

The within-group variation of an ASCA model has several sources of variability. The main source is variability between objects or subjects within that group (*e.g.*, biological variability), but there are also other sources such as sampling variability and measurement error. The latter can be related to repeatability and reproducibility, depending on the extent to which the measurement conditions change, *e.g.*, due to differences in the laboratory, equipment, and/or operator. This measurement error can be ascribed to different sources of variability, such as the measurement conditions or the physical properties of the sample, and it can become problematic for the estimation of the significance of effect sizes and their interpretation. There are methods to reduce such measurement variability where the obvious one is to perform more measurements and take averages. This is usually too expensive, and then other methods come into the picture and one of them is Repeatability-ASCA or REP-ASCA for short (Ryckewaert *et al.*, 2020).

The basic idea of REP-ASCA is to try to model the effect that possible sources of unwanted variability may have on the responses so as to be able to remove that contribution from the measured signal (through orthogonalization) prior to ASCA processing. Accordingly, REP-ASCA can be thought of as ASCA modeling on data appropriately preprocessed by a sort of External Parameter Orthogonalization (EPO) (Roger *et al.*, 2003). To do so, it is necessary to have a second set of data, made up of measurements of identical samples performed under similar measurement conditions mimicking the possible variability (*e.g.*, uncontrolled day-to-day temperature or humidity variations) experienced by those to be subjected to ASCA analysis.

Operationally, suppose that $\mathbf{Y}_s$ is a matrix containing a set of measurements performed on samples under the same or very similar repeatability conditions as the data $\mathbf{Y}$, to be subjected to ASCA. There can be multiple of such repeated sets, and each set is centered and then row-wise concatenated into the matrix $\mathbf{W}$. This matrix thus contains the differences among signals collected on samples that, otherwise, should be identical and which therefore could be ascribed to the varying measurement conditions, *i.e.*, to the contributions from the measurement errors due to (lack of) repeatability. Performing a PCA of $\mathbf{W}$ allows estimating loadings $\mathbf{P}$ which identify a subspace that reflects the part of the repeatability error due to the systematic but not controlled sources of variability. Subsequently, the rows of $\mathbf{Y}$ (the original data) are orthogonally projected onto the orthogonal complement of $\mathbf{P}$: $\mathbf{Y}_{\perp} = \mathbf{Y}(\mathbf{I} - \mathbf{P}\mathbf{P}^t)$ which should remove a part of the repeatability error. Next, $\mathbf{Y}_{\perp}$ is used for the subsequent ASCA analysis.

Several remarks are appropriate:

- The number of loadings in $\mathbf{P}$ has to be selected and is thus a meta-parameter.
- The underlying idea behind REP-ASCA is that the variability due to the measurement error is structured and can be fitted by a low-dimensional bilinear model, so it is not clear if and how REP-ASCA could work for cases in which the lack of repeatability is due to purely random error.
- There are some similarities between REP-ASCA and the approach proposed by (Climaco-Pinto *et al.*, 2009) to improve the detection of significant factors in ANOVA-PCA (APCA) (see Section 4.2) through reduction of residual variability. In the latter, after the ANOVA decomposition, the matrix $\mathbf{Y}_{\mathtt{Ind}}$ is subjected to PCA and iteratively deflated of its principal components ($\mathbf{Y}^{(r)}_{\mathtt{Ind}} = \mathbf{Y}_{\mathtt{Ind}} - \mathbf{T}_r\mathbf{P}^t_r$, $\mathbf{T}_r$ and $\mathbf{P}_r$ being the scores and loading matrices for the first r principal components of $\mathbf{Y}_{\mathtt{Ind}}$, respectively). APCA models are then calculated using $\mathbf{Y}^{(r)}_{\mathtt{Ind}}$ as the residual matrix, iteratively increasing the number of components r until factor effects become significant. On the other hand, in REP-ASCA the orthogonalization step is carried out before the ASCA analysis, while in the paper by (Climaco-Pinto *et al.*, 2009) it is an integral part of the APCA analysis. Moreover, in REP-ASCA the loadings for the orthogonal projection are calculated on a separate data set so as not to bias the estimation of effect significance and interpretation.
- An alternative procedure is to use Maximum Likelihood Scaling (MALS) (Hoefsloot *et al.*, 2006). That method uses maximum likelihood PCA to filter the data from noise *a priori*.

These types of general preprocessing of data focused on *a priori* removing noise from the data are not yet very much explored in ASCA-based methods.

4.1.9 ASCA as a Multivariate Multiple Regression Model

Up to now, ASCA has been presented in terms of "classical" ANOVA, where the factor estimates are obtained by calculating group means. In Chapter 3, it was explained how ANOVA models can be written as a regression problem using certain types of coding. Such an approach is more flexible and also works for the case of unbalanced designs (resulting in the ASCA+ method, see later). For simplicity, we will again indicate the total number of samples with the index I which should be understood as the summation across all levels of all factors and the replicates within these levels, which may be unequal in the case of unbalanced designs. The regression approach for a single response variable j can easily be expanded for J responses by concatenating these J responses into a matrix $\mathbf{Y}(I \times J)$ and postulating a multivariate multiple regression model:

$$\mathbf{Y} = \mathbf{X}\boldsymbol{\Theta} + \mathbf{Y}_{\mathtt{Ind}} \tag{4.13}$$

where $\mathbf{Y} = [\mathbf{y}_1 | \ldots | \mathbf{y}_J]$, $\mathbf{X}$ is as before, $\boldsymbol{\Theta}$ contains all parameters to be estimated, and $\mathbf{Y}_{\mathtt{Ind}}$ contains the individual contributions.[2] This is simply a partitioning of the single regression equations and the solution is $\widehat{\boldsymbol{\Theta}} = (\mathbf{X}^t\mathbf{X})^{-}\mathbf{X}^t\mathbf{Y}$. ASCA uses sum-coding (see Section 3.4.1) and in that case the matrix $\mathbf{X}$ has full column-rank; hence, the solution can also be written as $\widehat{\boldsymbol{\Theta}} = (\mathbf{X}^t\mathbf{X})^{-1}\mathbf{X}^t\mathbf{Y}$. The fitted part of Equation 4.13 can be written as $\widehat{\mathbf{Y}} = \mathbf{X}\widehat{\boldsymbol{\Theta}}$ and using a proper partitioning of the matrices $\mathbf{X}$ and $\widehat{\boldsymbol{\Theta}}$ this becomes

$$\widehat{\mathbf{Y}} = [\mathbf{1}_I | \mathbf{X}_{\mathtt{A}} | \mathbf{X}_{\mathtt{B}} | \mathbf{X}_{\mathtt{AB}}] \begin{bmatrix} \widehat{\theta}_{\mathtt{M}} \\ \widehat{\boldsymbol{\Theta}}_{\mathtt{A}} \\ \widehat{\boldsymbol{\Theta}}_{\mathtt{B}} \\ \widehat{\boldsymbol{\Theta}}_{\mathtt{AB}} \end{bmatrix} \tag{4.14}$$

where $\widehat{\theta}_{\mathtt{M}} = (\bar{y}_{\ldots 1}, \ldots, \bar{y}_{\ldots J})$, $\widehat{\boldsymbol{\Theta}}_{\mathtt{A}}(K_{\mathtt{A}} \times J)$ contains the estimated parameters of the Factor $\mathtt{A}$; $\widehat{\boldsymbol{\Theta}}_{\mathtt{B}}(K_{\mathtt{B}} \times J)$ contains the estimated parameters of the Factor $\mathtt{B}$; $\widehat{\boldsymbol{\Theta}}_{\mathtt{AB}}(K_{\mathtt{AB}} \times J)$ contains the estimated parameters of the interaction $\mathtt{AB}$; $\mathbf{X}$ is partitioned as $\mathbf{X}_{\mathtt{A}}(I \times K_{\mathtt{A}})$, $\mathbf{X}_{\mathtt{B}}(I \times K_{\mathtt{B}})$ and $\mathbf{X}_{\mathtt{AB}}(I \times K_{\mathtt{AB}})$. Hence, Equation 4.14 can be rewritten as

$$\begin{aligned} \widehat{\mathbf{Y}} &= \mathbf{1}_I\widehat{\theta}_{\mathtt{M}} + \mathbf{X}_{\mathtt{A}}\widehat{\boldsymbol{\Theta}}_{\mathtt{A}} + \mathbf{X}_{\mathtt{B}}\widehat{\boldsymbol{\Theta}}_{\mathtt{B}} + \mathbf{X}_{\mathtt{AB}}\widehat{\boldsymbol{\Theta}}_{\mathtt{AB}} \\ &= \widehat{\mathbf{Y}}_{\mathtt{M}} + \widehat{\mathbf{Y}}_{\mathtt{A}} + \widehat{\mathbf{Y}}_{\mathtt{B}} + \widehat{\mathbf{Y}}_{\mathtt{AB}} \end{aligned} \tag{4.15}$$

which is essentially the same as Equation 4.3 because of the use of sum-coding. The fitted values $\widehat{\mathbf{Y}}_f$ are sometimes called Least Squares means (LS-means) and are equivalent to the values $\overline{\mathbf{Y}}_f$ as shown in Equation 4.3. Note that this only holds for the case of balanced designs, and also note that this equation does not involve the individual level as does Equation 4.3. This shows the equivalence of the ANOVA and regression approaches. Thus, the next step in ASCA is then to perform PCA models on the factor estimates $\mathbf{X}_{\mathtt{A}}\widehat{\boldsymbol{\Theta}}_{\mathtt{A}}$, $\mathbf{X}_{\mathtt{B}}\widehat{\boldsymbol{\Theta}}_{\mathtt{B}}$, and $\mathbf{X}_{\mathtt{AB}}\widehat{\boldsymbol{\Theta}}_{\mathtt{AB}}$. A more elaborate treatment of the ASCA-regression approach will be given in Section 4.3.

4.1.10 Geometry of ASCA

The geometry of ASCA can be understood in terms of centering operations. Such centering operations have different geometrical interpretations in the column- and row-spaces of matrices. This will be explained first and is largely based on Smilde *et al.* (2008).

An illustration of the centering operation is already shown in Elaboration 2.1 in both the row- and column-space. A more formal treatment goes as follows. For the matrix $\mathbf{Y}(I \times J)$ the column

2 For simplicity we do not distinguish unobserved errors from (estimated) residuals.

means can be collected in $\bar{\mathbf{y}}^t = (\bar{y}_1, \dots, \bar{y}_J)$. It holds that $\bar{y}_j = \mathbf{1}^t\mathbf{y}_j/I$ where $\mathbf{y}_j$ is the jth column of $\mathbf{Y}$.[3] Hence, $\bar{\mathbf{y}}^t = \mathbf{1}^t\mathbf{Y}/I$ and thus the centered matrix $\mathbf{Y} - \mathbf{1}\bar{\mathbf{y}}^t$ can be written as $\mathbf{Y} - \mathbf{1}\mathbf{1}^t\mathbf{Y}/I = (\mathbf{I} - \mathbf{1}\mathbf{1}^t/I)\mathbf{Y}$ which is the basic equation for a centering operation (see Example 4.5).

This centering matrix $(\mathbf{I} - \mathbf{1}\mathbf{1}^t/I)$ is clearly symmetric, and it can easily be checked to be also idempotent. Hence, it is a projection matrix. To understand the projection it is easier to consider the matrix $\mathbf{1}\mathbf{1}^t/I$ which is also symmetric and idempotent. This matrix $\mathbf{1}\mathbf{1}^t/I$ projects orthogonally on the vector of ones and thus provides the means of each column. Hence, the centering operation projects on the orthogonal complement of this vector of ones and that is exactly what a centering operation is supposed to do: remove the constants. This is what is happening in the column-space of $\mathbf{Y}$ since the centering matrix works on this column-space (see $(\mathbf{I} - \mathbf{1}\mathbf{1}^t/I)\mathbf{Y}$).

To understand what is happening in the row-space of $\mathbf{Y}$ we can rewrite $\mathbf{Y}_C = (\mathbf{I} - \mathbf{1}\mathbf{1}^t/I)\mathbf{Y}$ as $\mathbf{Y}_C = \mathbf{Y} - \mathbf{A}$ where $\mathbf{A}$ is a matrix constructed from the column-means. Hence, from each row, $\mathbf{y}_1^t$ in $\mathbf{Y}$ a row of $\mathbf{A}$ is subtracted and this is a translation. The translation is such that afterward the values of the rows are spread around zero (see Elaboration 2.1).

The column-centering operation has a different meaning in the column- and row-spaces. This is also clear when considering the properties of $\mathbf{Y}_C$. It holds that $R(\mathbf{Y}_C^t) \subseteq R(\mathbf{Y}^t)$[4], but for the column-space this is different. If $\mathbf{1} \in R(\mathbf{Y})$ then $R(\mathbf{Y}_C) \subseteq R(\mathbf{Y})$ otherwise $R(\mathbf{Y}_C) \nsubseteq R(\mathbf{Y})$ (Smilde *et al.*, 2008). This is also understandable from a projection point of view: when vectors in $R(\mathbf{Y})$ are projected onto a vector in that same subspace, then the residuals are also in that subspace.

Example 4.5: Centering as a matrix operation

For the simple example as used in Elaboration 2.1 the original matrix was

$$\mathbf{Y} = \begin{bmatrix} 3 & 1 \\ 1 & 3 \\ 5 & 2 \end{bmatrix} \tag{4.16}$$

and the column means are 3 and 2, respectively. Now the centering matrix is

$$\begin{bmatrix} \frac{2}{3} & \frac{-1}{3} & \frac{-1}{3} \\ \frac{-1}{3} & \frac{2}{3} & \frac{-1}{3} \\ \frac{-1}{3} & \frac{-1}{3} & \frac{2}{3} \end{bmatrix} \tag{4.17}$$

and

$$\begin{bmatrix} \frac{2}{3} & \frac{-1}{3} & \frac{-1}{3} \\ \frac{-1}{3} & \frac{2}{3} & \frac{-1}{3} \\ \frac{-1}{3} & \frac{-1}{3} & \frac{2}{3} \end{bmatrix} \begin{bmatrix} 3 & 1 \\ 1 & 3 \\ 5 & 2 \end{bmatrix} = \begin{bmatrix} 0 & -1 \\ -2 & 1 \\ 2 & 0 \end{bmatrix} \tag{4.18}$$

which is the centered matrix (see Elaboration 2.1).

The geometry of ASCA can be explained in the row-space and in the column-space.

4.1.10.1 Geometry of ASCA in Row-Space

The geometry of ASCA in row-space will be explained with a small example in which rats are subjected to a treatment (Factor A) with two levels, and measurements of J metabolites in the

3 In this section the column of ones always has size I.

4 $\mathbf{Y}_C^t = \mathbf{Y}^t(\mathbf{I} - \mathbf{1}\mathbf{1}^t/I) = \mathbf{Y}^t\mathbf{V}$ where $\mathbf{V}$ is implicitly defined.

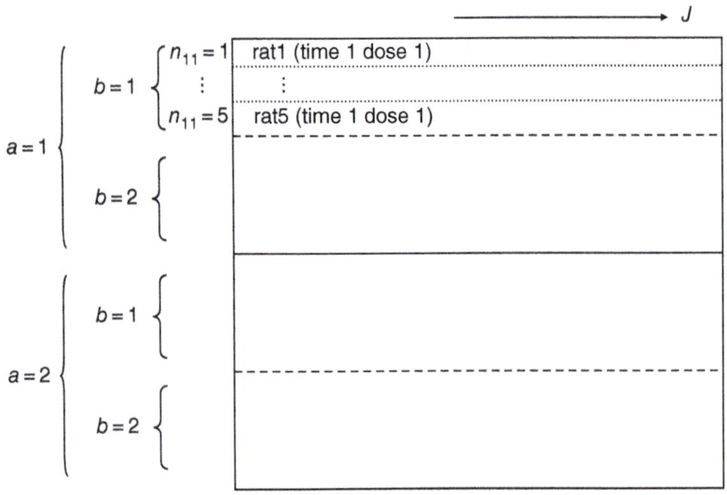

Figure 4.10 Arrangement of the data of the rat example in a matrix.

blood are performed at two time points (Factor B). Per group, there are five rats and these are sacrificed (hence, it is not a repeated measures design). The data set is then arranged as shown in Figure 4.10. For simplicity, we consider only two metabolites, but the argument carries over to the J metabolites. The data can now be plotted as in Figure 4.11. Each cross in this figure represents a measurement performed on one rat, and the treatment groups are indicated with the characters "a" and "b." There is clearly an effect of both factors and there is also an interaction. The crosses within a group show the biological correlation within a group. These correlations can be aligned between the groups (panel a) or can be different (panel b). Please note the resemblance with Figure 2.12 which shows (again) the problem of calculating correlations across all

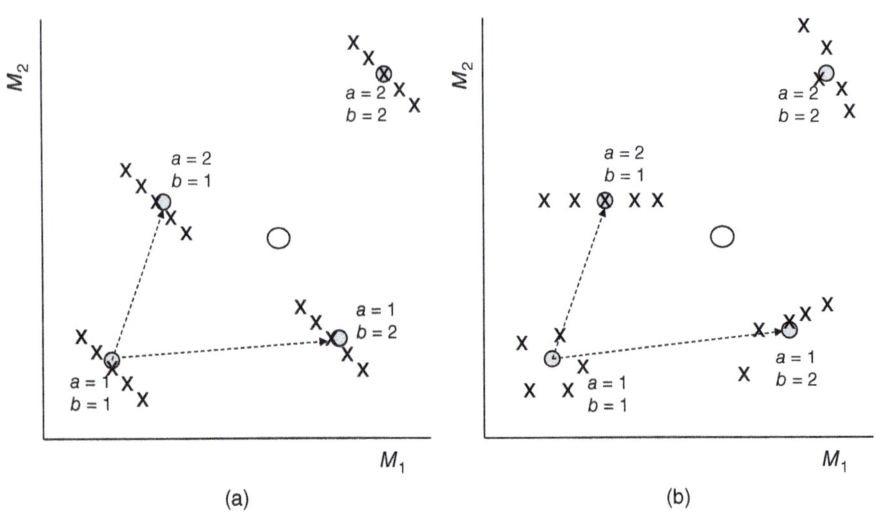

Figure 4.11 Plot of the data of the rat example. Each cross represents an individual rat; the open circle represents the overall mean of the data; the closed dots represent the group means. Panel (a): similar within-group correlation, panel (b): different within-group correlation. a and b are the levels of the Factors A and B, respectively. Source: Reproduced from Smilde *et al.* (2008)/with permission of John Wiley & Sons.

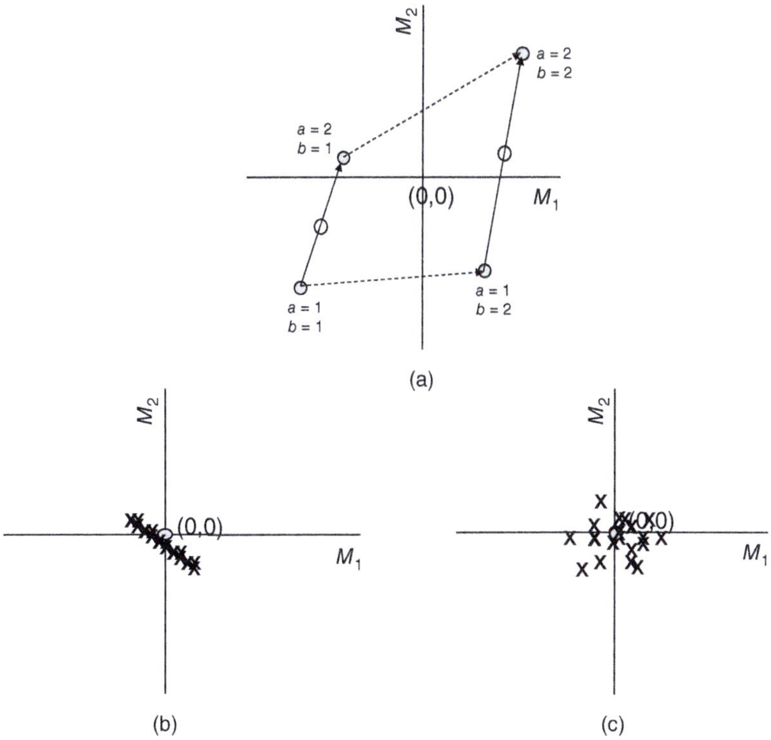

Figure 4.12 First step in ASCA: overall centering and centering per group. The group-centers (closed dots) are shown in panel (a) (the open circles in this panel show the means for the Factor B); the individual variation for the two cases is shown in panels (b) and (c). Source: Reproduced from Smilde *et al.* (2008)/with permission of John Wiley & Sons.

measurements. The data as shown in Figure 4.11 are the row-space representation, and the first step in ASCA is an overall mean-centering which translates the whole data configuration to the center of the data. Next, averages per treatment group are calculated, and the results are then visualized in Figure 4.12. This figure shows the separation of $\mathbf{Y}$ in $\hat{\mathbf{Y}}$ (panel a) and $\mathbf{Y}_{\texttt{Ind}}$ (panels b and c) where $\hat{\mathbf{Y}}$ contains the contributions of the factors. The situation of similar within-group correlations (Figure 4.12, panel b) shows that a PCA on $\mathbf{Y}_{\texttt{Ind}}$ is a sensible thing to do (White *et al.*, 2024).

The next step in ASCA is to separate the effects of the factors *Time* and *Treatment*. This is again a centering operation where the centering is performed across the levels of Factor B to arrive at the Factor A estimates and vice versa (see Figure 4.13). Note that these effects are the same for both situations of Figure 4.11.

The last part is estimating the interaction effect. This is shown in Figure 4.14 and is simply done by subtracting the factor estimates (right part of the figure) from $\hat{\mathbf{Y}}$ (left part of the figure). These figures and the explanation around them show that there is an intimate relationship between ASCA (and ANOVA), centering operations, projections, and separation of subspaces.

4.1.10.2 Geometry of ASCA in Column-Space

The geometry of ASCA in terms of column-spaces goes as follows. ASCA uses sum-coding, and then it can be shown that the hat matrix $\mathbf{X}(\mathbf{X}^t\mathbf{X})^{-1}\mathbf{X}^t$ of the regression model of Equation 4.13

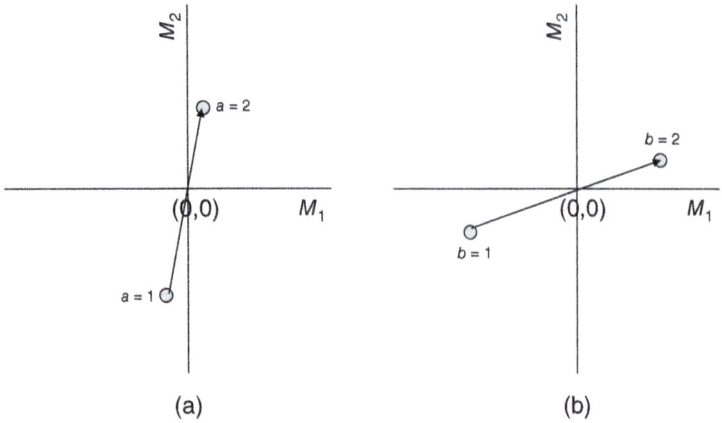

Figure 4.13 Estimating the main factor effects by taking averages across factors. Panel (a): estimated effects for Factor A (by averaging across Factor B); panel (b): estimated effects for Factor B (by averaging across Factor A). Source: Reproduced from Smilde *et al.* (2008)/with permission of John Wiley & Sons.

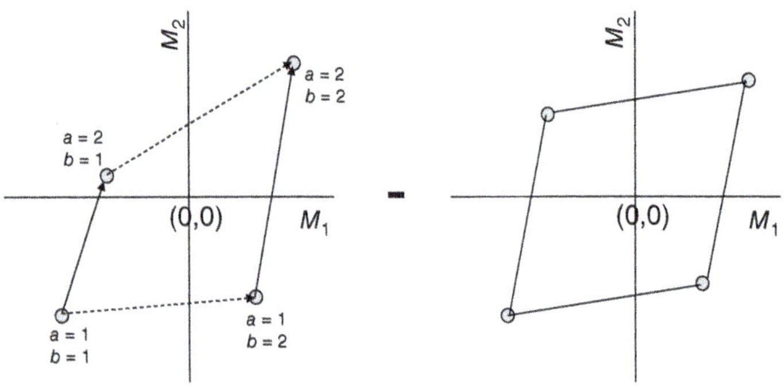

Figure 4.14 Estimating the interaction effect by subtracting the main effects from $\hat{\mathbf{Y}}$. Source: Reproduced from Smilde *et al.* (2008)/with permission of John Wiley & Sons.

can be written as

$$\mathbf{X}(\mathbf{X}^t\mathbf{X})^{-1}\mathbf{X}^t = \mathbf{Q}_{\mathrm{M}} + \mathbf{Q}_{\mathrm{A}} + \mathbf{Q}_{\mathrm{B}} + \mathbf{Q}_{\mathrm{AB}} \tag{4.19}$$

with

$$\begin{aligned}
\mathbf{Q}_{\mathrm{M}} &= \mathbf{1}_I(\mathbf{1}_I^t\mathbf{1}_I)^{-1}\mathbf{1}_I^t \\
\mathbf{Q}_{\mathrm{A}} &= \mathbf{X}_{\mathrm{A}}(\mathbf{X}_{\mathrm{A}}^t\mathbf{X}_{\mathrm{A}})^{-1}\mathbf{X}_{\mathrm{A}}^t \\
\mathbf{Q}_{\mathrm{B}} &= \mathbf{X}_{\mathrm{B}}(\mathbf{X}_{\mathrm{B}}^t\mathbf{X}_{\mathrm{B}})^{-1}\mathbf{X}_{\mathrm{B}}^t \\
\mathbf{Q}_{\mathrm{AB}} &= \mathbf{X}_{\mathrm{AB}}(\mathbf{X}_{\mathrm{AB}}^t\mathbf{X}_{\mathrm{AB}})^{-1}\mathbf{X}_{\mathrm{AB}}^t
\end{aligned} \tag{4.20}$$

and these matrices are symmetric, idempotent (thus projection matrices, see Section 2.1.4.2), and orthogonal to each other (see Section 4.7 and Smilde *et al.* (2012)) because of the properties of $\mathbf{X}$ using sum-coding. Thus $\mathbf{Y}$ can also be written as:

$$\begin{aligned}
\mathbf{Y} &= \mathbf{X}(\mathbf{X}^t\mathbf{X})^{-1}\mathbf{X}^t\mathbf{Y} + (\mathbf{I} - \mathbf{X}(\mathbf{X}^t\mathbf{X})^{-1}\mathbf{X}^t)\mathbf{Y} \\
&= \mathbf{Q}_{\mathrm{M}}\mathbf{Y} + \mathbf{Q}_{\mathrm{A}}\mathbf{Y} + \mathbf{Q}_{\mathrm{B}}\mathbf{Y} + \mathbf{Q}_{\mathrm{AB}}\mathbf{Y} + \mathbf{Q}_{\mathrm{Ind}}\mathbf{Y}
\end{aligned} \tag{4.21}$$

where $\mathbf{Q}_{\mathrm{Ind}} = (\mathbf{I} - \mathbf{X}(\mathbf{X}^t\mathbf{X})^{-1}\mathbf{X}^t)$, and this matrix is also symmetric, idempotent, and orthogonal to the other $\mathbf{Q}$ matrices. The interpretation of the $\mathbf{Q}$ matrices can be made in terms of certain centering operations. An example of the projection matrices is given in Example 4.6. Note that this analysis only works for the case of balanced data.

Example 4.6: ASCA in terms of projection matrices

As an example consider a simple case where $A = 2, B = 2, N = 2$, and $J = 2$. The matrix $\mathbf{X}$ for this case using sum-coding is:

$$\begin{bmatrix} 1 & -1 & -1 & 1 \\ 1 & -1 & -1 & 1 \\ 1 & -1 & 1 & -1 \\ 1 & -1 & 1 & -1 \\ 1 & 1 & -1 & -1 \\ 1 & 1 & -1 & -1 \\ 1 & 1 & 1 & 1 \\ 1 & 1 & 1 & 1 \end{bmatrix} \tag{4.22}$$

and $\mathbf{Y}$ is

$$\begin{bmatrix} y_{1111} & y_{1112} \\ y_{1121} & y_{1122} \\ y_{1211} & y_{1212} \\ y_{1221} & y_{1222} \\ y_{2111} & y_{2112} \\ y_{2121} & y_{2122} \\ y_{2211} & y_{2212} \\ y_{2221} & y_{2222} \end{bmatrix} \tag{4.23}$$

where the first index in y is for the level of Factor A; the second index is for the level of Factor B; the third index is for the replicate n; and the final index is for the response.

The projection matrix $\mathbf{Q}_A$ can be obtained by working out Equation 4.20 for this situation, and this becomes:

$$\frac{1}{8}\begin{bmatrix} 1 & 1 & 1 & 1 & -1 & -1 & -1 & -1 \\ 1 & 1 & 1 & 1 & -1 & -1 & -1 & -1 \\ 1 & 1 & 1 & 1 & -1 & -1 & -1 & -1 \\ 1 & 1 & 1 & 1 & -1 & -1 & -1 & -1 \\ -1 & -1 & -1 & -1 & 1 & 1 & 1 & 1 \\ -1 & -1 & -1 & -1 & 1 & 1 & 1 & 1 \\ -1 & -1 & -1 & -1 & 1 & 1 & 1 & 1 \\ -1 & -1 & -1 & -1 & 1 & 1 & 1 & 1 \end{bmatrix} \tag{4.24}$$

It is illustrative to work out $\mathbf{Q}_A\mathbf{Y}$:

$$\frac{1}{8}\begin{bmatrix} y_{1..1} - y_{2..1} & y_{1..2} - y_{2..2} \\ y_{1..1} - y_{2..1} & y_{1..2} - y_{2..2} \\ y_{1..1} - y_{2..1} & y_{1..2} - y_{2..2} \\ y_{1..1} - y_{2..1} & y_{1..2} - y_{2..2} \\ y_{2..1} - y_{1..1} & y_{2..2} - y_{1..2} \\ y_{2..1} - y_{1..1} & y_{2..2} - y_{1..2} \\ y_{2..1} - y_{1..1} & y_{2..2} - y_{1..2} \\ y_{2..1} - y_{1..1} & y_{2..2} - y_{1..2} \end{bmatrix} \tag{4.25}$$

which is equal to

$$\begin{bmatrix} \bar{y}_{1..1}-\bar{y}_{2..1} & \bar{y}_{1..2}-\bar{y}_{2..2} \\ \bar{y}_{1..1}-\bar{y}_{2..1} & \bar{y}_{1..2}-\bar{y}_{2..2} \\ \bar{y}_{1..1}-\bar{y}_{2..1} & \bar{y}_{1..2}-\bar{y}_{2..2} \\ \bar{y}_{1..1}-\bar{y}_{2..1} & \bar{y}_{1..2}-\bar{y}_{2..2} \\ \bar{y}_{2..1}-\bar{y}_{1..1} & \bar{y}_{2..2}-\bar{y}_{1..2} \\ \bar{y}_{2..1}-\bar{y}_{1..1} & \bar{y}_{2..2}-\bar{y}_{1..2} \\ \bar{y}_{2..1}-\bar{y}_{1..1} & \bar{y}_{2..2}-\bar{y}_{1..2} \\ \bar{y}_{2..1}-\bar{y}_{1..1} & \bar{y}_{2..2}-\bar{y}_{1..2} \end{bmatrix} \tag{4.26}$$

which is a centering operation, and this generates the (concatenated) solution of the sum-coding, see Section 3.4.1.

4.2 APCA

4.2.1 Basic Idea of APCA

The APCA approach will be explained using the example of a balanced design with two fixed factors. The basic starting model is the same as that of ASCA (see Equation 4.3) (Harrington *et al.*, 2005; Harrington *et al.*, 2006). After that, it starts to deviate from ASCA by performing a PCA on the factor matrices to which the matrix of individual variation is added. Hence, PCAs are performed on $\overline{\mathbf{Y}}_{\mathrm{A}} + \mathbf{Y}_{\mathtt{Ind}}$, $\overline{\mathbf{Y}}_{\mathrm{B}} + \mathbf{Y}_{\mathtt{Ind}}$, and $\overline{\mathbf{Y}}_{\mathrm{AB}} + \mathbf{Y}_{\mathtt{Ind}}$. This is different from the back-projection method as described in Section 4.3.1 since in APCA the individual variation is added *before* the PCA and in the back-projection approach *after* a PCA on $\overline{\mathbf{Y}}_{\mathrm{A}}$, $\overline{\mathbf{Y}}_{\mathrm{B}}$, and $\overline{\mathbf{Y}}_{\mathrm{AB}}$ (see also Section 4.2.2).

The advantage of APCA is that the PCA-score plot of the factor estimates made in this way can immediately give an impression of the variability and thus significance of the factor/interaction. This can be made a bit more precise by adding confidence bounds around the scores in each class (*i.e.*, factor levels for Factor A) in the factor estimate plots by using the Mahanalobis distance:

$$T^2_{an} = (\mathbf{t}_{an} - \bar{\mathbf{t}}_a)^t \mathbf{C}_a^{-1} (\mathbf{t}_{an} - \bar{\mathbf{t}}_a) \tag{4.27}$$

where $\mathbf{t}_{an}$ is the PCA score vector of replicate n at factor level a, $\bar{\mathbf{t}}_a$ are the mean score levels at factor level a (the factor estimates), and $\mathbf{C}_a$ is the covariance matrix of the scores on this level. A critical value for the value of T^2_{an} is

$$T^2_{crit} = \frac{2(N-1)(N+1)}{N(N-2)} F(\alpha, 2, N-2) \tag{4.28}$$

assuming a score plot of two principal components; N is the number of replicates per cell; F is the F-distribution; and α is the significance level. Note that Equation 4.28 is an approximation since the degrees of freedom for the F-distribution are not exactly known in this case. An example of APCA is given in Example 4.7.

Example 4.7: APCA applied to the plants science study of Arabidopsis

This example is a continuation of the ASCA example on the plant science study of *Arabidopsis* (Caldana *et al.* (2011)). *Time* and the combined *Light* + *Light:Time* interaction are included in the APCA, similar to the ASCA example and APCA also uses sum-coding for all effects. The variation explained by the combined effect (28.18%) is exactly the same as for ASCA. In Figure 4.15, loadings and scores for the combined effect are shown for APCA. We can see that the first APCA component is very similar to the first ASCA component (Figure 4.2), even though the loadings

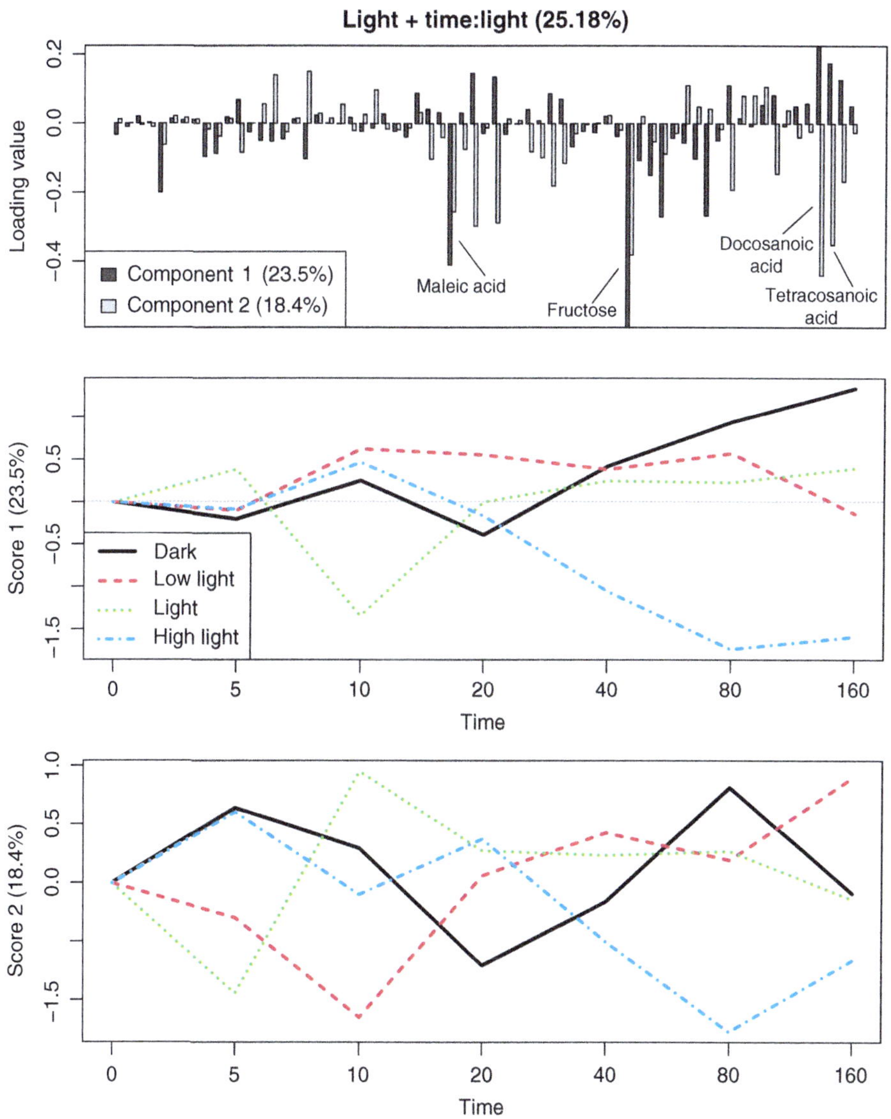

Figure 4.15 APCA analysis of metabolomics data. Source: Adapted from Caldana *et al.* (2011).

weigh the variables a bit differently. The first APCA component is dominated by fructose (like ASCA), but exchanges glucose with maleic acid. As in ASCA, the first APCA scores start at zero effect and show similar grouping as time progresses with *Dark* on top, *Low light* and *Light* in the middle, and *High light* clearly separated below. The explained variance of the first component of APCA is 23.5% which is considerably lower than that of ASCA (38.9%).

The second component of APCA (18.4% explained variance) clearly captures a different phenomenon than ASCA and is less interpretable than the second ASCA component (16.1% explained variance). The focus is shifted from *o*-acetyl-serine to docosanoic acid and tetracosanoic acid.

4.2.2 Comparing APCA with ASCA

The APCA and ASCA methods are very similar, but there are differences. A simulation study performed according to a balanced design and two fixed factors showed the following (Zwanenburg *et al.*, 2011):

1) The power of ASCA is higher than that of APCA.
2) ASCA retrieves true underlying factors better than APCA.

The first item shows that ASCA is better capable of finding small effects than APCA. This depends, of course, on the amount of individual variation and if this is small, then the differences are minor. For biological data, usually the amount of individual variation is large, and then ASCA might be advantageous. The second item is understandable since in ASCA individual variation is averaged within a group prior to performing PCA, whereas in APCA a PCA is performed on the factor estimates to which the individual variation is added. Some details are given in Elaboration 4.3.

ELABORATION 4.3

Comparing ASCA and APCA

In a simulation study, ASCA was compared with APCA in terms of power and retrieving underlying loadings (Zwanenburg *et al.*, 2011). A two-factor crossed ANOVA model was simulated with only the two main effects. Factor A was varied at four levels and Factor B at three levels; there were five replicates per cell. The factor effect differences for Factor A was 2 between all levels, and for Factor B this was only 1. There was only one component in the multivariate setting with 110 randomly generated loadings $\mathbf{p}_A$ and $\mathbf{p}_B$ where $\mathbf{p}_A^t\mathbf{p}_B \neq 0$. Random noise was added to the data to generate the replicates, and the level of this noise was manipulated by a variable δ where a value of $\delta = 1$ means that the signal-to-noise ratio for the effect of Factor A was 2 and for Factor B this was 1. Subsequently, ASCA and APCA models were fitted, and the significance of the factor effect estimates was assessed using permutation testing.

Figure 4.16 shows the performance of ASCA and APCA for the simulated example. For both factors, ASCA has a slightly better performance than APCA: for larger noise levels, ASCA has lower *p*-values for estimating effects than APCA and thus is better at finding small effects. This is due to the noise-averaging ability of ASCA (taking averages within cells).

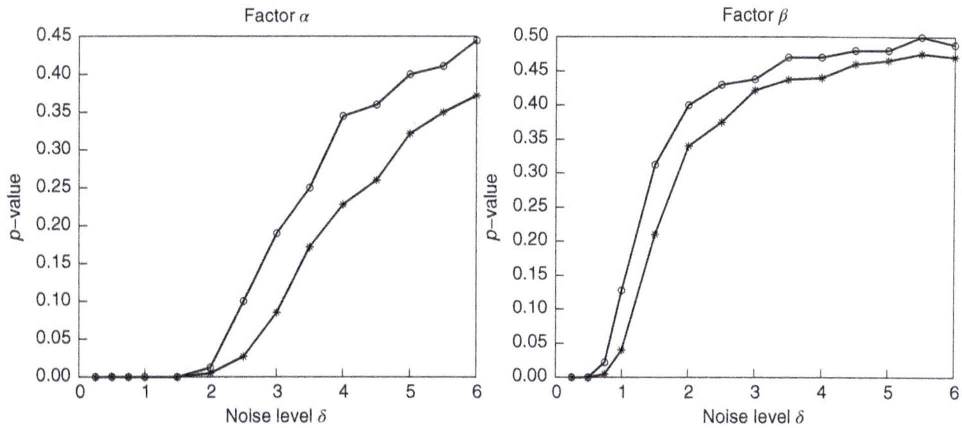

Figure 4.16 Performance plots for the Factors A (α, left) and B (β, right). On the x-axis the noise level δ and on the y-axis the p-values. Open circles: APCA; stars: ASCA. Source: Reproduced from Zwanenburg *et al.* (2011)/with permission of John Wiley & Sons.

4.3 ASCA+

In Section 4.1.9 we discussed ASCA in the context of a regression model. For the unbalanced case the standard ANOVA calculations of factor estimates using means as shown in Equation 4.2 give biased estimates of the true effects, and the more general approach based on regression is needed, see Elaboration 3.10.

The regression approach to ASCA models is called ASCA+, and this approach can also deal with the case of unbalanced data (Thiel *et al.*, 2017). The general ASCA+ algorithm is shown in Algorithm 4.5.

Algorithm 4.5

ASCA IN REGRESSION TERMS: ASCA+

The ASCA model in terms of a regression model (ASCA+) can be summarized in the following steps:

1) Setup the design matrix $\mathbf{X}$ for the regression using sum-coding.
2) Perform the multivariate multiple regression of Equation 4.13 for an appropriately centered and scaled response matrix $\mathbf{Y}$.
3) Perform PCAs on the effect matrices such as $\mathbf{X}_A\hat{\mathbf{\Theta}}_A$; plot and interpret the results.

The ASCA+ model then proceeds along the same line as ASCA: PCAs on factor estimates are performed. There are a few things that do not generalize that nicely from the balanced case. To start, the breakdown of sum-of-squares as shown in Equation 4.6 does not hold anymore. Subsequently, percentages of explained variation cannot be expressed easily. In line with the methodology of Type III sum of squares (see Elaboration 3.4), the following definition can be adopted:

$$Var_f = \frac{||\mathbf{E}_{/f}||^2 - ||\mathbf{E}_{full}||^2}{||\mathbf{Y} - \hat{\mathbf{Y}}_M||^2} \times 100\% \tag{4.29}$$

which gives the explained variance of factor f in percentages; $||\mathbf{E}_{full}||^2$ refers to the sum of squared residuals of the full model (*i.e.*, the individual variation); $||\mathbf{E}_{/f}||^2$ is the sum-of-squares of the full model excluding the factor f (which can be a main effect or an interaction). The sum of these percentages across all factors and interactions does not add up to 100% due to the unbalancedness. Equation 4.29 becomes Equation 4.8 in case of balanced data. Significance testing can be done using permutations much in the spirit of ASCA models.

Summarizing, we will use the term ASCA for an analysis based on the traditional ANOVA way of calculating effect sizes using means as can be done for balanced designs and the term ASCA+ for the regression formulation of such models. The latter then also includes analyses of data from unbalanced designs. Example 3.4 showed the use of WE-Coding; this can of course also be used in ASCA+. Note that the same remarks regarding factor codings in univariate models (see Section 3.A.3) hold in the case of multivariate data.

4.3.1 Confidence Ellipsoids for ASCA

Whereas the confidence ellipsoids based on back-projection are *data* ellipsoids, there are also *model* ellipsoids representing the uncertainty around factor effect estimates (Friendly *et al.*, 2013). This results from using the multivariate multiple regression formulation of the ASCA model, see Equation 4.13. We will first show the general theory and subsequently show how this can be applied in the ASCA context using the within correlation or correlation of the individual variation (considered as errors in the following). The next exposition is largely based on Mardia *et al.* (1979) and Liland *et al.* (2018). Upon assuming a multivariate normal distribution for the errors $\mathbf{E}$ (*i.e.*, $\mathbf{e}_i \sim N(\mathbf{0}, \mathbf{\Sigma})$ where $\mathbf{e}_i$ is a row of $\mathbf{E}$), the first step is to define the distribution of $\hat{\mathbf{\Theta}}$. It can be shown that for the jth and lth columns of $\hat{\mathbf{\Theta}}$ it holds that

$$cov(\hat{\theta}_j, \hat{\theta}_l) = \hat{\sigma}_{jl}(\mathbf{X}^t\mathbf{X})^{-1} \tag{4.30}$$

where $\hat{\sigma}_{jl}$ is the jlth element of $\hat{\mathbf{\Sigma}}$. The $\hat{\theta}_j$ and $\hat{\theta}_l$ contain the factor estimates of the responses j and l, respectively. This means that a linear combination of these estimated parameters has the following covariance:

$$cov(\mathbf{x}^t\hat{\theta}_j, \mathbf{x}^t\hat{\theta}_l) = \mathbf{x}^t\hat{\sigma}_{jl}(\mathbf{X}^t\mathbf{X})^{-1}\mathbf{x} \tag{4.31}$$

and if we consider the ith row of $\mathbf{X}$ (*i.e.*, $\mathbf{x}_i$) then it can be shown, due to the special block structure of $\mathbf{X}$ as dictated by the ANOVA model, that $\mathbf{x}_i^t\hat{\theta}_j = \hat{y}_{ij}$ and $\mathbf{x}_i^t\hat{\theta}_l = \hat{y}_{il}$. Hence, in that case the $cov(\mathbf{x}_i^t\hat{\theta}_j, \mathbf{x}_i^t\hat{\theta}_l) = cov(\hat{y}_{ij}, \hat{y}_{il})$. This holds for two responses but can be written also for all J responses simultaneously, by using $\hat{\mathbf{y}}_i^t = \mathbf{x}_i^t\hat{\mathbf{\Theta}}$. Then it holds that

$$cov(\hat{\mathbf{y}}_i) = \mathbf{x}_i^t(\mathbf{X}^t\mathbf{X})^{-1}\mathbf{x}_i\hat{\mathbf{\Sigma}} = c_i\hat{\mathbf{\Sigma}} \tag{4.32}$$

where c_i is implicitly defined. It can now be shown that

$$\frac{\mathbf{x}_i^t\hat{\mathbf{\Theta}}}{\sqrt{c_i}} \sim N\left(\frac{\mathbf{x}_i^t\mathbf{\Theta}}{\sqrt{c_i}}, \mathbf{\Sigma}\right) \tag{4.33}$$

which can thus be used to build confidence ellipsoids around $\widehat{\boldsymbol{\Theta}}$ for $\boldsymbol{\Theta}$:

$$\frac{I-K}{Ic_i}(\mathbf{x}_i^t\widehat{\boldsymbol{\Theta}} - \mathbf{x}_i^t\boldsymbol{\Theta})\widehat{\boldsymbol{\Sigma}}^{-1}(\mathbf{x}_i^t\widehat{\boldsymbol{\Theta}} - \mathbf{x}_i^t\boldsymbol{\Theta})^t \sim T^2(J, I-K) \tag{4.34}$$

where K is the number of columns in $\mathbf{X}$. Upon choosing a significance level α this equation can be turned into a confidence ellipsoid for $\boldsymbol{\Theta}$ and $T^2(J, I-K)$ is the Hotelling T^2 distribution.

To use this machinery to obtain confidence ellipsoids around the factor estimates we need to take four extra steps. First, we assume balanced designs. Secondly, note that the residuals $\mathbf{E}$ are now the individual contributions in the ASCA model, *i.e.*, $\mathbf{Y}_{\mathtt{Ind}}$ (note that we have to assume that the individual contributions all have the same covariance matrix). Thirdly, we have to use the loading matrices of the PCAs performed on the factor estimate matrices (*e.g.*, on $\widehat{\mathbf{Y}}_A$) to move from factor estimates to their scores in these PCAs. We will illustrate this for a Factor $\mathtt{A}$ which has loadings $\mathbf{P}_{\mathtt{A}}$ for the PCA of its effect matrix $\widehat{\mathbf{Y}}_A$. The confidence ellipsoid of Equation 4.34 then changes to

$$\frac{I-K}{Ic_i}(\mathbf{x}_i^t\widehat{\boldsymbol{\Theta}}\mathbf{P}_{\mathtt{A}} - \mathbf{x}_i^t\boldsymbol{\Theta}\mathbf{P}_{\mathtt{A}})(\mathbf{P}_{\mathtt{A}}^t\widehat{\boldsymbol{\Sigma}}\mathbf{P}_{\mathtt{A}})^{-1}(\mathbf{x}_i^t\widehat{\boldsymbol{\Theta}}\mathbf{P}_{\mathtt{A}} - \mathbf{x}_i^t\boldsymbol{\Theta}\mathbf{P}_{\mathtt{A}})^t \sim T^2(J, I-K) \tag{4.35}$$

and the next step is to focus on the level a of Factor $\mathtt{A}$ by using a $\mathbf{x}_i$ which has zeros everywhere except for the part corresponding to this level. We also need to choose the dimension of the ellipsoid, which is d (mostly $d = 2$ or 3). We translate from the Hotelling's T^2 distribution to the F-distribution using the general equation $T^2(p, m) = (mp/(m-p+1))F_{p,m-p+1}$. In our case, $m = I - K_{\mathtt{A}}$ (where $K_{\mathtt{A}}$ is the number of columns in $\mathbf{X}_{\mathtt{A}}$) and $p = d$. Then we must scale our ellipsoid with

$$g = \sqrt{\frac{(I-K_{\mathtt{A}})d}{I-K_{\mathtt{A}}-d+1}F(1-\alpha, d, I-K_{\mathtt{A}}-d+1)} \tag{4.36}$$

to arrive at a confidence ellipsoid of the desired level of confidence $1-\alpha$. Currently, these ellipsoids are only available for the balanced case, *i.e.*, ASCA and their behavior in the unbalanced case is not yet known.

An example of the use of data and model ellipsoids is given in Example 4.8. The crucial difference between a data and a model ellipsoid is that for a growing number of observations per group, the data ellipsoid remains almost the same whereas the model ellipsoid is shrinking. The reason is that for larger N, the F-distribution becomes more narrow resulting in a lower value for the F-cutoff value and thus in a lower value of g in Equation 4.36. This aligns with the intuition that more samples per group make the factor estimates more reliable.

Example 4.8: Confidence ellipsoids in candy assessment

The candy dataset was first introduced with a brief ASCA analysis in Example 1.6. It contains sensory assessments of 5 different candies by 11 trained assessors. Each assessor has judged 9 sensory attributes 3 times for each candy on a scale from 0 to 15. The dataset has 165 rows ($5 \times 11 \times 3$) and 2 factors: *Assessor* and *Candy*.

We run two models on the dataset, one with only main effects (assessment ~ *Assessor* + *Candy*) and one also including interactions (assessment ~ *Assessor* + *Candy* + *Assessor:Candy*). As expected, the assessor effect is small, while the candy effect is large with 4% and 74% explained variance, respectively. There is also a small interaction effect (8%), meaning that the assessors rate the candies differently (Table 4.1).

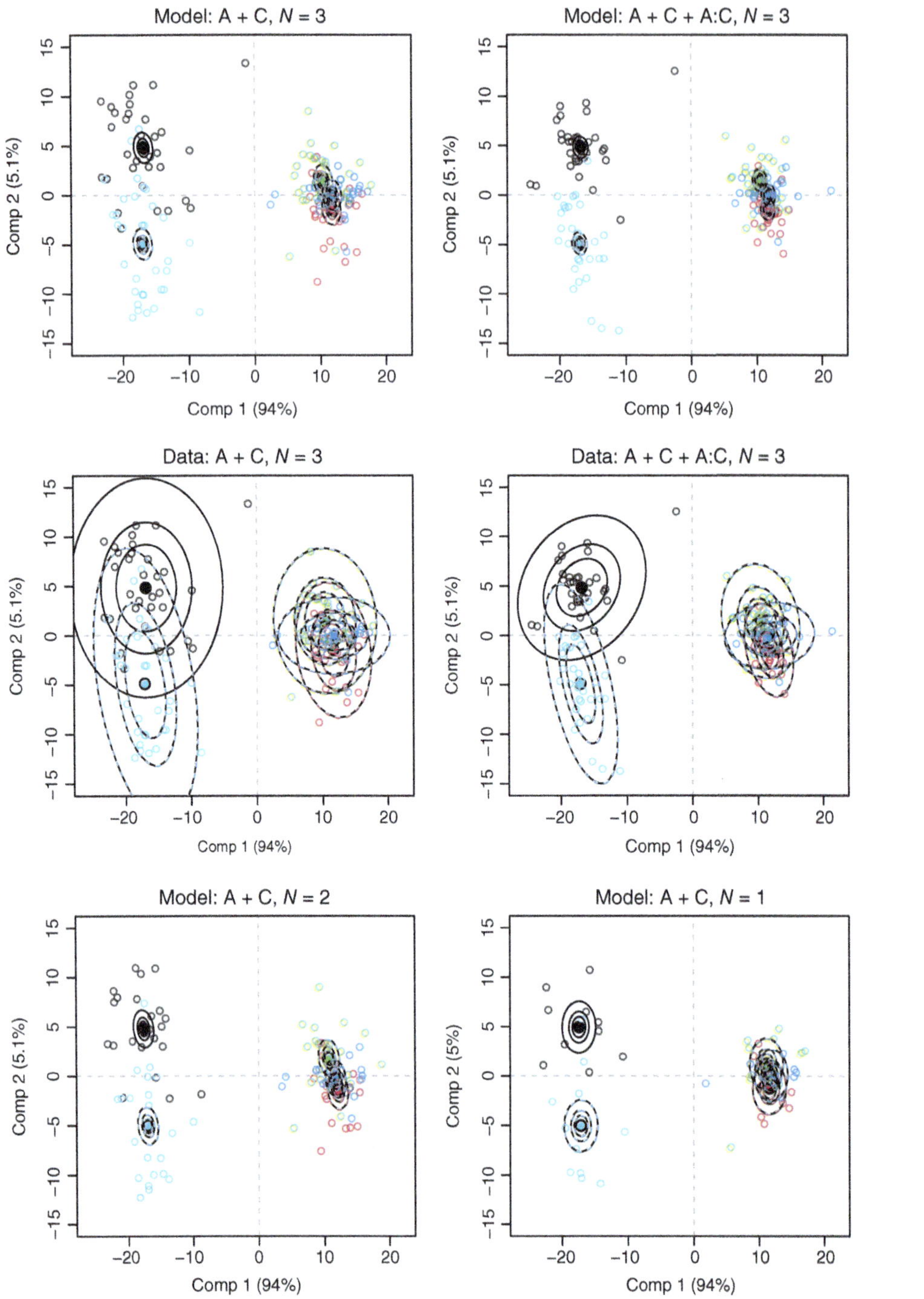

Figure 4.17 Score plot with back-projected samples, model ellipsoids, and data ellipsoids as analyzed by ASCA. The upper four sub-plots use are based on models using all three replicates, while the two lower sub-plots use $N = 2$ and $N = 1$ samples, respectively. A is *assessor* and C is *candy*.

Table 4.1 ASCA SS and explained variance per effect for the candy data.

Source	SS	R^2 (%)
Assessor	1961.37	4.37
Candy	33416.66	74.48
Assessor:Candy	3445.73	7.68
Error	6043.52	13.47

The score plots in Figure 4.17 show factor level means for the candy factor and back-projected observations for the two models. On top of these, we have added model ellipsoids and data ellipsoids.

The difference between data ellipsoids and model ellipsoids is evident as the model ellipsoids are all equal in shape and size while differing in location and do not cover the variation of the back-projected points. We observe that the two candies on the left are significantly different from the candies on the right due to their ellipsoids not overlapping each others's mean value. The two candies on the left are also easily distinguishable, while the three candies on the right are less obviously different. The effect of including the interaction term in the model can be observed by the reduced spread in points and reduced sizes of the ellipsoids. A reduction of spread is expected as the residuals are shrunken when the additional effect is included. As the calculations of the ellipsoids also rely on the residuals, this naturally manifests in smaller ellipsoids too. Changes in explained variances are so small that the rounding of percentages hides them. Finally, the effect of the number of replicates is illustrated by removing the 3rd or the 2nd and 3rd replicates from each factor level combination. This manifests in slight shifts in the mean values (especially the group on the right), larger confidence ellipsoids in the lower part of Figure 4.17, and slightly reduced explained variances.

4.3.2 ASCA and ASCA+ as RDA Models

ASCA can also be described as a special case of redundancy analysis (RDA) (see Section 2.2.2.7). For simplicity, we will first describe this for the one-way balanced fixed effect HD-ANOVA case using sum-coding. This model is then:

$$\mathbf{Y} = \mathbf{X}_{\mathrm{M}}\hat{\theta}_{\mathrm{M}} + \mathbf{X}_{\mathrm{A}}\hat{\mathbf{\Theta}}_{\mathrm{A}} + \mathbf{Y}_{\mathtt{Ind}} \tag{4.37}$$

where $\mathbf{X}_{\mathrm{M}} = \mathbf{1}_{AN}$. Now, since $\mathbf{X}_{\mathrm{M}}^t\mathbf{X}_{\mathrm{A}} = 0$ we can consider ASCA as an RDA model and focus on the part of $\mathbf{X}_{\mathrm{A}}\hat{\mathbf{\Theta}}_{\mathrm{A}}$ separately. The next step in the ASCA model is to perform a PCA on this matrix:

$$\mathbf{X}_{\mathrm{A}}\hat{\mathbf{\Theta}}_{\mathrm{A}} = \mathbf{T}_{\mathrm{A}}\mathbf{P}_{\mathrm{A}}^t + \mathbf{E}_{\mathrm{A}} \tag{4.38}$$

where only a limited number of principal components are taken (say, R_{A}), and thus $\mathbf{T}_{\mathrm{A}}(I \times R_{\mathrm{A}})$ are the scores and $\mathbf{P}_{\mathrm{A}}(J \times R_{\mathrm{A}})$ are the (orthonormal) loadings. Post-multiplying both sides of Equation 4.38 with $\mathbf{P}_{\mathrm{A}}$ gives:

$$\mathbf{X}_{\mathrm{A}}\hat{\mathbf{\Theta}}_{\mathrm{A}}\mathbf{P}_{\mathrm{A}} = \mathbf{T}_{\mathrm{A}} \tag{4.39}$$

since $\mathbf{E}_{\mathrm{A}}$ is orthogonal to $\mathbf{P}_{\mathrm{A}}$ because of Equation 4.38. Substituting $\mathbf{T}_{\mathrm{A}}$ in Equation 4.38 gives then

$$\mathbf{X}_{\mathrm{A}}\hat{\mathbf{\Theta}}_{\mathrm{A}} = \mathbf{X}_{\mathrm{A}}\hat{\mathbf{\Theta}}_{\mathrm{A}}\mathbf{P}_{\mathrm{A}}\mathbf{P}_{\mathrm{A}}^t + \mathbf{E}_{\mathrm{A}} \tag{4.40}$$

where now the term $\widehat{\boldsymbol{\Theta}}_{\mathrm{A}}\mathbf{P}_{\mathrm{A}}\mathbf{P}_{\mathrm{A}}^t$ contains the rank-reduced regression coefficients (see Section 2.2.2.7).

Similarly, for the balanced case with more than one factor, the different parts of the design matrix $\mathbf{X}$ related to the factors are orthogonal when using sum-coding. Thus, these can be written as separate RDA-problems per factor or interaction.

The ASCA+ model can also be cast in the framework of RDA, but that is more complicated than in the case of balanced data, see Elaboration 4.4.

ELABORATION 4.4

RDA for ASCA+

Suppose that a two-way HD-ANOVA model is under consideration with fixed effects and an unbalanced design. The model can again be written as

$$\widehat{\mathbf{Y}} = \mathbf{X}_{\mathrm{M}}\widehat{\theta}_{\mathrm{M}} + \mathbf{X}_{\mathrm{A}}\widehat{\boldsymbol{\Theta}}_{\mathrm{A}} + \mathbf{X}_{\mathrm{B}}\widehat{\boldsymbol{\Theta}}_{\mathrm{B}} + \mathbf{X}_{\mathrm{AB}}\widehat{\boldsymbol{\Theta}}_{\mathrm{AB}} \tag{4.41}$$

but in general the different sub-matrices in $\mathbf{X} = [\mathbf{X}_{\mathrm{M}}|\mathbf{X}_{\mathrm{A}}|\mathbf{X}_{\mathrm{B}}|\mathbf{X}_{\mathrm{AB}}]$ are not orthogonal anymore. Hence, we have to work with partial RDA, see Algorithm 2.2. This is not without problems, see Section 7.2.3.1.

4.4 Principal Response Curves

A method closely related to ASCA is Principal Response Curves (PRC) which is used in ecology (Van den Brink and Ter Braak, 1999; Smilde *et al.*, 2012). The basic idea of PRC is to express deviations relative to the control group. Hence, it is an instance of reference-coding where the control group is the reference. For a model with the Factor *Time* $a = 1, \dots, A$, *Treatment* $b = 1, \dots, B$ (with $b = 1$ for the control group) and $n = 1, \dots, N$ repeats per group, the PRC model for a single response is as follows:

$$y_{abn} = \alpha_{a1} + (\alpha\beta)_{ab} + \epsilon_{abn} \tag{4.42}$$

and in terms of coding this comes down to using full-coding for the Factor *Time* and reference-coding (relative to the control group) for the *Treatment*. The restrictions $\widehat{\mu} = 0$ and $(\widehat{\alpha\beta})_{a1} = 0;\ \forall a$ are used and the main effect for *Treatment* is incorporated in the term $(\alpha\beta)_{ab}$. The interpretation of the model is now that the estimates of the parameters α_{a1} are the effects of the control group at time points $a = 1, \dots, A$ and $(\alpha\beta)_{ab}$ is then the added effect due to treatment $b \neq 1$ at time point a. Elaboration 4.5 gives an example of the coding of a PRC model.

ELABORATION 4.5

Coding in PRC

We will illustrate the coding of PRC with a simple example with three time points ($a = 1, 2, 3$), two treatments ($b = 1, 2$), and no repeats. For a univariate model the response vector is then $\mathbf{y} = (y_{11}, y_{21}, y_{31}, y_{12}, y_{22}, y_{32})^t$ with time the fastest running index. The full-coding part for the Factor *Time* is

$$\mathbf{X}_{\mathrm{A}} = \begin{bmatrix} 1 & 0 & 0 \\ 0 & 1 & 0 \\ 0 & 0 & 1 \\ 1 & 0 & 0 \\ 0 & 1 & 0 \\ 0 & 0 & 1 \end{bmatrix} \tag{4.43}$$

and the reference-coding for the *Treatment* is

$$\mathbf{X}_{\mathrm{B}} = \begin{bmatrix} 0 \\ 0 \\ 0 \\ 1 \\ 1 \\ 1 \end{bmatrix} \tag{4.44}$$

where the first treatment level (control group) is the reference level. After taking the Hadamard products of the columns of $\mathbf{X}_{\mathrm{A}}$ and $\mathbf{X}_{\mathrm{B}}$ this results in the coding for the *Treatment:Time* interaction:

$$\mathbf{X}_{\mathrm{AB}} = \begin{bmatrix} 0 & 0 & 0 \\ 0 & 0 & 0 \\ 0 & 0 & 0 \\ 1 & 0 & 0 \\ 0 & 1 & 0 \\ 0 & 0 & 1 \end{bmatrix} \tag{4.45}$$

and the full design matrix is now $\mathbf{X} = [\mathbf{X}_{\mathrm{A}}|\mathbf{X}_{\mathrm{AB}}]$ and the model can be written as $\mathbf{y} = \mathbf{X}_{\mathrm{A}}\hat{\boldsymbol{\theta}}_1 + \mathbf{X}_{\mathrm{AB}}\hat{\boldsymbol{\theta}}_2 + \mathbf{y}_{\mathtt{Ind}}$. The parameter vectors can be written as $\hat{\boldsymbol{\theta}}_1 = (\hat{\theta}_{11}, \hat{\theta}_{12}, \hat{\theta}_{13})^t$ and $\hat{\boldsymbol{\theta}}_2 = (\hat{\theta}_{21}, \hat{\theta}_{22}, \hat{\theta}_{23})^t$. The matrix $\mathbf{X}^t\mathbf{X}$ is:

$$\mathbf{X}^t\mathbf{X} = \left[\begin{array}{ccc|ccc} 2 & 0 & 0 & 1 & 0 & 0 \\ 0 & 2 & 0 & 0 & 1 & 0 \\ 0 & 0 & 2 & 0 & 0 & 1 \\ \hline 1 & 0 & 0 & 1 & 0 & 0 \\ 0 & 1 & 0 & 0 & 1 & 0 \\ 0 & 0 & 1 & 0 & 0 & 1 \end{array}\right] \tag{4.46}$$

which is highly structured and full rank but not orthogonal. The matrix $\mathbf{X}^t\mathbf{y}$ equals

$$\mathbf{X}^t\mathbf{y} = \begin{bmatrix} y_{11} + y_{12} \\ y_{21} + y_{22} \\ y_{31} + y_{32} \\ y_{12} \\ y_{22} \\ y_{32} \end{bmatrix} \tag{4.47}$$

Solving the normal equations $(\mathbf{X}^t\mathbf{X})\hat{\theta} = \mathbf{X}^t\mathbf{y}$ where the parameter vector $\hat{\theta}$ collects $\hat{\theta}_1$ and $\hat{\theta}_2$ gives:

$$\begin{aligned}
\hat{\theta}_{11} &= y_{11} \\
\hat{\theta}_{12} &= y_{21} \\
\hat{\theta}_{13} &= y_{31} \\
\hat{\theta}_{21} &= y_{12} - y_{11} \\
\hat{\theta}_{22} &= y_{22} - y_{21} \\
\hat{\theta}_{23} &= y_{32} - y_{31}
\end{aligned} \tag{4.48}$$

which shows that indeed the $\hat{\mu} = 0$, and the effects are expressed relative to the control treatment ($b = 1$).

In the multivariate context using J variables the PRC model then becomes

$$y_{abnj} = \alpha_{a1j} + (\alpha\beta)_{abj} + \epsilon_{abnj} \tag{4.49}$$

and constraints must be invoked to obtain unique estimates. These constraints in this case are $\widehat{(\alpha\beta)}_{a1j} = 0; (\forall a, j)$; and $\hat{\mu}_j = 0; (\forall j)$ and assuming a balanced design this generates the decomposition:

$$y_{abnj} = \bar{y}_{a1.j} + (\bar{y}_{ab.j} - \bar{y}_{a1.j}) + (\bar{y}_{abnj} - \bar{y}_{ab.j}) \tag{4.50}$$

and thus the subsequent effect estimates can be collected in the matrices $\overline{\mathbf{Y}}_{\mathrm{A}}$ and $\overline{\mathbf{Y}}_{\mathrm{AB}}$ giving:

$$\mathbf{Y} = \overline{\mathbf{Y}}_{\mathrm{A}} + \overline{\mathbf{Y}}_{\mathrm{AB}} + \mathbf{Y}_{\mathtt{Ind}} \tag{4.51}$$

which is in the spirit of ASCA. Note that the matrices $\overline{\mathbf{Y}}_{\mathrm{A}}$ and $\overline{\mathbf{Y}}_{\mathrm{AB}}$ are not orthogonal. Analysis can now proceed by performing a PCA on $\overline{\mathbf{Y}}_{\mathrm{AB}}$ and plotting the results as time curves for each treatment, the so-called principal response curves (PRCs).

The PRC model can also be written as a multivariate regression model which then also holds for the case of unbalanced designs:

$$\mathbf{Y} = \mathbf{X}_{\mathrm{A}}\hat{\mathbf{\Theta}}_{\mathrm{A}} + \mathbf{X}_{\mathrm{AB}}\hat{\mathbf{\Theta}}_{\mathrm{AB}} + \mathbf{Y}_{\mathtt{Ind}} \tag{4.52}$$

where $\hat{\mathbf{\Theta}}_{\mathrm{A}}$ carries all the parameters related to the main effect time (for all responses) and $\hat{\mathbf{\Theta}}_{\mathrm{AB}}$ has all the parameters related to interactions except the ones for the control treatment (for all responses). Thus, the PRC model can also be written as an RDA problem while reducing the dimension of $\mathbf{X}_{\mathrm{AB}}\hat{\mathbf{\Theta}}_{\mathrm{AB}}$; but in this case, a partial RDA is required since $\mathbf{X}_{\mathrm{A}}$ is not orthogonal to $\mathbf{X}_{\mathrm{AB}}$. The PRC model is a least squares solution, but since $\mathbf{X}_{\mathrm{A}}$ is not orthogonal to $\mathbf{X}_{\mathrm{AB}}$ the explained variance of the term associated with $\mathbf{X}_{\mathrm{AB}}$ is an extra sum of squares after having accounted for the $\mathbf{X}_{\mathrm{A}}$ term (see also Elaboration 3.4).

There are several diagnostics to validate the PRC model. The first one is the amount of explained variation of the Factor `A` and the (combined) interaction `AB` using the extra sum-of-squares principle explained above. It is also possible to perform a permutation test for assessing the significance of the interaction `AB` (Van den Brink and Ter Braak, 1999). Moreover, bootstrap confidence intervals can be calculated for the principal response curves (Timmerman *et al.*, 2007). An example of PRC is given in Example 4.9.

Example 4.9: Principal response curves

We applied the PRC model to the same plant science data as earlier in the ASCA and APCA examples. The results are shown in Figure 4.18, and we can see that PRC uses the first level of light as a baseline to contrast the other levels. Except for this (and a random sign flip in the first component), the results of PRC are very similar to the ASCA results (Figure 4.2), including the loading values. The same compounds dominate the components, and component-wise explained variances are the same as for ASCA, though with a slightly lower explained variance for the combined effect (21.92% vs 25.18% for ASCA). Also the second component is very similar after having postprocessed the second ASCA component by expressing the profiles relative to the first level of *Light* as baseline.

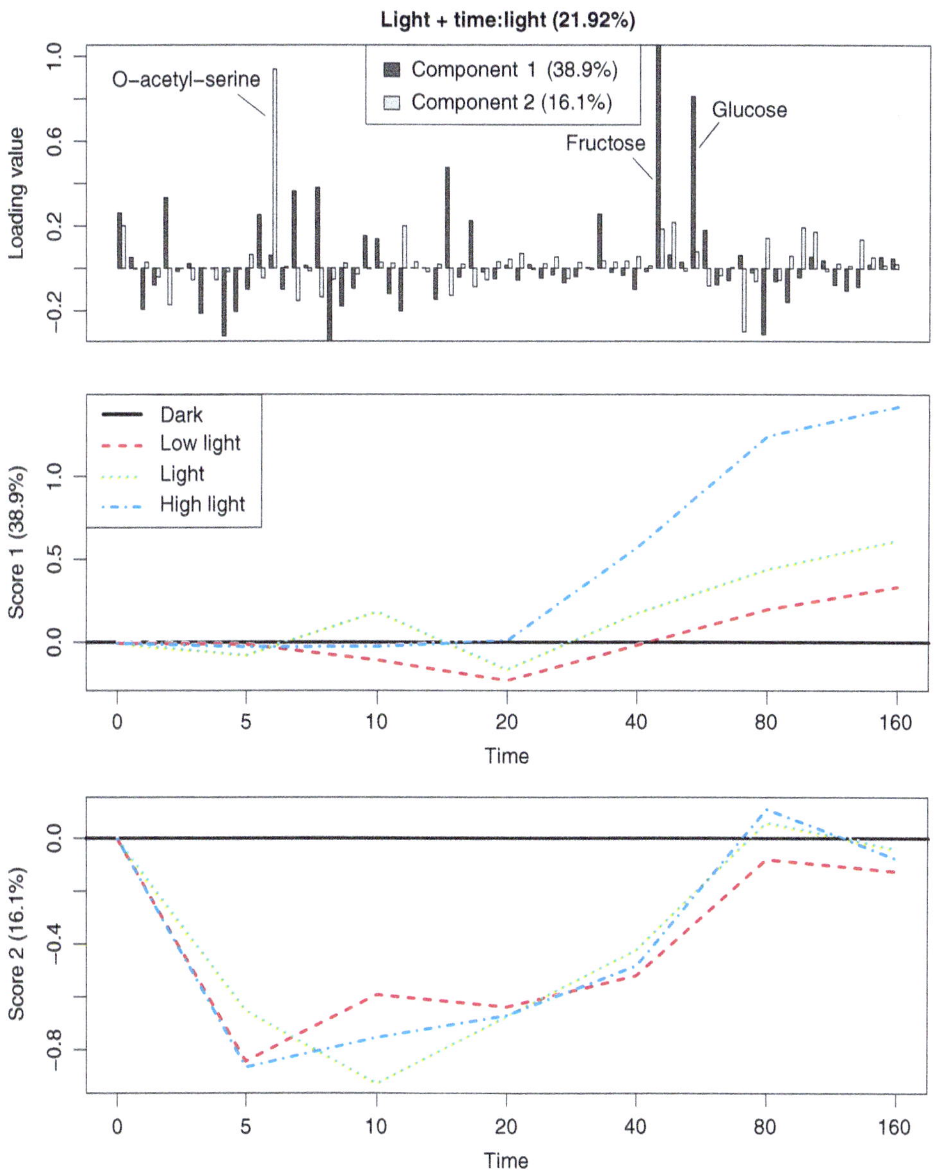

Figure 4.18 PRC analysis of metabolomics data. Source: Adapted from Caldana *et al.* (2011).

4.5 SMART

In metabolomics, another version of an HD-ANOVA has been published and is called scaled-to-maximum and aligned reduced trajectories or SMART for short (Keun *et al.*, 2004). Using again the example of a model with the Factor *Time* $a = 1, \ldots, A$, *Treatment* $b = 1, \ldots, B$ (with $b = 1$ for the control group), and $n = 1, \ldots, N$ repeats per group, the basic model of SMART assuming a balanced design is

$$y_{abn} = \beta_{1b} + (\alpha\beta)_{ab} + \epsilon_{abn} \tag{4.53}$$

and in terms of coding this comes down to using full-coding for the Factor *Treatment* and reference-coding (relative to the first time point) for the Factor *Time*. It is assumed that $\widehat{(\alpha\beta)}_{1b} = 0; (\forall b)$ and $\hat{\mu} = 0$ and the main effect for time is absorbed in the terms $(\alpha\beta)_{ab}$. The interpretation of the model is now that the estimates of the parameters β_{1b} are the effects at time point 1 for the treatments $b = 1, \ldots, B$, and $(\alpha\beta)_{ab}$ is then the added effect at time point $a \neq 1$ due to treatment b. Elaboration 4.6 gives an example of the coding of a SMART model.

ELABORATION 4.6

Coding in SMART

We will illustrate the SMART model using the same simple example as for PRC. The full-coding part for the Factor *Treatment* is

$$\mathbf{X}_\text{B} = \begin{bmatrix} 1 & 0 \\ 1 & 0 \\ 1 & 0 \\ 0 & 1 \\ 0 & 1 \\ 0 & 1 \end{bmatrix} \tag{4.54}$$

and the reference-coding for the *Time* is

$$\mathbf{X}_\text{A} = \begin{bmatrix} 0 & 0 \\ 1 & 0 \\ 0 & 1 \\ 0 & 0 \\ 1 & 0 \\ 0 & 1 \end{bmatrix} \tag{4.55}$$

where the first time point is the reference level. After taking the Hadamard products of the columns of $\mathbf{X}_\text{A}$ and $\mathbf{X}_\text{B}$ this results in the coding for the *Treatment:Time* interaction:

$$\mathbf{X}_\text{AB} = \begin{bmatrix} 0 & 0 & 0 & 0 \\ 1 & 0 & 0 & 0 \\ 0 & 1 & 0 & 0 \\ 0 & 0 & 0 & 0 \\ 0 & 0 & 1 & 0 \\ 0 & 0 & 0 & 1 \end{bmatrix} \tag{4.56}$$

which then results in the design matrix

$$\mathbf{X} = [\mathbf{X}_{\mathrm{B}}|\mathbf{X}_{\mathrm{AB}}] = \left[\begin{array}{cc|cccc} 1 & 0 & 0 & 0 & 0 & 0 \\ 1 & 0 & 1 & 0 & 0 & 0 \\ 1 & 0 & 0 & 1 & 0 & 0 \\ 0 & 1 & 0 & 0 & 0 & 0 \\ 0 & 1 & 0 & 0 & 1 & 0 \\ 0 & 1 & 0 & 0 & 0 & 1 \end{array}\right] \tag{4.57}$$

and the model now becomes $\mathbf{y} = \mathbf{X}_{\mathrm{B}}\hat{\boldsymbol{\theta}}_1 + \mathbf{X}_{\mathrm{AB}}\hat{\boldsymbol{\theta}}_2 + \mathbf{y}_{\mathtt{Ind}}$. The parameter vectors can be written as $\hat{\theta}_1 = (\hat{\theta}_{11}, \hat{\theta}_{12})$ and $\hat{\theta}_2 = (\hat{\theta}_{21}, \hat{\theta}_{22}, \hat{\theta}_{23}, \hat{\theta}_{24})^t$. The matrix $\mathbf{X}^t\mathbf{X}$ is:

$$\mathbf{X}^t\mathbf{X} = \left[\begin{array}{cc|cccc} 3 & 0 & 1 & 1 & 0 & 0 \\ 0 & 3 & 0 & 0 & 1 & 1 \\ \hline 1 & 0 & 1 & 0 & 0 & 0 \\ 1 & 0 & 0 & 1 & 0 & 0 \\ 0 & 1 & 0 & 0 & 1 & 0 \\ 0 & 1 & 0 & 0 & 0 & 1 \end{array}\right] \tag{4.58}$$

and the matrix $\mathbf{X}^t\mathbf{y}$ equals

$$\mathbf{X}^t\mathbf{y} = \begin{bmatrix} y_{11} + y_{21} + y_{31} \\ y_{12} + y_{22} + y_{32} \\ y_{21} \\ y_{31} \\ y_{22} \\ y_{32} \end{bmatrix} \tag{4.59}$$

and, finally, the solution for the normal equations is:

$$\begin{aligned} \hat{\theta}_{11} &= y_{11} \\ \hat{\theta}_{12} &= y_{12} \\ \hat{\theta}_{21} &= y_{21} - y_{11} \\ \hat{\theta}_{22} &= y_{31} - y_{11} \\ \hat{\theta}_{23} &= y_{22} - y_{12} \\ \hat{\theta}_{24} &= y_{32} - y_{12} \end{aligned} \tag{4.60}$$

which shows that the effects are expressed relative to the first time point.

In the multivariate context using J variables the SMART model then becomes

$$y_{abnj} = \beta_{1bj} + (\alpha\beta)_{abj} + \epsilon_{abnj} \tag{4.61}$$

and using the constraints $(\widehat{\alpha\beta}_{1bj}) = 0; (\forall b, j)$ and $\hat{\mu}_j = 0; (\forall j)$ thereby generates the decomposition (assuming a balanced design):

$$y_{abnj} = \bar{y}_{1b.j} + (\bar{y}_{ab.j} - \bar{y}_{1b.j}) + (\bar{y}_{abnj} - \bar{y}_{ab.j}) \tag{4.62}$$

and thus the subsequent effect estimates can be collected in the matrices $\overline{\mathbf{Y}}_{\mathtt{B}}$ and $\overline{\mathbf{Y}}_{\mathtt{AB}}$ giving:

$$\mathbf{Y} = \overline{\mathbf{Y}}_{\mathtt{B}} + \overline{\mathbf{Y}}_{\mathtt{AB}} + \mathbf{Y}_{\mathtt{Ind}} \tag{4.63}$$

which is again in the spirit of ASCA. Note again that the matrices $\overline{\mathbf{Y}}_{\mathtt{B}}$ and $\overline{\mathbf{Y}}_{\mathtt{AB}}$ are not orthogonal. Analysis can now proceed by performing a PCA on $\overline{\mathbf{Y}}_{\mathtt{AB}}$ and plotting the results as time curves for each treatment.

The SMART model can also be written as a multivariate regression model which also holds for the general case of an unbalanced design:

$$\mathbf{Y} = \mathbf{X}_{\mathtt{B}}\widehat{\mathbf{\Theta}}_{\mathtt{B}} + \mathbf{X}_{\mathtt{AB}}\widehat{\mathbf{\Theta}}_{\mathtt{AB}} + \mathbf{Y}_{\mathtt{Ind}} \tag{4.64}$$

where $\widehat{\mathbf{\Theta}}_{\mathtt{B}}$ carries all the parameters related to the main effect treatment (for all responses) and $\widehat{\mathbf{\Theta}}_{\mathtt{AB}}$ has all the parameters related to interactions except the ones for the first time point (for all responses). Thus, the SMART model can also be written as an RDA problem while reducing the dimension of $\mathbf{X}_{\mathtt{AB}}\widehat{\mathbf{\Theta}}_{\mathtt{AB}}$; but in this case, a partial RDA is required since $\mathbf{X}_{\mathtt{B}}$ is not orthogonal to $\mathbf{X}_{\mathtt{AB}}$. The same comments about variance explained as in PRC apply in this case.

4.6 ASCA, PRC, and SMART Compared

There are relationships among ASCA, PRC, and SMART, which is already clear from the fact that they can all be written as multivariate multiple regression models using different types of coding schemes. Although for the univariate response case, all the effect estimates using different coding schemes can be converted into each other (see Chapter 3) this is not the case anymore in the multivariate case due to the PCA-dimensionality reduction step. Yet, there are relationships among the methods and the subspaces spanned by the factor/interaction estimates (Smilde *et al.*, 2012) which will be explained briefly.

We will explain the commonalities and differences of ASCA, PRC, and SMART using a simple example, visualized in Figure 4.19. In this example, we have two metabolites measured under three treatment regimes where treatment one serves as the control. There are also biological repeats, but these are not shown since they would clog the figure; only group means are shown. Factor A represents *Time* and Factor B represents *Treatment* as before. The full ASCA decomposition of the resulting data is now $\mathbf{Y} = \overline{\mathbf{Y}}_{\mathtt{M}} + \overline{\mathbf{Y}}_{\mathtt{A}} + \overline{\mathbf{Y}}_{\mathtt{B}} + \overline{\mathbf{Y}}_{\mathtt{AB}} + \mathbf{Y}_{\mathtt{Ind}}$ which is a direct generalization of the (univariate) ANOVA model $y_{abnj} = \overline{y}_{...j} + (\overline{y}_{a..j} - \overline{y}_{...j}) + (\overline{y}_{.b.j} - \overline{y}_{...j}) + (\overline{y}_{ab.j} - \overline{y}_{a..j} - \overline{y}_{.b.j} + \overline{y}_{...j}) + (\overline{y}_{abnj} - \overline{y}_{ab.j})$(see Chapter 3). We assume balanced data for simplicity.

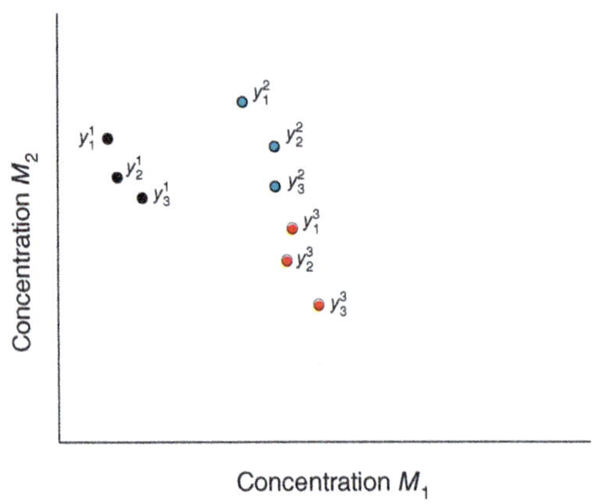

Figure 4.19 Small example of two metabolites. Treatment groups are indicated by color and superscript; time points are indicated by subscript. Source: Reproduced from Smilde *et al.* (2012)/with permission of Oxford University Press.

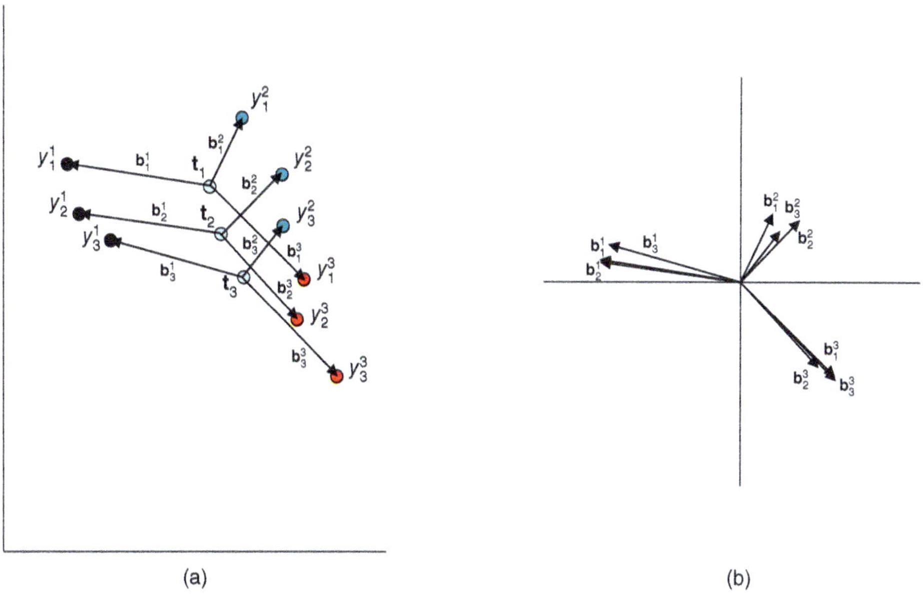

Figure 4.20 Small example of two metabolites. Representation of the ASCA model making the time effect explicit; (a) the whole data set (concentration M_1 versus concentration M_2); (b) translation of the deviations $\mathbf{b}_a^b$ to the origin. A value y_a^b is a row in $\mathbf{Y}$. For more explanation, see the main text. Source: Reproduced from Smilde *et al.* (2012)/with permission of Oxford University Press.

An alternative way of representing this data is shown in Figure 4.20. In this figure, the addition of the mean and the main time effect $(\overline{\mathbf{Y}}_{\text{M}} + \overline{\mathbf{Y}}_{\text{A}})$ is made explicit and indicated by the open dots with labels t_1, t_2, and t_3. The original data points (rows of $\mathbf{Y}$) are indicated by the dots labeled y and the deviations of these original data points relative to $\mathbf{t}_1, \mathbf{t}_2$, and $\mathbf{t}_3$ are indicated with the dots labeled as $\boldsymbol{b}$ with obvious sub- and superscripts. The vectors $\mathbf{t}_1, \mathbf{t}_2$, and $\mathbf{t}_3$ can be subtracted from the y_a^b thereby inducing a translation of the vectors y_a^b to the origin, as shown in the right panel of Figure 4.20. An ASCA model could now proceed by performing a PCA on the values of $\overline{\mathbf{Y}}_{\text{B}} + \overline{\mathbf{Y}}_{\text{AB}}$ (*Treatment* + *Treatment:Time*) the rows of which are the data as shown in Figure 4.20b.

For PRC, the situation is different. The PRC decomposition is $\mathbf{Y} = \overline{\mathbf{Y}}_{\text{A1}} + \overline{\mathbf{Y}}_{\text{PRC}} + \mathbf{Y}_{\text{Ind}}$. This is based on the (univariate) decomposition $y_{abnj} = \overline{y}_{a1.j} + (\overline{y}_{ab.j} - \overline{y}_{a1.j}) + (\overline{y}_{abnj} - \overline{y}_{ab.j})$ where the matrices $\overline{\mathbf{Y}}_{\text{A1}}$ and $\overline{\mathbf{Y}}_{\text{PRC}}$ collect the appropriate terms. Thus, the effects are described relative to the first level of Factor B which serves as the control. Note that the residuals are equal to the ones in the ASCA model. The matrices $\overline{\mathbf{Y}}_{\text{A1}}$ and $\overline{\mathbf{Y}}_{\text{PRC}}$ are not orthogonal, so calculating explained variance is more complicated and has to be done by using constraints on the PRC model. The visual representation of the PRC model of our simple example is shown in Figure 4.21. Figure 4.21b shows the rows of the matrix $\mathbf{Y}_{\text{PRC}}$, and a subsequent analysis would now be a PCA on these data. The time scores of this PCA are then the Principal Response Curves. A simple geometric illustration of the relation between ASCA and PRC is now shown in Figure 4.22 for time point one. The other time points work in the same way.

A similar type of analysis can be made for the SMART model; this will not be explained in detail. It suffices to say that the SMART model comes with the decomposition $\mathbf{Y} = \overline{\mathbf{Y}}_{\text{1B}} + \overline{\mathbf{Y}}_{\text{SMART}} + \mathbf{Y}_{\text{Ind}}$ where again the residuals are the same as in ASCA and PRC; and this model takes everything relative to the first time point ($a = 1$).

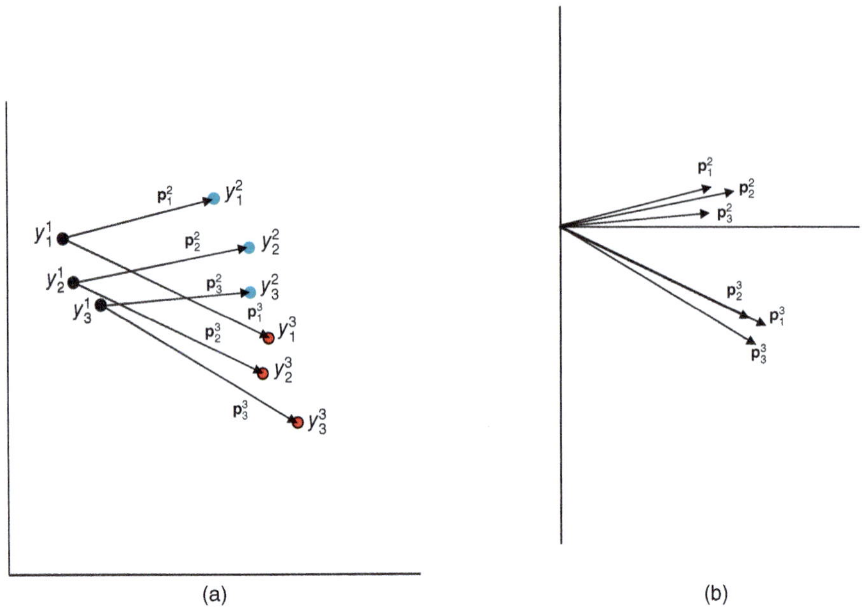

Figure 4.21 Small example of two metabolites. Representation of the PRC model; (a) the whole data set with the control made explicit (concentration M1 versus concentration M2); (b) translation of the deviations $\mathbf{p}_a^b$ relative to the control. For more explanation, see the main text. Source: Reproduced from Smilde *et al.* (2012)/with permission of Oxford University Press.

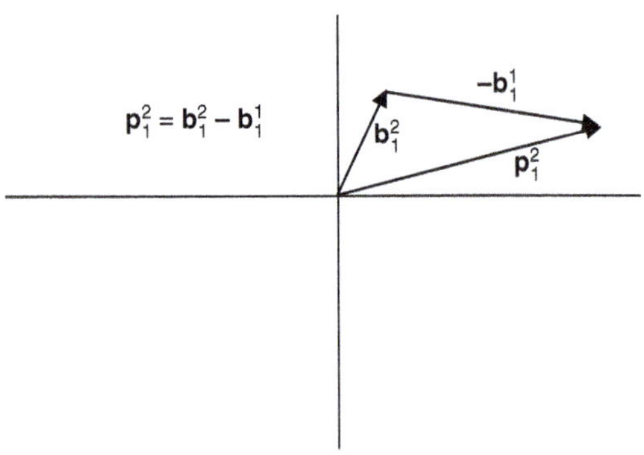

Figure 4.22 Comparing ASCA and PRC for time point 1. The vectors ***b*** and **p** are explained in the previous figures. For more explanation, see the main text. Source: Reproduced from Smilde *et al.* (2012)/with permission of Oxford University Press.

This small example and its geometric illustration indicate that there are also mathematical relationships between the matrices of the decompositions of ASCA, PRC, and SMART. These relationships are:

1) $R(\overline{\mathbf{Y}}^t_{\mathrm{PRC}}) = R((\overline{\mathbf{Y}}_{\mathrm{B}} + \overline{\mathbf{Y}}_{\mathrm{AB}})^t)$
2) $R(\overline{\mathbf{Y}}^t_{\mathrm{SMART}}) = R((\overline{\mathbf{Y}}_{\mathrm{A}} + \overline{\mathbf{Y}}_{\mathrm{AB}})^t)$
3) $R(\overline{\mathbf{Y}}^t_{\mathrm{PRC}}) \neq R(\overline{\mathbf{Y}}^t_{\mathrm{SMART}})$
4) $R(\overline{\mathbf{Y}}^t_{\mathrm{PRC}}) \cap R(\overline{\mathbf{Y}}^t_{\mathrm{SMART}}) \neq \emptyset$ if $J < 2AB - A - B$

where the last property should be understood in terms of vectors other than the zero vector which is always in the range of a subset (for proofs, see Smilde *et al.*, 2012). From relationship 1, we can conclude that $\overline{\mathbf{Y}}^t_{\mathrm{PRC}}$ and $(\overline{\mathbf{Y}}_{\mathrm{B}} + \overline{\mathbf{Y}}_{\mathrm{AB}})^t$ can be expressed on the same basis $\mathbf{V}$. Hence, we can express $\overline{\mathbf{Y}}_{\mathrm{PRC}}$ as $\mathbf{T}_{\mathrm{PRC}}\mathbf{V}^t$ and likewise $\overline{\mathbf{Y}}_{\mathrm{B}} + \overline{\mathbf{Y}}_{\mathrm{AB}}$ as $\mathbf{T}_{\mathrm{ASCA}}\mathbf{V}^t$. Thus $\mathbf{T}_{\mathrm{ASCA}} = \mathbf{Q}\mathbf{T}_{\mathrm{PRC}}$ where $\mathbf{Q}$

is a non-singular matrix. This means that if no dimension reduction is performed on either $\overline{\mathbf{Y}}_{\mathtt{PRC}}$ or $\overline{\mathbf{Y}}_{\mathtt{B}} + \overline{\mathbf{Y}}_{\mathtt{AB}}$ then both methods (ASCA and PRC) give essentially the same results which were already obvious from the Examples 4.2 and 4.9. A similar reasoning holds for $\overline{\mathbf{Y}}_{\mathtt{SMART}}$ and $\overline{\mathbf{Y}}_{\mathtt{A}} + \overline{\mathbf{Y}}_{\mathtt{AB}}$. Once dimensionality reduction is performed (*e.g.*, by using a PCA) then these results may differ.

It is illustrative to give a real-life example of ASCA, PRCA, and SMART on the same data set to show similarities and differences, see Example 4.10.

Example 4.10: A metabolomics example of the use of ASCA, PRC, and SMART

This example uses the same data as in Example 4.4 (van Velzen *et al.*, 2014) and concerns measurements of 11 metabolites in the blood of 20 individuals at regular timepoints in a cross-over study of three treatments of which one treatment is a control. Hence, it is possible to use ASCA, PRC, and SMART on this data. The data was meancentered but not scaled. The scores of the three methods on the first component are shown in Figure 4.23. This figure shows that the percentage of explained variation of the first component in the respective matrices $\overline{\mathbf{Y}}_{\mathtt{A}} + \overline{\mathbf{Y}}_{\mathtt{AB}}$,

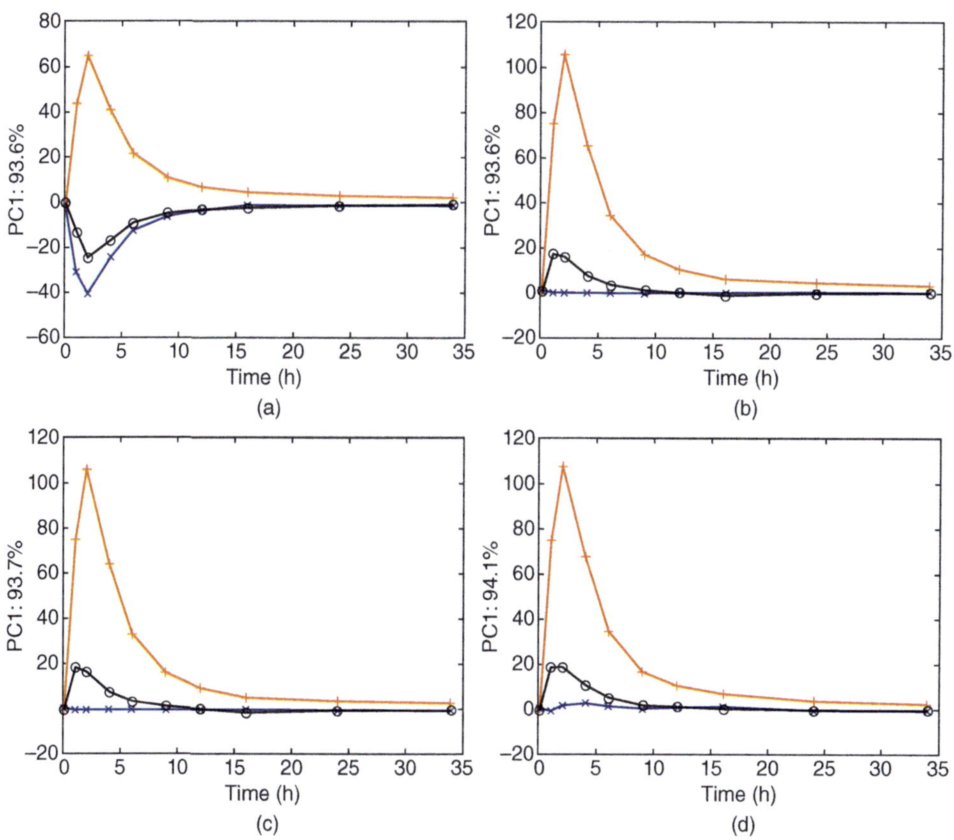

Figure 4.23 The scores of the different methods on the first component; (a) ASCA; (b) post-processed ASCA; (c) PRC; and (d) SMART. Colors: blue is placebo (control group), red is tea extract, and black is wine extract. On the y-axis the amount of explained variation of that PC in the respective matrix is shown. Source: Reproduced from Smilde *et al.* (2012)/with permission of Oxford University Press.

$\overline{\mathbf{Y}}_{\mathrm{PRC}}$, and $\overline{\mathbf{Y}}_{\mathrm{SMART}}$ is very high and that thus this component reveals the major variation. For ease of comparison, the ASCA scores were post-processed by expressing them as deviations from the estimated scores of the control group. The results for the three methods are very similar.

Deviations of the methods start to show up in the third component, see Figure 4.24. Although the explained variances are small, the patterns are not random and clearly show a different behavior for ASCA on the one hand and PRC and SMART on the other hand. The loadings on these third components are shown in Figure 4.25 and show that these differences can be attributed to variables 10 and 11. These variables are special and have a deviating behavior relative to the others since these are gut-mediated metabolites that have their peaks at a later point in time. This is reflected in the PRC and SMART scores on the third component and to some extent also in the ASCA scores.

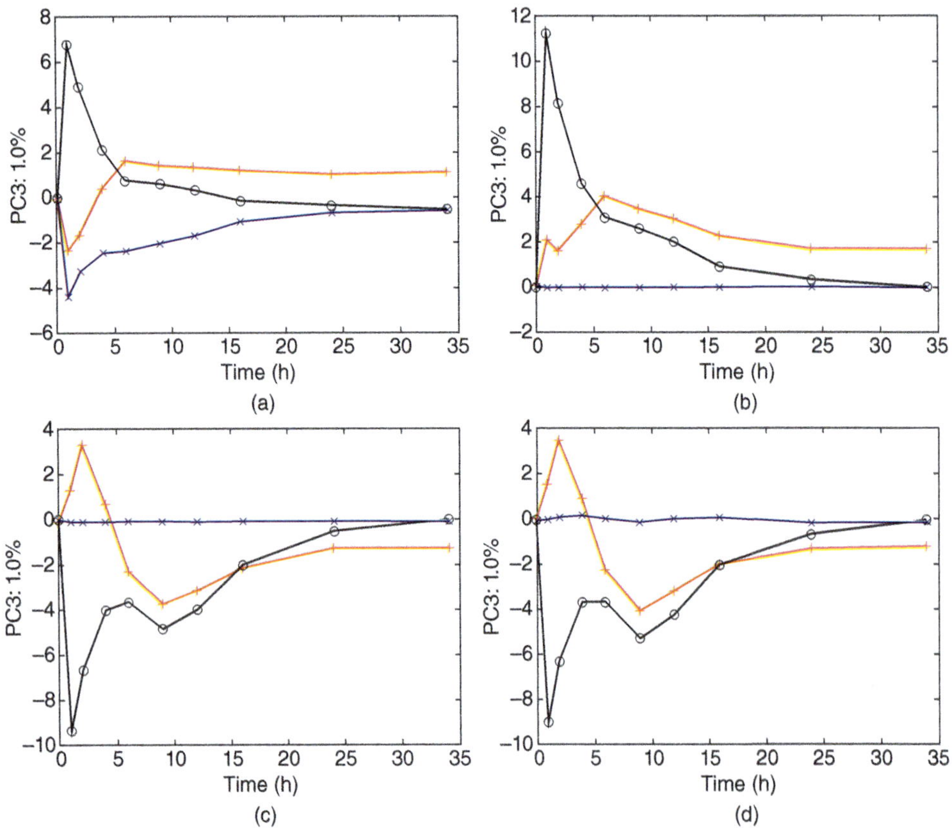

Figure 4.24 The scores of the different methods on the third component; (a) ASCA; (b) post-processed ASCA; (c) PRC; and (d) SMART. Colors: blue is placebo, red is tea extract, and black is wine extract. On the *y*-axis the amount of explained variation of that PC in the respective matrix. Source: Reproduced from Smilde *et al.* (2012)/with permission of Oxford University Press.

This example shows that the results of ASCA, PRC, and SMART can be very similar, but this is data-dependent and not a general rule.

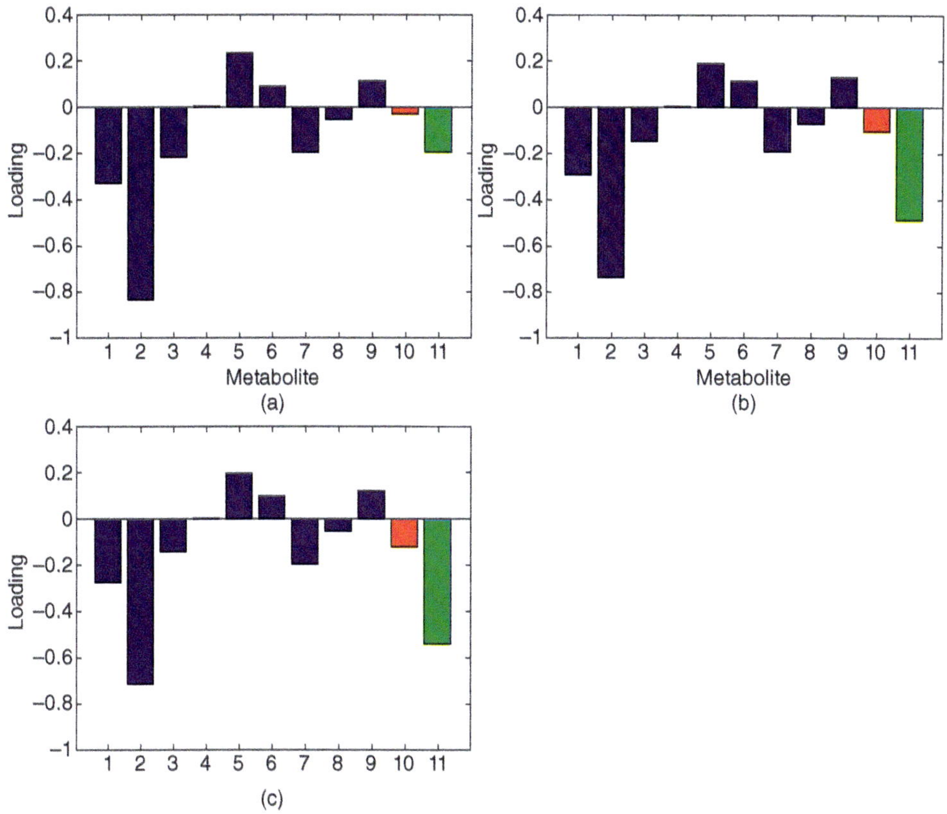

Figure 4.25 The loadings of the different methods on the third component; (a) ASCA; (b) PRC; (c) SMART. Colors: red is 5-(3′-methoxy - 4′-hydroxyphenyl)-g-valerolactone; green is 5-(3′,4′-dihydroxyphenyl)-g-valerolactone. Source: Reproduced from Smilde *et al.* (2012)/with permission of Oxford University Press.

4.7 MSCA

A special case of HD-ANOVA is Multilevel Component Analysis (Timmerman, 2006) and its derivative Multilevel Simultaneous Component Analysis or MSCA for short. This method has been applied in metabolomics (Jansen *et al.*, 2005; Saccenti *et al.*, 2014) and in process chemometrics (de Noord and Theobald, 2005). It is basically an extension of nested ANOVA models (see Section 3.2.2.2) to $j = 1, \ldots, J$ variables:

$$y_{abnj} = \mu_j + \alpha_{aj} + \beta_{b(a)j} + \epsilon_{abnj} \tag{4.65}$$

where the Factor B is nested within Factor A, and we assume the same number of replicates per factor combination. We will call the Factor A the between-factor and the Factor B the within-factor. An example from metabolomics is the measurement of metabolites in the urine of monkeys at several points in time (Jansen *et al.*, 2005). These time points are not exactly the same but – more importantly – time cannot be considered a crossed-factor since a specific time point t does not have the same biological meaning for all the monkeys: they are not synchronized (see Example 4.11). Hence, the model in Equation 4.65 is the proper model with the Factor $\mathtt{A}$ the specific monkey and the Factor $\mathtt{B}$ the time points of the measurements.

We will first discuss such a model based on the example in Section 3.4.2. In that example, the Factor $\mathtt{A}$ had two levels and Factor $\mathtt{B}$ had two levels for $a = 1$ and three levels for $a = 2$. We assume two replicates per factor combination. The full-coding for this model is already given in Equation 3.88, and thus the model can be written as:

$$\mathbf{Y} = \overline{\mathbf{Y}}_M + \mathbf{X}_{\mathtt{A}}\widehat{\boldsymbol{\Theta}}_{\mathtt{A}} + \mathbf{X}_{\mathtt{B(A)}}\widehat{\boldsymbol{\Theta}}_{\mathtt{B(A)}} + \mathbf{Y}_{\mathtt{Ind}} \tag{4.66}$$

where $\overline{\mathbf{Y}}_M$ contains the means of the J variables as before; $\widehat{\boldsymbol{\Theta}}_{\mathtt{A}}(2 \times J)$ contains the estimates for all variables related to Factor $\mathtt{A}$; and $\widehat{\boldsymbol{\Theta}}_{\mathtt{B(A)}}(5 \times J)$ contains the estimates for all variables related to Factor $\mathtt{B}$. The term $\mathbf{X}_{\mathtt{A}}\widehat{\boldsymbol{\Theta}}_{\mathtt{A}}$ is now summarized by a PCA with scores $\mathbf{T}_{\mathtt{A}}(10 \times R_{\mathtt{A}})$ and loadings $\mathbf{P}_{\mathtt{A}}(J \times R_{\mathtt{A}})$ with $R_{\mathtt{A}}$ the number of components. Likewise, $\mathbf{X}_{\mathtt{B(A)}}\widehat{\boldsymbol{\Theta}}_{\mathtt{B(A)}}$ is summarized by a PCA with scores $\mathbf{T}_{\mathtt{B(A)}}(10 \times R_{\mathtt{B(A)}})$ and loadings $\mathbf{P}_{\mathtt{B(A)}}(J \times R_{\mathtt{B(A)}})$ with $R_{\mathtt{B(A)}}$ the number of components.

Several restrictions are now imposed on the scores $\mathbf{T}_{\mathtt{A}}$ and $\mathbf{T}_{\mathtt{B(A)}}$ to arrive at an orthogonal decomposition. Due to the special structure, the matrix $\mathbf{T}_{\mathtt{A}}$ can also be written as:

$$\begin{bmatrix} \mathbf{1}_4\mathbf{t}_1^t \\ \mathbf{1}_6\mathbf{t}_2^t \end{bmatrix} \tag{4.67}$$

where $\mathbf{t}_1$ contains the $R_{\mathtt{A}}$ scores for the first level of Factor $\mathtt{A}$ and $\mathbf{t}_2$ the $R_{\mathtt{A}}$ scores for the second level of Factor $\mathtt{A}$. Moreover, the scores matrix $\mathbf{T}_{\mathtt{B(A)}}$ can be written as $[\mathbf{T}^t_{\mathtt{B(A)}_1}|\mathbf{T}^t_{\mathtt{B(A)}_2}]^t$ with $\mathbf{T}_{\mathtt{B(A)}_1}(4 \times R_{\mathtt{B(A)}})$ and $\mathbf{T}_{\mathtt{B(A)}_2}(6 \times R_{\mathtt{B(A)}})$ referring, respectively, to the estimates of the Factor $\mathtt{B(A)}$ levels for the first and second levels of Factor $\mathtt{A}$. The restrictions on the score matrices are now:

$$\begin{aligned} \mathbf{1}_4^t\mathbf{T}_{\mathtt{B(A)}_1} &= 0 \\ \mathbf{1}_6^t\mathbf{T}_{\mathtt{B(A)}_2} &= 0 \\ 2\mathbf{t}_1 + 3\mathbf{t}_2 &= 0 \end{aligned} \tag{4.68}$$

and this assures that all the parts in the RHS of the equation

$$\mathbf{Y} = \overline{\mathbf{Y}}_M + \mathbf{T}_{\mathtt{A}}\mathbf{P}^t_{\mathtt{A}} + \mathbf{T}_{\mathtt{B(A)}}\mathbf{P}^t_{\mathtt{B(A)}} + \tilde{\mathbf{E}} \tag{4.69}$$

are orthogonal to each other (Timmerman, 2006) and $\tilde{\mathbf{E}}$ contains the contributions from $\mathbf{Y}_{\mathtt{Ind}}$ and the not-used parts in the PCAs of Factors $\mathtt{A}$ and $\mathtt{B(A)}$. The scores $\mathbf{T}_{\mathtt{A}}$ and loadings $\mathbf{P}_{\mathtt{A}}$ can now be plotted and studied as between-variation (levels of Factor $\mathtt{A}$), and likewise $\mathbf{T}_{\mathtt{B(A)}}$ and $\mathbf{P}_{\mathtt{B(A)}}$ represent the within-variation (levels of Factor $\mathtt{B(A)}$ within Factor $\mathtt{A}$).

A formal treatment of MSCA goes as follows (where we drop the explicit indication of Factor $\mathtt{B(A)}$ being nested in Factor $\mathtt{A}$ to simplify notation). Suppose we have a Factor $\mathtt{A}$ with levels $a = 1, \ldots, A$ and a nested Factor $\mathtt{B}$ with levels $b = 1, \ldots, B_a$ where the subscript a in B_a indicates that the nested Factor $\mathtt{B}$ has B_a levels within level a of Factor $\mathtt{A}$. For simplicity, we do not consider replicated measurements. If all values pertaining to the measurements performed at level a of

Factor A are collected in $\mathbf{Y}_a(B_a \times J)$ then the MSCA model is:

$$\mathbf{Y}_a = \overline{\mathbf{Y}}_{Ma} + \mathbf{1}_{B_a}\mathbf{t}_a^t\mathbf{P}_{\mathrm{A}}^t + \mathbf{T}_{\mathrm{B}_a}\mathbf{P}_{\mathrm{B}}^t + \tilde{\mathbf{E}}_a \tag{4.70}$$

where $\overline{\mathbf{Y}}_{Ma}$ contains the overall means of the data across all samples at level a of Factor A and the restrictions on the parameters are:

$$\mathbf{1}_{B_a}^t\mathbf{T}_{\mathrm{B}_a} = \mathbf{0}_{R_{\mathrm{B}}}^t$$
$$\sum_{a=1}^{A} B_a\mathbf{t}_a^t = \mathbf{0}_{R_{\mathrm{A}}}^t \tag{4.71}$$

which ensures orthogonality between the different parts of the decomposition as explained above. The total MSCA model is now as presented in Equation 4.69. There is rotational freedom in this model in the parts $\mathbf{T}_A\mathbf{P}_A^t$ and $\mathbf{T}_B\mathbf{P}_B^t$, and thus $\mathbf{P}_A$ and $\mathbf{P}_B$ can be chosen to have orthonormal columns (Timmerman, 2006) which simplifies calculating explained variances (see Elaboration 4.7). In fact, the within part of the MSCA model (Factor B) is an SCA of the stacked matrices $\mathbf{Y}_a; a = 1, \ldots, A$ and different versions of SCA exist depending on the type of restrictions put on the scores (Timmerman and Kiers, 2003); the version as explained above is the least restricted one and is called SCA-P. An example of MSCA is given in Example 4.11.

ELABORATION 4.7

Explained variances in MSCA

Explained variances can be calculated at different levels in an MSCA model. At the highest level, the explained variances of the between and within parts of the model can be calculated on the basis of the equation:

$$||\mathbf{Y}||^2 = ||\overline{\mathbf{Y}}_M||^2 + ||\mathbf{T}_{\mathrm{A}}\mathbf{P}_{\mathrm{A}}^t||^2 + ||\mathbf{T}_{\mathrm{B}}\mathbf{P}_{\mathrm{B}}^t||^2 + ||\tilde{\mathbf{E}}||^2 \tag{4.72}$$

which is due to the orthogonality of the different terms on the RHS of Equation 4.69. This easily leads to the standard definitions of explained variances for the within and between parts. For the within part this would amount to:

$$\text{VarExplWithin} = \frac{||\mathbf{T}_{\mathrm{B}}\mathbf{P}_{\mathrm{B}}^t||^2}{||\mathbf{Y}||^2 - ||\overline{\mathbf{Y}}_M||^2} \times 100\% \tag{4.73}$$

The second level of explained variation is the amount of explained within-variation per subject which starts with Equation 4.70 and for each within-subject block the following holds:

$$||\mathbf{Y}_a||^2 = ||\overline{\mathbf{Y}}_{Ma} + \mathbf{1}\mathbf{t}_a^t\mathbf{P}_{\mathrm{A}}^t||^2 + ||\mathbf{T}_{\mathrm{B}a}\mathbf{P}_{\mathrm{B}}^t||^2 + ||\tilde{\mathbf{E}}_a||^2 \tag{4.74}$$

due to the restrictions on the MSCA model. The residuals $\tilde{\mathbf{E}}_a$ are orthogonal to $\mathbf{P}_{\mathrm{A}}$, $\mathbf{P}_{\mathrm{B}}$, and $\boldsymbol{m}$ which follows from $\tilde{\mathbf{E}}\mathbf{P}_{\mathrm{A}} = \tilde{\mathbf{E}}\mathbf{P}_{\mathrm{B}} = \tilde{\mathbf{E}}\boldsymbol{m} = 0$ with $\overline{\mathbf{Y}}_M = \mathbf{1}\boldsymbol{m}^t$. Hence, the amount of explained variation within level a of Factor B is:

$$\text{VarExplWithin}_a = \frac{||\mathbf{T}_{\mathrm{B}a}\mathbf{P}_{\mathrm{B}}^t||^2}{||\mathbf{Y}||^2 - ||\overline{\mathbf{Y}}_{Ma} + \mathbf{1}\mathbf{t}_a^t\mathbf{P}_{\mathrm{A}}^t||^2} \times 100\% \tag{4.75}$$

and since the loading matrix $\mathbf{P}_{\mathrm{B}}$ has orthonormal columns this variance explained within block a can also be calculated per SCA-component.

Example 4.11: MSCA analysis of monkey-urine metabolomics

An extensive treatment of this example is given elsewhere (Jansen *et al.*, 2005), but a short description is given here. A related study compares these results with a data set of humans (Saccenti *et al.*, 2014). The current exposition concerns a so-called normality analysis. Urine was obtained from rhesus monkeys at 29 non-equidistant days over a time period of 57 days. Of the monkeys, five were male and five were female. Note that there are two reasons for considering this experimental setup as an MSCA model. First, the time points at which measurements were obtained are not the same for all monkeys, and, secondly, even if the time points would have been the same, the monkeys are not "synchronized": measurements at time point t for two monkeys do not have the same meaning. This is a problem of comparability (Van Mechelen and Smilde, 2011). An MSCA model was fitted to the data where 24% of the variation was explained by the between part and 76% of the variation by the within part. A score plot of the between-monkey variation is shown in Figure 4.26 and reveals a separation between male and female monkeys.

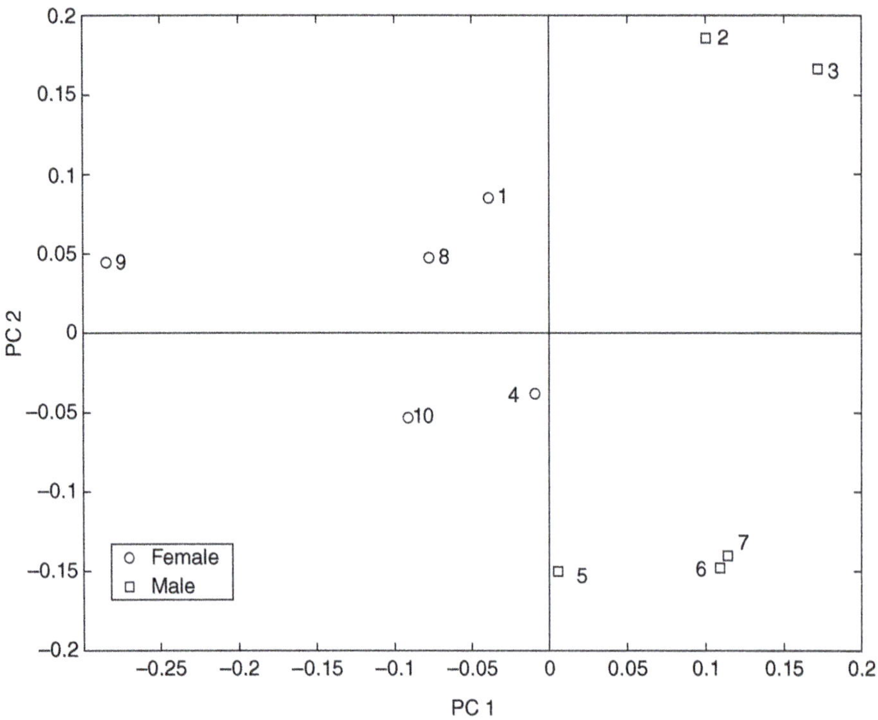

Figure 4.26 Score plot of the between-monkey variation of the normality study. PC1 explains 33% of the between-variation and PC2 explains 29%. Source: Reproduced from Jansen *et al.* (2005)/with permission of Elsevier.

The within-individual variation explained per individual monkey ranges from 49 to 77% (of the 76% within-variation) where monkey 9 is associated with the 49%. This shows that monkey 9 has different biorhythms compared to the other monkeys; note that this same monkey also has a deviating between-level score.

The within-level scores for monkey 6 (a representative monkey) are shown in Figure 4.27, and these are compared with the scores of a PCA analysis separately performed for monkey 6. The left panel in this figure clearly shows dynamic features. The right panel shows that the difference between the MSCA and PCA component 1 is almost constant, and this relates to the between-level variation of this monkey, whereas the latter components show deviations between the PCA and within-level MSCA model scores.

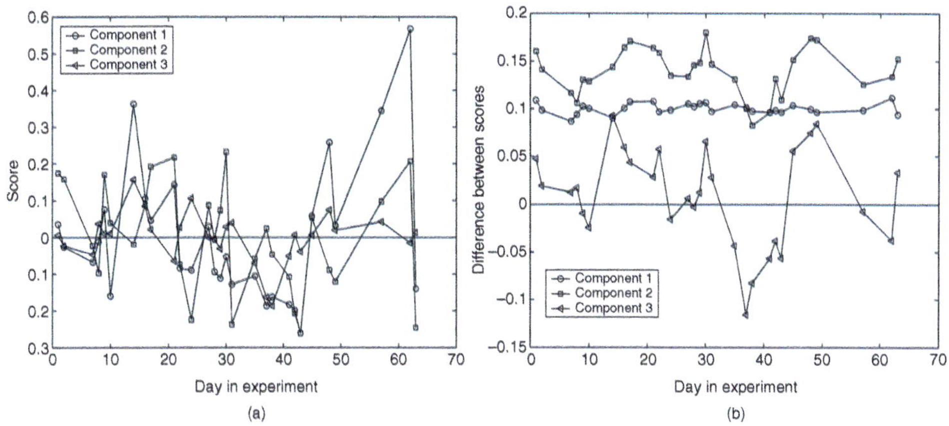

Figure 4.27 Score plot of the within-monkey variation of monkey 6; (a) the within-scores; (b) the difference between within-level MSCA scores and those of a separate PCA on the monkey 6 data. Source: Reproduced from Jansen *et al.* (2005)/with permission of Elsevier.

The concept of multilevel modeling of which MSCA is an example can also be used fruitfully in other contexts, *e.g.*, combined with multiway analysis (Rubingh *et al.*, 2011) or combined with PLSDA (Westerhuis *et al.*, 2010).

4.A Appendix

4.A.1 Proof of Equation 4.20

Equations 4.19 and 4.20 can be found by using $\mathbf{X} = [\mathbf{1}_I|\mathbf{X}_\mathrm{A}|\mathbf{X}_\mathrm{B}|\mathbf{X}_\mathrm{AB}]$ and subsequently working out $\mathbf{X}(\mathbf{X}^t\mathbf{X})^{-1}\mathbf{X}^t$ using the orthogonality properties in $\mathbf{X}$. Then the matrix $\mathbf{X}^t\mathbf{X}$ is

$$\begin{bmatrix} \mathbf{1}_I^t\mathbf{1}_I & \mathbf{0} & \mathbf{0} & \mathbf{0} \\ \mathbf{0} & \mathbf{X}_\mathrm{A}^t\mathbf{X}_\mathrm{A} & \mathbf{0} & \mathbf{0} \\ \mathbf{0} & \mathbf{0} & \mathbf{X}_\mathrm{B}^t\mathbf{X}_\mathrm{B} & \mathbf{0} \\ \mathbf{0} & \mathbf{0} & \mathbf{0} & \mathbf{X}_\mathrm{AB}^t\mathbf{X}_\mathrm{AB} \end{bmatrix} \tag{4.A.1}$$

which makes calculating the inverse and thus the hat matrix particularly easy. Straightforward calculations lead then to Equation 4.20. The proof of orthogonality follows then by considering, *e.g.*, $\mathbf{Q}_\mathrm{A}^t\mathbf{Q}_\mathrm{B} = \mathbf{X}_\mathrm{A}(\mathbf{X}_\mathrm{A}^t\mathbf{X}_\mathrm{A})^{-1}\mathbf{X}_\mathrm{A}^t\mathbf{X}_\mathrm{B}(\mathbf{X}_\mathrm{B}^t\mathbf{X}_\mathrm{B})^{-1}\mathbf{X}_\mathrm{B}^t$ which is zero since $\mathbf{X}_\mathrm{A}^t\mathbf{X}_\mathrm{B}$ is zero because of the sum-coding.

4.A.2 Proof of Equation 4.6

Working out term $||\mathbf{Y}||^2$ in terms of traces becomes:

$$\begin{aligned} ||\mathbf{Y}||^2 &= tr(\mathbf{Y}^t\mathbf{Y}) \\ &= tr(\mathbf{Y}_{\mathrm{M}} + \mathbf{Y}_{\mathrm{A}} + \mathbf{Y}_{\mathrm{B}} + \mathbf{Y}_{\mathrm{AB}} + \mathbf{Y}_{\mathrm{Ind}})^t(\mathbf{Y}_{\mathrm{M}} + \mathbf{Y}_{\mathrm{A}} + \mathbf{Y}_{\mathrm{B}} + \mathbf{Y}_{\mathrm{AB}} + \mathbf{Y}_{\mathrm{Ind}}) \end{aligned} \quad (4.A.2)$$

which amounts to terms like $tr(\mathbf{Y}_{\mathrm{A}}^t\mathbf{Y}_{\mathrm{A}}) = ||\mathbf{Y}_{\mathrm{A}}||^2$ and many cross-products, *e.g.*, $tr(\mathbf{Y}_{\mathrm{A}}^t\mathbf{Y}_{\mathrm{B}})$. The latter terms can be written using the **Q** matrices, *e.g.*, $tr(\mathbf{Y}_{\mathrm{A}}^t\mathbf{Y}_{\mathrm{B}}) = tr(\mathbf{Q}_{\mathrm{A}}\mathbf{Y})^t(\mathbf{Q}_{\mathrm{B}}\mathbf{Y})$ using Equation 4.21. Now $tr(\mathbf{Q}_{\mathrm{A}}\mathbf{Y})^t\mathbf{Q}_{\mathrm{B}}\mathbf{Y}$ can be written as $tr(\mathbf{Y}^t\mathbf{Q}_{\mathrm{A}}^t\mathbf{Q}_{\mathrm{B}}\mathbf{Y})$ which is zero due to the orthogonality of $\mathbf{Q}_{\mathrm{A}}$ and $\mathbf{Q}_{\mathrm{B}}$. Hence, all cross-products disappear and the result is Equation 4.6.

5

Alternative Methods

5.1 General Introduction

In this chapter, we will describe alternative methods. These methods are either not yet much used or rather advanced relative to the methods treated in Chapter 4. Broadly speaking, the alternative methods can be divided into PLS-based methods, methods based on extensions of Linear Mixed Models, and miscellaneous methods.

5.2 PLSR-Based Methods

Some alternative methods that have been proposed in the literature, start from the same multivariate ANOVA decomposition described, *e.g.*, in Equation 4.3 or, more generally, from the regression formulation of the model (see Equation 4.15) to use approaches based on PLS regression both to test for the statistical significance of the effect of the design terms and for the successive interpretation of the results and data display. Among those methods, the first and most used is ANOVA Target Projection (ANOVA-TP). Moreover, taking the lead from ANOVA-TP, and exploiting the toolkit of multi-block data analysis, another method, namely ANOVA multiblock OPLS (AMOPLS) was also proposed in the literature.

5.2.1 ANOVA-TP

ANOVA Target Projection (ANOVA-TP) (Marini *et al.*, 2015, 2017; Marini and Walczak, 2020) starts with decomposing the experimental matrix $\mathbf{Y}$, according to the ANOVA scheme, as already discussed in Chapter 4 and schematized, *e.g.*, in Equation 4.15. For instance, in the case of a three-factor problem, the corresponding decomposition would be:

$$\mathbf{Y} = \widehat{\mathbf{Y}}_{\mathrm{M}} + \widehat{\mathbf{Y}}_{\mathrm{A}} + \widehat{\mathbf{Y}}_{\mathrm{B}} + \widehat{\mathbf{Y}}_{\mathrm{C}} + \widehat{\mathbf{Y}}_{\mathrm{AB}} + \widehat{\mathbf{Y}}_{\mathrm{AC}} + \widehat{\mathbf{Y}}_{\mathrm{BC}} + \widehat{\mathbf{Y}}_{\mathrm{ABC}} + \mathbf{Y}_{\mathtt{Ind}} \tag{5.1}$$

where, as in Chapter 4, $\widehat{\mathbf{Y}}_{\mathrm{M}}$, also indicated as $\overline{\mathbf{Y}}$, is a matrix where all rows contain the mean of $\mathbf{Y}$, $\widehat{\mathbf{Y}}_{\mathrm{A}}, \widehat{\mathbf{Y}}_{\mathrm{B}}, \dots, \widehat{\mathbf{Y}}_{\mathrm{ABC}}$ are the effect matrices for factors and interactions and $\mathbf{Y}_{\mathtt{Ind}}$ collects the residuals. In ANOVA-TP, the decomposition is estimated using a regression approach where the design matrix is usually sum-coded. This allows to deal also with unbalanced data similar to what was already discussed for ASCA+ (see Section 4.3).

Based on the complete ANOVA decomposition of $\mathbf{Y}$, significance testing and interpretation of the effect of the specific design term is carried out by building individual PLS models, where the independent and dependent blocks are defined in the following way. The dependent block

Analysis of Variance for High-Dimensional Data: Applications in Life, Food and Chemical Sciences,
First Edition. Age K. Smilde, Federico Marini, Johan A. Westerhuis, and Kristian H. Liland.

is made of the properly coded (*e.g.*, by sum-coding) design submatrix corresponding to the effect to be tested ($\mathbf{X}_f$ with $f = \mathtt{A, B, C, AB}, \ldots$), while the independent matrix block, here indicated as $\widehat{\mathbf{Y}}_f^*$ ($f = \mathtt{A, B, C, AB}, \ldots$), is defined as the sum of all the design terms in Equation 5.1, that would contribute to the expected mean squares (EMS) of the factor or interaction under investigation. The latter choice, *i.e.*, to use the concept of EMS, in principle allows to have a common framework to deal with fixed, random, and mixed effect models in crossed or nested configurations. When dealing with a fixed-effects crossed design, then the above definition of the independent block boils down to adding the effect matrix to the residual matrix:

$$\widehat{\mathbf{Y}}_f^* = \widehat{\mathbf{Y}}_f + \mathbf{Y}_{\mathtt{Ind}} \qquad f = \mathtt{A, B, C, AB}, \ldots \tag{5.2}$$

When dealing with random or mixed effect models, it may not be straightforward to identify which design terms contribute to the EMS for a specific factor or interaction; in such cases, Hasse diagrams (see Section 3.6) can provide a valid help for the purpose. As said, the core of ANOVA-TP relies on calculating a PLS model relating $\mathbf{X}_f$ as dependent block to $\widehat{\mathbf{Y}}_f^*$ as predictors, according to:

$$\mathbf{X}_f = \widehat{\mathbf{Y}}_f^* \mathbf{B}_{f,PLS} + \mathbf{E}_f = \hat{\mathbf{X}}_f + \mathbf{E}_f \tag{5.3}$$

$\mathbf{E}_f$ being the matrix collecting the residuals of the PLS regression, whose coefficients are gathered in $\mathbf{B}_{f,PLS}$. The use of $\widehat{\mathbf{Y}}_f^*$ instead of $\widehat{\mathbf{Y}}_f$ as predictor matrix allows to account not only for the between-levels but also for the within-levels variability in the successive steps of the analysis. The optimal complexity of the PLS model is chosen through a cross-validation scheme, where cancelation groups take as much as possible into account the underlying treatment design (*e.g.*, by withdrawing one unit at a time from each design cell). As for ASCA and other models presented in this book, the statistical significance of the effect of the individual design terms (factors or interactions) is non-parametrically estimated through permutation testing (Anderson and Ter Braak, 2003). In this context, a possible test statistic is represented by the sum-of-squares of the residuals of the PLS model:

$$SSQR_f = ||\mathbf{E}_f||_F^2 = ||\mathbf{X}_f - \hat{\mathbf{X}}_f||_F^2 \tag{5.4}$$

Based on this definition, in order for a design term to have a significant effect, $SSQR_f$ should be as low as possible. Operationally, the value of $SSQR_f$ is compared to its null distribution estimated by a permutation test (see Section 3.7.3), where the rows of the design matrix $\mathbf{X}_f$ are randomly permuted (taking into account the exchangeable units, according to the rules already explained in Section 3.7.3) and consequently, new predictions of the response are calculated by PLS, leading to estimates of the sum-of-squares of the residuals under the permutation $SSQR_{f,perm}$. Accordingly, a p-value can be calculated by counting the fraction of the number of total permutations (N_{perm}) leading to a value of $SSQR_{f,perm}$ lower than or equal to that obtained on the unpermuted data:

$$p_{f,SSQR} = \frac{\#(SSQR_{f,perm} \leq SSQR_f) + 1}{N_{perm} + 1} \tag{5.5}$$

For each design term, interpretation of the corresponding PLS model is accomplished through post-processing of the solution by Target Projection (TP) (see Elaboration 2.7), a transformation that provides a more straightforwardly response-related latent structure. In particular, for each column of the response matrix (in the case of ANOVA-TP, corresponding to the coded design matrix $\mathbf{X}_f$), the target projection summarizes the related variability in the independent block on a single latent variable. When factors or interactions have more than two levels, *i.e.*, when the design matrix $\mathbf{X}_f$ is multivariate, TP scores and loadings are calculated as follows. Target projection scores are individually calculated for each column of the predicted response matrix:

$$\mathbf{t}_{k,TP}^{(f)} = \frac{\hat{\mathbf{x}}_k^{(f)}}{\|\mathbf{b}_k^{(f)}\|} \tag{5.6}$$

where $\hat{\mathbf{x}}_k^{(f)}$ and $\mathbf{b}_k^{(f)}$ are the kth column of $\hat{\mathbf{X}}_f$ and $\mathbf{B}_{f,PLS}$, respectively. The score vectors are then collected into the matrix $\mathbf{T}_{TP}^{(f)}$, which is not necessarily orthogonal:

$$\mathbf{T}_{TP}^{(f)} = [\mathbf{t}_{1,TP}^{(f)}, \ldots, \mathbf{t}_{K_f,TP}^{(f)}] \tag{5.7}$$

K_f being the number of columns in $\hat{\mathbf{X}}_f$. TP loadings are then calculated as:

$$\mathbf{P}_{TP}^{(f)} = \mathbf{X}_f^t \mathbf{T}_{TP}^{(f)} (\mathbf{T}_{TP}^{(f)t} \mathbf{T}_{TP}^{(f)})^{-1} \tag{5.8}$$

In ANOVA-TP, differences among the levels of the individual factors or interaction terms are highlighted by inspecting the scores along the target projection components corresponding to the related PLS model, and the interpretation is analogous to the one of the SCA models in ASCA after back projection of the residuals. The contribution of the measured variables to the observed differences can be ascertained by examining the target projection loadings of the same PLS model, possibly after having estimated their confidence intervals through bootstrapping or jack-knifing (Efron, 1979). However, for complex designs, with factors controlled at many levels, interpretation may not be straightforward, since there is a TP component for each of the many columns in the corresponding design matrix $\mathbf{X}_f$.

Example 5.1: Example of ANOVA-TP on the pasta data

The use of ANOVA-TP will be illustrated through the analysis of a dataset resulting from the investigation of the effect of production-related factors on the NIR spectroscopic signature of egg-pasta samples (Bevilacqua *et al.*, 2013). In the study, that was already briefly introduced in Example 1.5, egg-pasta samples were prepared according to a full factorial design where three factors were controlled, namely *Egg content* (at 6 levels: 20.25%, 22.53%, 25.15%, 27.00%, 29.92%, and 33.50%), *Drying temperature* (at 3 levels: 40 °C, 52 °C, and 65 °C) and *Drying time* (at 3 levels: 4, 6, and 8 hours). Ten samples were prepared and analyzed by NIR spectroscopy in reflectance mode, for each combination of factor levels, giving a total of 540 spectra. To remove the effect of spurious spectral variability (due, *e.g.*, to scattering), the signals were pre-treated by Standard Normal Variate transform (Barnes *et al.*, 1989). The first step in ANOVA-TP, as in many HD-ANOVA approaches, is the ANOVA decomposition of the variability in the original data matrix, as described in Equation 5.1. More specifically, since the data were collected according to a fully crossed design with three factors (*Temperature* (T), *Egg concentration* (C), and *Time* (t)), the following partitioning is obtained through regression of the spectroscopic data $\mathbf{Y}$ on the sum-coded design matrix $\mathbf{X}$:

$$\mathbf{Y} = \hat{\mathbf{Y}}_{\mathtt{M}} + \hat{\mathbf{Y}}_{\mathtt{T}} + \hat{\mathbf{Y}}_{\mathtt{C}} + \hat{\mathbf{Y}}_{\mathtt{t}} + \hat{\mathbf{Y}}_{\mathtt{TC}} + \hat{\mathbf{Y}}_{\mathtt{Tt}} + \hat{\mathbf{Y}}_{\mathtt{Ct}} + \hat{\mathbf{Y}}_{\mathtt{TCt}} + \mathbf{Y}_{\mathtt{Ind}} \tag{5.9}$$

Then, for each design term (*i.e.*, the main effect of the factors or interactions), significance testing and interpretation rely on the calculation of separate PLS models and target-projection post-processing. At first, the effect of *Drying temperature* will be considered. After the ANOVA decomposition in Equation 5.9, the independent matrix for building the PLS model is obtained by combining the factor effect matrix $\hat{\mathbf{Y}}_{\mathrm{T}}$ with the residual matrix $\mathbf{Y}_{\mathtt{Ind}}$ as described in Equation 5.2. The resulting spectroscopic matrix ($\hat{\mathbf{Y}}_{\mathrm{T}}^{*}$) is graphically displayed in Figure 5.1.

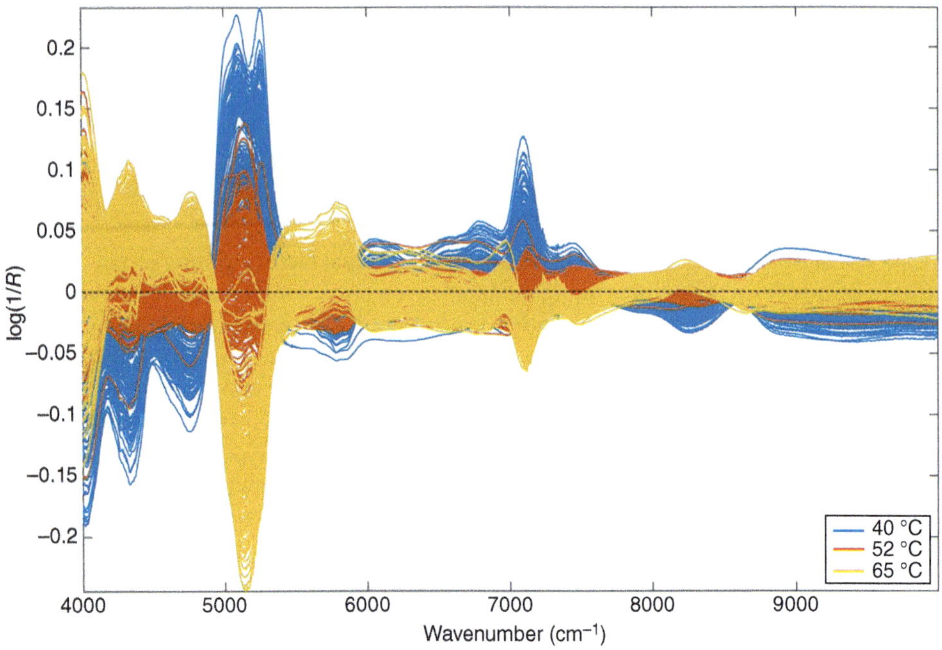

Figure 5.1 Residual-augmented effect matrix for *Temperature*, which is used as the independent block for the PLS regression that constitutes the engine of ANOVA-TP.

A PLS model is then built between the design submatrix $\mathbf{X}_{\mathrm{T}}$ as dependent block and $\hat{\mathbf{Y}}_{\mathrm{T}}^{*}$ as predictors, the optimal complexity being selected based on the lowest cross-validation error. Specifically, in the case of the model for the effect of *Temperature*, 2 latent variables are selected. Significance of the effect of the factor is then evaluated by comparing the sum-of-squares of the $\mathbf{E}_{\mathrm{T}}$ residuals of the full model with its null distribution estimated by a permutation test (in the present case, with 10,000 randomizations). The results are graphically displayed in Figure 5.2.

Interpretation of the effect of the factor on the spectroscopic signal can be then accomplished by first inspecting the scores of the samples on the two target projection vectors (one for each of the columns of the design matrix; see Figure 5.3), and successively by examining the corresponding target projection loadings (Figure 5.4).

The scores plot in Figure 5.3 shows that the samples corresponding to the three *Temperature* levels are well separated from one another: the separation between the level means (filled symbols) is larger than the within-level variability. More specifically, a decreasing trend along both

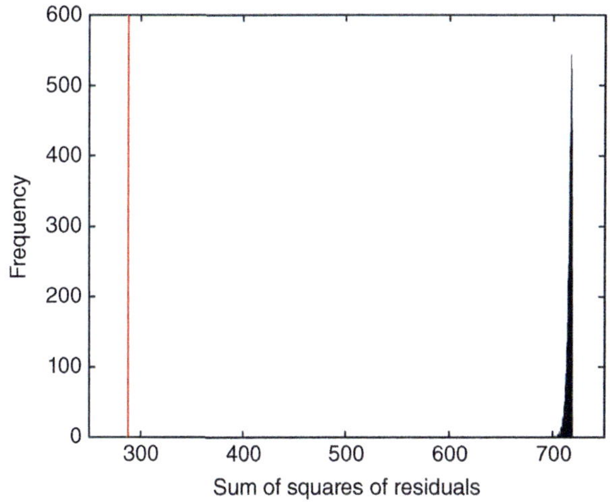

Figure 5.2 Comparison of the sum-of-squares of the $\mathbf{E}_T$ residuals of the PLS model for the effect of *Temperature* (red vertical bar) with its null distribution estimated by a permutation test with 10,000 randomizations (black histogram). A corresponding p-value < 0.0001 is obtained.

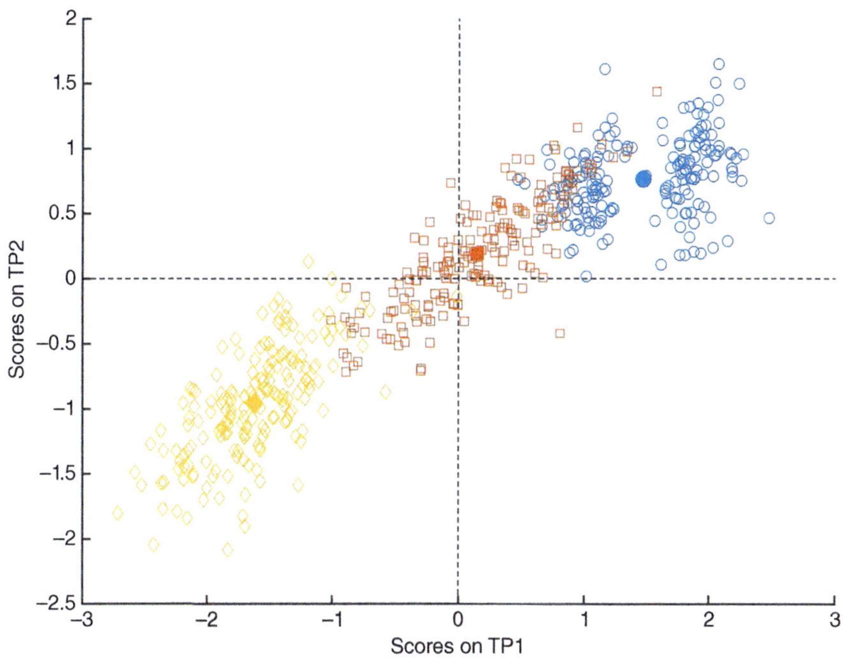

Figure 5.3 Scores of the samples onto the two target projection vectors of the PLS model for the effect of *Temperature* (empty symbols), together with their mean values for the three levels of the factor (filled symbols). Blue circles – 40 °C; orange squares – 52 °C; yellow diamonds – 65 °C.

components can be observed: scores along both TP1 and TP2 decrease as *Temperature* increases. Inspection of the loadings in Figure 5.4 provides the interpretation of the observed differences in terms of the spectroscopic bands that are mostly involved.

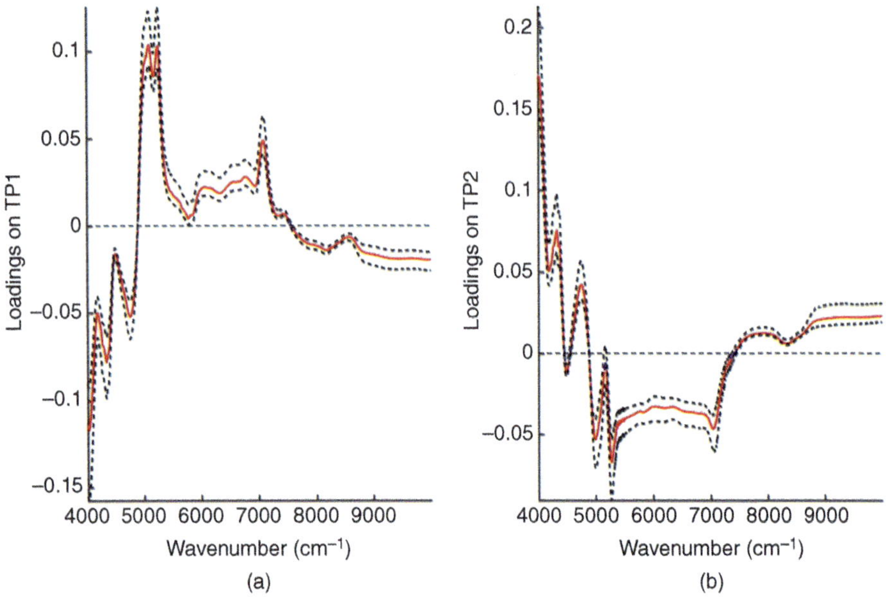

Figure 5.4 Variable loadings for the first (a) and the second (b) target projection vectors (continuous red line), together with their 95% confidence interval, estimated by a bootstrap procedure with 1000 repetitions.

Here, it should be stressed that, depending on the data, the TPs may become heavily correlated, as evident from the score plot for the effect of temperature shown in Figure 5.3. In such cases, it is possible to obtain uncorrelated TP scores spanning the same subspace, by SVD of the TP scores matrix. The corresponding TP loadings are then obtained as in Equation 5.8 using the orthogonal scores. The results for the effect of *Temperature* are shown in Figure 5.5, where the orthogonal scores calculated by SVD of the scores in Figure 5.3 and the corresponding TP loadings are displayed.

When imposing orthogonality, the score plot in Figure 5.5 looks very similar to the one obtained with ASCA (see Bevilacqua *et al.*, 2013). Most of the between-level variability is accounted for by TP1, whereas TP2 is mainly associated with the within-level variation. Since an increase of *Temperature* corresponds to a decrease of the score along TP1, inspection of the loadings along the same component suggests that increasing the *Temperature* results in a higher spectral intensity between 4000 and 4900 cm^{-1} and between 5400 and 5940 cm^{-1} and a lower intensity between 4900 and 5400 cm^{-1} and between 6400 and 7600 cm^{-1}.

Similar considerations can be made about the effect of the other factors and interactions. In particular, based on the permutation tests conducted on the residuals analogously to what already discussed for the case of temperature, the following p-values can be estimated (Table 5.1).

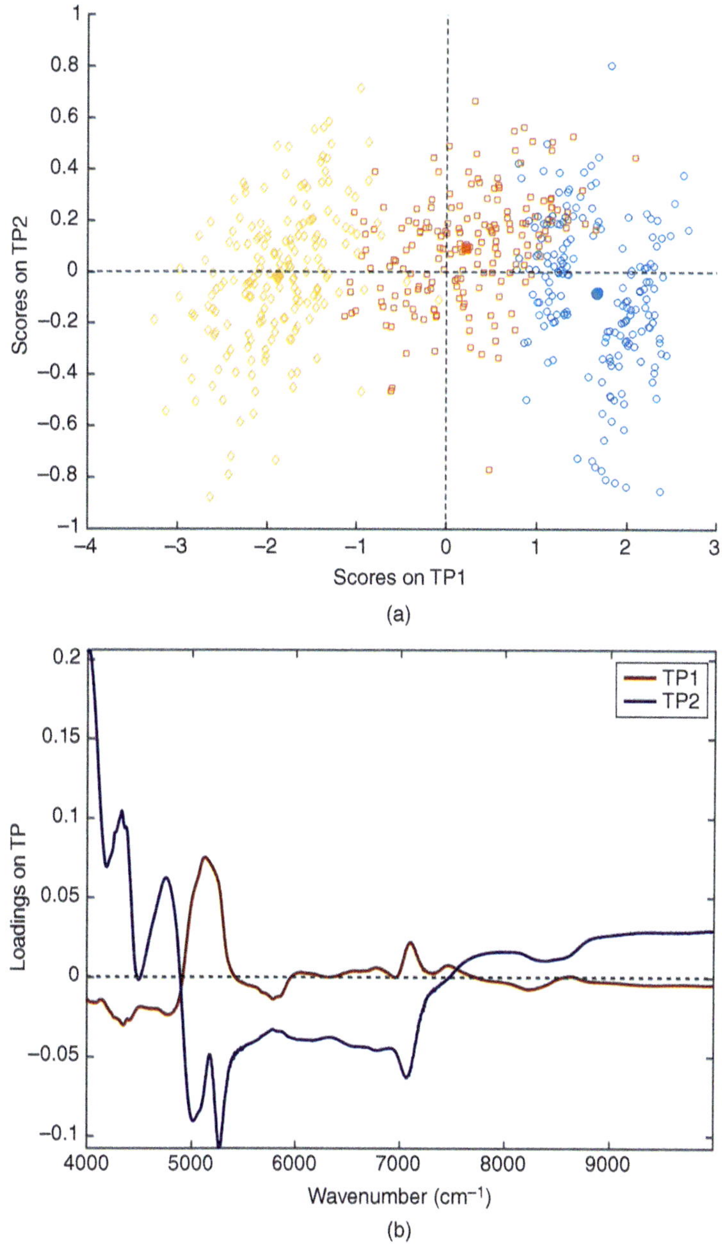

Figure 5.5 (a) Orthogonal samples scores onto the two target projection components of the PLS model for the effect of *Temperature* (empty symbols), together with their mean values for the three levels of the factor (filled symbols). Blue circles – 40 °C; orange squares – 52 °C; yellow diamonds – 65 °C. (b) Variable loadings for the two TP components, calculated based on orthogonal TP scores.

Table 5.1 Results of permutation testing of the $\mathbf{X}_f$ residuals for estimating the effect significance.

Source	*p*-value
Temperature (T)	<0.0001
Concentration (C)	<0.0001
Time (t)	<0.0001
T:C	<0.0001
T:t	0.0072
C:t	0.9629
T:C:t	0.9998

5.2.2 ANOVA Multiblock Orthogonal Partial Least Squares (AMOPLS)

ANOVA Multiblock Orthogonal Partial Least Squares (AMOPLS) (Boccard and Rudaz, 2016) also starts from the same ANOVA-based partitioning of the sources of variation, already described in Equation 5.1, and analogously to ANOVA-TP considers a supervised approach to analyze the resulting submatrices for investigating the effect of the controlled factors and interaction and for further interpretation. However, effect submatrices are not individually analyzed, but they are jointly processed through a multiblock algorithm, namely, Multiblock Orthogonal Partial Least Squares (MB-OPLS), which is implemented through its kernel formulation. The main steps of AMOPLS are described in Algorithm 5.1.

Algorithm 5.1

AMOPLS

1) Perform the ANOVA decomposition of the matrix **Y** as described in Equation 5.1.
2) Perform a PCA decomposition of each of the effect matrices resulting from the ANOVA decomposition and collect the scores of all the components with non-zero eigenvalues:

$$\hat{\mathbf{Y}}_f = \mathbf{T}_f \mathbf{P}_f^t \qquad f = \mathtt{A}, \mathtt{B}, \ldots \tag{5.10}$$

3) Collect all the scores resulting from the previous step into a matrix. For instance, in the case of a two-factor crossed design:

$$\mathbf{T}_{all} = [\mathbf{T}_{\mathtt{A}} | \mathbf{T}_{\mathtt{B}} | \mathbf{T}_{\mathtt{AB}}] \tag{5.11}$$

This matrix will be used as the dependent variable for the successive multiblock predictive modeling.
4) A multiblock data set of predictors is built by adding the residual matrix to each of the effect matrices of the ANOVA decomposition and including the residual matrix itself as an additional block. By defining:

$$\begin{aligned} \hat{\mathbf{Y}}_{\mathtt{A}}^* &= \hat{\mathbf{Y}}_{\mathtt{A}} + \mathbf{Y}_{\mathtt{Ind}} \\ \hat{\mathbf{Y}}_{\mathtt{B}}^* &= \hat{\mathbf{Y}}_{\mathtt{B}} + \mathbf{Y}_{\mathtt{Ind}} \\ \hat{\mathbf{Y}}_{\mathtt{AB}}^* &= \hat{\mathbf{Y}}_{\mathtt{AB}} + \mathbf{Y}_{\mathtt{Ind}} \end{aligned} \tag{5.12}$$

Then the multiblock set of predictors $\mathbf{Y}_{mb}$ is constituted of the matrices $\widehat{\mathbf{Y}}^*_{\mathtt{A}}$, $\widehat{\mathbf{Y}}^*_{\mathtt{B}}$, $\widehat{\mathbf{Y}}^*_{\mathtt{AB}}$, and $\mathbf{Y}_{\mathtt{Ind}}$. This step is similar to the one that is used in AComDim (see Section 5.4.6).

5) A multiblock OPLS model is built between $\mathbf{T}_{all}$ as dependent block and $\mathbf{Y}_{mb}$ as predictor blocks using the kernel OPLS algorithm, according to the following steps:
   a) For each block in $\mathbf{Y}_{mb}$, a kernel matrix $\mathbf{W}_f$ ($f = \mathtt{A}, \mathtt{B}, \mathtt{AB}, \mathtt{Ind}$) is built as:

$$\begin{aligned}\mathbf{W}_{\mathtt{A}} &= \widehat{\mathbf{Y}}^*_{\mathtt{A}}\widehat{\mathbf{Y}}^{*t}_{\mathtt{A}}\\ \mathbf{W}_{\mathtt{B}} &= \widehat{\mathbf{Y}}^*_{\mathtt{B}}\widehat{\mathbf{Y}}^{*t}_{\mathtt{B}}\\ \mathbf{W}_{\mathtt{AB}} &= \widehat{\mathbf{Y}}^*_{\mathtt{AB}}\widehat{\mathbf{Y}}^{*t}_{\mathtt{AB}}\\ \mathbf{W}_{\mathtt{Ind}} &= \mathbf{Y}_{\mathtt{Ind}}\mathbf{Y}^{t}_{\mathtt{Ind}}\end{aligned} \tag{5.13}$$

   b) A consensus kernel $\mathbf{W}_{mb}$ is built by a weighted sum of the individual kernels:

$$\mathbf{W}_{mb} = c_{\mathtt{A}}\mathbf{W}_{\mathtt{A}} + c_{\mathtt{B}}\mathbf{W}_{\mathtt{B}} + c_{\mathtt{AB}}\mathbf{W}_{\mathtt{AB}} + c_{\mathtt{Ind}}\mathbf{W}_{\mathtt{Ind}} \tag{5.14}$$

   the weights c_f ($f = \mathtt{A}, \mathtt{B}, \mathtt{AB}, \mathtt{Ind}$) being the RV coefficients between the corresponding predictor block and the response $\mathbf{T}_{\mathtt{all}}$. The RV coefficient can be considered as a multivariate generalization of the squared correlation coefficient, and it provides a measure of the similarity between two matrices.
   c) A kernel OPLS (KOPLS) model is calculated between $\mathbf{T}_{\mathtt{all}}$ and $\mathbf{W}_{\mathrm{mb}}$

$$\begin{aligned}\mathbf{W}_{mb} &= \mathbf{T}_{p\mathtt{A}}\mathbf{P}^t_{p\mathtt{A}} + \mathbf{T}_{p\mathtt{B}}\mathbf{P}^t_{p\mathtt{B}} + \mathbf{T}_{p\mathtt{AB}}\mathbf{P}^t_{p\mathtt{AB}} + \mathbf{T}_o\mathbf{P}_o + \mathbf{E}_{W_{mb}}\\ \mathbf{T}_{all} &= \mathbf{T}_{p\mathtt{A}}\mathbf{Q}^t_{p\mathtt{A}} + \mathbf{T}_{p\mathtt{B}}\mathbf{Q}^t_{p\mathtt{B}} + \mathbf{T}_{p\mathtt{AB}}\mathbf{Q}^t_{p\mathtt{AB}} + \mathbf{E}_{T_{all}}\end{aligned} \tag{5.15}$$

   where the subscripts p and o indicate the predictive and orthogonal components, while $\mathbf{E}_{W_{mb}}$ and $\mathbf{E}_{T_{all}}$ are the KOPLS residuals for the predictors and the responses, respectively. Accordingly, $\mathbf{T}_{p\mathtt{A}}$ and $\mathbf{P}_{p\mathtt{A}}$ are the scores and loadings, respectively, for the predictive components related to the effect of Factor A, $\mathbf{T}_{p\mathtt{B}}$ and $\mathbf{P}_{p\mathtt{B}}$ those associated with the effect of Factor B, and $\mathbf{T}_{p\mathtt{AB}}$ and $\mathbf{P}_{p\mathtt{AB}}$ the ones related to the interaction A : B.

The characteristics of multiblock OPLS are that it allows the decomposition of the predictor blocks into response-predictive and response-orthogonal components and, in the context of the present method, that each predictive component is related to a specific and individual term in the ANOVA decomposition. Accordingly, interpretation of the effect of factor or interaction terms relies on the inspection of the predictive components. In particular the contribution of the fth block to the rth component $\mathbf{t}_r$ (also referred to as its weight), $\lambda_r^{(f)}$, can be calculated as:

$$\lambda_r^{(f)} = \mathbf{t}_r^t\mathbf{W}_f\mathbf{t}_r \tag{5.16}$$

The block weights are exploited to match each predictive component to the factor or interaction it refers to, and also as a criterion to assess whether the corresponding term is worth to be interpreted or not. Indeed, as the only common source of variation among the effect submatrices are the ANOVA residuals ($\mathbf{Y}_{\mathtt{Ind}}$) that have been added to all of them when multiple blocks have high contributions on a component, the latter will be very likely to be dominated by noise. On the other hand, a relevant source of variation will provide a dominant contribution to the corresponding component. After having paired the components with the corresponding design terms, as in other HD-ANOVA approaches, the score plot allows us to evaluate the distribution of the samples associated with the different factor levels and to evaluate whether the design structure emerges from the plot.

Significance of the effect of the factors or interactions can be evaluated by inspecting the first orthogonal component (here denoted by the subscript $oc1$), which should summarize the main contribution of the ANOVA residuals. In particular, it is suggested to compare the contribution of each effect submatrix to this component, with the contribution of the residual submatrix, by introducing the so-called Residual Structure Ratio (RSR):

$$RSR_f = \frac{\lambda_{oc1}^{(\mathrm{Ind})}}{\lambda_{oc1}^{(f)}} \tag{5.17}$$

This statistic resembles very closely the F-ratio used for significance testing in AComDim (see Equation 5.57). When a factor or interaction has a relevant effect, its contribution to the submatrix will be dominant with respect to the added residuals and, correspondingly, its contribution to the first orthogonal component associated with residuals will be low, leading to a high RSR value. On the other hand, when dealing with a factor or interaction inducing a low variation, comparable with that of the residuals, the RSR will be close to one.

Example 5.2: Example of AMOPLS on the pasta data

The use of AMOPLS will be illustrated through the analysis of the same pasta data set already discussed in Example 5.1. However, for the AMOPLS modeling the decomposition in Equation 5.9 includes only the main terms and the binary interactions, while the ternary interaction is included in $\mathbf{Y}_{\mathrm{Ind}}$. To prepare the multivariate response for the multiblock OPLS model, each of the 6 effect matrices $\hat{\mathbf{Y}}_f$ is subjected to SVD, and the scores vectors for all the components with non-zero eigenvalues are retained and concatenated:

$$\mathbf{T}_{all} = [\mathbf{T}_{\mathrm{T}}|\mathbf{T}_{\mathrm{C}}|\mathbf{T}_{\mathrm{t}}|\mathbf{T}_{\mathrm{TC}}|\mathbf{T}_{\mathrm{Tt}}|\mathbf{T}_{\mathrm{Ct}}] \tag{5.18}$$

where each of the matrices $\mathbf{T}_f$ collects the scores extracted from $\hat{\mathbf{Y}}_f$ (with $f = \mathrm{T, C, t, TC, Tt, Ct}$). At the same time, to build the multiblock dataset of predictors, the matrix $\mathbf{Y}_{\mathrm{Ind}}$ is added to each of the matrices $\hat{\mathbf{Y}}_f$ (with $f = \mathrm{T, C, t, TC, Tt, Ct}$) and the kernel matrix $\mathbf{W}$ is built as in Equation 5.14.

A kernel OPLS model is then calculated between $\mathbf{T}_{all}$ as the multivariate response matrix (with 33 columns) and the consensus kernel $\mathbf{W}$ as predictor block. In AMOPLS, optimal model complexity is evaluated by including as many predictive components as the number of dependent variables (the columns of $\mathbf{T}_{all}$; in this case, 33) and testing how the model changes with the addition of one or more orthogonal components. This is accomplished by evaluating how the overall R^2 for the prediction of the responses changes with the addition of an increasing number of orthogonal components. Since no significant change in fitting was achieved with the addition of further components, only one orthogonal component was included in the final model.

To relate each of the predictive components to the terms in the experimental design, it is necessary to inspect the weights of the different blocks (each corresponding to a factor or interaction, or to the residuals) on those components (see Figure 5.6).

By looking at Figure 5.6, it is evident how the first predictive component accounts for the effect of *Temperature*, the second predictive component for the effect of *Time*, the third for that of *Egg concentration* and the fourth and fifth components for the interaction between *Temperature* and *Egg concentration*. For all the remaining components, since the weights of the various blocks are almost equal, it is difficult to assign them to the effect of a specific factor or interaction, and this is one of the most critical aspects of the method.

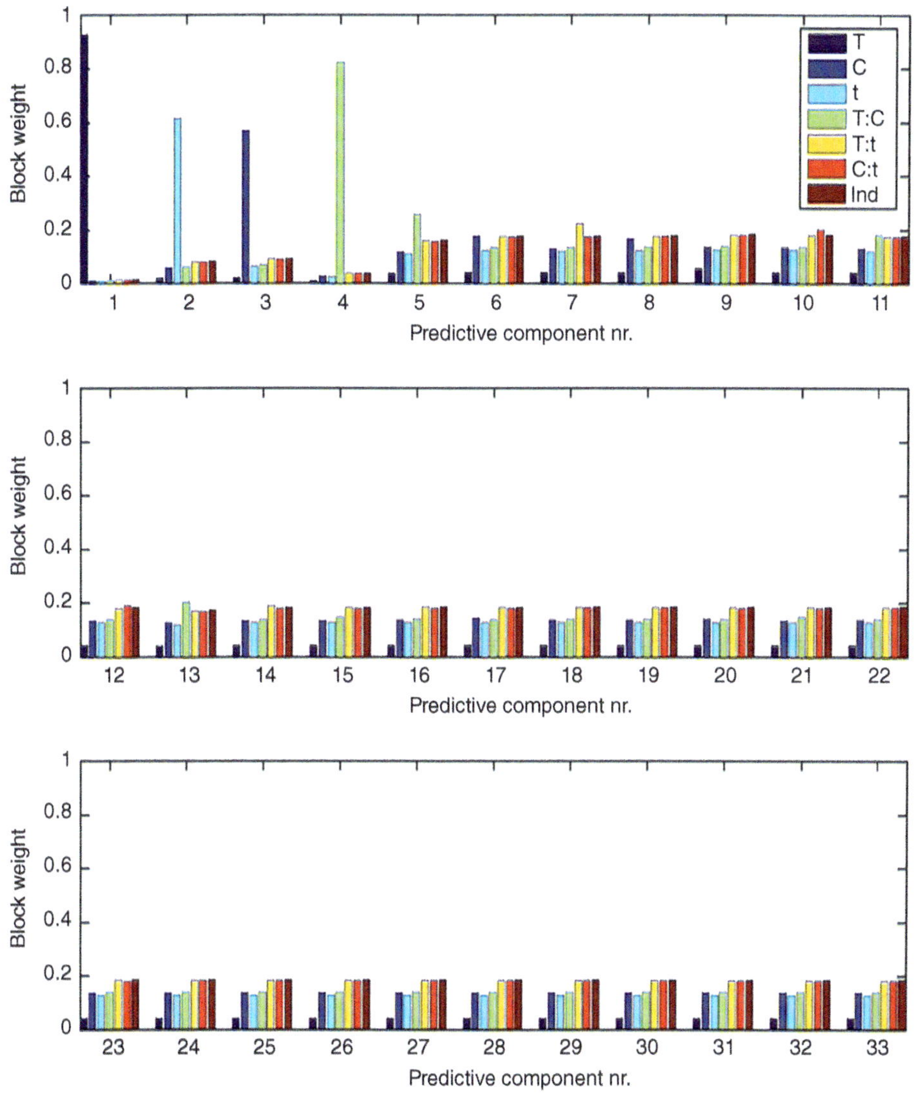

Figure 5.6 Block weights on the 33 predictive components for the AMOPLS model on the pasta data set.

Significance of the effect of the factors or interactions can be evaluated by calculation of the Residual Structure Ratio (RSR), according to Equation 5.17. The idea behind the definition of the RSR is that, if a factor or interaction has a significant effect, the weight of the corresponding block on the orthogonal component should be low, leading to a high RSR value. Operationally, for each factor or interaction, the RSR value is compared to its null distribution evaluated by permutation testing to obtain a non-parametric estimate of a p-value. For the pasta data set, the values of the RSR and the corresponding p-values (estimated with 1000 randomizations) are summarized in Table 5.2.

Table 5.2 RSR values and results of permutation testing for estimating the effect significance.

Source	RSR	*p*-value
Temperature (*T*)	4.333	0.001
Concentration (*C*)	1.338	0.001
Time (*t*)	1.434	0.001
T:*C*	1.313	0.001
T:*t*	1.010	0.310
C:*t*	1.018	0.560

According to the results in Table 5.2, all the three factors and the interaction between *Temperature* and *Egg concentration* have a significant effect on the spectral profile, consistent with what was reported in the original paper (Bevilacqua *et al.*, 2013).

Interpretation of the effect of a factor or interaction can then be accomplished by inspecting the scores and loadings on the corresponding predictive component. As an example, the effect of *Temperature*, which is associated with the first predictive component, will be briefly discussed.

The scores and loadings for the first predictive component of the AMOPLS model, which is associated with the effect of *Temperature*, are displayed in Figure 5.7. The score plot clearly shows

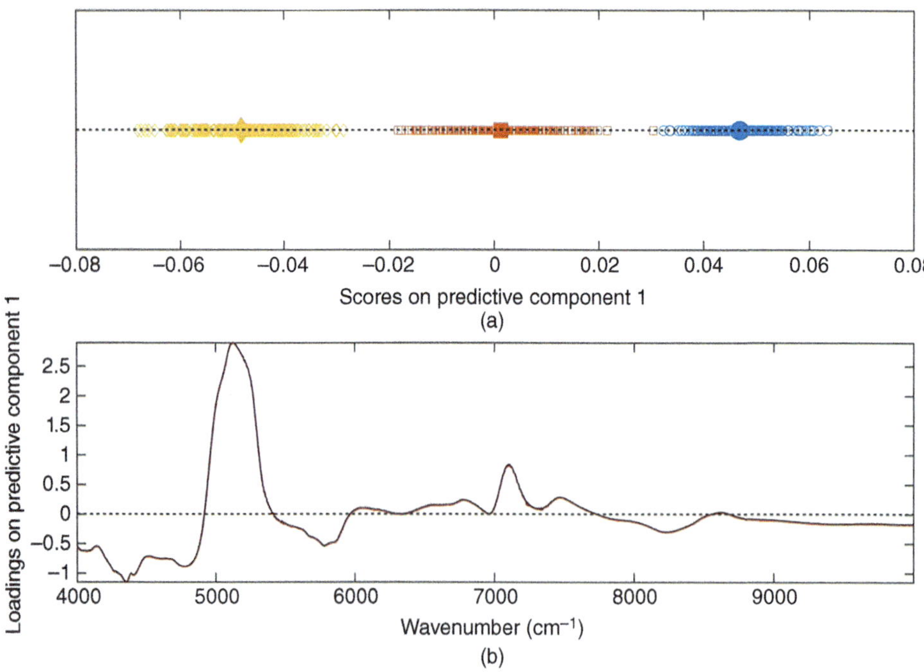

Figure 5.7 (a) Samples scores onto the first predictive components of the AMOPLS model, corresponding to the effect of temperature (empty symbols), together with their mean values for the three levels of the factor (filled symbols). Blue circles – 40 °C; orange squares – 52 °C; yellow diamonds – 65 °C. (b) Variable loadings for first predictive components (5.57).

that the three levels are well separated along the component, the scores progressively decreasing with increasing *Temperature*. The loadings on the components are consistent with what was already discussed for the ANOVA-TP model (see Example 5.1): increasing the *Temperature* results in a higher spectral intensity between 4000 and 4900 cm^{-1} and between 5400 and 5940 cm^{-1} and a lower intensity between 4900 and 5400 cm^{-1} and between 6400 and 7600 cm^{-1}.

5.3 LMM-Based Methods

In this section, we will discuss ASCA-type models based on linear mixed models, with a focus on Repeated Measures ASCA+ (RM-ASCA+, Madssen *et al.*, 2021) and Linear Mixed Models-PCA (LiMM-PCA, Martin and Govaerts, 2020). Univariate mixed models for longitudinal studies in which the effect of different treatments is followed over time were introduced already in Section 3.5.2. Although the linear mixed models are well established for single-response variables, their extension toward high-dimensional data structures with correlated variables is still in its infancy. Recently, two new approaches that combine linear mixed models and high-dimensional data analysis methods such as PCA have been introduced. These two methods will be discussed in this section.

5.3.1 RM-ASCA+ with Qualitative Time Models

Repeated measures ASCA+ extends the linear mixed models to the multivariate domain (Madssen *et al.*, 2021). The linear mixed models are estimated for each variable separately, and then the corresponding model parameters are combined for all J variables. RM-ASCA+ was applied to studies with short time series of 3 and 5 time points and used the categorical approach for the *Time* factor.

To illustrate the multivariate RM-ASCA+ method the example of Section 3.5.2.5 where $I(i = 1, \ldots, I)$ individuals take part in a study with $A = 3$ different treatments ($a = 0, \ldots, 2$) will be continued. Each individual only receives a single treatment, and treatment 0 is considered the reference. After the treatment has been provided, measurements are obtained at $T = 4$ different time points ($t = 0, \ldots, 3$) for each individual and J variables ($j = 1, \ldots, J$) are measured. The first timepoint $t = 0$ is measured before the start of the treatment. The LMM model for measurement y_{aitj} is:

$$y_{aitj} = (\mu_j + \gamma_{ij}) + \alpha_{aj} + \beta_{tj} + (\alpha\beta)_{atj} + \epsilon_{aitj} \tag{5.19}$$

The model can also be defined in regression style. Here, both factors *Time* and *Treatment* are reference coded, and the 0 levels for both *Time* and *Treatment* are used as the reference levels. Then the RM-ASCA+ model can be defined as

$$\begin{aligned} y_{aitj} = {} & (\theta_{0j} + z_{ij}\gamma_{ij}) + A_{1a}\theta_{1aj} + A_{2a}\theta_{2aj} + T_{1t}\theta_{3tj} + T_{2t}\theta_{4tj} + T_{3t}\theta_{5tj} + T_1A_{1at}\theta_{6atj} \\ & + T_2A_{1at}\theta_{7atj} + T_3A_{1at}\theta_{8atj} + T_1A_{2at}\theta_{9atj} + T_2A_{2at}\theta_{10atj} + T_3A_{2at}\theta_{11atj} + \epsilon_{aitj} \end{aligned} \tag{5.20}$$

where the indicator variables are defined as was described in Tables 3.8–3.11. Aggregating over all samples and variables the multivariate mixed model can be written as:

$$\mathbf{Y} = \mathbf{X}\hat{\boldsymbol{\Theta}} + \mathbf{Z}\hat{\boldsymbol{\Gamma}} + \mathbf{E} \tag{5.21}$$

where $\mathbf{Y}$ now contains J variables and IT samples, $\mathbf{X} = [\mathbf{1}|\mathbf{X}_{\mathrm{A}}|\mathbf{X}_{\mathrm{T}}|\mathbf{X}_{\mathrm{AT}}]$, $\hat{\boldsymbol{\Theta}}$ contains the regression coefficients for each of the 12 $\boldsymbol{\theta}$ values for all J variables. Furthermore, $\mathbf{Z}$ is the indicator variable for the individual which in this example only contains a contribution for the random intercept, and $\hat{\boldsymbol{\Gamma}}$ can be predicted as discussed in Equation 3.100. $\hat{\boldsymbol{\Gamma}}$ contains the best linear unbiased predictor (BLUP) for the random intercept for each variable for each individual. The fixed part of the high-dimensional LMM, $\mathbf{X}\hat{\boldsymbol{\Theta}}$, can be separated into parts for each of the factors, their interaction, and for the intercept:

$$\mathbf{X}\hat{\boldsymbol{\Theta}} = \left[\mathbf{1}|\mathbf{X}_{\mathrm{A}}|\mathbf{X}_{\mathrm{T}}|\mathbf{X}_{\mathrm{AT}}\right]\begin{bmatrix}\hat{\theta}_0^t\\ \hat{\boldsymbol{\Theta}}_{\mathrm{A}}\\ \hat{\boldsymbol{\Theta}}_{\mathrm{T}}\\ \hat{\boldsymbol{\Theta}}_{\mathrm{AT}}\end{bmatrix} = [\mathbf{1}\hat{\theta}_0^t] + [\mathbf{X}_{\mathrm{A}}\hat{\boldsymbol{\Theta}}_{\mathrm{A}}] + [\mathbf{X}_{\mathrm{T}}\hat{\boldsymbol{\Theta}}_{\mathrm{T}}] + [\mathbf{X}_{\mathrm{AT}}\hat{\boldsymbol{\Theta}}_{\mathrm{AT}}] \tag{5.22}$$

Here $\mathbf{1}$ represents a column of ones $(IT \times 1)$ and $\hat{\theta}_0(J \times 1)$ contains the mean value of each of the J variables, $\mathbf{X}_{\mathrm{A}}(IT \times 2)$ contains the columns $A1$ and $A2$ for each measurement, and $\hat{\boldsymbol{\Theta}}_{\mathrm{A}}(2 \times J)$ contains the two coefficients $[\hat{\theta}_1|\hat{\theta}_2]^t$ for each variable, see Equation 5.20. $\mathbf{X}_{\mathrm{T}}(IT \times 3)$ contains the columns $T1$, $T2$, and $T3$ for each measurement, and $\hat{\boldsymbol{\Theta}}_{\mathrm{T}}(3 \times J)$ contains the three coefficients $[\hat{\theta}_3|\hat{\theta}_4|\hat{\theta}_5]^t$ for each variable, and $\mathbf{X}_{\mathrm{AT}}(IT \times 6)$ contains the interactions $[T_1A_1|T_2A_1|T_3A_1|T_1A_2|T_2A_2|T_3A_2]$ for each of the measurements, and $\hat{\boldsymbol{\Theta}}_{\mathrm{AT}}(6 \times J)$ contains the coefficients $[\hat{\theta}_6|\hat{\theta}_7|\hat{\theta}_8|\hat{\theta}_9|\hat{\theta}_{10}|\hat{\theta}_{11}]^t$ for each variable. In balanced situations, with an equal number of replicates for each experimental setting, the variance explained by the full model $\mathbf{X}\hat{\boldsymbol{\Theta}}$ can be separated into its contributions to each of the factors and their interactions.

$$||\mathbf{X}\hat{\boldsymbol{\Theta}}||^2 = ||\mathbf{1}\hat{\theta}_0^t||^2 + ||\mathbf{X}_{\mathrm{A}}\hat{\boldsymbol{\Theta}}_{\mathrm{A}}||^2 + ||\mathbf{X}_{\mathrm{T}}\hat{\boldsymbol{\Theta}}_{\mathrm{T}}||^2 + ||\mathbf{X}_{\mathrm{AT}}\hat{\boldsymbol{\Theta}}_{\mathrm{AT}}||^2 \tag{5.23}$$

and these effect matrices can be explored using PCA.

$$\mathbf{X}\hat{\boldsymbol{\Theta}} = \mathbf{M}_0 + (\mathbf{T}_{\mathrm{A}}\mathbf{P}_{\mathrm{A}}^t + \mathbf{E}_{\mathrm{A}}) + (\mathbf{T}_{\mathrm{T}}\mathbf{P}_{\mathrm{T}}^t + \mathbf{E}_{\mathrm{T}}) + (\mathbf{T}_{\mathrm{AT}}\mathbf{P}_{\mathrm{AT}}^t + \mathbf{E}_{\mathrm{AT}}) \tag{5.24}$$

If reference coding is used, $\mathbf{M}_0$ represents the estimated value for the samples at the reference levels for both *Treatment* $(a = 0)$ and *Time* $(t = 0)$. The effect matrices $\mathbf{X}_{\mathrm{A}}\hat{\boldsymbol{\Theta}}_{\mathrm{A}}$, $\mathbf{X}_{\mathrm{T}}\hat{\boldsymbol{\Theta}}_{\mathrm{T}}$ and $\mathbf{X}_{\mathrm{AT}}\hat{\boldsymbol{\Theta}}_{\mathrm{AT}}$ are therefore not mean-centered, and could be mean-centered before applying PCA.

In our example, there is only a single random factor, which is the individual intercept γ_{ij} for each variable j, such that $\mathbf{Z}\hat{\boldsymbol{\Gamma}}$ is represented by $\mathbf{Z}$ as indicated in Table 3.11 and $\hat{\boldsymbol{\Gamma}}$ is an $I \times J$ matrix with BLUPs for all individuals for variable j in column $\hat{\gamma}_j$.

5.3.2 Validation of the RM-ASCA+ Model

There are different approaches to validate the RM-ASCA+ model, or specific terms within the model. Here we will discuss the nonparametric bootstrap approach and the permutation approach. A disadvantage of both approaches is that the calculation of mixed models is rather time-consuming and thus does not apply well to high-dimensional data as linear mixed models have to be calculated for each variable for all bootstrap or permutation calculations. The LiMM-PCA model offers a solution in these cases because it reduces the number of variables using an initial PCA step (see Section 5.3.3).

5.3.2.1 Validation of RM-ASCA+ Models with Nonparametric Bootstrap

The nonparametric bootstrap was introduced in the RM-ASCA+ paper (Madssen *et al.*, 2021). The main idea of the nonparametric bootstrap is to create datasets similar to the original one by resampling of each group in the data with replacement. In Example 5.3 (see later), there are three treatment groups that consist of 60 distal Roux-en-Y gastric bypass (RYGB), 27 proximal RYGB, and 19 gastric sleeve individuals. As each of these individuals is measured 5 times, the data has a repeated measures structure. To create a similar dataset we randomly sample 60 individuals *with replacement* from the distal RYGB group, 27 from the proximal RYGB group, and 19 from the gastric sleeve group. Because of the possible replacement, some individuals will be selected twice or even more often, while others will not be selected. The RM-ASCA+ model will be applied to all these randomly created datasets and provide model parameters. If, *e.g.*, 1000 bootstrap samples are created then for each parameter in the model there will be 1000 estimates. If the model parameter is very important to describe the variation in the data, then the 1000 estimates will all be similar and different from zero. From the 1000 estimates, an average and a 95% confidence interval can be calculated. If the interval does not include 0, the model parameter is considered significant. As the RM-ASCA+ model also includes a PCA step, there are three sources of unintended variation.

- The sign of the scores and loadings can be inverted without affecting the fit of the model.
- If two components almost have equal explained variation, their order can easily change in the different bootstrap samples.
- Due to the rotational freedom of PCA, groups of scores and loadings can be rotated while still providing the same fit of the data.

These three sources of variation in the model parameters have to be checked and corrected for using Procrustes rotation (Martens and Martens, 2000; Timmerman *et al.*, 2007) before the different estimates of the model parameters can be compared. The nonparametric bootstrap provides confidence intervals around each level of a factor or interaction, comparable to a post-hoc test. It thus provides an attractive visualization of differences between time profiles of the different treatments as provided in Example 5.3.

5.3.2.2 Validation of RM-ASCA+ Models with Permutation Testing

Validation of the treatment effect can also be performed using permutation testing. Permutation testing was introduced for univariate ANOVA models in Section 3.7.3, where it was discussed how to permute the samples in a permutation test, and whether an exact or an approximate test can be applied. The main idea of a permutation test is to create a H_0 distribution for the appropriate test statistic by reshuffling samples that are exchangeable under the H_0 distribution for the hypothesis being tested. In this example, we focus on testing the statistical significance of the complete *Treatment* effect including the interaction with *Time* (*Treatment* + *Time:Treatment* interaction). To guarantee the exchangeability, permutation is performed on the residuals of a reduced regression model. The permutation approach to test the significance of the fixed factor f is performed in these steps:

1) Fit the full model $\mathbf{Y} = \mathbf{X}\hat{\mathbf{\Theta}} + \mathbf{Z}\hat{\mathbf{\Gamma}} + \hat{\mathbf{E}}$ with all fixed and random factors. Calculate the test statistic S^{obs} of the unpermuted data. S^{obs} can be $||\mathbf{X}_f\hat{\mathbf{\Theta}}_f||^2$ or the approximation of $||\mathbf{X}_f\hat{\mathbf{\Theta}}_f||^2$ by its first PCs.
2) Fit the reduced model $\mathbf{Y} = \mathbf{X}_{-f}\hat{\hat{\mathbf{\Theta}}}_{-f} + \mathbf{Z}\hat{\hat{\mathbf{\Gamma}}} + \hat{\hat{\mathbf{E}}}$ where a fixed factor (f) is removed from $\mathbf{X}$.
3) For each permutation $p = 1 \ldots P$
   a) $\widetilde{\hat{\mathbf{E}}}^p$ is $\hat{\hat{\mathbf{E}}}$ after randomly permuting its rows.
   b) Create the permuted data: $\mathbf{Y}^p = \mathbf{X}_{-f}\hat{\hat{\mathbf{\Theta}}}_{-f} + \mathbf{Z}\hat{\hat{\mathbf{\Theta}}} + \widetilde{\hat{\mathbf{E}}}^p$.
   c) Fit the full model on the permuted data: $\mathbf{Y}^p = \mathbf{X}\hat{\mathbf{\Theta}}^p + \mathbf{Z}\hat{\mathbf{\Gamma}}^p + \mathbf{E}^p$ and calculate the test statistic S^p.
4) Calculate the p-value as: $p < \frac{(\#S^p \geq S^{obs})+1}{P+1}$

5.3.2.3 Visualization

The random part $\mathbf{Z\Gamma}$ can also be used to indicate the variation in the fixed scores due to the random effect. In ASCA score plots, the back projection approach is used in which the residual variation $\mathbf{E}$ is added to the score plot to indicate the residual variation of the replicates $\mathbf{T}_f^a = (\mathbf{X}_f\boldsymbol{\Theta}_f + \mathbf{E})\mathbf{P}_f$, see Section 4.1.4. Using the random effects, the total residual variation can now be highlighted by adding various random effects: $\mathbf{T}_f^a = (\mathbf{X}_f\boldsymbol{\Theta}_f + \mathbf{Z}_r\boldsymbol{\Gamma}_r + \mathbf{E})\mathbf{P}_f$. Thus the effect of some selected $\mathbf{Z}_r\boldsymbol{\Gamma}_r$ or all of the random effects can be used to show their contribution to the variability of the replicates in the score plots.

Example 5.3: Bariatric surgery

In this example serum metabolites of $I = 105$ patients are measured before and after one of three types ($a = 1 : 3$) of bariatric surgery, the proximal Roux-en-Y gastric bypass (RYGB) (26), distal RYGB (60), and gastric sleeve (GS) (19) (Gralka *et al.*, 2015). The selected type of surgery was based on factors such as obesity and comorbidities. Therefore, the baseline levels can be expected to be different for the individuals, and an unconstrained longitudinal model has to be used, see Equation 5.19. After surgery, four additional measurements were obtained after 3, 6, 9, and 12 months ($t = 0 : 4$). Data is missing for some post-baseline timepoints 7%, 8%, 12%, and 32%, respectively. NMR spectra were obtained and $J = 21$ quantified metabolites were included in the study. The data is available from MetaboLights (MTBLS242). The example is discussed in more detail in Madssen *et al.* (2021).

The unconstrained mixed model used to describe the data (see Section 3.5.2.5) was as follows:

$$\begin{aligned} y_{aitj} = {} & (\theta_{0j} + \gamma_{0ij}) + \theta_{1j}T_{1t} + \theta_{2j}T_{2t} + \theta_{3j}T_{3j} + \theta_{4j}T_{4t} + \theta_{5j}(Distal_a) + \theta_{6j}(Proximal_a) \\ & + \theta_{7j}(Distal_a * T_{1t}) + \theta_{8j}(Distal_a * T_{2t}) + \theta_{9j}(Distal_a * T_{3t}) + \theta_{10j}(Distal_a * T_{4t}) \\ & + \theta_{11j}(Proximal_a * T_{1t}) + \theta_{12j}(Proximal_a * T_{2t}) + \theta_{13j}(Proximal_a * T_{3t}) \\ & + \theta_{14j}(Proximal_a * T_{4t}) + e_{aitj} \end{aligned} \tag{5.25}$$

Time was reference-coded with baseline (T_0) being the reference and the *Treatment* factor was sum-coded, with GS being coded −1. $T_1..T_4$ thus represents the difference between the baseline value and the timepoints 3, 5, 9, 12 months. We will only look at the *Treatment* + *Time:Treatment* interaction effect. The result is visualized in Figure 5.8.

In this example, the combined effect of *Treatment* and the *Time:Treatment* interaction, estimated with the repeated measures model, is analyzed with PCA. The (a) plot shows a strong first PC with over 60% explained variation. The scoreplot (b) shows the group scores for the three *Treatment* levels at different time points with their 95% confidence intervals. The distal RYGB treatment shows a clearly deviating profile from the other two *Treatment* levels. It can also be observed that at baseline the *Treatment* levels are not exactly the same, which is allowed due to the use of the unconstrained model. The loading plot (c) shows a large contribution of the metabolite methylsulfonylmethane, where the group receiving distal RYGB showed more pronounced increases compared with proximal RYGB or Gastric Sleeve.

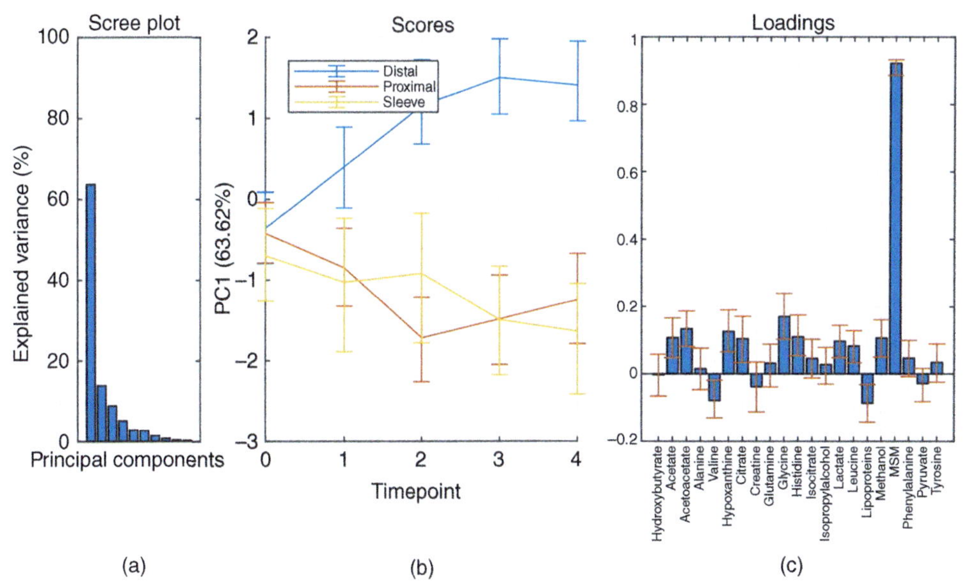

Figure 5.8 RM-ASCA+ decomposition of *Treatment* + *Time:Treatment* interaction effect on the metabolites of the Bariatric surgery data; (a) explained variation of the principal components; (b) score values and confidence intervals of each *Treatment* level for the 5 time points; (c) loading plot of the first principal component. Source: Madssen *et al.* (2021)/PLOS/CC BY 4.0.

5.3.2.4 RM-ASCA+ with Quantitative Time Models

The extension of LMM with quantitative time models to the multivariate domain follows the same steps as described in Section 3.5.2.5. Extending Equation 3.116 for $j = 1 \ldots J$ variables

$$y_{aitj} = (\theta_{0j} + \gamma_{0ij}) + (\theta_{1j} + \gamma_{1ij})T_{it} + (\theta_{2j} + \gamma_{2ij})T_{it}^2 + \theta_{3j}A_{1a} + \theta_{4j}A_{2a} + (\theta_{5j} + \gamma_{5ij})T_{it}A_{1a} + (\theta_{6j} + \gamma_{6ij})T_{it}^2A_{1a} + (\theta_{7j} + \gamma_{7ij})T_{it}A_{2a} + (\theta_{8j} + \gamma_{8ij})T_{it}^2A_{2a} + \epsilon_{aitj} \tag{5.26}$$

represents the quantitative time model for each variable with multiple random effects for intercept γ_{0ij} and slopes $\gamma_{1ij}, \ldots, \gamma_{8ij}$. Equation 5.26 can be extended for all samples and variables of a single individual i simultaneously.

$$\mathbf{Y}_i = \mathbf{X}_i\boldsymbol{\Theta} + \mathbf{Z}_i\boldsymbol{\Gamma}_i + \mathbf{E}_i \tag{5.27}$$

$\mathbf{X}_i$ and $\mathbf{Z}_i$ are then defined as already presented in Tables 3.12 and 3.13, which show the $\mathbf{X}_i$ values for individual i for time points (1.5, 2.8, 4.3, 6.7, …). Instead of coded values, the actual time values are used in the quantitative model. Each individual can have different time values.

Extending Equation 5.27 for all individuals ($i = 1..I$), $\mathbf{X}$ is formed by stacking all $\mathbf{X}_i$ matrices under each other, $\mathbf{X} = [\mathbf{X}_1^t|\mathbf{X}_2^t|\mathbf{X}_i^t|\mathbf{X}_I^t]^t$. The size of $\mathbf{X}$ is $\sum_{i=1}^{I} T_i \times (K_f + 1)$. $\mathbf{Z}$ will be formed by placing all $\mathbf{Z}_i$ matrices in the diagonal of a huge matrix, as was already shown in Equation 3.119. Both $\boldsymbol{\Gamma}$ and $\mathbf{Y}$ are obtained by placing all their individual $\boldsymbol{\Gamma}_i$ and $\mathbf{Y}_i$ matrices below each other, similar to $\mathbf{X}$.

The systematic variation present in the multivariate response $\mathbf{Y}$ is separated into a fixed part ($\mathbf{X\Theta}$) describing the variation due to time and treatment and a random individual $\mathbf{Z\Gamma}$ part. A Linear Mixed Model based on the design matrices $\mathbf{X}$ and $\mathbf{Z}$ is applied to each variable separately, and the coefficients are collected in $\mathbf{\Theta}$ and $\mathbf{\Gamma}$.

In Section 3.5.2.5 it was already discussed that splines can be favorable over polynomial time models because of the improved interpolation properties. In the next example, natural cubic splines were used to model the time profiles. The basis functions $\mathbf{S}$ of a natural cubic spline can be used to model the longitudinal response $\mathbf{y}$ in such a way that the model is linear beyond the boundaries of the measured time points (Hastie *et al.*, 2009).

Example 5.4: RM-ASCA+ with quantitative time model

In this example, metabolomics data from a double-blind randomized controlled trial is analyzed with RM-ASCA+ with a quantitative time model based on natural cubic splines (Erdős *et al.*, 2023). Twenty healthy male volunteers consumed one of two liquid diet meals containing vegetable fat or a mixture of vegetable and bovine fat. After a washout period of a week, the other drink is consumed by the same individual according to a cross-over design. Just before and until 5 hours after the drink, blood samples were obtained of the volunteers (every 30 minutes) such that 11 timepoints were available for each volunteer after both drinks. From the blood samples, 35 different lipoprotein fractions (LDL and VLDL, XXL – S subgroups) were collected and Esterified Cholesterol (CE), Free Cholesterol (FC), Phospholipids (PL), and Triglycerides (TG) from these fractions were measured. The measurements were scaled by the standard deviation of the variable at baseline across individuals. A constrained model was used as no group average differences were expected at the baseline of the two drinks. The model for each metabolite j is defined as:

$$\begin{aligned} y_{aitj} &= (\theta_{0j} + \gamma_{i0j}) + ((\theta_{1j} + \gamma_{i1j})S_{1t}) + \cdots + ((\theta_{3j} + \gamma_{i3j})S_{3t}) \\ &\quad + ((\theta_{4j} + \gamma_{i4j})S_{1t}A_{1a}) + \cdots + ((\theta_{9j} + \gamma_{i9j})S_{3t}A_{2a}) + \epsilon_{aitj} \end{aligned} \tag{5.28}$$

Here $S_{1t}, \ldots, S_{3t}$ represent the basis functions for the natural cubic spline for time t, A_{1a}, A_{2a} represent the two treatments, and θ_j and γ_j represent the fixed and random model parameters for each of the *Time* contributions and the *Time:Treatment* interactions for metabolite j. This model has many random effects which might be reduced, as was also performed in Example 3.6, where the same random effects γ_{2i} and γ_{3i} were used for multiple time points. Combining all *Time* and *Time:Treatment* interactions for all metabolites simultaneously, the model can be written as:

$$\mathbf{Y} = \mathbf{M}_0^f + \mathbf{M}_T^f + \mathbf{M}_{TA}^f + \mathbf{M}_0^r + \mathbf{M}_T^r + \mathbf{M}_{TA}^r + \mathbf{E}_Y \tag{5.29}$$

Each effect matrix separately or in combination can be explored using PCA, and score and loading plots can help in interpreting the effect profiles. The fixed effect matrices ($\mathbf{M}_T^f + \mathbf{M}_{TA}^f$) contain the systematic variation over all individuals where all the random individual effects $\mathbf{M}_0^r$, $\mathbf{M}_T^r$, and $\mathbf{M}_{TA}^r$ have been removed. PCA analysis of this variation is shown below.

The scores of PC1 of the systematic ($\mathbf{M}_T^f + \mathbf{M}_{TA}^f$) variation in the fixed effects show the average time profiles of both diets (A and B) together with their confidence intervals estimated by bootstrap resampling. The bootstrap procedure was applied 100 times in which individuals were

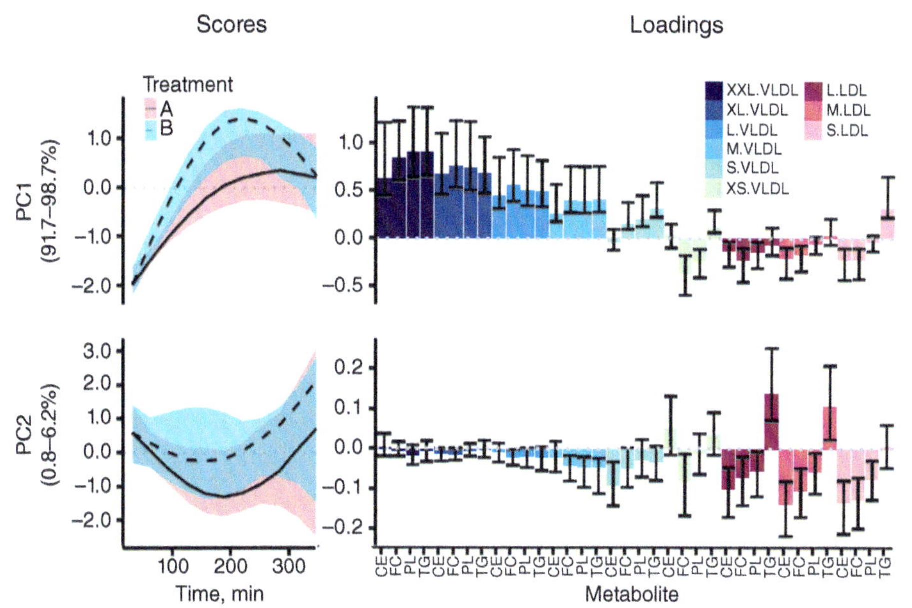

Figure 5.9 PCA Score and loading plots *Time* +*Time:Treatment* effect variation. Scores contain prominent patterns over time, while the loadings show the association of the scores with the metabolite time-courses. Metabolites are shown in the axis label of the loadings with the colors indicating the various subclasses. FC: free cholesterol, CE: esterified cholesterol, PL: phospholipids, TG: triglycerides. Resampling-based 95% confidence intervals are shown as shaded areas for the scores and error bars for the loadings. Source: Erdős *et al.* (2023).

selected with replacement, and the analysis of the data was repeated. The amounts of variation explained by the first PC ranged between 91.7% and 98.7%. The average and a 95% confidence interval of the 100 score profiles are shown in Figure 5.9. Diet B shows a faster increase than diet A. This faster increase is especially related to all CE, FC, PL and TG in the large (XXL, XL, X, M) VLDL particles as shown in the loading plot of PC1, where the confidence intervals do not include 0. The rather small contribution of the 2nd PC does not seem significant as all loading value confidence intervals include 0.

The random individual variation due to the *Time:Treatment* interaction, $\mathbf{M}^r_{TA}$, can also be studied to explore individual differences, whether there are clearly deviating individuals or whether groups of similar individuals exist in the cohort. The $AIT \times J$ matrix $\mathbf{M}^r_{TA}$ can be reshaped in an $I \times ATJ$ matrix where each row contains all variations related to a single individual. PCA on this matrix gives a score matrix $I \times R$ containing score values for all individuals.

Figure 5.10a shows the score plot of the individual variation of the *Time:Treatment* interaction over all timepoints, *Treatments* and metabolites. It can be observed that individuals 17 and 18 are on either side of PC1 of this score plot indicating deviating patterns for these two individuals. Figure 5.10b shows some of the compounds measured where these individuals are highlighted for their two time profiles after drinking the two different diets.

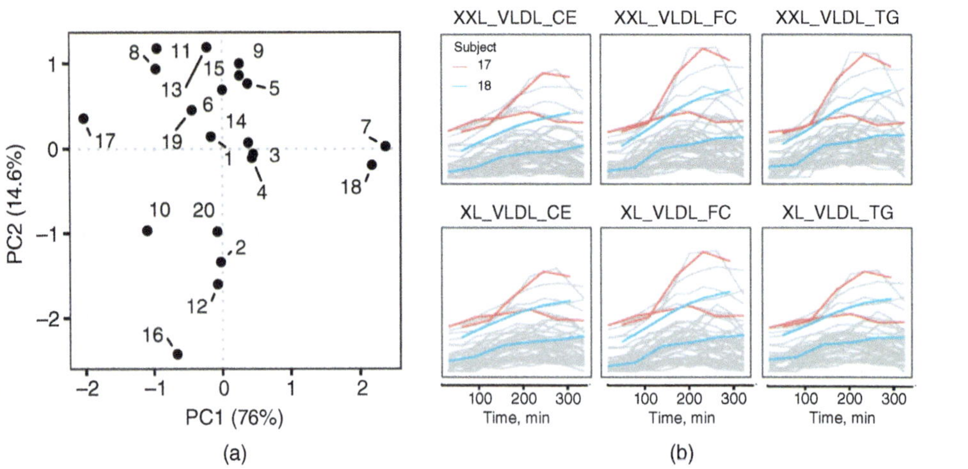

Figure 5.10 (a) PCA Score of the *Time:Treatment* random individual variation. (b) *Time* profiles of esterified cholesterol (CE), free cholesterol (FC) and triglycerides (TG) in XXL-VLDL and XL-VLDL particles. Profiles of individuals 17 (red) and 18 (cyan) for both diets are indicated and show clear individual differences reflected in the position in the score plot. Source: Erdős *et al.* (2023)/PLOS/CC BY 4.0.

5.3.3 LiMM-PCA

A second method for analyzing high-dimensional data using a mixed model that includes fixed and random effects is called Linear Mixed Models-PCA, abbreviated as LiMM-PCA. After an optional initial step of data preprocessing (*e.g.*, transformation, normalization, scaling of the response matrix) the LiMM-PCA approach chooses a different route than RM-ASCA+. As a first step, a PCA of the response matrix $\mathbf{Y}$ is performed to create a score matrix $\widetilde{\mathbf{Y}}$ that consists of independent non-correlated latent responses and a loading matrix $\mathbf{Q}$ that mainly serves as a link between the measured variables $\mathbf{Y}$ and the orthogonal responses $\widetilde{\mathbf{Y}}$ that will be analyzed using the linear mixed models.

$$\mathbf{Y} = \widetilde{\mathbf{Y}}\mathbf{Q}^t + \mathbf{E} \tag{5.30}$$

There are two good reasons for the PCA step before applying the linear mixed model estimation. Many data sets, especially in omics research, have many more variables than samples. After the PCA step, the number of principal components is much lower than the initial number of variables, and as the LMM modeling is computationally costly, reducing the number of variables is rather useful, especially if resampling strategies are used for validation.

The second reason for the initial PCA step is that the principal component scores are orthogonal and independent if the original responses do follow a multivariate normal distribution. This is a necessary assumption for the Global Log-Likelihood ratio test that is used in LiMM-PCA as a validation step to test whether the effects are significant.

The number of components to retain $(L, l = 1, \ldots, L)$ in $\widetilde{\mathbf{Y}}$ should be high to retain almost all variation. The main goal is the projection toward orthogonal variables in a reduced space. Depending on the effect size and nature of the applied factor the induced variation will appear in earlier or later components. Now similar to the model in Equation 5.19, the linear mixed model can be applied to each of the independent variables in $\widetilde{\mathbf{Y}}$.

$$\tilde{\mathbf{y}}_l = \mathbf{X}\tilde{\boldsymbol{\theta}}_l + \mathbf{Z}\tilde{\boldsymbol{\gamma}}_l + \tilde{\boldsymbol{\epsilon}}_l \tag{5.31}$$

$\tilde{\mathbf{y}}_l$, the lth column of $\tilde{\mathbf{Y}}$, can be modeled in a similar way as in the RM-ASCA+ approach using the fixed effect factors in $\mathbf{X}$ and the random effects in $\mathbf{Z}$ but now with corresponding coefficients $\tilde{\theta}_l$ and $\tilde{\gamma}_l$ as was explained in Section 3.5.2.5.

The fixed factor parameter $\tilde{\theta}_l$ is estimated using weighted least squares. $\hat{\tilde{\theta}}_l = (\mathbf{X}^t\hat{\mathbf{V}}_l^{-1}\mathbf{X})^{-1}\mathbf{X}^t\hat{\mathbf{V}}_l^{-1}\tilde{\mathbf{y}}_l$, while $\hat{\tilde{\gamma}}_l = \hat{\mathbf{G}}_l\mathbf{Z}^t\hat{\mathbf{V}}_l^{-1}(\tilde{\mathbf{y}}_l - \mathbf{X}\hat{\tilde{\theta}}_l)$. Here $\hat{\mathbf{G}}_l$ is the estimated random effects covariance matrix for $\tilde{\mathbf{y}}_l$. Furthermore, the variance $\hat{\sigma}^2_{rl}$ of the R random effects for $\tilde{\mathbf{y}}_l$ and the residual variance for column l $\hat{\sigma}^2_{el}$ of $\tilde{\boldsymbol{\epsilon}}_l$ are estimated by the REML procedure, see Elaboration 2.4. Combining Equation 5.31 for all columns $\tilde{\mathbf{y}}_l$ leads to the complete decomposition of $\tilde{\mathbf{Y}}$.

$$\tilde{\mathbf{Y}} = \mathbf{X}\hat{\tilde{\boldsymbol{\Theta}}} + \mathbf{Z}\hat{\tilde{\boldsymbol{\Gamma}}} + \hat{\tilde{\mathbf{E}}} \tag{5.32}$$

Now $\hat{\tilde{\boldsymbol{\Theta}}}$ is a matrix of estimated fixed parameters for all columns in $\mathbf{X}$ and $\hat{\tilde{\boldsymbol{\Gamma}}}$ the matrix of random effects coefficients for each of the random effects $\mathbf{Z}$. Similar to RM-ASCA+, here also $\mathbf{X}\hat{\tilde{\boldsymbol{\Theta}}}$ can be decomposed into separate effects for each factor and their interactions provided the design is balanced.

$$||\mathbf{X}\hat{\tilde{\boldsymbol{\Theta}}}||^2 = ||\mathbf{1}\hat{\tilde{\theta}}_0^t||^2 + ||\mathbf{X}_A\hat{\tilde{\boldsymbol{\Theta}}}_A||^2 + ||\mathbf{X}_T\hat{\tilde{\boldsymbol{\Theta}}}_T||^2 + ||\mathbf{X}_{AT}\hat{\tilde{\boldsymbol{\Theta}}}_{AT}||^2 \tag{5.33}$$

5.3.3.1 Validation

The estimation of significance of effects is more cumbersome for multivariate mixed models. The approach used for LiMM-PCA calculates significance for every single column $\tilde{\mathbf{y}}_l$ using a log-likelihood ratio test, and then the results are combined for all variables in $\tilde{\mathbf{Y}}$ using the global log-likelihood ratio (GLLR) test. The GLLR test assumes independence between the different columns in $\tilde{\mathbf{Y}}$ and that has been achieved by the initial PCA step in Equation 5.30, although this only holds for multivariate data that is normally distributed.

For fixed effects, the variance due to effect f for each variable $\tilde{\mathbf{y}}_l$ is $\hat{\sigma}^2_{fl} = var(\mathbf{X}_f\hat{\tilde{\boldsymbol{\Theta}}}_{fl})$. The total variance of $\tilde{\mathbf{Y}}$ is the sum of the variance components of the fixed and random effects.

$$var(\tilde{\mathbf{Y}}|\mathbf{X}) = \sum_{l=1}^{L} var(\tilde{\mathbf{y}}_l) = \sum_{l=1}^{L}\left(\sum_{f=1}^{F}\hat{\tau}^2_{fl} + \sum_{r=1}^{R}\hat{\sigma}^2_{rl} + \hat{\sigma}^2_{el}\right) \tag{5.34}$$

In the case of a balanced design, the method is equal to the least squares approach. However, for an unbalanced design where the factors are not orthogonal anymore, the least squares approach does not work. Thiel *et al.* (2017) showed that in unbalanced cases the Type III sum-of-squares can be used instead, adapted according to the Type III sum-of-squares approach that was used in ASCA+. The percentage of explained variation due to a single fixed factor f can be addressed as:

$$\%Var_f = \frac{||\hat{\tilde{\mathbf{E}}}_{-f}||^2 - ||\hat{\tilde{\mathbf{E}}}_{full}||^2}{||\tilde{\mathbf{Y}} - \mathbf{1}\hat{\tilde{\theta}}_0^t||^2} \tag{5.35}$$

where $\hat{\tilde{\mathbf{E}}}_{-f}$ is the error matrix of a full model without effect f included, while $\hat{\tilde{\mathbf{E}}}_{full}$ is the error matrix of the full model with all effects included. The denominator represents the error matrix of the model with only the intercept. The type III sum-of-squares equals the normal sum-of-squares when the orthogonality between f and the other effects holds, such as in balanced situations. In the high-dimensional LMM framework, the variance described by fixed factor f can be obtained from

$$\%Var_f = \frac{1}{I}(\mathbf{X}_f\hat{\tilde{\boldsymbol{\Theta}}}_f)^t\left(\mathbf{X}_f(\mathbf{X}^t\mathbf{X})^{-1}\mathbf{X}_f^t\right)^{-1}(\mathbf{X}_f\hat{\tilde{\boldsymbol{\Theta}}}_f) \tag{5.36}$$

The statistical significance of the variance described by factor f over all variables $\tilde{\mathbf{y}}_l$ can be tested with the Global Log-Likelihood ratio test, which tests the improvement of the likelihood of the alternative hypothesis H_1 with the extended model compared to the likelihood of the reduced model without factor f in H_0.

- H_0: Variation in $\mathbf{Y}$ is described optimally by the reduced model.
- H_1: Variation is $\mathbf{Y}$ is described optimally by the extended model.

To quantify the significance of the fixed effect in the multivariate case for all columns of $\tilde{\mathbf{Y}}$ simultaneously, the log-likelihood ratios of all columns in $\tilde{\mathbf{Y}}$ are summed.

$$GLLR = 2\left[\sum_{l=1}^{L}\left(log(L_{H_{1,l}})\right) - log\left(L_{H_{0,l}}\right)\right] \tag{5.37}$$

As an example we will use the estimate of the significance of the interaction between *Time* and *Treatment* of Equation 5.33. The models under the H_0 and H_1 hypotheses are defined as:

- Reduced model: $\tilde{\mathbf{Y}} = [\mathbf{x}_0\hat{\theta}_0^t] + [\mathbf{X}_A\hat{\boldsymbol{\Theta}}_A] + [\mathbf{X}_T\hat{\boldsymbol{\Theta}}_T] + \mathbf{Z}\hat{\boldsymbol{\Gamma}} + \hat{\mathbf{E}}$
- Extended model: $\tilde{\mathbf{Y}} = [\mathbf{x}_0\hat{\theta}_0^t] + [\mathbf{X}_A\hat{\boldsymbol{\Theta}}_A] + [\mathbf{X}_T\hat{\boldsymbol{\Theta}}_T] + [\mathbf{X}_{AT}\hat{\boldsymbol{\Theta}}_{AT}] + \mathbf{Z}\hat{\boldsymbol{\Gamma}} + \hat{\mathbf{E}}$

The GLLR statistic can then be calculated according to Algorithm 5.2.

Algorithm 5.2

CALCULATION OF THE GLLR STATISTIC

1) Fit the reduced model on the data, and obtain the model parameters $\hat{\theta}_0^t$, $\hat{\boldsymbol{\Theta}}_A$, and $\hat{\boldsymbol{\Theta}}_T$, and the variances of the random effects and the residual variance $\hat{\sigma}_{rm}^2$ and $\hat{\sigma}_{em}^2$.
2) Produce bootstrap response $\tilde{\mathbf{Y}}_m^b$, $(b = 1 \ldots B)$.
   a) Based on the H_0 estimates draw $\hat{\boldsymbol{\Gamma}}$ and $\hat{\mathbf{E}}$ randomly from Normal distributions with zero mean and variances $\hat{\sigma}_{rm}^2$ for the random effect part and $\hat{\sigma}_{em}^2$ for the residual variance part.
   b) Recalculate the likelihoods of both the H_0 and H_1 models for the bootstrap response and calculate $GLLR^b$.
3) The p-value is defined as the number of times the bootstrap sample provides a better result than the original sample divided over the number of bootstrap samples:

$$p_{interaction}^b = \frac{\sum_{b=1}^{B} I(GLLR^b > GLLR) + 1}{B+1} \tag{5.38}$$

The addition of 1 in both the numerator and denominator comes from the fact that the original data is also considered one of the bootstrap samples and forced to have a p-value unequal to 0 when all GLLRb are worse than the original GLLR.

The bootstrap approach of Algorithm 5.2 can be applied for each fixed effect or combinations thereof and also to the random effects to obtain an estimate of their significance.

5.3.3.2 Visualization of Effects in LiMM-PCA

The visualization of important effects in ASCA is usually done using score and loading plots. The scores and loadings from a PCA model of each effect $\mathbf{X}_f\hat{\boldsymbol{\Theta}}_f = \tilde{\mathbf{T}}_f\tilde{\mathbf{P}}_f^t + \tilde{\mathbf{E}}_f$ however are based on $\tilde{\mathbf{Y}}$ and not on $\mathbf{Y}$ as in ASCA of RM-ASCA+. To obtain loadings for the measured variables, $\tilde{\mathbf{P}}_f^t$ has to be transformed back to its original space by multiplication with the transpose of $\mathbf{Q}$: $\mathbf{P}_f^t = \widetilde{\mathbf{P}}_f^t\mathbf{Q}^t$. For better visualization of the effect, the PCA scores are augmented with residual variation between the individuals by projecting the residuals on the model plain defined by the loadings of the effect: $\mathbf{T}_f^{aug} = (\mathbf{X}_f\widehat{\widehat{\mathbf{\Theta}}}_f + \widehat{\widehat{\mathbf{E}}})\widetilde{\mathbf{P}}_f^t$. Now $\widehat{\widehat{\mathbf{E}}}$ can include variation from both residuals and the random effect.

Example 5.5: Bariatric surgery with LiMM-PCA

In this example, the same Bariatric surgery data was used as described in Example 5.3 where RM-ASCA+ was used. The example is discussed in more detail in (Madssen *et al.*, 2021). For the first PCA step of LiMM-PCA, 10 components were used.

The results, which are summarized in Figure 5.11, are very similar to the RM-ASCA+ approach. There is a slight difference in score trajectories, explained variation, and confidence intervals of the loadings. The amount of variation explained in the first PC is larger in the LiMM-PCA model. Note that if all components had been used, the score plots would be exactly equal.

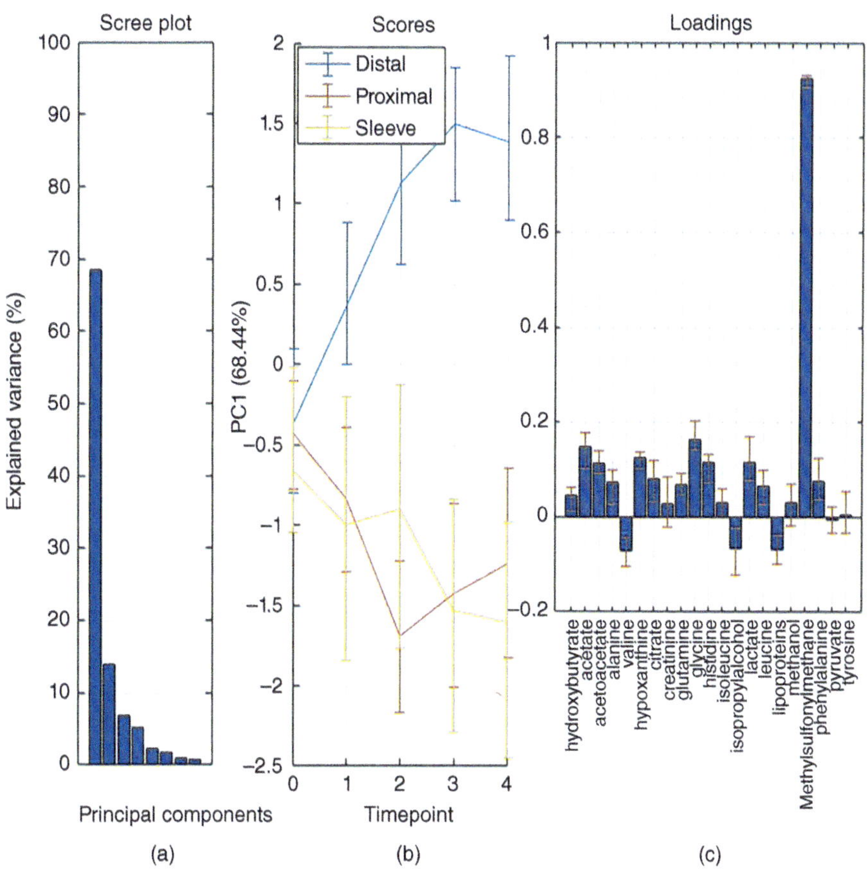

Figure 5.11 LiMM-PCA decomposition of *Treatment* and *Time:Treatment* interaction of the Bariatric surgery data; (a) explained variation of the principal components; (b) score values and confidence intervals of each *Treatment* for the 5 time points; (c) loading plot of the first principal component. Source: Madssen *et al.* (2021)/PLOS/CC BY 4.0.

5.4 Miscellaneous Methods

5.4.1 PC-ANOVA

5.4.1.1 Basic Idea of PC-ANOVA

Another way of combining two building blocks of data analysis – PCA and ANOVA – is by first performing a PCA on the data and subsequently an ANOVA on the scores of this PCA (Ellekjær *et al.*, 1996). This is a simple and straightforward method that uses visualization tools to understand the effects of the factors and is similar in reasoning to LiMM-PCA, already discussed in Section 5.3.3. Formal significance tests from ANOVA can be used on the principal component scores, but that would require having the proper degrees of freedom for these tests which is not trivial.

Example 5.6: Example of PC-ANOVA

To illustrate the main features of PC-ANOVA, the metabolomic measurements of the Caldana dataset (Caldana *et al.*, 2011) will be used once again, considering only the set of experiments conducted at 21 °C, which constitute a fully crossed design with 4 *Light conditions* (*Dark (D), Low light (LL), Medium light (ML),* and *High light (HL)*) and 7 *Exposure times* (0, 5, 10, 20, 40, 80, and 160 minutes), with 5 replicates of each design point (for a total of 140 experiments).

As discussed, PC-ANOVA operates by first performing PCA on the original data and then by calculating individual ANOVA models on the first extracted PCs. PCA scores and loadings can then be further exploited for interpreting the significant effects. After filtering the data by removing variables with a high fraction of missing values and suspect outliers a matrix with 140 experiments and 67 metabolites is obtained. The matrix is then subjected to PCA, after autoscaling. The scores and loadings along the first two components, which together account for about 28% of the total variance, are displayed in Figures 5.12 and 5.13, respectively.

Once the PCA model has been calculated, PC-ANOVA proceeds by fitting separate ANOVA models on each principal component scores vector (the analysis is usually limited to the first PCs, usually up to four or five). Here, the ANOVA models are calculated on the first 5 PCs, which jointly account for more than 50% of the original variance. The results of the ANOVA modeling are summarized in Table 5.3

By looking at Table 5.3, it can be observed that there is always at least one of the design terms (main effect or interactions) having a significant effect on the variability observed along the investigated principal components. In particular, the effect of *Time* is dominating along PC1 and PC3, that of *Light* along PC4, while all three terms contribute significantly to PC2 and PC5. Interpretation can then be carried out by plotting model estimates (the $\hat{\alpha}_i$, the $\hat{\beta}_j$, or the $(\widehat{\alpha\beta}_{ij})$) for the components of interest in a score plot and then resorting to the corresponding loadings. For instance, the main effect of light could be interpreted by plotting its model estimates along PC2 and PC4 (components for which the factor has a significant contribution). The results are shown in Figure 5.14.

By looking at Figure 5.14, it is apparent how the scores along PC progressively reflect the amount of light the sample is exposed to moving from extremely positive to extremely negative values, the scores follow the sequence *D, LL, ML, HL*. Following what was reported by (Luciano and Næs, 2009), the square root of the residual error variances for the ANOVA models of the two components is also shown in the plot (dotted lines) to provide a straightforward indication of the

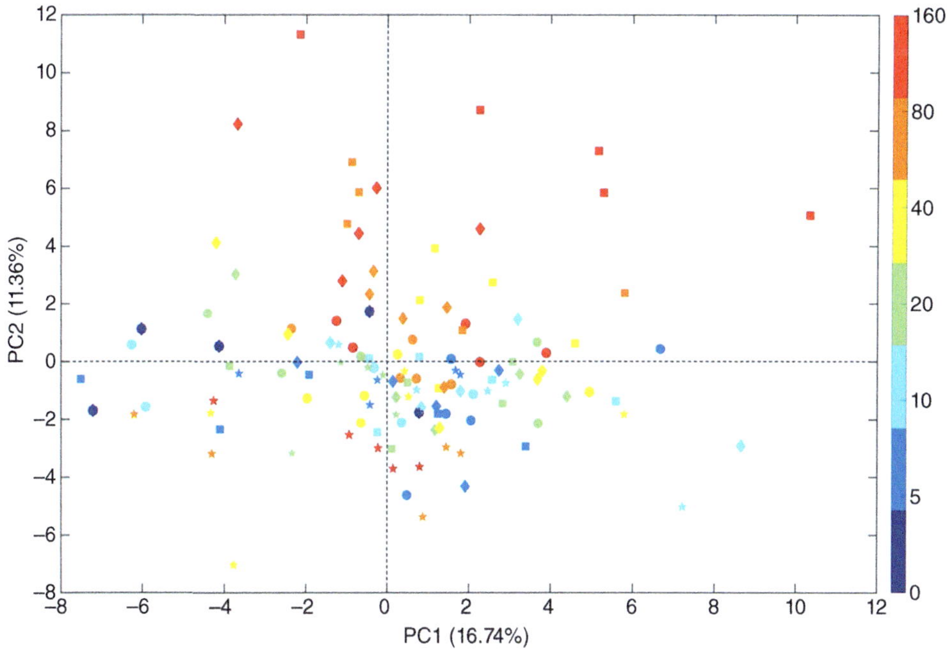

Figure 5.12 Scores of the 140 experiments along the first two PCs extracted from the Caldana metabolomic dataset. The different symbols correspond to the *Light* conditions (square = *D*; diamond = *LL*; circle = *ML*; star = *HL*), while the color reflects the exposure time (see color bar for reference).

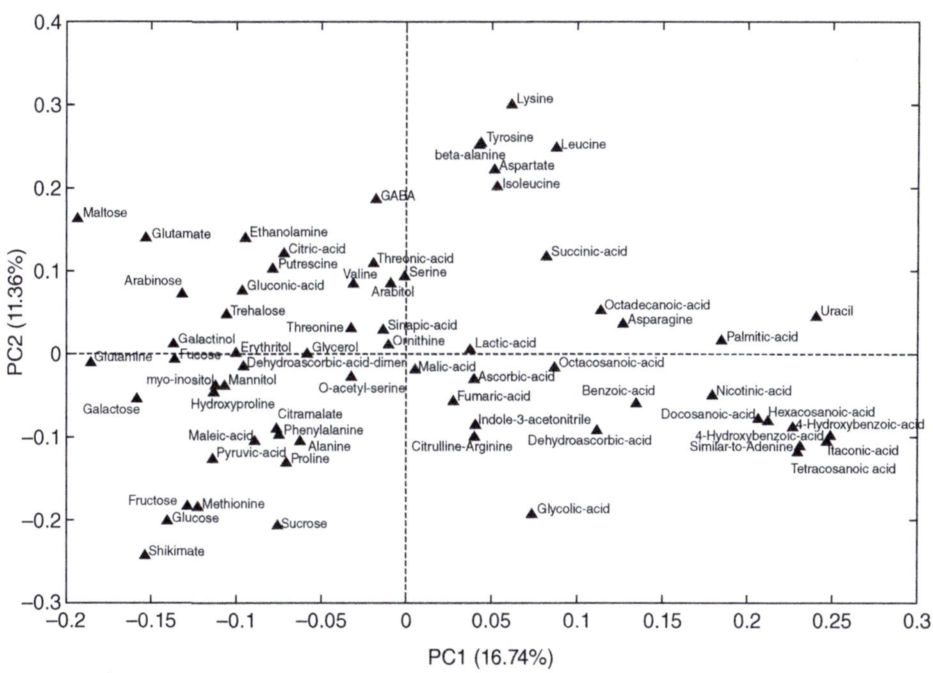

Figure 5.13 Loading of the 67 metabolites along the first two PCs extracted from the metabolomic dataset.

Table 5.3 Overview of two-way ANOVA models on the first five PCs extracted from the Caldana dataset.

Results of ANOVA on PC1

Source	*SS*	*df*	*MS*	*F*	*p*-value
Light	28.93	3	9.642	1.048	0.3744
Time	289.2	6	48.21	5.239	8.6×10^{-5}
Light:Time	210.1	18	11.67	1.269	0.2222
Error	1030.6	112	9.201		
Total	1558.8	139			

Results of ANOVA on PC2

Source	*SS*	*df*	*MS*	*F*	*p*-value
Light	192.7	3	64.24	23.14	9.9×10^{-12}
Time	214.9	6	35.82	12.91	4.8×10^{-11}
Light:Time	339.6	18	18.87	6.797	3.1×10^{-11}
Error	310.9	112	2.775		
Total	1058.1	139			

Results of ANOVA on PC3

Source	*SS*	*df*	*MS*	*F*	*p*-value
Light	212.53	3	4.176	0.6249	0.6003
Time	111.4	6	18.57	2.780	0.0148
Light:Time	80.30	18	4.461	0.6676	0.8361
Error	748.4	112	6.682		
Total	952.7	139			

Results of ANOVA on PC4

Source	*SS*	*df*	*MS*	*F*	*p*-value
Light	87.70	3	29.23	6.711	3.3×10^{-4}
Time	5.599	6	0.9332	0.2142	0.9716
Light:Time	85.86	18	4.770	1.095	0.3666
Error	487.9	112	4.356		
Total	667.0	139			

Table 5.3 (Continued)

Results of ANOVA on PC5					
Source	*SS*	*df*	*MS*	*F*	*p*-value
Light	84.53	3	28.18	11.66	1.0×10^{-6}
Time	107.4	6	17.90	7.408	1.1×10^{-6}
Light:Time	82.58	18	4.588	1.898	0.0227
Error	270.7	112	2.417		
Total	545.2	139			

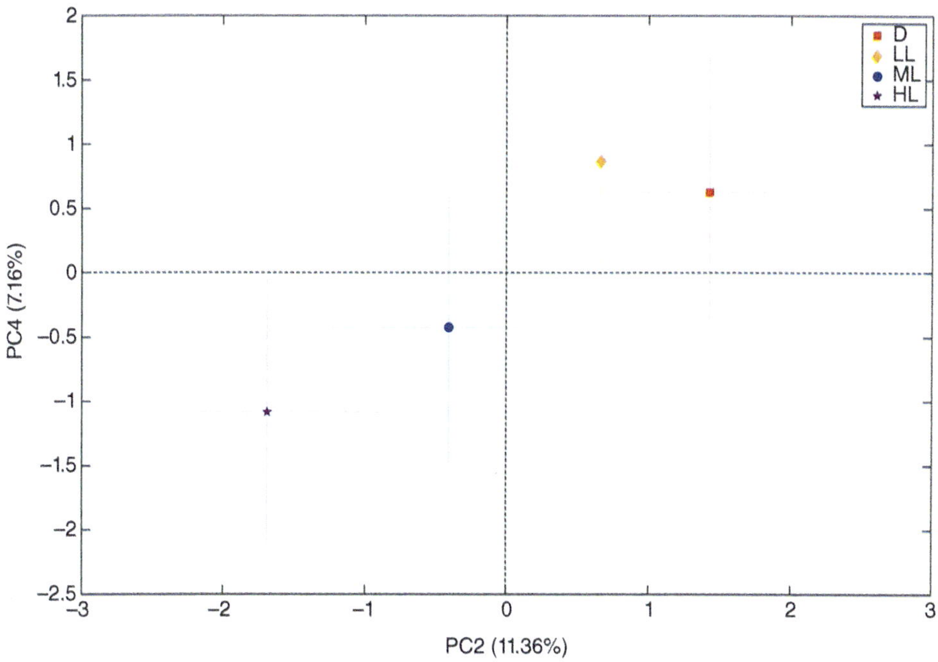

Figure 5.14 Scores of the 4 *Light* levels along PC2 and PC4 extracted from the metabolomic dataset. The different symbols correspond to the *Light* conditions (square = *D*; diamond = *LL*; circle = *ML*; star = *HL*), while dotted vertical and horizontal lines correspond to the square root of the MS for the error term of the corresponding ANOVA decomposition, as suggested. Source: Luciano and Næs (2009)/with permission of Elsevier.

noise level. Inspection of the corresponding loadings (not shown) indicates that increasing the amount of light the samples are exposed to, on average, results mainly in a decrease of GABA, lysine, leucine, isoleucine, tyrosine, aspartate, and β-alanine, and a corresponding increase of the levels of glycolic acid, methionine, shikimate, sucrose, fructose, glucose, pyruvate, alanine, and phenylalanine. Analogously, it is also possible to inspect and interpret the combination of multiple terms, such as a main factor together with an interaction. To do so, for each component the mean scores corresponding to the factor or interaction levels are summed. The resulting scores can then be plotted. An example is given in Figure 5.15, where the effect of *Light* and of its interaction with *Time* ($\overline{\mathbf{Y}}_{Light} + \overline{\mathbf{Y}}_{Light:Time}$) on PC2 is considered.

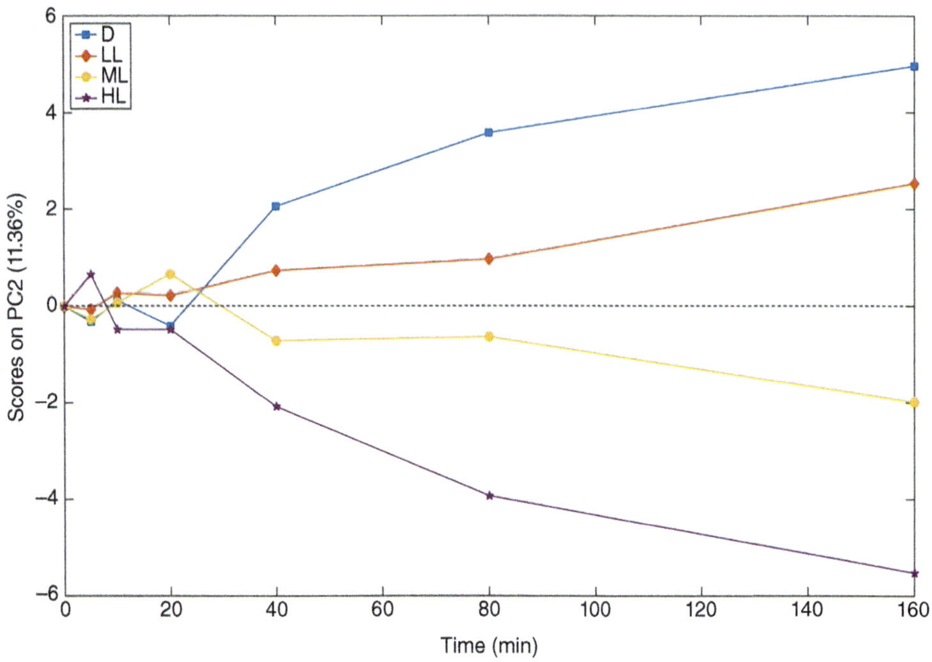

Figure 5.15 Scores of $\overline{\mathbf{Y}}_{Light} + \overline{\mathbf{Y}}_{Light:Time}$ along PC2 plotted longitudinally against *Time*.

5.4.1.2 Comparing PC-ANOVA with ASCA

A comparison of PC-ANOVA with ASCA is also available (Luciano and Næs, 2009). In a sensory example, the results of PC-ANOVA and ASCA were comparable. It was also concluded that ASCA may have an advantage if the principal components of the different factor estimate matrices are very different. In such a case, PC-ANOVA may have to use many components to describe all these different variations and thus would end up having to interpret all these components.

5.4.2 PARAFASCA

For some types of ANOVA models and subsequent ASCA generalizations, it is possible to rearrange the effect matrices in a way that makes them amenable to a multiway decomposition. Assuming a crossed-effect ASCA model where the Factor A represents the *Time* (at A levels); the Factor B the *Treatment* (at B levels) and N repeats per cell (for simplicity we assume balanced data) the ASCA model is

$$\mathbf{Y} = \overline{\mathbf{Y}}_{\mathtt{M}} + \overline{\mathbf{Y}}_{\mathtt{A}} + \overline{\mathbf{Y}}_{\mathtt{B}} + \overline{\mathbf{Y}}_{\mathtt{AB}} + \mathbf{Y}_{\mathtt{Ind}} \tag{5.39}$$

where the term $\mathbf{Y}_{\mathtt{Ind}}$ collects the individual or residual contributions $(y_{abnj} - \bar{y}_{ab.j})$. This is the basic decomposition in the ASCA approach and is a repeat of Equation 4.3. In many examples, it is useful to consider $\overline{\mathbf{Y}}_{\mathtt{B}} + \overline{\mathbf{Y}}_{\mathtt{AB}}$ simultaneously, *e.g.*, in cases where the treatment kicks in after a certain amount of time since a main treatment effect is then difficult to interpret. The structure of the matrix $\overline{\mathbf{Y}}_{\mathtt{B}} + \overline{\mathbf{Y}}_{\mathtt{AB}}$ is now as follows:

$$\overline{\mathbf{Y}}_{\mathtt{B}} + \overline{\mathbf{Y}}_{\mathtt{AB}} = \begin{bmatrix} \bar{\mathbf{y}}_{.1.1} + \bar{\mathbf{y}}_{11.1} & \cdots & \bar{\mathbf{y}}_{.1.J} + \bar{\mathbf{y}}_{11.J} \\ \vdots & \ddots & \vdots \\ \bar{\mathbf{y}}_{.B.1} + \bar{\mathbf{y}}_{1B.1} & \cdots & \bar{\mathbf{y}}_{.B.J} + \bar{\mathbf{y}}_{1B.J} \\ \vdots & \ddots & \vdots \\ \bar{\mathbf{y}}_{.1.1} + \bar{\mathbf{y}}_{A1.1} & \cdots & \bar{\mathbf{y}}_{.1.J} + \bar{\mathbf{y}}_{A1.J} \\ \vdots & \ddots & \vdots \\ \bar{\mathbf{y}}_{.B.1} + \bar{\mathbf{y}}_{AB.1} & \cdots & \bar{\mathbf{y}}_{.B.J} + \bar{\mathbf{y}}_{AB.J} \end{bmatrix} \tag{5.40}$$

where $\bar{\mathbf{y}}_{ab.j}(N \times 1)$ is the constant vector containing estimates of Factor A at level a and Factor B at level b for response j (we assume that n is the fastest running index and a the slowest). Considering that this is a vector of constants, we can also represent this as the scalar constant $y_{ab.j}$. Since the design is fully crossed, we can also conceive these values as elements of the three-way array $\underline{\mathbf{Y}}_{\mathtt{B+AB}}$. Hence, we can use three-way decomposition methods to study the variability in this array, *e.g.*, using a PARAFAC model (Smilde *et al.*, 2005). This then results in a PARAFASCA approach which is illustrated in Example 5.7 (Jansen *et al.*, 2008).

Example 5.7: Example of PARAFASCA in toxicology

The example concerns a toxicity study among rats. These were administered a toxic compound in three *Dose* levels (*zero*, *low*, and *high*), and their urine was collected at several time points among which a pre-dose point and analyzed by NMR for their metabolites. Each treatment group contains five rats, and the measurements were performed on the same rats. So, effectively this is a repeated measurement design but for simplicity, we ignore this aspect.

Our analysis starts with a PCA on the total column-centered data set, and no scaling was applied. The scores on the first PC are visualized in Figure 5.16 and explain 81% of the variation in the data. These scores clearly show a *Dose* effect and also a higher spread in the *high-dose* group, pointing to a possible need for homogenizing the variances.

The next step in the analysis is to run an ASCA model where we combine the main effect *Dose* and the *Dose:Time* interaction (see the main text). This part of the ASCA model explains 64% of the variation, and the first PC on the *Dose* + *Dose:Time* interaction factor estimates describes 93% of its variation and the second component describes 4%. These scores are shown in Figure 5.17. Component one in this figure clearly describes a contrast between control and low dose on the one hand, and high dose on the other hand. Component two describes a differentiation between high dose and control.

By considering the estimates of the combined *Dose* and *Dose:Time* interaction as elements of a three-way array, we can apply a PARAFAC model to this data. This model explains 95% of the variation of these factor effect estimates in two PARAFAC components. The results are shown in Figure 5.18. The (a) component shows a late response mainly as a contrast between the high-dose group and the other groups. The (b) component shows an early response mainly as a contrast between the control- and high-dose groups.

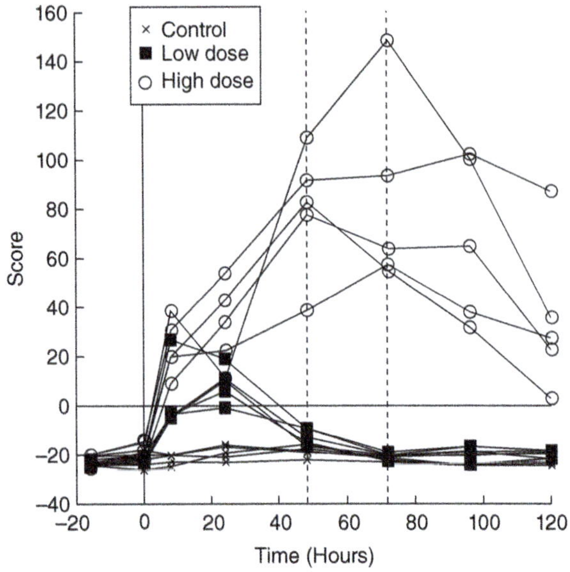

Figure 5.16 Scores on the first PC of a PCA on the toxicology data showing the *Time* evolvement and the *Dose* groups. The toxic compound is administered at time zero and the dashed lines show different stages in the process. Source: Jansen *et al.* (2008)/with permission of John Wiley & Sons.

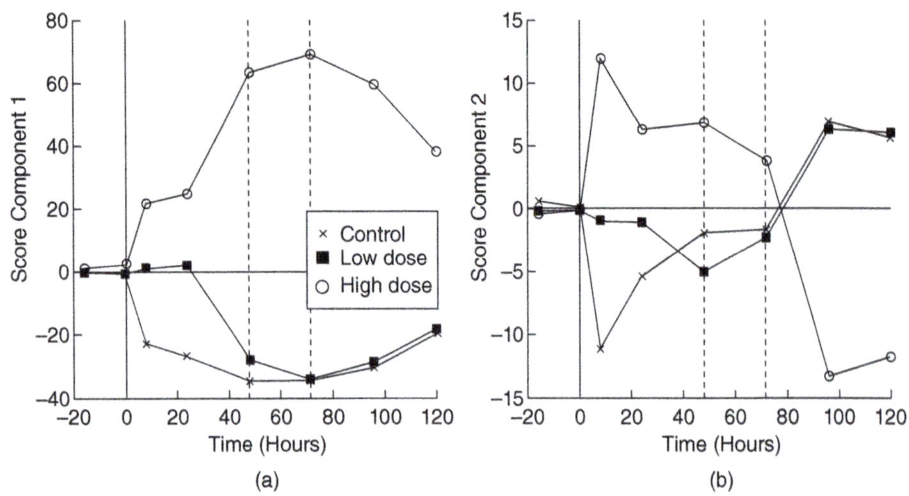

Figure 5.17 ASCA scores on the first (a) and second (b) PARAFAC component of the combined effect of *Dose* and *Dose:Time* interaction. The toxic compound is administered at time zero, and the dashed lines show different stages in the process. Source: Jansen *et al.* (2008)/with permission of John Wiley & Sons.

This analysis shows that a PARAFAC model can be used to interpret ASCA decompositions. In this case, instead of an ASCA model, also a PRC model could have been used since there is a control group available. Also, the results may point to the necessity to homogenize the variance among the groups.

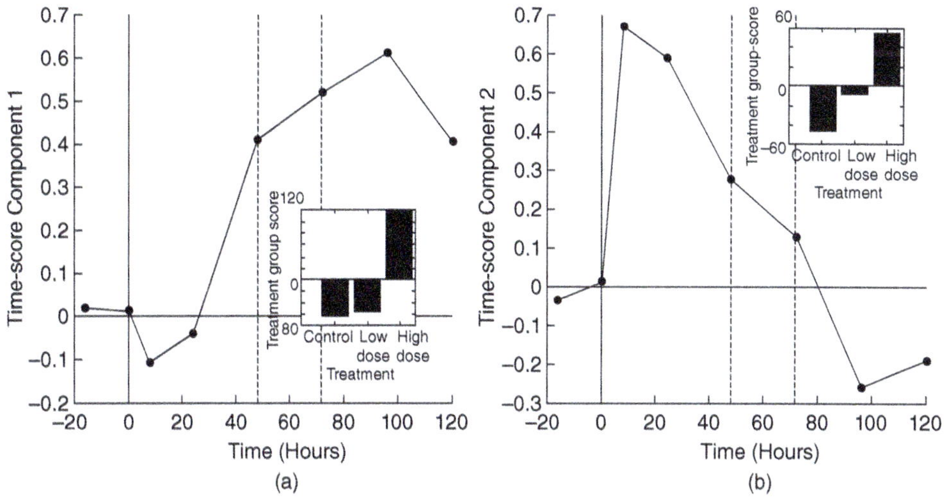

Figure 5.18 PARAFASCA scores on the first (a) and second (b) factor estimates of the combined effect of *Dose* and *Dose:Time* interaction. The toxic compound is administered at time zero and the dashed lines show different stages in the process. The insert shows the loadings for the different treatment groups. Source: Jansen *et al.* (2008)/with permission of John Wiley & Sons.

5.4.3 PE-ASCA

The PE-ASCA model is based on the P-ESCA (exponential family SCA) model for data fusion (Song *et al.*, 2020). The method combines the ideas of ASCA and data fusion. Its main goal is to find common and distinct components in multiple sets of data that are all measured on the same sets of samples, and of which the samples were varied according to an experimental design. The common and distinct structure is defined by using group-wise sparseness penalties on the loading matrices for each set of data. Assuming two sets of data $\mathbf{Y}_1(I \times J_1)$ and $\mathbf{Y}_2(I \times J_2)$ that provide different sets of measurements on the same set of samples, a P-ESCA model is defined as:

$$[\mathbf{Y}_1|\mathbf{Y}_2] = [\mathbf{Y}_{M1}|\mathbf{Y}_{M2}] + \mathbf{T}[\mathbf{P}_1^t|\mathbf{P}_2^t] + [\mathbf{E}_1|\mathbf{E}_2] \tag{5.41}$$

with $\mathbf{T} = [\mathbf{T}_c|\mathbf{T}_{d1}|\mathbf{T}_{d2}]$ and $[\mathbf{P}_1^t|\mathbf{P}_2^t] = \begin{bmatrix} \mathbf{P}^t_{1,c} & \mathbf{P}^t_{2,c} \\ \mathbf{P}^t_{1,d1} & 0 \\ 0 & \mathbf{P}^t_{2,d2} \end{bmatrix}$.

There are three sets of principal components to describe the systematic variation in the two datasets. $\mathbf{T}_c(I \times R_c)$ represents the common principal components that describe systematic variation that is present in both $\mathbf{Y}_1$ and $\mathbf{Y}_2$. $\mathbf{T}_{d1}(I \times R_{d1})$ describes additional systematic variation that is only present in $\mathbf{Y}_1$, and $\mathbf{T}_{d2}(I \times R_{d2})$ describes additional variation that is only present in $\mathbf{Y}_2$. $[\mathbf{P}_1^t|\mathbf{P}_2^t]$ represents the loadings of both data sets next to each other, and both are also separated into parts for loadings that correspond to the common variation $\mathbf{P}^t_{1,c}(R_c \times J_1)$, and $\mathbf{P}^t_{2,c}(R_c \times J_2)$, loadings that correspond to the distinctive variation for $\mathbf{Y}_1$, ($\mathbf{P}^t_{1,d1}(R_{d1} \times J_1)$) while $\mathbf{P}^t_{2,d1} = 0\,(R_{d1} \times J_2)$, as such variation is not present in $\mathbf{Y}_2$. Furthermore, the loadings for $\mathbf{T}_{d2}$ are $\mathbf{P}^t_{1,d2} = 0\,(R_{d2} \times J_1)$ and $\mathbf{P}^t_{2,d2}(R_{d2} \times J_2)$. The estimation of the loading matrix $\mathbf{P}^t$ is obtained by applying structured sparsity penalization such that groups of loadings are forced to zero.

Note that the structured loading is not predefined, but a result of the sparseness penalization. If a loading vector is forced to 0, *e.g.*, $\mathbf{P}^t_{1,d2}$, then that component is defined to be a distinctive component $\mathbf{T}_{d2}$ for dataset $\mathbf{Y}_2$ using group-wise penalties, as discussed in Elaboration 5.1.

ELABORATION 5.1

Group-wise penalties

In Section 2.2.2.4 penalized regression was introduced. Penalties are used a lot in the estimation of model parameters to avoid issues of overfitting and to obtain sparser models. A much-used example is the LASSO penalty to induce sparse solutions in which many of the parameters are forced to 0 (Hastie *et al.*, 2009), as was already discussed in Section 2.2.2.4. Such penalties can often also be imposed on a group of parameters simultaneously, such as the group-LASSO (Yuan and Lin, 2006). For a group-LASSO penalty on the P-ESCA model described in Equation 5.41, the penalty is defined as $\lambda \sum_{m=1}^{M} \sum_{r=1}^{R} ||\mathbf{p}_{mr}||_2$. Here, $\mathbf{p}_{mr}$ represents the loading of the rth component of block $\mathbf{Y}_m$, and $||\mathbf{p}_{mr}||_2$ is its L_2 norm. This group-LASSO penalty shrinks the whole loading vector $\mathbf{p}_{mr}$ as a LASSO penalty, and the elements inside the loading $\mathbf{p}_{mr}$ as a ridge penalty. The tuning parameter λ defines the strength of the penalty. The group-LASSO penalty shrinks all elements of the loading matrix in the same way; thus, also the nonzero parameters are shrunken. Although a re-estimation procedure can be used where only the non-zero parameters are re-estimated without a penalty, this may lead to estimation problems, especially for binary data (Song *et al.*, 2021).

Alternative penalties were introduced as improvements over the group-LASSO. If $\sigma_{mr} = ||\mathbf{p}_{mr}||_2$, then the bridge penalty (Fu, 1998) is defined as $\lambda_m \sqrt{J_m} \sum_{r=1}^{R} (\sigma_{mr}^q)$, with $0 < q \leq 1$. The number of features in block $\mathbf{Y}_m$, J_m, is added to correct for a different number of features in different blocks. The Generalized Double Pareto (GDP) penalty (Armagan *et al.*, 2013) is defined as $\lambda_m \sqrt{J_m} \sum_{r=1}^{R} (log(1 + \frac{\sigma_{mr}}{\gamma})$ with $\gamma = 1$. The GDP penalty has been adopted by Song *et al.* (2021) as the penalty of choice in the P-ESCA method.

The P-ESCA was developed as a data fusion approach for data sets of different distributions, *i.e.*, binary data, counts or quantitative data. Here we assume both datasets to be quantitative. In that case, the decomposition of both data sets can be defined as follows:

$$\begin{aligned}\mathbf{Y}_1 &= \mathbf{Y}_{1M} + \mathbf{T}_c\mathbf{P}^t_{1,c} + \mathbf{T}_{1d}\mathbf{P}^t_{1,1d} + \mathbf{E}_1 \\ \mathbf{Y}_2 &= \mathbf{Y}_{2M} + \mathbf{T}_c\mathbf{P}^t_{2,c} + \mathbf{T}_{2d}\mathbf{P}^t_{2,2d} + \mathbf{E}_2\end{aligned} \tag{5.42}$$

If the samples of two data sets are varied according to a predefined experimental design with Factors `A` and `B` and interaction `AB`, the ASCA model for both data sets then can be written as:

$$\begin{aligned}\mathbf{Y}_1 &= \mathbf{Y}_{1M} + \mathbf{Y}_{1,\mathtt{A}} + \mathbf{Y}_{1,\mathtt{B}} + \mathbf{Y}_{1,\mathtt{AB}} + \mathbf{Y}_{1,\mathtt{Ind}} \\ \mathbf{Y}_2 &= \mathbf{Y}_{2M} + \mathbf{Y}_{2,\mathtt{A}} + \mathbf{Y}_{2,\mathtt{B}} + \mathbf{Y}_{2,\mathtt{AB}} + \mathbf{Y}_{2,\mathtt{Ind}}\end{aligned} \tag{5.43}$$

The PE-ASCA represents the combination of the ASCA and P-ESCA models for multiset data of which the samples are varied according to an experimental design with Factors `A` (*Time*), `B` (*Treatment*), and their interaction (Alinaghi *et al.*, 2020). The combination `B + AB` provides information on how each of the *Treatments* changes in time, after correction of the main *Time* effect. Both the common parts and the distinctive parts can be induced by the experimental factors or their interaction. Also systematic individual variation could be induced by the factors. This leads to the following general model:

$$\begin{aligned}\mathbf{Y}_m = \mathbf{Y}_{mM} &+ \mathbf{T}_{c,\mathtt{A}}\mathbf{P}^t_{mc,\mathtt{A}} + \mathbf{T}_{c,\mathtt{B+AB}}\mathbf{P}^t_{mc,\mathtt{B+AB}} + \mathbf{T}_{c,\mathtt{Ind}}\mathbf{P}^t_{mc,\mathtt{Ind}} + \mathbf{T}_{md,\mathtt{A}}\mathbf{P}^t_{md,\mathtt{A}} \\ &+ \mathbf{T}_{md,\mathtt{B+AB}}\mathbf{P}^t_{md,\mathtt{B+AB}} + \mathbf{T}_{md,\mathtt{Ind}} + \mathbf{P}^t_{md,\mathtt{Ind}} + \mathbf{E}_m\end{aligned} \tag{5.44}$$

where the common scores and the distinctive scores now all have contributions from the main Factor A, the combination B+AB, and the individual level. All scores $\mathbf{T}$ in this equation, and thus also their corresponding loadings $\mathbf{P}$ can have multiple components. The full score and loading matrices of Equation 5.41 can thus be described as:

$$\mathbf{T} = [\mathbf{T}_{c\mathrm{A}}|\mathbf{T}_{c,\mathrm{B+AB}}|\mathbf{T}_{c,\mathrm{Ind}}|\mathbf{T}_{d1,\mathrm{A}}|\mathbf{T}_{d1,\mathrm{B+AB}}|\mathbf{T}_{d1,\mathrm{Ind}}|\mathbf{T}_{d2,\mathrm{A}}|\mathbf{T}_{d2,\mathrm{B+AB}}|\mathbf{T}_{d2,\mathrm{Ind}}] \tag{5.45}$$

and

$$[\mathbf{P}_1^t|\mathbf{P}_2^t] = \begin{bmatrix} \mathbf{P}^t_{1c,\mathrm{A}} & \mathbf{P}^t_{2c,\mathrm{A}} \\ \mathbf{P}^t_{1c,\mathrm{B+AB}} & \mathbf{P}^t_{2c,\mathrm{B+AB}} \\ \mathbf{P}^t_{1c,\mathrm{Ind}} & \mathbf{P}^t_{2c,\mathrm{Ind}} \\ \mathbf{P}^t_{1d,\mathrm{A}} & 0 \\ \mathbf{P}^t_{1d,\mathrm{B+AB}} & 0 \\ \mathbf{P}^t_{1d,\mathrm{Ind}} & 0 \\ 0 & \mathbf{P}^t_{2d,\mathrm{A}} \\ 0 & \mathbf{P}^t_{2d,\mathrm{B+AB}} \\ 0 & \mathbf{P}^t_{2d,\mathrm{Ind}} \end{bmatrix}$$

Again the loading matrix is estimated here using structured sparseness penalties to force loading vectors toward 0. The use of the method is illustrated in Example 5.8.

Example 5.8: Simulated example PE-ASCA

A simulated example will be used to show the performance of PE-ASCA compared to SCA and P-ESCA. The simulation is based on the structure of the real data that is described in Example 5.9. The goal of the simulated example is to show that PE-ASCA is able to estimate common and distinctive profiles in situations where SCA and P-ESCA fail. On the left of Figure 5.19 the systematic variation to simulate the data for two datasets $\mathbf{Y}_1$ and $\mathbf{Y}_2$ is provided. Each dataset has 100 variables. The samples in the data sets are coming from an experimental design with two Factors A and B. Factor A represents the *Litter* effect (3 groups of 18 piglets from 3 sows), Factor B represents the *Treatment* effect (*Control, Infected1, Infected2*). For each setting of the design factors, 6 piglets were used. The systematic variation between the samples is induced by 1 common component induced by Factor A ($\mathbf{t}_{c,\mathrm{A}}$), another common component induced by Factor B ($\mathbf{t}_{c,\mathrm{B}}$). The common variation is present in both datasets, each with its own specific loading. A distinctive component for dataset 1 induced by Factor B ($\mathbf{t}_{d,\mathrm{B}}\mathbf{p}^t_{1,d,\mathrm{B}}$), and another distinctive component for dataset 2 that shows systematic variation between the individuals in the study ($\mathbf{t}_{d,\mathrm{Ind}}\mathbf{p}^t_{2,d,\mathrm{Ind}}$).

$$\begin{aligned} \mathbf{Y}_1 &= \mathbf{Y}_{M1} + \mathbf{t}_{c,\mathrm{A}}\mathbf{p}^t_{1,c,\mathrm{A}} + \mathbf{t}_{c,\mathrm{B}}\mathbf{p}^t_{1,c,\mathrm{B}} + \mathbf{t}_{d,\mathrm{B}}\mathbf{p}^t_{1,d,\mathrm{B}} + \mathbf{E}_1 \\ \mathbf{Y}_2 &= \mathbf{Y}_{M2} + \mathbf{t}_{c,\mathrm{A}}\mathbf{p}^t_{2,c,\mathrm{A}} + \mathbf{t}_{c,\mathrm{B}}\mathbf{p}^t_{2,c,\mathrm{B}} + \mathbf{T}_{d,\mathrm{Ind}}\mathbf{p}^t_{2,d,\mathrm{Ind}} + \mathbf{E}_2 \end{aligned} \tag{5.46}$$

The scores and loadings to create the datasets $\mathbf{Y}_1$ and $\mathbf{Y}_2$ are visualized on the left in Figure 5.19. The top row shows the loadings for each of the two datasets. The common loadings $\mathbf{p}^t_{1,c,\mathrm{A}}$, $\mathbf{p}^t_{1,c,\mathrm{B}}$, $\mathbf{p}^t_{2,c,\mathrm{A}}$, and $\mathbf{p}^t_{2,c,\mathrm{B}}$ both in light and dark blue show a signal at variable 20 in both datasets. The distinctive component $\mathbf{p}^t_{1,d,\mathrm{B}}$ for $\mathbf{Y}_1$ is shown in red, and the distinctive loading for the individual variation $\mathbf{p}^t_{2,d,\mathrm{Ind}}$ for $\mathbf{X}_2$ is shown in green. The score $\mathbf{t}_{c,\mathrm{A}}$ just shows three different levels for the three litters, while $\mathbf{t}_{c,\mathrm{B}}$ shows a different level only for infected group.

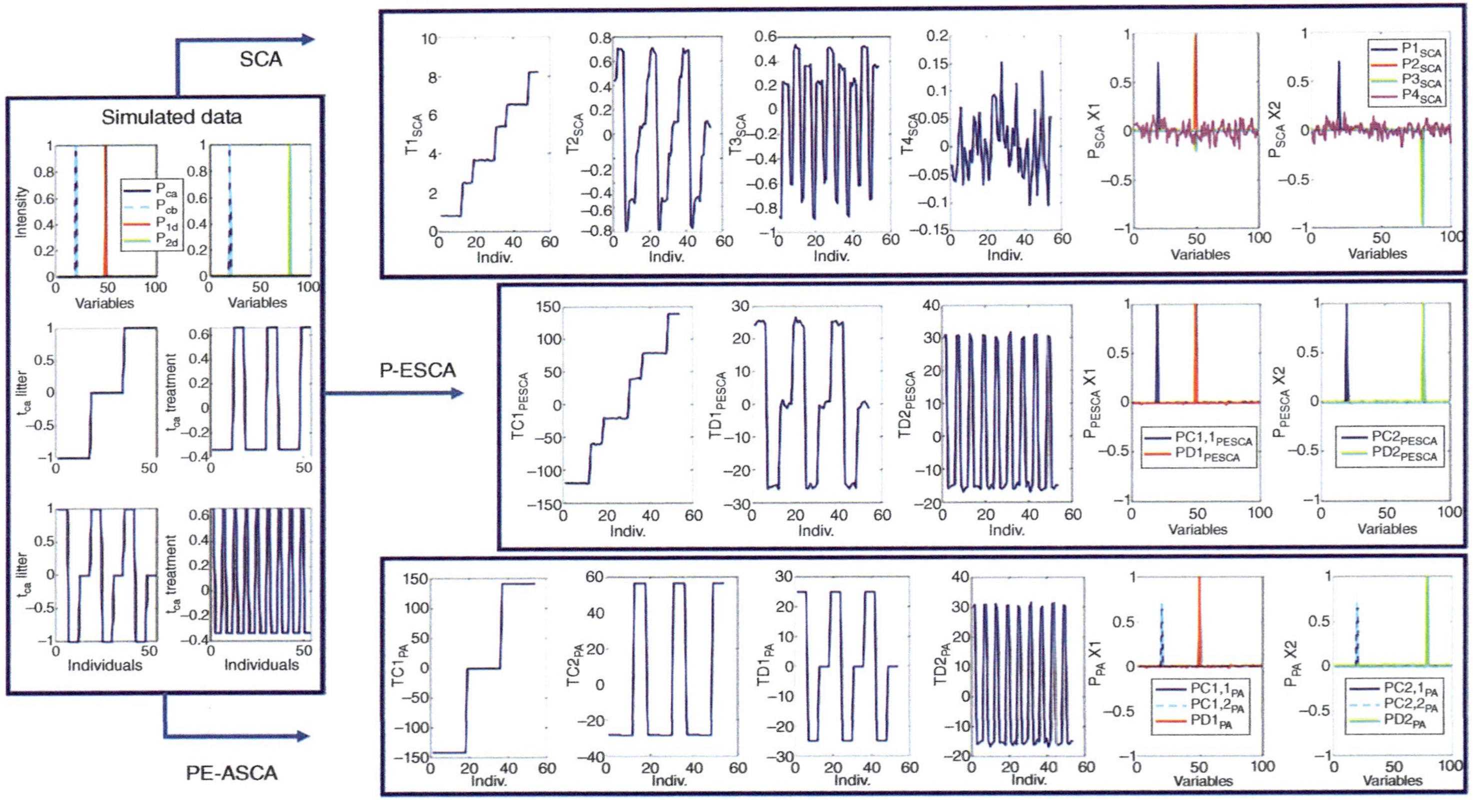

Figure 5.19 Simulated example of PE-ASCA. The simulation of the data (left) shows loadings (upper row) and scores (middle and lower rows) to create data for two data sets. On the right, the decomposition solutions were obtained using SCA, P-ASCA, and PE-ASCA. Only the latter approach is able to exactly reproduce the underlying components in both data sets.

The distinctive treatment component, $\mathbf{t}_{d,\mathtt{B}}$, shows a difference between all treatment groups, and the final distinctive component $\mathbf{t}_{d,\mathtt{Ind}}$ shows a difference between some of the piglets and the rest. The error matrices $\mathbf{E}_1$ and $\mathbf{E}_2$ are from a normal distribution with an average of zero and a standard deviation of 0.01. Signal-to-noise ratios for both $\mathbf{Y}_1$ and $\mathbf{Y}_2$ are 35. On the right-hand side, the results of simultaneous component analysis (SCA), P-ESCA and PE-ASCA are provided. The SCA can only find three significant components (the 4th one only contains noise), but the scores and loadings do not represent the induced variation. The P-ESCA method is not able to use the designed structure of the data and therefore cannot distinguish between the two common components that both have a contribution to the same variable, which is variable 20. Its first score is therefore a combination of the simulated $\mathbf{t}_{c,\mathtt{A}}$ and $\mathbf{t}_{c,\mathtt{B}}$. The PE-ASCA model is able to find all the simulated profiles underlying the data.

The real data example is provided in Example 5.9. Details can be found in (Alinaghi *et al.*, 2020).

Example 5.9: Brain tissue analysis with PE-ASCA

The PE-ASCA model was applied to a subset of NMR metabolomics data from two brain tissues (hypothalamus and midbrain) of piglets in an intervention study of experimental bloodstream infection (Alinaghi *et al.*, 2020). Twenty-two piglets obtained from two mothers (*Litter* 1 and *Litter* 2) were randomly allocated to one of three *Treatment* groups: a control group receiving intra-arterial saline and total parenteral nutrition (CON+TPN, $N = 7$), an infected group receiving Staphylococcus epidermidis (SE) inoculation and total parenteral nutrition (SE+TPN, $N = 5$) or a colostrum supplementation group receiving the SE inoculation and bovine colostrum and supplementary parenteral nutrition (SE+COL, $N = 10$). To balance the groups, four samples from the SE+COL group and one sample from CON+TPN were removed from the data sets. Some of the removed samples were outliers, while others were randomly selected. Furthermore, one sample of the SE+TPN group was reconstructed by adding normally distributed noise to the mean response litter 1 piglets in the group. The final NMR data set of the hypothalamus and midbrain (18×330) containing the three samples in every *Treatment* group of each litter, was used for the integrative analysis with PE-ASCA. A schematic overview of the treatment design is depicted on the left part in Figure 5.20.

After appropriate preprocessing of the NMR data (see (Alinaghi *et al.*, 2020) for details), PE-ASCA was applied and both the common *Litter* and *Treatment* effect were found significant using permutation testing. Two components can be disentangled for the common *Treatment* effect (see Figure 5.20, upper part right). The score plot shows a separation of the three different *Treatment* groups. The loading plot reveals that lactate is increased in the SE+TPN group in both compartments, and choline-containing metabolites are decreased in the CON+TPN group, forcing discrimination between the other groups. Specifically, for the midbrain compartment, glycerophosphocholine (GPC) and phosphocholine (PC) show a higher level in CON+TPN.

On top of the common variation, a distinct component for the hypothalamus ($p = 0.06$) also shows increased levels of choline in the infected group as well as some other metabolites such as γ-aminobutyric acid (GABA), N-acetylaspartate (NAA), and acetate in the infected group (SE+TPN) compared to the control group (CON+TPN) (see Figure 5.20, lower part right).

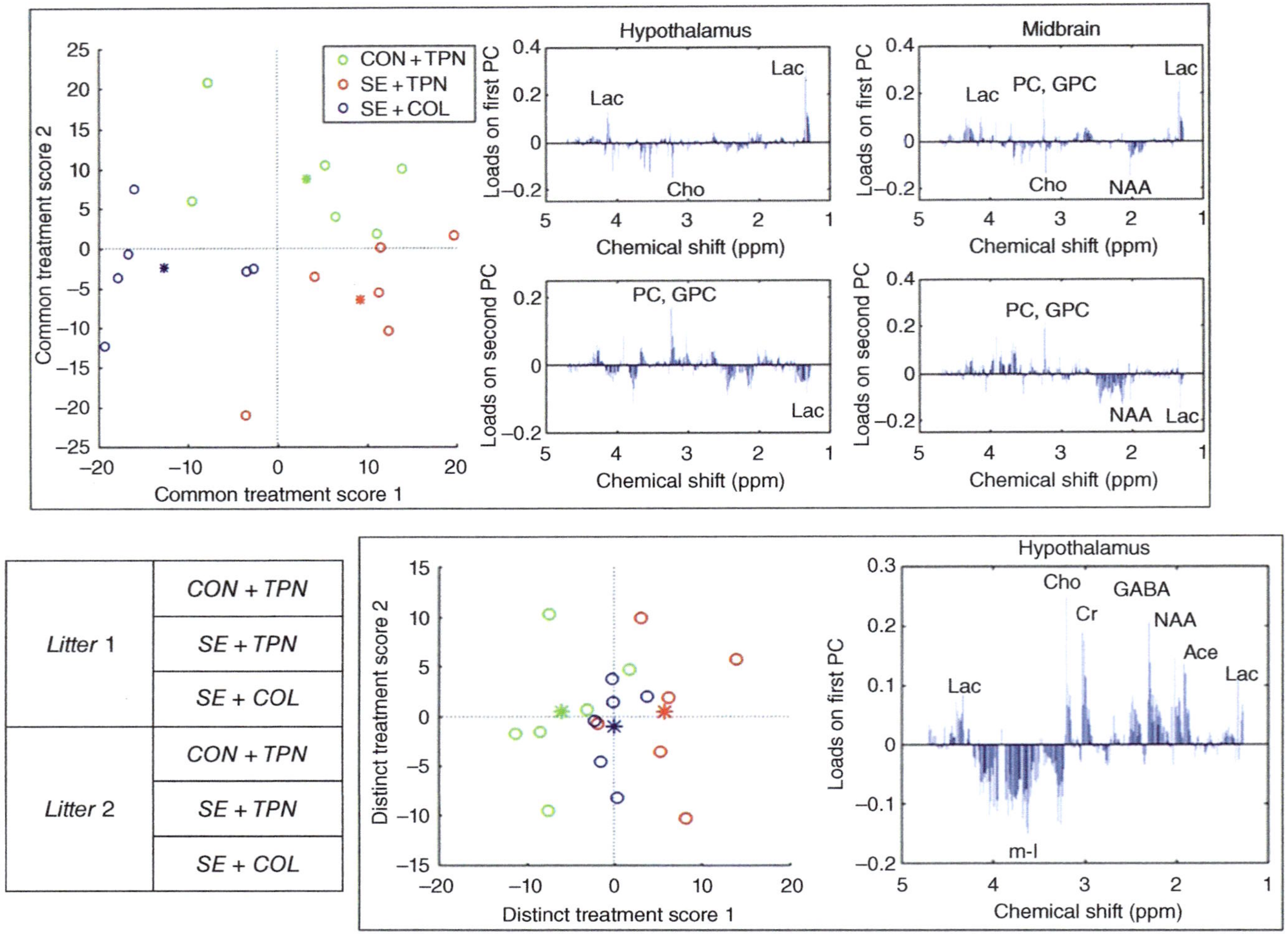

Figure 5.20 PE-ASCA result of the experimental NMR-metabolomics data, (bottom left) schematic overview of the balanced experimental design of the SE study, (top) common score and loading plots of hypothalamus and midbrain, (bottom right) distinct score and loading plots of hypothalamus. CON+TPN (green circles) represents the control group, SE+TPN (red circles) represents the infected group by Staphylococcus epidermidis, and SE+COL (blue circles) represents the colostrum supplementation group. The stars represent the mean value of each treatment group. In the loading plots: Ace: acetate; GABA: γ-aminobutyric acid; GPC: glycerophosphocholine; Cho: choline; Cr: creatine; Lac: lactate; m-I: myo-inositol; NAA: N-acetylaspartate; PC: phosphocholine. Source: Alinaghi *et al*. (2020)/Springer Nature/CC BY 4.0.

5.4.4 rMANOVA

Regularized MANOVA (rMANOVA) (Engel *et al.*, 2015, 2020) was introduced in the literature to overcome the problems posed by ill-conditioned matrices to the application of classical multivariate analysis of variance (MANOVA), whose fundamental concepts were reviewed in Section 3.8.1. As discussed in that section, classical MANOVA requires the inversion of the within-levels scatter matrix $\mathbf{W}$ which cannot be performed when the experimental matrix is ill-conditioned or singular (*i.e.*, when the number of variables is higher than that of experiments and/or when the variables are highly correlated). Regularized MANOVA was proposed as a possible solution to this limitation. As the name suggests, rMANOVA is based on exploiting a regularization of the within-levels scatter matrix $\mathbf{W}$, which is meant to allow its inversion also in the case of high-dimensional data. More specifically, in rMANOVA, the "ratio" matrix is defined as:

$$\mathbf{J}_f^* = [(1-\delta)\mathbf{W} + \delta\boldsymbol{\Xi}]^{-1}\mathbf{B}_f \qquad f = \mathtt{A}, \mathtt{B}, \ldots \tag{5.47}$$

where $\boldsymbol{\Xi}$ is a target matrix, which imposes a specific structure on $\mathbf{W}$ while $\delta \in [0,1]$ is an adjustable parameter tuning the extent to which the regularized within-levels scatter matrix is more similar to $\boldsymbol{\Xi}$ or to $\mathbf{W}$ and, consequently, providing a trade-off between bias and variance in the estimate of the inverse of the scatter matrix itself. Equation 5.47 shows that regularization is achieved by shrinking the covariance matrix toward a target matrix $\boldsymbol{\Xi}$ which, according to the proposers of the method, should have lots of structure imposed (*i.e.*, few parameters to be estimated). In particular, among the possible targets, the developers of the method identify two special cases, that they call *unique variance* and *average variance*:

$$\begin{aligned} \boldsymbol{\Xi}_{unique} &= diag(\mathbf{W}) \\ \boldsymbol{\Xi}_{average} &= \tfrac{trace(\mathbf{W})}{J}\mathbf{I} \end{aligned} \tag{5.48}$$

J being the number of measured variables. The differences between the two targets are in the implicit assumptions they make about the variance of the individual variables: $\boldsymbol{\Xi}_{average}$ assumes the same variance for each variable, while $\boldsymbol{\Xi}_{unique}$ hypothesizes the variance of each variable to be different. Here, it is worth highlighting that it is reported that, by setting $\delta = 1$ and using $\boldsymbol{\Xi} = \mathbf{I}$ as a target, the results of rMANOVA should be equal to performing ASCA on the data. As for the adjustable parameter δ, the authors suggest finding its optimal parameter with a data-driven procedure based on the Ledoit–Wolf theorem (Ledoit and Wolf, 2004), which, consequently, leads to different estimates when the average or the unique variance targets are adopted. Indeed, for the average variance target, the suggested optimal value of δ is:

$$\delta_{opt}^{ave} = \frac{\sum_{i\neq j} Var(w_{ij}) + \sum_i Var(w_{ii})}{\sum_{i\neq j} w_{ij}^2 + \sum_i (w_{ii}^2 - v)} \tag{5.49}$$

with $v = trace(\mathbf{W})/J$, whereas, in the case of the unique variance target, it is:

$$\delta_{opt}^{un} = \frac{\sum_{i\neq j} Var(w_{ij})}{\sum_{i\neq j} w_{ij}^2} \tag{5.50}$$

As far as interpretation is concerned, visualization of the multivariate effect of a design term can be carried out by plotting the scores of the experiments along the canonical variates, *i.e.*, the eigenvectors of $\mathbf{J}_f^*$ ($f = \mathtt{A}, \mathtt{B}, \ldots$). Identification of the relevant variables relies on inspection of the corresponding weights on the canonical variates. At the same time, it should be stressed that the extension to more complicated designs is not straightforward and it has not been explored yet in the literature, but it can be done (see Example 5.10).

Example 5.10: Example of rMANOVA

The use of rMANOVA will be illustrated by means of an example of its application to the pasta data set, already used for ANOVA-TP, where NIR spectra were collected on egg-pasta samples prepared by varying *Egg concentration*, drying *Time* and drying *Temperature* (at 6, 3, and 3 levels, respectively). Here both the use of average and unique variance as targets will be considered.

When using the *unique* variance target, all factors and interactions appear to be significant ($p < 0.05$) according to all four statistics used in MANOVA (Wilks' Lambda, Pillai–Bartlett trace, Hotellings–Lawley trace, and Roy's largest root, see Section 3.8.1), their null distributions having been estimated by permutation testing, If the *average* variance target is chosen, only the effects of the three factors and that of the *Temperature:Egg concentration* interaction are found to be significant. In particular, the effect of *Temperature* and that of *Egg concentration* are significant according to all four statistics; the effect of *Time* for all statistics except for Pillai–Bartlett trace and the one of the interaction according to all except Roy's largest root. In both cases, the optimal value of δ was found to be 0.0069, indicating that the largest part of the regularized variance is still due to the original within-group scatter matrix **W**.

For each significant factor or interaction, interpretation is then carried out by inspecting the scores and weights along the corresponding canonical variates (CVs). As an example, interpretation of the effect of *Temperature* will be discussed.

The plots in Figure 5.21 show how the use of a particular target affects the scores pattern. More specifically, *unique* variance target results in more compact groups corresponding to the factor levels, and there is a clear separation both along CV1 and CV2. On the other hand, when *average*

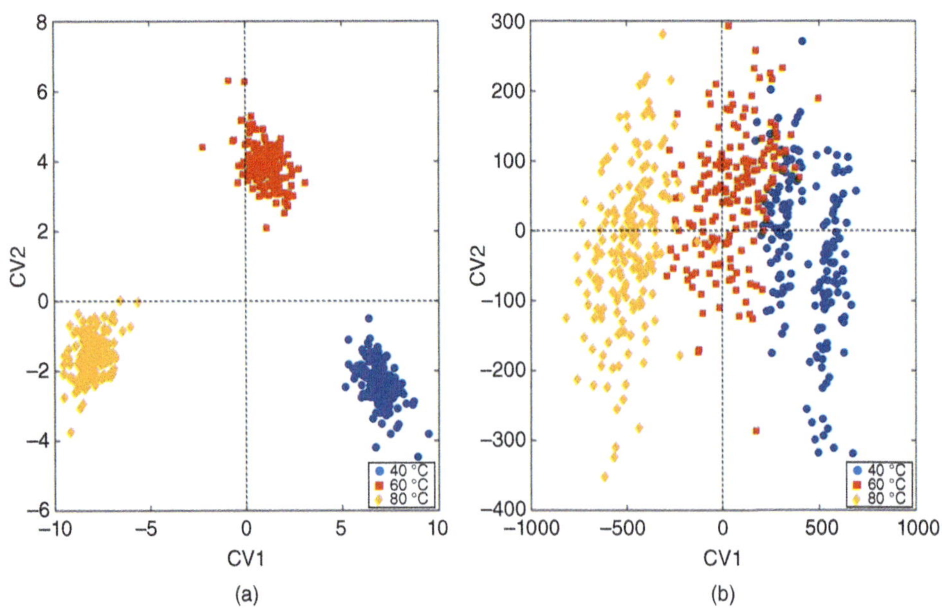

Figure 5.21 Scores along the two Canonical Variates for the effect of *Temperature* calculated by rMANOVA using as target the *unique* (a) and the *average* (b) variance.

variance target is used, the factor levels are mostly separated along CV1, and the corresponding groups are more scattered, in particular along CV2. Moreover, two subgroups emerge for 40 °C, as also observed in the original paper (Bevilacqua *et al.*, 2013), where ASCA was used.

When looking at the corresponding weights, displayed in Figure 5.22, they appear very noisy, especially in the case of *unique* variance target. When considering *average* variance target, all wavenumbers below 7800 cm^{-1} (with the exception of the interval 4910–4966 cm^{-1}) have positive weights on CV1, suggesting that these bands decrease in intensity with increasing temperature. On the other hand, in the case of *unique* variance target, many bands in the same interval have negative weights (*e.g.*, 4574–4820, 5593–5801, 5993–6279, 6646–6827 cm^{-1}).

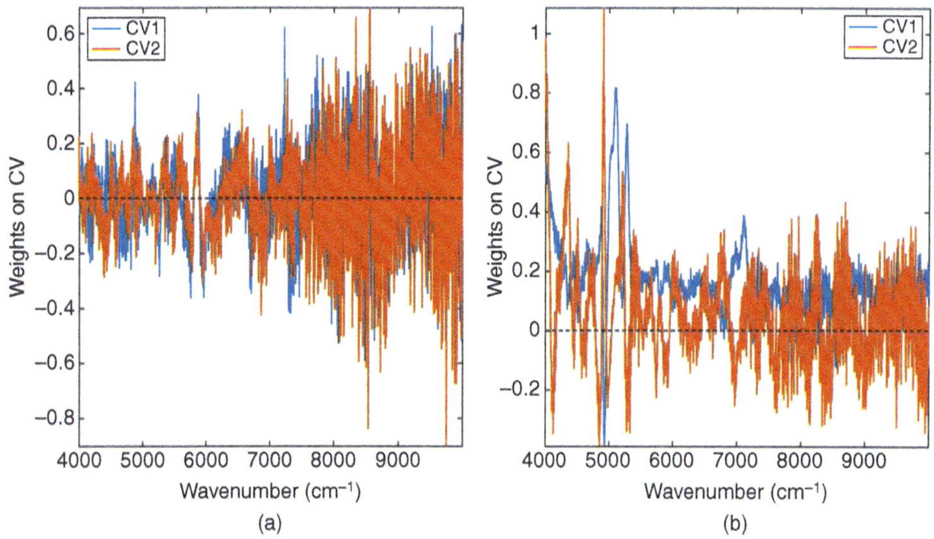

Figure 5.22 Variable loadings along the two Canonical Variates for the effect of *Temperature* calculated by rMANOVA using as target the *unique* (a) and the *average* (b) variance.

5.4.5 Fifty–Fifty MANOVA

An approach very much related to PC-ANOVA is Fifty–Fifty MANOVA (FF-MANOVA) (Langsrud, 2002). The basic idea will be explained for a one-way MANOVA model:

$$\mathbf{Y} = \mathbf{1}\boldsymbol{\theta}_M^t + \mathbf{x}\boldsymbol{\theta}^t + \mathbf{E}_Y \tag{5.51}$$

where $\mathbf{Y}(I \times J)$ contains the J responses; $\boldsymbol{\theta}_M(J \times 1)$ is vector of means; $\mathbf{x}(I \times 1)$ is a vector of zeros and ones indicating the two groups; $\boldsymbol{\theta}(J \times 1)$ is a vector of regression coefficients; and $\mathbf{E}_Y$ is the matrix of residuals. Testing the hypothesis $H_0 : \boldsymbol{\theta} = \mathbf{0}$ versus $H_1 : \boldsymbol{\theta} \neq \mathbf{0}$ can be done using Hotelling's T^2 test, but this breaks down when $J > I$.

Connected to the model in Equation 5.51 is the reverse regression model:

$$\mathbf{x} = \beta_0 + \mathbf{Y}\boldsymbol{\beta} + \mathbf{e}_x \tag{5.52}$$

where β_0 is the offset term; $\boldsymbol{\beta}(J \times 1)$ a vector of regression coefficients; and $\mathbf{e}_x$ a generic error term. The earlier-mentioned hypothesis testing is now the same as testing the hypothesis

$H_0 : \boldsymbol{\beta} = \mathbf{0}$ versus $H_1 : \boldsymbol{\beta} \neq \mathbf{0}$ which can be done with ordinary F-test for regression. Also this breaks down, however, in the case of $J > I$.

One solution for this problem is to work with the principal components of $\mathbf{Y}$ thereby invoking Principal Component Regression. Suppose that the SVD of the centered matrix $\mathbf{Y}_c$ equals $\mathbf{Y}_c = \mathbf{USV}^t$. Now the testing of Equation 5.52 can be done per principal component $\mathbf{u}_l$ where $\mathbf{u}_l$ is the lth column of $\mathbf{U}$. This is, however, problematic if there are more than one PCs relevant for $\mathbf{x}$ since then tests are not independent of each other. Thus, we have to use multiple PCs simultaneously, and the suggestion is then to define a test statistic based on sums-of-squares $SS_l = (\mathbf{u}_l^t\mathbf{x})^2$ to arrive at

$$F = \frac{(SS_1 + \cdots + SS_L)/L}{(SS_{L+1+d} + \cdots + SS_{I-1})/(I - 1 - L - d)} \tag{5.53}$$

which follows an F distribution with L, $(I - 1 - L - d)$ degrees of freedom. There are two crucial elements in this test statistic. First, the number of principal components in the denominator (L) should be chosen independent of the regression Equation 5.52; otherwise, the test is not valid. This is comparable to performing variable selection and subsequently testing the regression coefficients. Hence, the L has to be established *a priori*, and the recommendation is to use the smallest number L such that 50% of the variation in $\mathbf{Y}_c$ is explained. Secondly, in the numerator, not all remaining PCs (after the first L) are used, but the components $L + 1$ to $L + d$ are skipped. These d components are called buffer components. The reason is that if there is still some systematic variation in these components and if they are included in the numerator, this numerator will be too large thereby hampering the power of the test. The recommendation now is to use 50% of the remaining PCs as buffer components. This then constitutes the F-test for a significant effect of the one-way factor on the (PCs) of the responses.

Several remarks are appropriate:

- For obvious reasons, this method is called Fifty–Fifty MANOVA (FF-MANOVA).
- This test can be extended to more than one-way MANOVA.
- The values of L and d are meta-parameters. Although the authors suggest the 50/50% rule, they also indicate that these values may be chosen differently depending on the specific data set.
- This method only gives a formal test for significance and no facilities for visualization and plotting of the results. That may be the reason why it is not used very often. It would be worthwhile to develop methods combining FF-MANOVA with ASCA-type methods.
- Note the resemblance in the way of reasoning of FF-MANOVA and ANOVA-PLS, *i.e.*, using the similarity between testing based on Equations 5.51 and 5.52 on the one hand, and ASCA and ANOVA-PLS (or ANOVA-TP) on the other hand.

5.4.6 AComDim

ANOVA-Common Dimensions (AComDim) (Bouveresse *et al.*, 2011) starts from the same decomposition of the matrix $\mathbf{Y}$ which constitutes the basis of most HD-ANOVA generalizations and which is shown in Equation 5.1. AComDim shares some similarities with APCA but, instead of using PCA for the analysis of the individual effect matrices, it exploits a multi-block approach to jointly process all of them. In particular, once the overall variability in $\mathbf{Y}$ is partitioned according to the ANOVA scheme, analogously to what already discussed for APCA, new effect matrices are built by adding the residuals $\mathbf{Y}_{\mathtt{Ind}}$ to the corresponding pure effect terms $\widehat{\mathbf{Y}}_f$

$$\widehat{\mathbf{Y}}_f^* = \widehat{\mathbf{Y}}_f + \mathbf{Y}_{\mathtt{Ind}} \qquad f = \mathtt{A}, \mathtt{B}, \mathtt{C}, \mathtt{AB}, \ldots \tag{5.54}$$

All these matrices, including also the residual matrix $\mathbf{Y}_{\mathtt{Ind}}$, are treated as a multi-block data set, and they are jointly analyzed through a multi-block exploratory analysis method, namely Common components and specific weight analysis (CCSWA, also known as Common dimensions, ComDim (Qannari *et al.*, 2000), see Elaboration 5.2).

ELABORATION 5.2

ComDim

ComDim is an exploratory multi-block method that can be considered as one of the possible generalizations of PCA to multiple data tables. As the name suggests, it focuses on extracting components that explain as much as possible of the joint variability among the different data matrices. Operationally, common components are extracted by maximizing a weighted sum of cross-product matrices, where the weights, also called *saliences*, are estimated during the decomposition. Therefore, ComDim hypothesizes that the blocks of variables ($\mathbf{Y}_f, f = 1, \ldots, F$) share a common column space and that each dimension of this space may be differentially weighted for each block through the corresponding salience:

$$\mathbf{W}_f = \mathbf{Y}_f\mathbf{Y}_f^t = \mathbf{Q}\boldsymbol{\Lambda}_f\mathbf{Q}^t \tag{5.55}$$

where $\mathbf{Q}$ (of size $I \times R$) and $\boldsymbol{\Lambda}_f = diag([\lambda_1^{(f)}\lambda_2^{(f)} \cdots \lambda_R^{(f)}])$ are the matrices collecting the common components and the saliences of the fth block for those components, respectively. More in detail, ComDim works as illustrated in Algorithm 5.3.

Algorithm 5.3

CALCULATING THE COMDIM SOLUTION

Assuming that each block $\mathbf{Y}_f$ is mean-centered and normalized through division by its Frobenius' norm, the algorithm extracts the components sequentially, according to the following steps

1) For each block f, initialize the saliences for the rth component, setting it equal to one: $\lambda_r^{(f)} = 1$ for $f = 1 : F$.
2) Repeat until convergence of $\sum_{f=1}^{F} \|\mathbf{W}_f - \lambda_r^{(f)}\mathbf{q}_r\mathbf{q}_r^t\|^2$:
   a) Define $\mathbf{W} = \sum_{f=1}^{F} \lambda_r^{(f)}\mathbf{W}_f = \sum_{f=1}^{F} \lambda_r^{(f)}\mathbf{Y}_f\mathbf{Y}_f^t$.
   b) Estimate $\mathbf{q}_r$ as the first singular vector of $\mathbf{W}$.
   c) Update the values of the saliences for the component as: $\lambda_r^{(f)} = \mathbf{q}_r^t\mathbf{W}_f\mathbf{q}_r$.
3) Perform deflation on each block: $\mathbf{W}_f^{upd} = \mathbf{W}_f - \lambda_r^{(f)}\mathbf{q}_r\mathbf{q}_r^t$.
4) Proceed with the calculation of the successive component by setting $r = r + 1$ and iterating steps (1)–(3).

Once the ComDim solution is calculated according to what is described in Algorithm 5.3, if needed, loadings for each block can be obtained as:

$$\mathbf{P}_f = \mathbf{Y}_f^t\mathbf{Q} \tag{5.56}$$

since the scores are constrained to: $\mathbf{Q}^t\mathbf{Q} = \mathbf{I}$.

As anticipated above, the main idea behind AComDim is to replace the multiple PCAs which constitute the second step of APCA, with a simultaneous analysis of the multiple blocks resulting from the ANOVA decomposition, to determine their common components using ComDim. Therefore, after the ANOVA decomposition, all the matrices prepared as described in Equation 5.54 together with the matrix $\mathbf{Y}_{\mathtt{Ind}}$ are jointly subjected to ComDim analysis. Once the model is calculated, inspection of the saliences allows us to determine which common component is related to which factor.

Since by construction, all blocks contain a contribution from $\mathbf{Y}_{\mathtt{Ind}}$, it is very likely that the first common component will be associated with the added residuals. This assumption constitutes the basis for statistical testing. According to the authors, since the first common component (CC1) is mostly associated with the residual variability (*i.e.*, that is not accounted for by the design), blocks having high salience on that component should contain mainly noise, and therefore their effect should not be significant. On the other hand, blocks that contribute less to CC1 should probably contain other sources of variability (the effect of the corresponding design term). Based on these considerations, the significance of each design term is evaluated through an F-test, where the salience of the corresponding block on CC1 is compared to that of the residual block on the same component:

$$F_f = \frac{\lambda_1^{(Ind)}}{\lambda_1^{(f)}} = \frac{\mathbf{q}_1^t \mathbf{W}_{Ind} \mathbf{q}_1}{\mathbf{q}_1^t \mathbf{W}_f \mathbf{q}_1} \tag{5.57}$$

For testing, $I - 1$ degrees of freedom are considered for both the numerator and the denominator, I being the total number of rows in each of the matrices. Based on the considerations above, a large value of F indicates that the effect of the factor or interaction is significant; if this is the case, then interpretation is accomplished by inspecting the corresponding scores and loadings for the associated common component(s). The use of AComDim is illustrated in Example 5.11.

Example 5.11: Example of AComDim on the pasta data

The use of AComDim will be illustrated through the analysis of the already mentioned pasta dataset (see Example 5.1). As in the previously discussed examples, the SNV-preprocessed spectral data were at first subjected to the ANOVA decomposition described in Equation 5.9. Successively, the residual matrix $\mathbf{Y}_{\mathtt{Ind}}$ is added to each of the effect matrices, as described in Equation 5.54. Accordingly, a multiblock dataset made up of the eight matrices shown in Figure 5.23 is obtained.

The multiblock dataset, made up of the residual-augmented effect matrices and the residual matrix, is then analyzed by ComDim. The plot in Figure 5.24 shows the saliences on the 8 effect blocks on the first 20 common components (CCs). It is evident from the plot how not only the first common component, as expected by construction, but also the second one is dominated by the contribution of the residuals. On the other hand, other components can be ascribed almost completely to specific main effects or interactions. For instance, the effect of *Temperature* contributes almost exclusively to the definition of CC3 and CC19, that of *Time* to CC5, the effect of *Egg concentration* to CC6, CC11, and CC14, and the binary interaction to CC7, CC9, and CC20.

To evaluate whether the effect of a factor or of an interaction term on the spectroscopic signature of the samples is significant, AComDim relies on performing an F test, where the salience of

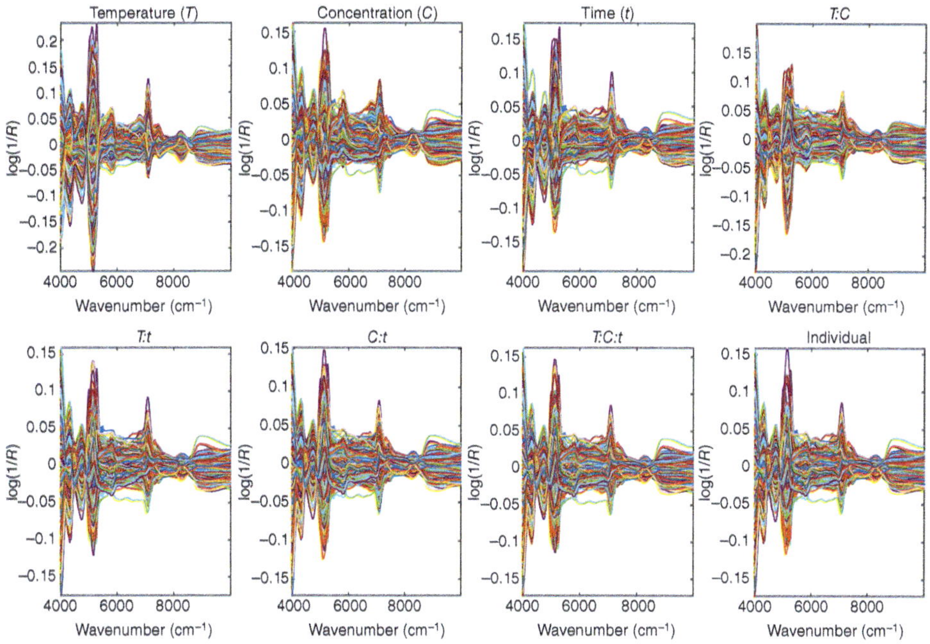

Figure 5.23 Effect matrices resulting from the ANOVA decomposition of the pasta dataset after addition of the residual matrix to each of them (except, of course, of the residual matrix itself). The resulting 8 matrices make up the multiblock data set which will be then subjected to ComDim analysis.

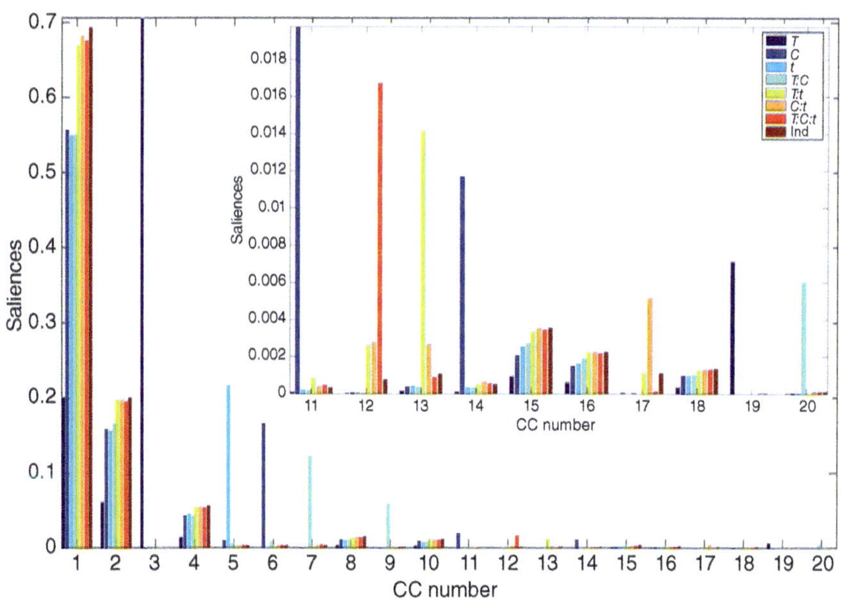

Figure 5.24 Saliences of the different residual-augmented data blocks resulting from the ANOVA decomposition on the first 20 common components (CCs) extracted by the ComDim model (a zoom of the values for the lower variance common components (11 to 12) is displayed in the inset).

each effect block on the first component is compared to that of the residual block on the same common component, as described in Equation 5.57, the degrees of freedom for both the numerator and the denominator being the total number of observations- minus one (*i.e.*, 539). The results are summarized in Table 5.4.

Table 5.4 Results of the F test on the block saliences on CC1 for testing the effect significance.

Source	Salience	*F*	*p*-value
Temperature (*T*)	0.2007	3.451	<0.0001
Concentration (*C*)	0.5576	1.242	0.0059
Time (*t*)	0.5502	1.259	0.0038
T:C	0.5510	1.257	0.0040
T:t	0.6720	1.031	0.3629
C:t	0.6832	1.014	0.4369
T:C:t	0.6769	1.023	0.3944
Error	0.6927		

The results summarized in Table 5.4 indicate a significant effect of all three factors and of the binary interaction between *Temperature* and *Egg concentration*. These findings are in agreement with the original publication (Bevilacqua *et al.*, 2013), where the HD-ANOVA analysis was performed using ASCA.

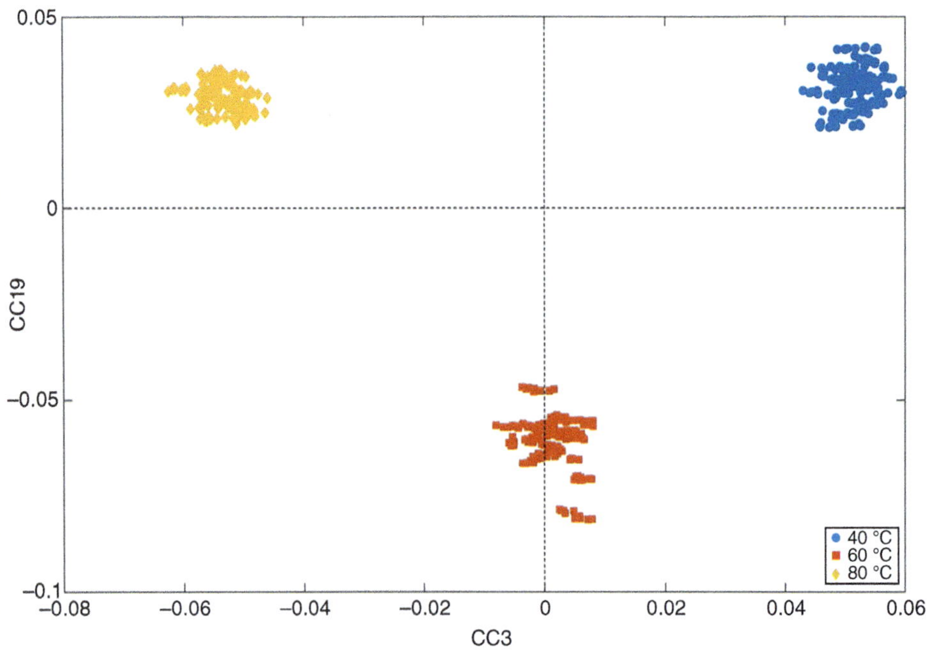

Figure 5.25 Sample scores along CC3 and CC19 which, based on the values of their saliences, can be attributed to the effect of *Temperature*.

Once the significance of the effect of the different design terms has been evaluated, interpretation can be carried out by inspecting the scores and loadings along the corresponding common components. For example, to investigate the effect of *Temperature*, one should inspect the scores and loadings along CC3 and CC19. The score plot, displayed in Figure 5.25, shows a clear separation among the three factor levels. In particular, the scores along CC3 are inversely proportional to the *Temperature* levels, with the higher temperature (80 °C) showing negative values, the intermediate (60 °C) falling around zero and the lowest (40 °C) having positive values. On the other hand, CC19 further differentiates the intermediate temperature (at negative scores) from the other two (at positive scores).

The interpretation of the observed differences can be achieved by inspecting the loadings on the corresponding common components. In ComDim a separate set of loadings is calculated for each of the modeled blocks, according to Equation 5.56. Since CC3 and CC19 reflect almost exclusively the contribution of the *Temperature* effect matrix, interpretation can be carried out by considering only the loadings calculated for that particular block, which are displayed in Figure 5.26.

When looking at Figure 5.26, at first it should be stressed that, in ComDim, the scores are normalized, so that all the variance explained by the components is reflected in the loadings. Accordingly, it is evident how the largest part of the variability on the signal induced by the effect of temperature is captured by CC3. In particular, by comparing the loading plot with the scores plot in Figure 5.25, it is possible to affirm that increasing the roasting temperature results in a progressive decrease of the spectral intensity of the signals at 4915–5411 cm^{-1} and 5950–7700 cm^{-1}, and a corresponding increase of the absorbance in the range 4000–4915 cm^{-1}, 5411–5900 cm^{-1}, and 7700–8540 cm^{-1}. These results are similar to those obtained by applying AMOPLS on the same data (see Example 5.2).

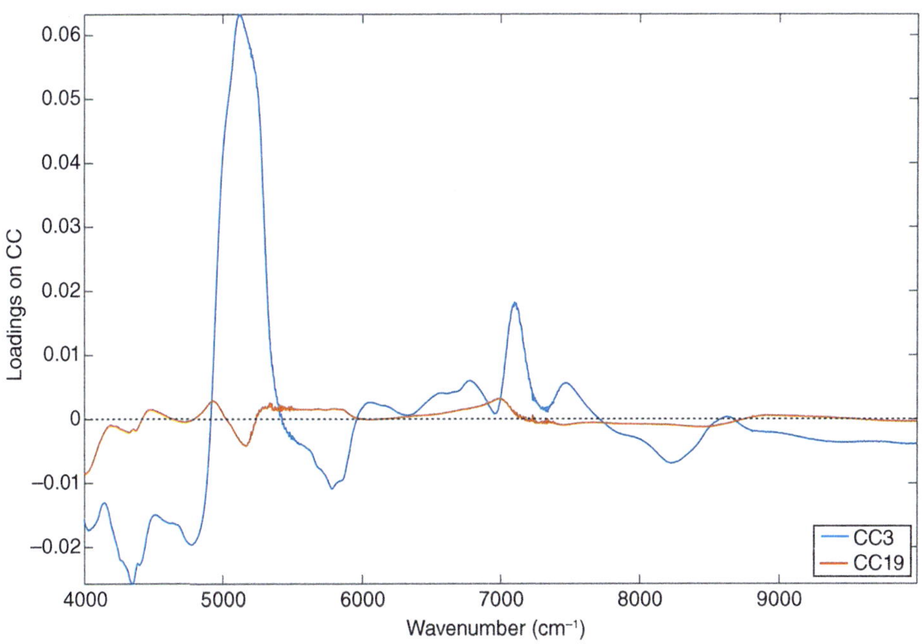

Figure 5.26 Loadings of the block corresponding to the residual-augmented effect of *Temperature* along CC3 and CC19.

5.4.7 General Effect Modeling (GEM)

Mosleth et al. have developed a framework related to APCA (Mosleth *et al.*, 2020). The original formulation uses regression-based factor estimation as in APCA, creating so-called Effect+Residual (ER) matrices that are the basis for subsequent unsupervised or supervised analyses. ER matrices are defined the same way as effect matrices in APCA, where the residual errors are added to the least squares estimates of each factor:

$$\mathrm{ER}_f = \mathbf{X}_f\widehat{\Theta}_f + \mathbf{Y}_{\mathtt{Ind}} = LS_f + \mathbf{Y}_{\mathtt{Ind}} \tag{5.58}$$

If PCA was applied to the ER matrices, this would be exactly APCA. In GEM, PCA is only one of many options. Other examples of analyses applied to the ER matrices include univariate and bivariate visualizations of variables, PLS discriminant analysis with variable selection, and Elastic net regularization, but many other analyses could be useful. These can either be used as a single method, or multiple methods can be applied in parallel to look for a consensus and common patterns or variables that have an effect across methodologies.

Later, the framework was extended and renamed General Effect Modeling (Mosleth and Liland, 2024). The underlying regression model has been extended to include covariates, and the subsequent analysis follows two main paths. Terminology is borrowed from the knock-out and knock-in of genes in biology, where genes are deactivated or activated to study their effect on an organism or pathway. The first path of GEM is called *knock-in* – analyzing a single, isolated effect as in Equation 5.58, and the second is called *knock-out* – removing one or more effects and analyzing the remaining patterns:

$$\mathrm{ER}_{-f} = \mathbf{Y} - \mathbf{X}_f\widehat{\Theta}_f \tag{5.59}$$

Here, all effects except effect f have been retained. *Knock-in* coincides with the principal strategy of APCA (and extensions), *i.e.*, analyses of effect matrices (with added residuals), though with more flexibility in the subsequent analysis of the ER matrices. *Knock-out* is an alternative to orthogonalizing on a factor but rather doing this through regression.

Although GEM so far has only leveraged basic (unbalanced) fixed effect regression-based ANOVA for generating effect matrices, we could easily plug any of the more advanced variants shown in this book into the framework. This would open up for analyzing more types of designs combined with other (un)supervised analyses in *knock-in* or *knock-out* mode.

Notable applications so far have been the analysis of data from proteomics and transcriptomics in connection to human cohort and intervention studies on multiple sclerosis (Mosleth *et al.*, 2021, 2024b) and diabetes (Mosleth *et al.*, 2024a), and the study of metabolic pathways in bacteria (Mosleth *et al.*, 2020). Common for the GEM studies is that the methodology is used to peel away interfering or confounding effects and to assist and steer biomarker selection using domain knowledge about the problem under study.

6

Distance-based Methods

In Chapters 3–5, the theory of linear models, linear mixed models, and ANOVA-Simultaneous Component Analysis (ASCA) methods have been introduced. In this chapter, we will leave the linear models and go to a generalization of analysis of variance (ANOVA) based on between-sample distances or dissimilarities. Distance-based data analysis is a form of indirect data analysis, meaning that analysis is performed on a reduced version of the data set in which only properties of one of the modes (samples or variables) are used. Distances between samples can be calculated from the variables in the data set, but they could also have been obtained directly, *e.g.*, from a projective mapping experiment (Risvik *et al.*, 1994). Indirect data analysis is often divided into Q-mode analysis, which analyzes pairs of samples or subjects, and R-mode analysis which focuses on pairs of descriptors or species (Legendre and Legendre, 2012). A formal definition of distances and dissimilarities was already provided in Section 2.1.4, where also various examples of distances, and their relationships were shown and discussed. In this chapter, we focus on Q-mode analysis using distance or dissimilarity measures. Distance-based methods are often used when the measured data are not normally distributed, compositional, have many zeros, and are overdispersed (*e.g.*, abundance data in ecology or metagenomics) and thus disobey many requirements of statistical tests. Furthermore, they are selected if the chosen distance metric offers better interpretability than the distances that are implicitly considered in direct methods. PCoA, see Elaboration 1.5 and permutational multivariate analysis of variance (PERMANOVA) (Legendre and Legendre, 2012) are two approaches that use such distance-based analysis.

6.1 Introduction

In many microbiome applications, abundance measurements of various species are used to compare different geographic units, or the same geographic unit under different conditions. In microbiome studies, counts of bacterial species members are used to quantify the microbial community. Such communities can consist of hundreds to thousands of different species that are usually indicated as Operational Taxonomic Units (OTUs) or as Amplicon Sequence Variants (ASVs), see, *e.g.*, Section 1.4.3. Reads per species, ASV, or OTU are counted for each sample. Their presence and abundance can be rather different across the various samples.

In the ASCA framework, linear models separating variation in high dimensional data into groups, and explorative analysis using Principal Component Analysis (PCA) of those groups are combined. This approach usually fares well for quantitative data with symmetric error distributions. The multivariate or high-dimensional abundance data discussed above has some specific distributional properties that are not well matched with traditional data analysis

Analysis of Variance for High-Dimensional Data: Applications in Life, Food and Chemical Sciences,
First Edition. Age K. Smilde, Federico Marini, Johan A. Westerhuis, and Kristian H. Liland.

methods such as PCA or linear models. The main issues are that there are many zeros in the data (sparsity), the counts should be interpreted as proportions (compositionality), and there is complex heteroscedasticity (overdispersion). Solutions for these problems can be found in using special statistical distributions in generalized linear models, by using transformation methods toward more normal-like distributions of the data, or by the use of distance-based data analysis methods, as we see a lot in ecological data analysis literature (Zuur *et al.*, 2007; Legendre and Legendre, 2012). Examples of distance-based methods are PCoA (Borg and Groenen, 2010), PERMANOVA (Anderson, 2001), distance-based redundancy analysis (RDA) (Legendre and Legendre, 2012), and many methods based on Multidimensional Scaling (Borg and Groenen, 2010).

6.1.1 Double Zeros

In microbiome studies, the number of species detected in a full data set can be large although most samples will lack most species, *i.e.*, the data set is sparse. The total number of zeros in such data sets can be as large as 80–90%, where some species have almost no zeros but other species are only present in very few samples. In the microbiome research field, where variables represent the abundance of species, a zero value can have different meanings, being biological, *e.g.*, the species is not present in the sample due to whatever reason, a sampling zero where the species might be present, but the sample was too small to detect it, or a technical zero where an error occurred in the measurement (Jiang *et al.*, 2022). In the first situation such zeros are often called structured or structural zeros, meaning that there are biological reasons that the species cannot exist in such condition and thus can never be found in that sample. In the second situation of the sampling zeros, a larger sample size might have shown the presence of that species. It is important to distinguish between these two situations, as the interpretation is very different. The real reason for the zero can often not be clarified by the data alone and needs biological experts to be found.

In the calculation of the dissimilarity between two samples, the contribution of each species is summed to a total dissimilarity. For many of the species, however, both samples will have 0 counts. The general interpretation is that the presence of a species in both samples decreases the dissimilarity more than the common absence of a species because there can be many different reasons for the absence of a species in a sample. Dissimilarity measures that rate the double absence equally important as the double presence are called symmetrical, while dissimilarity measures that treat them differently are indicated as asymmetrical (Legendre and Legendre, 2012).

In situations with many double zeros in microbiome data, asymmetrical dissimilarity measures, which ignore the double zeros, are preferred over symmetrical ones that take the double zeros into account. A good example of an asymmetrical binary dissimilarity is the Jaccard dissimilarity between two samples $\mathbf{y}_1$ and $\mathbf{y}_2$ that both have J species measured. Here, $\mathbf{y}_1(j)$ is 1 if species j is present in sample $\mathbf{y}_1$, while it is 0 when it is absent. The Jaccard dissimilarity measures the fraction of species the two samples share, ignoring the features that are absent in both samples. For binary distances, observations are usually summarized in a frequency table where the combined presence or absence of species in samples $\mathbf{y}_1$ and $\mathbf{x}_2$ is indicated by $a, b, c,$ and d.

In Table 6.1, a represents the number of species that are present in both samples $\mathbf{y}_1$ and $\mathbf{y}_2$, and d represents the number of species that are absent in both samples (the double zeros). The Jaccard dissimilarity is then defined as: $d_J(\mathbf{y}_1, \mathbf{y}_2) = 1 - \frac{a}{a+b+c}$, where the double zeros are ignored and do not change the distance d_J. A symmetrical distance such as the *simple matching distance* which is defined as $d_{smc}(\mathbf{y}_1, \mathbf{y}_2) = 1 - \frac{a+d}{a+b+c+d}$ would decrease when the number of double zeros increases and in general will not be used in microbiome applications.

Table 6.1 Frequency table for variables in samples $\mathbf{y}_1$ and $\mathbf{y}_2$.

	$\mathbf{y}_2(j) = 1$	$\mathbf{y}_2(j) = 0$
$\mathbf{y}_1(j) = 1$	a	b
$\mathbf{y}_1(j) = 0$	c	d

6.1.2 Horseshoe Effect

When samples only share a few of the measured species, they also share the absence of many other species in the dataset. Figure 6.1a shows an example where the black diagonal represents the presence of the species, while the white part means the absence of the species. Samples 1–5 share at least one species, but samples 6–20 do not share any species with sample 1. Such data can be expected when the samples represent a gradient (Morton *et al.*, 2017). However, when mapping all samples in only two dimensions, it is not possible to position each of these samples (6–20) equally far away from sample 1. The dataset has a much higher rank than two, and mapping all samples and variables in two dimensions leads to distorted PCA score plots as can be seen in Figure 6.1. This phenomenon is often referred to as the horseshoe or arch effect (Figure 6.1b) if the data is not centered or the heart effect (Figure 6.1c) when the data is centered before analysis. The horseshoe shows that a low-dimensional approximation of the data does not represent its information well and that more dimensions carry important information on the differences between the samples.

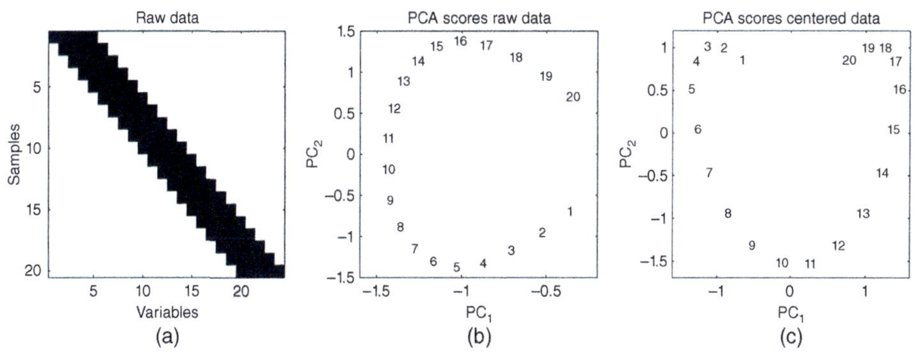

Figure 6.1 The horseshoe effect; (a) layout of the data (black is nonzero); (b) PCA scores of raw data; (c) PCA scores of centered data.

6.1.3 Compositionality

High-throughput count data of DNA or RNA sequences such as microbiome or gene expression data are compositional. Compositional microbiome data consists of counts for each species that add up to a total number of counts. Compositional data is subject to the "closure problem" that occurs when species compete to make up the constant sum constraint (Aitchison, 1986). The high-throughput instrument has a specific capacity and can only provide reads up to a certain number. There is no information on the absolute number of bacteria in the sample, only about their relative amounts. The library size (total number of reads, see Section 1.4.2) for a given measurement of a microbiome is only a random value and is not constant for different samples. This complicates the comparison of counts per species for different samples, and thus

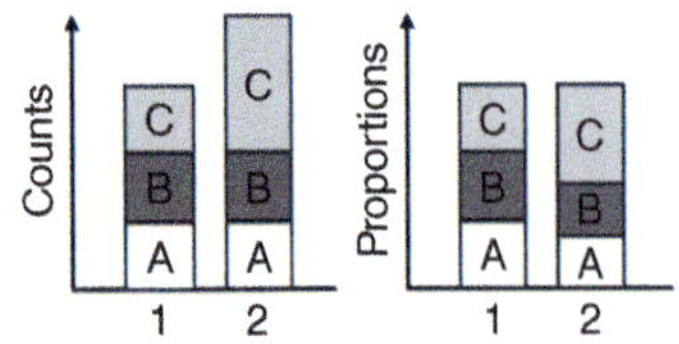

Figure 6.2 The compositionality issue. In sample 2, species C is doubled in size compared to sample 1, while species A and B are unchanged. The proportion of species C is increased, but proportions for species A and B seem to have decreased. However, the ratio between species A and B is the same in samples 1 and 2 for both the count data and the proportions.

proportions values are needed (Gloor *et al.*, 2017). However, even from these proportions it is not possible to explore the difference of a species in different samples due to the compositionality (see Elaboration 1.4). If one species doubles in relative abundance while the other species do not change and the microbiome is measured again, the proportions will show an increase of that species and a decrease of all other species as shown in Figure 6.2.

In Figure 6.2, the total counts of the samples are not representative of the total number of bacteria in the sample, it only represents an unknown fraction of that. The proportions show the relative amounts. The total number of reads of a microbiome sample is irrelevant for the relative abundance of each species, but it does contain information on the precision of the estimated relative abundance or proportions (Aitchison, 1994). The relevant information is in the ratios of the species. The ratio between species A and B is the same when looking at counts and looking at the proportions.

The specific number of reads of the high-throughput instrument is actually not the same for each sample and can depend on a different amount of DNA provided from each microbiome, or on its quality. The total number of reads per sample is thus irrelevant to the relative abundance, but it does affect many data analysis methods. Therefore, various normalization approaches have been suggested to remove the effect of the different library sizes, *e.g.*, rarefying (or subsampling), or count normalization, see Section 1.5.1. Normalization removes the library size differences and makes the counts comparable between samples. However, normalization does not remove the compositionality in the data. After normalization, the counts (in case of subsampling) still add up to a constant value, or the proportions (in case of count normalizations) still add up to 1 for each sample. Thus if a single species increases in abundance, the others still decrease. Interpretation of changes in the composition still has to be done in terms of ratios of species.

Aitchison's work on compositional data provides a much-used transformation of compositional data (Aitchison, 1994). The Centered Log-Ratio (CLR) transforms each sample of compositional data separately by correcting for library size and centering over the features within each sample, after which it can be subjected to a PCA to explore the differences between the samples.

$$CLR(\mathbf{y}) = \left(log\frac{y_1}{g(\mathbf{y})}, \dots, log\frac{y_j}{g(\mathbf{y})}, \dots, log\frac{y_J}{g(\mathbf{y})} \right) \tag{6.1}$$

where $\mathbf{y} = [y_1, \dots, y_j, \dots, y_J]^t$ is a sample with compositional values and $g(\mathbf{y}) = \sqrt[J]{y_1 \cdot \dots \cdot y_J}$ is the geometric mean of $\mathbf{y}$. The geometric mean is often used to represent values of populations where doubling or halving the size is more meaningful than the addition or subtraction of a fixed amount. The CLR equals a row-centering after a log transform of the proportions (Filzmoser *et al.*, 2018). The log transform makes sense as the data is usually right skewed (*i.e.*, most features are low abundant and only a few are highly abundant). The Euclidean distance of CLR-transformed data is called the Aitchison distance and is useful for compositional data.

The interpretation of CLR-transformed data still has to be done in terms of log-ratios. The advantage of log-ratios over proportions is that the scale now goes from $-\infty$ to $+\infty$ instead of a constrained scale between 0 and 1. However, the interpretation of a single feature alone is still not possible.

The CLR transformation uses the log of the counts. As microbiome data has many zeros, and the log of zero does not exist, all zeros are replaced by pseudo counts. A pseudo count is a small positive value that replaces the 0. However, as microbiome data are sparse with many zeros, the effect of imputation of so many pseudo counts can become problematic. There are several different strategies to address the problem of zero inflation in the transformation of microbiome data. Different values for the pseudo-counts, replacement of zeros using multivariate imputation, and even deep learning methods are being used. Lubbe *et al.* (2021) provided a detailed comparison of these strategies in compositional methods. They recommended that the selection of the appropriate method should be based on the goal of the analysis and on the dimension and amount of zero inflation of the data.

Correlation analysis of compositional data can suffer from a correlation bias (see Example 6.1). In the example provided in Figure 6.2, where only a single species was doubled, the relative abundance data shows a nonexisting decrease of the other species. Gloor *et al.* (2017) discussed alternative methods that can be used after the CLR transformation to explore correlations, *e.g.*, SPARCC and SPieCeasi, that both assume a sparse data matrix. Although these methods provide greater precision in estimation for higher correlation values, the estimation of correlation in compositional datasets remains an open research problem.

Example 6.1: Correlation bias

Correlation analysis of compositional data is cumbersome and often leads to incorrect interpretations (Filzmoser and Hron, 2009). The correlation between two species can be affected by the change in the level of another species. A small example is shown to explain the correlation bias in compositional data. In this example, four samples were measured for three species as presented in Table 6.2 with their counts, proportions, and CLR-transformed values.

Table 6.2 Counts, proportions, and CLR-transformed values of example data.

	Counts			Proportions			CLR-transformed values		
	A	*B*	*C*	*A*	*B*	*C*	*A*	*B*	*C*
s1	10	5	10	0.40	0.20	0.40	0.10	−0.20	0.10
s2	10	5	20	0.29	0.14	0.57	0.00	−0.30	0.30
s3	10	5	30	0.22	0.11	0.66	−0.06	−0.36	0.42
s4	10	5	40	0.18	0.09	0.73	−0.10	−0.40	0.50

It can be observed that in the four samples only the counts in feature *C* increase, thus the library size changes. The count data can be normalized by expressing the counts as proportions, *i.e.*, dividing the counts by the total counts of each sample. Both the proportions and the CLR-transformed values also change for features A and B. The CLR-transformed values are equal when calculated from the raw counts or from the proportions. The CLR-transformed values are also equal to the row-centered log-transformed proportions. These CLR-transformed values are not constrained anymore between 0 and 1. The correlation between features A and C is not defined when using the count data, as there is no variation in feature A. However, when going to proportions or to CLR-transformed values, the correlation between features A and C has become −1. Furthermore, a positive correlation of +1 is observed for the proportions between features A and B, only due to a change in feature C. Correlations between species from compositional data have to be interpreted with great care.

Finally, differential abundance analyses of specific features are often performed using tools that are developed for differential gene expression using RNAseq data such as DESeq2 (Love *et al.*, 2014) and edgeR (Robinson *et al.*, 2010). Unfortunately, these methods lead to a high number of falsely detected features due to the very sparse nature of metagenomics data, and alternative approaches taking these many zeros into account are preferred (Abegaz *et al.*, 2024).

6.2 Methods

In this section, we will discuss two distance-based methods. PCoA, a generalization of PCA, explores the relationship between samples and provides a low-dimensional mapping of the samples (Borg and Groenen, 2010). PERMANOVA is a generalization of ANOVA applied on a sample-to-sample distance matrix to study the effect of experimental or observational factors on the multivariate or high-dimensional (HD) response (Anderson, 2017). Although PCoA is not an HD-ANOVA method, it is discussed here to serve as an introduction to PERMANOVA.

6.2.1 Principal Coordinate Analysis

PCoA, also called Classical Multidimensional Scaling (MDS), is an ordination method that aims to visualize the similarity of samples in a low-dimensional space similar to PCA. The main difference between the two is that PCoA uses a distance matrix to define this low-dimensional space, where PCA is calculated directly from the measurement data itself. The special link between the two is that if the distance matrix is calculated using Euclidean distances, then PCoA and PCA provide the same results, *i.e.*, they provide the same low-dimensional mapping of the samples. So we could say that PCA is a special case of PCoA or that PCoA is a generalization of PCA to any distance or dissimilarity measure.

Consider high-dimensional data of a set of subjects $\mathbf{Y}(I \times J)$. The distance $d^f_{ii'}$ between two samples $\mathbf{y}_i$ and $\mathbf{y}_{i'}$ in the J-dimensional space can be expressed as:

$$d^f_{ii'} = f(\mathbf{y}_i, \mathbf{y}_{i'}) \tag{6.2}$$

where f is a function or algorithm to calculate the dissimilarity between the two samples, and these dissimilarities are collected in $\mathbf{D}_f(\mathbf{Y})$. There are many different dissimilarity measures that can be selected based on the properties of the data and the goal of the analysis, some of which are true distances,[1] see Section 2.1.4.1. The goal of PCoA is the inverse problem, *i.e.*, to find coordinates $\mathbf{Z}(I \times M)$ with $M < J$ for the samples in $\mathbf{Y}$ such that the Euclidean distance between the rows (samples) of $\mathbf{Z}$, $\mathbf{D}_E(\mathbf{Z})$ approximates $\mathbf{D}_f(\mathbf{Y})$ as good as possible. We will start with the small Example 6.2.

1 We will use the generic term distance matrix, although in some cases the elements of this matrix are dissimilarities.

Example 6.2: Small example of PCoA

A small example of a data matrix $\mathbf{Y}$ with three samples and three variables is used as a guide through the equations and calculations.

$$\mathbf{Y} = \begin{bmatrix} 1 & 3 & 2 \\ 3 & 0 & 3 \\ 3 & 2 & 1 \end{bmatrix} \tag{6.3}$$

Each sample has the same library size (six counts), and thus normalization is not necessary. We use the Bray–Curtis dissimilarity (see Equation 2.16) as the dissimilarity measure: $\mathbf{D}_{BC}$ matrix of $\mathbf{Y}$ is:

$$\mathbf{D}_{BC}(\mathbf{Y}) = \begin{bmatrix} 0 & 0.500 & 0.333 \\ 0.500 & 0 & 0.333 \\ 0.333 & 0.333 & 0 \end{bmatrix} \tag{6.4}$$

Figure 6.3a shows the three samples in the three-dimensional space. The PCoA solution is not unique as both the solutions $\mathbf{V}$ with coordinates $\mathbf{v}_1$ and $\mathbf{v}_2$ (Figure 6.3b) and $\mathbf{Z}$ with coordinates $\mathbf{z}_1$ and $\mathbf{z}_2$ provide the same Euclidean distance matrix as the Bray–Curtis dissimilarity matrix of Equation 6.4; $\mathbf{D}_E(\mathbf{V}) = \mathbf{D}_E(\mathbf{Z}) = \mathbf{D}_{BC}(\mathbf{Y})$. This group of samples could be located anywhere in the two-dimensional space without changing the distances. PCoA is looking for that set of coordinates $\mathbf{Z}$ that is centered around the origin, which is represented in Figure 6.3c.

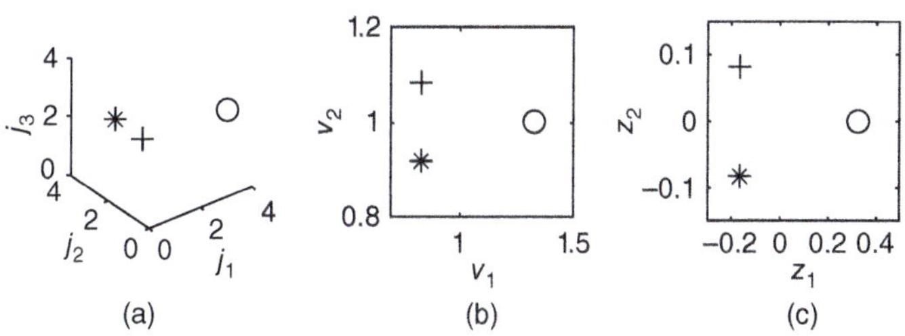

Figure 6.3 (a) Three samples in space of variables j_1, j_2, and j_3; (b) samples in non-centered configuration with coordinates v_1 and v_2; (c) samples in centered configuration with coordinates z_1 and z_2.

The matrix $\mathbf{Z}$ is now:

$$\mathbf{Z} = \begin{bmatrix} -0.2500 & -0.0735 \\ 0.2500 & -0.0735 \\ 0.0000 & 0.1470 \end{bmatrix} \tag{6.5}$$

which is indeed centered. The corresponding Euclidean distance matrix of $\mathbf{Z}$ is:

$$\mathbf{D}_E(\mathbf{Z}) = \begin{bmatrix} 0 & 0.500 & 0.333 \\ 0.500 & 0 & 0.333 \\ 0.333 & 0.333 & 0 \end{bmatrix} \tag{6.6}$$

which is exactly $\mathbf{D}_{BC}(\mathbf{Y})$. In this example, $\mathbf{Z}$ could be represented exactly in the two-dimensional space, and no dimension reduction was needed. This is not always true: for larger datasets with more samples and features, more dimensions will generally be necessary for $\mathbf{Z}$ such that $\mathbf{D}_E(\mathbf{Z})$ can fully represent $\mathbf{D}_{BC}(\mathbf{Y})$. In such situations, usually, only the eigenvectors corresponding to the largest eigenvectors will be visualized. Moreover, an exact equivalence between $\mathbf{D}_{BC}(\mathbf{Y})$ and $\mathbf{D}_E(\mathbf{Z})$, even when taking many components, is not always possible which will be explained below.

The general description of PCoA goes as follows (Mardia *et al.*, 1979). We first need a definition regarding the nature of the matrix $\mathbf{D}_f(\mathbf{Y})$. This distance matrix is called Euclidean if there exists a configuration of points in the Euclidean space whose interpoint distances are given by $\mathbf{D}_f(\mathbf{Y})$. Hence, there are points $\mathbf{z}_i; i = 1, \ldots, I$ in $\mathbb{R}^M$ such that $(\mathbf{z}_i - \mathbf{z}_{i'})^t(\mathbf{z}_i - \mathbf{z}_{i'}) = (d^f_{ii'})^2$ (thus $\mathbf{D}^2_f(\mathbf{Y}) = \mathbf{D}^2_E(\mathbf{Z})$). Note that $\mathbf{D}^2_f$ is a Hadamard (elementwise) product of $\mathbf{D}_f * \mathbf{D}_f$, and this is also the case for $\mathbf{D}^2_E$. The central theorem underpinning PCoA is now: the dissimilarity matrix $\mathbf{D}_f(\mathbf{Y})$ is Euclidean if and only if the Gower matrix $\mathbf{G} = \frac{-1}{2}\mathbf{J}\mathbf{D}^2_f\mathbf{J}$ is positive semi-definite where $\mathbf{J}$ is the centering matrix $\mathbf{J} = \mathbf{I} - \mathbf{1}\mathbf{1}^t/I$. Elaboration 6.1 provides more information on the interpretation of the Gower matrix.

ELABORATION 6.1

Interpretation of the Gower matrix

The Gower matrix $\mathbf{G} = \frac{-1}{2}\mathbf{J}\mathbf{D}^2_f\mathbf{J}$ has a clear interpretation. Recall from Example 6.2 that many configurations of coordinates $\mathbf{z}_i$ can represent a configuration of $\mathbf{Y}$. The one with coordinates centered around the origin is special. If $\mathbf{D}^2_E(\mathbf{Z})$ is the matrix of interpoint distances $(\mathbf{z}_i - \mathbf{z}_{i'})^t(\mathbf{z}_i - \mathbf{z}_{i'})$ then $\mathbf{G}$ is the matrix with elements $g_{ii'} = (\mathbf{z}_i - \overline{\mathbf{z}})^t(\mathbf{z}_{i'} - \overline{\mathbf{z}})$ where $\overline{\mathbf{z}}$ is the centerpoint of the configuration. The double centering in the definition of $\mathbf{G}$ comes from the need to center both $\mathbf{z}_i$ and $\mathbf{z}_{i'}$ and $\frac{-1}{2}$ is a correction factor to make $tr(\mathbf{G})$ equal to the sum-of-squares of the mean-centered $\mathbf{Z}$. In other words, $\mathbf{D}^2_E(\mathbf{Z})$ represents the squared interpoint distances and $\mathbf{G}$ represents these squared distances relative to the center of gravity of the original configuration.

The full algorithm for PCoA is described in Algorithm 6.1.

Algorithm 6.1

PRINCIPAL COORDINATE ANALYSIS

The general algorithm for PCoA is:

1) Calculate the distance matrix from matrix $\mathbf{Y}$ using any distance or dissimilarity function: $\mathbf{D}_f(\mathbf{Y})$.
2) Calculate the Gower matrix, which is the double-centered squared distance matrix: $\mathbf{G} = \frac{-1}{2}\mathbf{J}\mathbf{D}^2_f\mathbf{J}$, where $\mathbf{J}$ is the centering matrix $\mathbf{J} = \mathbf{I} - \mathbf{1}\mathbf{1}^t/I$.
3) Calculate the eigenvector decomposition of the Gower matrix to obtain the eigenvectors $\mathbf{V}$ and eigenvalues $\mathbf{\Lambda}$.
4) The coordinates of the samples $\mathbf{Z}$ can be obtained from multiplication $\mathbf{Z} = \mathbf{V}\mathbf{\Lambda}^{1/2}$.
5) $\mathbf{Z}$ will have Euclidean distance $\mathbf{D}_E(\mathbf{Z})$ equal to the distance $\mathbf{D}_f(\mathbf{Y})$, if $\mathbf{D}_f(\mathbf{Y})$ is Euclidean.
6) A low-dimensional visualization of $\mathbf{Z}$ will only approximate $\mathbf{D}_f(\mathbf{Y})$.

Distance measures that obey all four axioms of Equation 2.11 are called metric (see Section 2.1.4.1). For metric distance matrices, the coordinates $\mathbf{Z}$ perfectly preserve all the distances of $\mathbf{D}_f(\mathbf{Y})$ if all L components are used. These measures are also called Euclidean to represent that their distances can be embedded in Euclidean space. Distance measures that can disobey the triangle inequality are called semi-metric. Such measures are usually called dissimilarities, *e.g.*, the Bray–Curtis dissimilarity. The eigenvector analysis of the Gower matrix $\mathbf{G}$ of a semi-metric dissimilarity matrix may provide negative eigenvalues, and a full representation of the distances between the samples in Euclidean space is still not possible if all L eigenvectors are used (Krzanowski, 2000). Such distance matrices are called non-Euclidean. Although semi-metric dissimilarities cannot guarantee the triangular inequality to hold, they can produce distance matrices where the triangular inequality does hold. These distance matrices are called Euclidean, and their Gower matrix $\mathbf{G}$ does not have negative eigenvalues.

In case a dissimilarity matrix is non-Euclidean, the severity of the "non-Euclideanity" depends on the dissimilarity measure used and on the data itself (Legendre and Anderson, 1999). It can, however, be shown that even in those cases PCoA has optimality properties in terms of approximating $\mathbf{D}_f^2(\mathbf{Y})$ in a least squares sense (Mardia *et al.*, 1979). Note that the usual goal of PCoA is to approximate the dissimilarity matrix in low dimensions and not to provide a perfect Euclidean representation of the dissimilarity matrix. In Example 6.2, the Gower matrix $\mathbf{G}$ is positive semi-definite since it has two positive eigenvalues and one zero eigenvalue. Note that the matrix $\mathbf{D}_{BC}(\mathbf{Y})$ in that example obeys the triangle inequality since $0.333 + 0.333 > 0.500$.

Some possible corrections exist to adjust the non-Euclidean dissimilarity matrix $\mathbf{D}_f$ somewhat such that a full Euclidean representation of the adjusted $\mathbf{D}_f$ is possible. This adjustment then removes the negative eigenvalues of the $\mathbf{G}$ matrix, which is necessary for PCoA to be able to present the samples in a Euclidean space. For many dissimilarity measures, $\sqrt{\mathbf{D}_f}$ becomes Euclidean. In this case, the PCoA approximates the square root transformed dissimilarity matrix. Another approach is to add a small constant to each of the distances (except on the diagonal elements which should stay 0). The small constant c equals the absolute value of the largest negative eigenvalue. Thus the corrected distance matrix is then calculated as: $\tilde{d}_{ii'} = (d_{ii'} + 2c)^{0.5}$ (Gower and Legendre, 1986; Legendre and Anderson, 1999). This also leads to $\mathbf{G}$ matrices that only have non-negative eigenvalues.

After correction for the negative eigenvalues as discussed above, the amount of explained variation can be calculated from the eigenvalues of $\mathbf{G}$ (Mardia *et al.*, 1979). Suppose these eigenvalues are $\lambda_1 \geq \cdots \geq \lambda_I$ and $\lambda_L > 0$:

$$ExplVar_L = \frac{\sum_{i=1}^{L} \lambda_i}{\sum_{i=1}^{I} \lambda_i} \times 100\% \tag{6.7}$$

Example 6.3 provides a PCoA analysis of 16S rRNA extracts of biofilm samples from different positions along the Loue river in France.

Example 6.3: PCoA applied to biofilm samples

Biofilm samples were harvested during the summer of 2012 in the Loue River at two different sites (Zancarini *et al.*, 2017). DNA was extracted from the different samples, and the 16S rRNA gene was amplified and sequenced using 454-pyrosequencing to describe bacterial communities within the Loue River biofilms. There are nine samples harvested for each of the two *Sites: Cléron* (located in the upstream area of the river) and *Parcey* (located in the downstream area of the river). These samples were harvested at different times during the year, but here we ignore the time differences and consider the 9 samples of each *Site* to be replicates. Features that have less than 2 counts for at least half of the samples are removed from the data set. To correct for different library sizes, the data was rarified to 5466 counts for each sample. Features that did not have any counts left after rarifying were removed, and only 372 features remained. Bray Curtis dissimilarity was used to calculate the dissimilarities between the 18 samples, and PCoA was used to visualize the samples in two dimensions, see Figure 6.4. In total 53.4% of the variation between the samples was explained in the two PCoA dimensions.

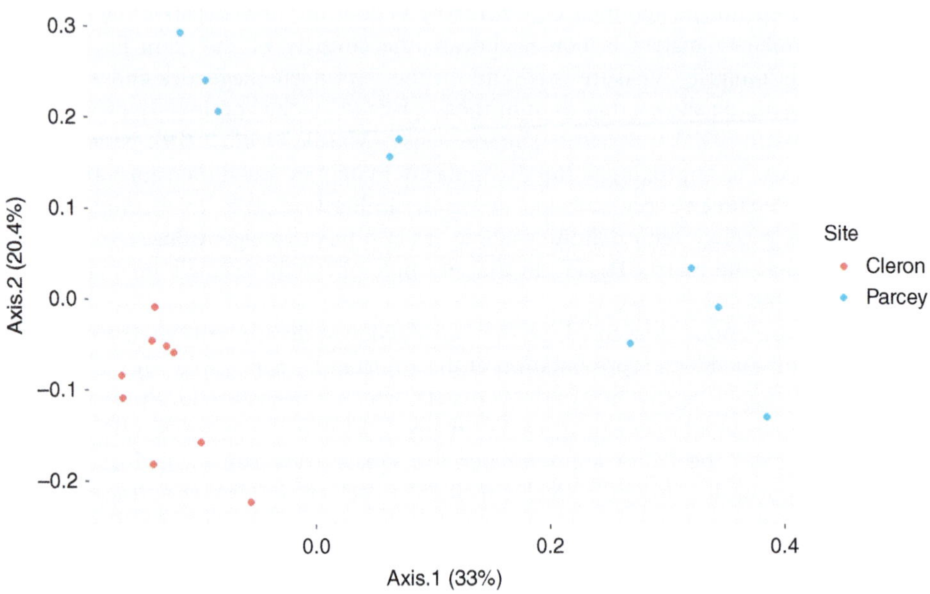

Figure 6.4 PCoA plot of the Loue river example.

The two-dimensional PCoA plot shows the samples as a dot, colored according to their sampling *Site* (*Cléron* in red and *Parcey* in blue). There seems to be a clear difference between the samples from the two locations, and the intra-location variability between the *Cléron* samples is lower than the intra-location variability between the *Parcey* samples. Figure 6.4 provides a low-dimensional visualization of the between-sample distances in the Euclidean space using the first two eigenvectors.

6.2.2 PERMANOVA

PERMANOVA is a distance-based approach to HD-ANOVA with a sample structure that is defined by an underlying statistical design. It is described as "a geometric partitioning of multivariate variation in the space of a chosen dissimilarity measure" (Anderson, 2017).

Briefly, the dissimilarities in high-dimensional space between all sample pairs are calculated using a predefined dissimilarity measure similar to what was shown in the PCoA. Next, the sources of these dissimilarities are related to the experimental factors underlying the samples and the interactions of these factors. Validation of factor and interaction significance is performed using random permutation of samples with respect to their factor settings.

The Null hypothesis of the PERMANOVA test is: "The centroids of the groups, as defined in the space of the chosen (dis-)similarity measure, are equivalent for all groups." This means that if H_0 is not rejected, the observed differences between the centroids are similar to what would be obtained under random permutation of the subjects of the different groups. The only assumption for PERMANOVA is that samples are exchangeable under the H_0 hypothesis. It tests the difference in location. PERMANOVA is quite robust against heterogeneity in multivariate dispersions for balanced designs but less robust for unbalanced designs (Anderson, 2017).

PERMANOVA combines the ANOVA approach in which main effects, interactions, nested structures, and repeated measures can be tested for significance with the flexibility of choosing the most appropriate distance measure for the problem at hand, without distributional assumptions of the data. In ANOVA, the between-group variation and within-group variation are quantified using the sum of squared difference between the samples and their group mean. The sum of squared differences based on the Euclidean distance is defined between the samples and their group mean. For many semi-metric dissimilarity measures, however, the group mean is not representative of the group as it does not have the minimum sum of all squared distances with its group members, see Example 6.2.

ELABORATION 6.2

Bray–Curtis dissimilarity and group means

As an example to show that the group mean does not always have to be representative of the group, we will use the Bray–Curtis dissimilarity between three samples ($S1 \dots S3$) of which five species ($F1 \dots F5$) are measured. The total number of counts for each sample is 18, see Table 6.3.

Table 6.3 Simple example with three samples and five features.

Sample	F1	F2	F3	F4	F5
S1	6	6	3	3	0
S2	3	0	15	0	0
S3	3	3	3	3	6
Mean	4	3	7	2	2
Centroid	3	3	9	3	0

The Bray–Curtis dissimilarity between the three samples and the mean is: $BC_{S1,Mean} = 0.333$, $BC_{S2,Mean} = 0.444$, and $BC_{S3,Mean} = 0.278$; thus, the sum of the squared dissimilarities between each sample and the mean is: $0.333^2 + 0.444^2 + 0.278^2 = 0.386$. To see whether the mean is a good representative for the three samples, a grid search was performed and revealed many positions in the 5-dimensional space that have a smaller sum of squared distances with the three samples than the 0.386. The lowest sum of squared BC-dissimilarities with the three samples was found to be 0.333. There were many centroids with this sum, but only one that also had a sample size of 18. This centroid sample is also indicated in Table 6.3.

Thus the simple mean across the samples is not a good representation of the samples if the Bray–Curtis dissimilarity is used, but it is not straightforward to calculate the location of the centroid sample directly. An interesting observation is that the minimum sum of squared dissimilarities (0.333) equals the sum of squared dissimilarity between all sample combinations in the group divided by the number of group members. The BC-dissimilarity between each pair of samples is: $BC_{12} = 0.667; BC_{13} = 0.333; BC_{23} = 0.667$ and the sum of squared inter-sample dissimilarities divided by the number of samples: $(0.667^2 + 0.333^2 + 0.667^2)/3 = 0.333$.

The phenomenon that for a dissimilarity function satisfying $d_{ii'} = d_{i'i}$ and $d_{ii} = 0$ the within-group sum-of-squares relative to the centroid can be obtained from the sum of the squared inter-sample dissimilarities, divided by the number of samples (see Elaboration 6.2 and Figure 6.5), was already indicated by Gower and Krzanowski (1999). This is very useful for dissimilarity measures where the average location does not provide the minimum sum of squared dissimilarities with its group members, such as the Bray–Curtis dissimilarity. This is called Huygens' theorem (Christian Huygens (1629–1695)) and is nicely illustrated in Anderson (2001).

$$\sum_{i=1}^{I} d^2_{i\text{-}centroid} = \frac{1}{I} \sum_{i=1}^{I-1} \sum_{i'=i+1}^{I} d^2_{ii'} \tag{6.8}$$

where $d^2_{i\text{-}centroid}$ represents the squared distance between point i and the centroid. To study differences among groups, the sum of squared distances between samples that do not belong to the same groups is compared to the sum of squared distances between samples that do belong to the same group. As an example, we will use a small set of nine samples that belong to three groups. The samples are represented by two variables $\mathbf{y}_1$ and $\mathbf{y}_2$. Columns $\mathbf{x}_1$ and $\mathbf{x}_2$ are the sum-coded indicator variables that indicate the group of each sample, see Table 6.4. This example does not consider compositional data.

The Euclidean distance will be used to define the distance matrix of the 9 samples. The squared Euclidean distance matrix is represented in Table 6.5.

The total sum-of-squares for all I samples is the sum of all squared distances between all combinations of samples indicated in Table 6.5 divided by the number of samples I:

$$SS_T = \frac{1}{I} \sum_{i=1}^{I-1} \sum_{i'=i+1}^{I} d^2_{ii'} = 39.77 \tag{6.9}$$

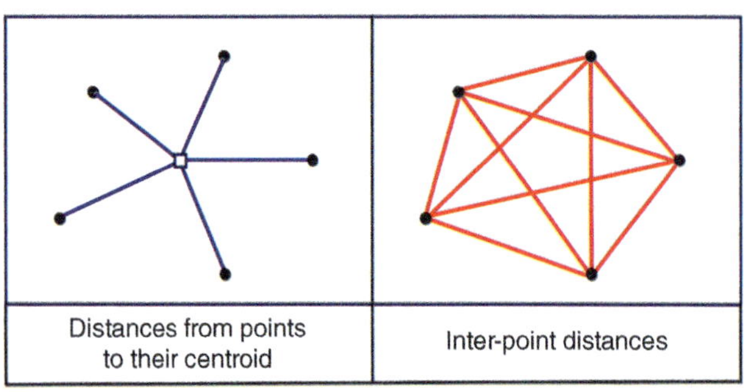

Figure 6.5 Equivalence between the sum of squared distances between points and centroid and the sum of all squared inter-point distances divided over the number of points.

Table 6.4 Design and samples for the simple PERMANOVA example.

Group	x_1	x_2	y_1	y_2
A_1	1	0	1	1
A_2	1	0	2	3
A_3	1	0	1	4
B_1	0	1	4	1
B_2	0	1	5	3
B_3	0	1	6	2
C_1	−1	−1	3	3
C_2	−1	−1	2	5
C_3	−1	−1	4	4

The within-group (or residual) sum-of-squares is the sum of all squared distances of samples in the same group divided by the number of samples in that group I_g. In the distance matrix of Table 6.5 they are indicated as the distances in the colored parts of the matrix.

$$SS_W = \frac{1}{I_g}\sum_{i=1}^{I-1}\sum_{i'=i+1}^{I} w_{ii'} \cdot d_{ii'}^2 = 13.33 \tag{6.10}$$

where $w_{ii'}$ is 1 if samples i and i' belong to the same group, while it is 0 if samples belong to different groups, and I_g is the number of samples within group g. The between-group sum-of-squares is then obtained from the difference between the two: $SS_B = SS_T - SS_W = 39.77 - 13.33 = 26.44$. The ratio between SS_B and SS_W is often indicated as the F-ratio because of its resemblance with the ANOVA F-test. In ANOVA the F-test is formally defined as:

$$F = \frac{SS_B/(g-1)}{SS_W/(I-g)} \tag{6.11}$$

where SS_B represents the variation among the group, SS_W the variation within the groups, g is the number of groups, and I the total number of samples. As the F-test compares variances, SS_B and SS_W are divided by their degrees of freedom.

Table 6.5 Euclidean distances between the samples of the simple PERMANOVA example.

D_E^2	A_1	A_2	A_3	B_1	B_2	B_3	C_1	C_2	C_3
A_1	0	5	9	9	20	26	8	17	18
A_2	5	0	2	8	9	17	1	4	5
A_3	9	2	0	18	17	29	5	2	9
B_1	9	8	18	0	5	5	5	20	9
B_2	20	9	17	5	0	2	4	13	2
B_3	26	17	29	5	2	0	10	25	8
C_1	8	1	5	5	4	10	0	5	2
C_2	17	4	2	20	13	25	5	0	5
C_3	18	5	9	9	2	8	2	5	0

PERMANOVA is developed for situations where the normality does not hold, and thus the F-distribution cannot be used. In that case, a p-value is obtained from permutation testing in which the exchangeable samples are randomly permuted over the groups. In the case of permutations, the degrees of freedom for both within and between group differences can be ignored as they are equal for the original F value and the permutation F^p-value and thus do not affect whether F is smaller or larger than F^p. The "pseudo F-ratio" can be obtained by simply comparing the ratio of $F = \frac{SS_B}{SS_W}$ from the original groups with $F^p = \frac{SS_B^p}{SS_W^p}$, where SS_B^p and SS_W^p are obtained after random permutation p of the samples in the group.

The p-value can then be obtained from counting the frequency of F^p values being equal or larger than the F-value:

$$p_P = \frac{(\#F^p \geq F) + 1}{P + 1} \tag{6.12}$$

where P is the total number of permutations. As the total number of possible permutations is often too large to be executed, generally a large number of permutations are applied (*e.g.*, 9999). To prevent that the p_P value can become 0, when none of the $F^p > F$, it is customary to add 1 to both the numerator and denominator of the equation. Under H_0 the original labeling can also be considered as one of the permutations.

6.2.2.1 PERMANOVA Calculated from the Gower Matrix

A more general approach to calculating PERMANOVA models with a more complex statistical design is described by Anderson (2017). As was shown in the PCoA section, the dissimilarity matrix of the samples can be transformed into the Gower matrix $\mathbf{G} = \frac{-1}{2}(\mathbf{J}\mathbf{D}_f^2(\mathbf{Y})\mathbf{J})$, where $\mathbf{D}_f^2$ in Table 6.5 is the squared Euclidean matrix $\mathbf{D}_E^2$ and $\mathbf{J}$ is the centering matrix. To estimate whether there is a difference between the three groups, the effect of the Group Factor represented by the design matrix $\mathbf{X} = [\mathbf{x}_1|\mathbf{x}_2]$ in Table 6.4 on the Gower matrix $\mathbf{G}$ is calculated using a linear regression model, $\mathbf{G} = \mathbf{X}\mathbf{b}_X$ with $\mathbf{b}_X = (\mathbf{X}^t\mathbf{X})^{-1}\mathbf{X}^t\mathbf{G}$. The part of $\mathbf{G}$ that can be explained by $\mathbf{X}$, can be obtained by projecting $\mathbf{G}$ on the space described by $\mathbf{X}$, $\mathbf{X}(\mathbf{X}^t\mathbf{X})^{-1}\mathbf{X}^t\mathbf{G} = \mathbf{H}\mathbf{G}$, where $\mathbf{H}$ represents the projection matrix for the matrix $\mathbf{X}$. The variation equals the between group sum-of-squares of $\mathbf{y}$, while the residual variation $(\mathbf{I} - \mathbf{H})\mathbf{G}$ represents the within group sum-of-squares of $\mathbf{y}$, as is detailed below.

In PERMANOVA, matrix $\mathbf{G}$ is projected on the space defined by the model matrix of the experimental factors $\mathbf{X}$. To explain this in more detail we use the example data $\mathbf{Y}$ of Table 6.4 in which nine samples are appointed to three different levels of a treatment factor. Sum-coding was used for this factor to define the model matrix $\mathbf{X}$, where $\mathbf{X}$ combines the columns $\mathbf{x}_1$ and $\mathbf{x}_2$. The Gower matrix $\mathbf{G}$ is represented in Table 6.6. As the Gower matrix is symmetric, the upper triangle is omitted.

From the Gower matrix, $\mathbf{G}$, the coordinates $\mathbf{Z}$ can be obtained as was explained in the PCoA section above. Figure 6.6 shows positions of data points of example data set $\mathbf{Y}$ in the original space and their PCoA coordinates $\mathbf{Z}$. Note that no dimension reduction was performed: $\mathbf{G} = \mathbf{Z}\mathbf{Z}^t$. In this case, PCoA provided a centering and a small rotation such that the direction $\mathbf{z}_1$ corresponds to the largest eigenvalue. The total sum-of-squares described by $\mathbf{Z}$ of the three group example can now be found by taking the trace of $\mathbf{G}$ which is the sum of its diagonal elements: $SS_T = tr(\mathbf{G}) = 39.77$, which equals the value that was found in Equation 6.9.

The samples are visualized in Figure 6.6a as points in the 2-dimensional space of variables y_1 and y_2, and in Figure 6.6b as PCoA coordinates in the z_1 and z_2 space.

The model matrix $\mathbf{X}$ represents the basis for the space on which the data $\mathbf{G}$ will be projected. Columns $\mathbf{x}_1$ and $\mathbf{x}_2$ form the basis vectors of this space. Matrix $\mathbf{H}$ is the hat matrix that projects the data onto the space defined by $\mathbf{X}$. The hat matrix: $\mathbf{H} = \mathbf{X}(\mathbf{X}^t\mathbf{X})^{-1}\mathbf{X}^t$ is a linear projector

Table 6.6 The matrix **G** in the simple PERMANOVA example.

G	A_1	A_2	A_3	B_1	B_2	B_3	C_1	C_2	C_3
A_1	8.02								
A_2	2.14	1.25							
A_3	2.36	2.47	5.69						
B_1	1.69	−1.20	−3.98	4.36					
B_2	−4.20	−2.09	−3.86	1.47	3.58				
B_3	−4.42	−3.31	−7.09	4.25	5.36	9.14			
C_1	0.025	−0.14	0.36	−0.31	−0.20	−0.42	0.025		
C_2	−1.64	1.47	4.69	−4.98	−1.86	−5.09	0.3622	5.69	
C_3	−3.98	−0.86	−0.64	−1.31	1.80	1.58	0.025	1.36	2.02

(see Section 2.1.4.2); thus, PERMANOVA applies a linear model between the transformed dissimilarity values and the different levels of the experimental factor indicating the three groups. The hat matrix is shown in Table 6.7. As the hat matrix is symmetric, the upper triangle is omitted.

Matrix **H** is double-centered, which is the case for sum-coded design matrices. As **X** has rank two, the projection matrix **H** is also of rank two, and thus all projections on **X** will also be in the two-dimensional space. The projection of the Gower matrix **HG** (see Table 6.8) provides a matrix in which the values for all samples in the same group are equal (in each column) and represents the group average. **HG** occupies a two-dimensional subspace in the 9-dimensional space of the samples. Within that space, it can be observed that **HG** only has three distinct values representing the centroids of each of the groups. A PCA of **HG** will show that it has rank two and a PCA score plot provides the relative positions of the centroids in two dimensions. The between-group sum-of-squares described by the variation between the group centroids can be obtained from the trace of **HG**: $SS_B = trace(\mathbf{HG}) = 26.44$, which equals the value calculated earlier.

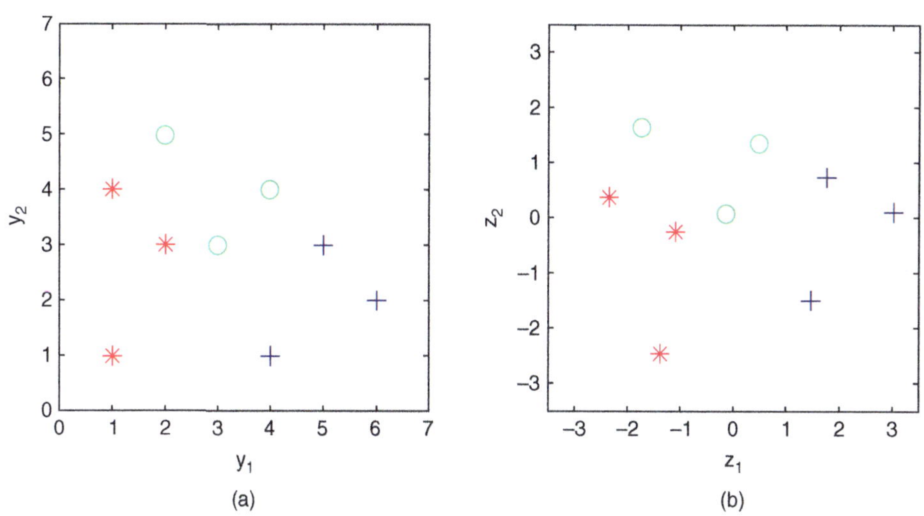

Figure 6.6 (a) Original samples; (b) z-coordinates of PCoA analysis.

Table 6.7 The matrix **H** in the PERMANOVA example.

H	A_1	A_2	A_3	B_1	B_2	B_3	C_1	C_2	C_3
A_1	0.22								
A_2	0.22	0.22							
A_3	0.22	0.22	0.22						
B_1	−0.11	−0.11	−0.11	0.22					
B_2	−0.11	−0.11	−0.11	0.22	0.22				
B_3	−0.11	−0.11	−0.11	0.22	0.22	0.22			
C_1	−0.11	−0.11	−0.11	−0.11	−0.11	−0.11	0.22		
C_2	−0.11	−0.11	−0.11	−0.11	−0.11	−0.11	0.22	0.22	
C_3	−0.11	−0.11	−0.11	−0.11	−0.11	−0.11	0.22	0.22	0.22

The part of the variation not described is the variation within groups or residual variation, visualized in Table 6.9. This variation can be obtained by projecting the matrix **G** in a direction orthogonal to the space defined by **X**. The projection matrix for this direction equals (**I** − **H**), where **I** is the identity matrix. (**I** − **H**)**G** provides the difference between the actual values and their group differences.

The within-group variation can be obtained from the trace of matrix (**I** − **H**)**G** = 13.33, which is equal to the value calculated in Equation 6.10. The rank of (**I** − **H**)**G** is 6. For visualization, we can perform a PCA on both **HG** and (**I** − **H**)**G**. Figure 6.7a shows the location of the three groups' means **HG** around the origin, while the difference of each sample to its own group mean (**I** − **H**)**G** is located around the origin in Figure 6.7b.

Figure 6.7 shows the differences between group centroids (a) and the differences between each sample compared to their own group mean (Figure 6.7b). The right-hand figure can give information on whether outlying samples are present and whether within-group correlations exist between the replicates in a group. The pseudo F-value is estimated as

$$F = \frac{SS_B}{SS_W} = \frac{26.44}{13.33} = 1.98 \tag{6.13}$$

The significance of this pseudo-F value is obtained using permutation analysis where the samples are randomly permuted over the three groups. For each of the permutations, the

Table 6.8 The matrix **HG** in the PERMANOVA example.

HG	A_1	A_2	A_3	B_1	B_2	B_3	C_1	C_2	C_3
A_1	4.17	1.95	3.51	−1.16	−3.38	−4.93	0.17	1.50	−1.83
A_2	4.17	1.95	3.51	−1.16	−3.38	−4.93	0.17	1.50	−1.83
A_3	4.17	1.95	3.51	−1.16	−3.38	−4.93	0.17	1.50	−1.83
B_1	−2.31	−2.20	−4.98	3.36	3.47	6.25	−0.31	−3.98	0.69
B_2	−2.31	−2.20	−4.98	3.36	3.47	6.25	−0.31	−3.98	0.69
B_3	−2.31	−2.20	−4.98	3.36	3.47	6.25	−0.31	−3.98	0.69
C_1	−1.86	0.25	1.47	−2.20	−0.086	−1.31	0.14	2.47	1.14
C_2	−1.86	0.25	1.47	−2.20	−0.086	−1.31	0.14	2.47	1.14
C_3	−1.86	0.25	1.47	−2.20	−0.086	−1.31	0.14	2.47	1.14

Table 6.9 The matrix $(\mathbf{I}-\mathbf{H})\mathbf{G}$ in the PERMANOVA example.

(I – H)G	A_1	A_2	A_3	B_1	B_2	B_3	C_1	C_2	C_3
A_1	3.85	0.19	−1.15	2.85	−0.81	0.52	−0.15	−3.15	−2.15
A_2	−2.04	−0.70	−1.04	−0.037	1.30	1.63	−0.037	−0.037	0.96
A_3	−1.81	0.52	2.19	−2.81	−0.48	−2.15	0.19	3.19	1.19
B_1	4.00	1.00	1.00	1.00	−2.00	−2.00	0.00	−1.00	−2.00
B_2	−1.89	0.11	1.11	−1.89	0.11	−0.89	0.11	2.11	1.11
B_3	−2.11	−1.11	−2.11	0.89	1.89	2.89	−0.11	−1.11	0.89
C_1	1.89	−0.11	−1.11	1.89	−0.11	0.89	−0.11	−2.11	−1.11
C_2	0.22	1.22	3.22	−2.78	−1.78	−3.78	0.22	3.22	0.22
C_3	−2.11	−1.11	−2.11	0.89	1.89	2.89	−0.11	−1.11	0.89

F-value is obtained, and in this example, none of the F-values of the permutations was higher than the original: $p < 0.001$. Permutation analysis in PERMANOVA is discussed in Section 6.2.4.

6.2.2.2 PERMANOVA of Non-Euclidean Dissimilarity Matrices

For any response matrix $\mathbf{Y}$ and a design matrix $\mathbf{X}$, the response matrix can be decomposed using a linear model as $\mathbf{Y} = \hat{\mathbf{Y}} + \mathbf{E}$, which is equal to $\mathbf{Y} = \mathbf{HY} + (\mathbf{I}-\mathbf{H})\mathbf{Y}$, where $\mathbf{H}$ is the projection matrix $\mathbf{X}(\mathbf{X}^t\mathbf{X})^{-1}\mathbf{X}^t$, assuming $\mathbf{X}$ is of full column rank. Since $\mathbf{H}$ and $(\mathbf{I}-\mathbf{H})$ are orthogonal to each other, the total sum-of- squares of $\mathbf{Y}$ is decomposed as:

$$trace(\mathbf{YY}^t) = trace(\mathbf{HYY}^t\mathbf{H}^t) + trace((\mathbf{I}-\mathbf{H})\mathbf{YY}^t(\mathbf{I}-\mathbf{H})^t) \tag{6.14}$$

Thus, even if $\mathbf{Y}$ is unknown, any outer product $\mathbf{YY}^t$ of $\mathbf{Y}$ can be decomposed into a part related to the design matrix $\mathbf{X}$, and the residuals $\mathbf{E}$. Such an outer product can be obtained from any symmetric matrix of dissimilarities. The Gower matrix $\mathbf{G}$, as discussed in Elaboration 6.1, is such an outer product as $\mathbf{G} = \frac{-1}{2}\mathbf{JD}_f^2\mathbf{J}$. Thus $\mathbf{G}$ can be decomposed into $\mathbf{HGH}$ and $(\mathbf{I}-\mathbf{H})\mathbf{G}(\mathbf{I}-\mathbf{H})^t$.

For distance matrices that are Euclidean, obtained from a centered data matrix $\mathbf{Y}$, $\mathbf{G} = \mathbf{YY}^t$, so $\mathbf{HGH}^t$ equals $(\mathbf{HYY}^t\mathbf{H}^t) = \hat{\mathbf{Y}}\hat{\mathbf{Y}}^t$ and $(\mathbf{I}-\mathbf{H})\mathbf{G}(\mathbf{I}-\mathbf{H})^t = (\mathbf{I}-\mathbf{H})\mathbf{YY}^t(\mathbf{I}-\mathbf{H})^t = \mathbf{EE}^t$. However,

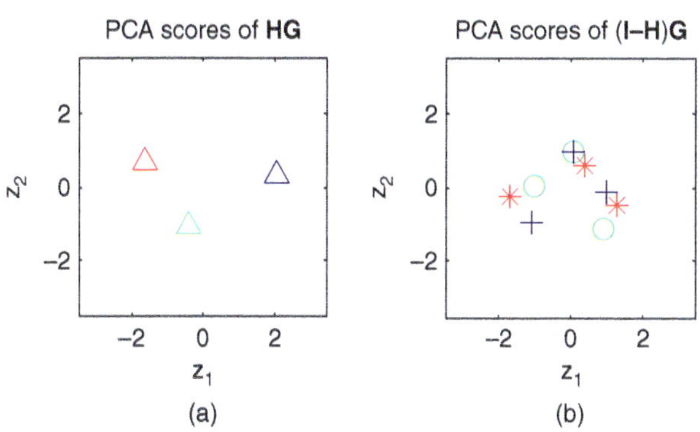

Figure 6.7 (a) PCA scores of $\mathbf{HG}$ indicating group centroids of the three groups of Figure 6.6; (b) PCA scores of $(\mathbf{I}-\mathbf{H})\mathbf{G}$ indicating within-group variation of each sample around their group origins. For both plots, the colors indicate their group membership.

G can also be calculated from any symmetric non-Euclidean dissimilarity matrix, and Equation 6.14 still holds. Even if **G** has negative eigenvalues, then still the sum of all eigenvalues of **G** equals *trace*(**G**), representing the total sum-of-squares. Thus for PERMANOVA, the correction approaches for negative eigenvalues, introduced for PCoA in Section 6.2.1, are not necessary to estimate the effect of the factors (McArdle and Anderson, 2001). These corrections are only necessary to place all samples in a Euclidean space. An example of PERMANOVA to biofilm samples of the Loue river is provided in Example 6.4.

Example 6.4: PERMANOVA applied to biofilm samples of Loue river

Continuing the example of the 18 samples from the two *Sites* of the Loue river, *Cleron* and *Parcey*, a statistical test with respect to the centroids can be performed using PERMANOVA. The PERMANOVA uses the same Bray–Curtis dissimilarity matrix as the PCoA analysis based on the 372 features that remained after filtering and rarifying. The R-function adonis2 of the Vegan-package is used to calculate the PERMANOVA test, and the significance is tested using 9999 permutations. Both the *Site* and the *Month* and their interaction *Site:Month* were used as factors. The results are presented in Table 6.10.

Table 6.10 Results of the PERMANOVA analysis of the Loue data, in which the amount of explained variation by the factors *Site, Month* and their interaction are provided.

Source	SS	df	MS	F	R^2	p
Site	0.49392	1	0.49392	9.9318	25.520%	1e−04
Month	0.47030	2	0.23515	4.7285	24.299%	1e−04
Site:Month	0.37445	2	0.18723	3.7648	19.347%	1e−04
Error	0.59677	12	0.04973		30.834%	
Total	1.93545	17			100.000%	

The p-value of the test whether the between-group dissimilarities are equal to the within-group dissimilarities is lower than the predefined cutoff value of 0.05, $(Pr(> F) = 1e - 04)$ for both factors and their interaction. We can reject the H_0 and conclude that there is a significant effect of the both the *Site* and the *Month* and their interaction. The PERMANOVA results can also be calculated by projecting the Gower matrix on the hat matrix for each of the factors and their interaction. The sum-of-squares can then be obtained as the trace of the projected matrices.

6.2.3 Effect Sizes in PERMANOVA

PERMANOVA calculates p-values for each of the factors and their interaction; it does not provide effect sizes for the levels of the different factors. If the distance matrix **D** would have been obtained from Euclidean distances, then the distances between the centroids of each cell in the design are equal to the distances between the means of each cell. For other dissimilarity measures, this is not the case. However, an *averaged* Gower matrix $\overline{\mathbf{G}}$ can be calculated which can provide the distances between means in terms of Euclidean distances also for other dissimilarity measures. For Factor A with levels $a = 1, \ldots, A$ and N_a replicates for level a, and

$N = \sum_{a=1}^{A} N_a$. $\overline{g}_{a,a'} = \frac{1}{N_a \cdot N_{a'}} \sum_{i \in a}^{N} \sum_{i' \in a'}^{N} g_{i,i'}$, where $i \in a$ represents an indicator vector of length N that is 1 for the elements related to level a and 0 for all the other levels. The distance between the centroids can then obtained from $d_{a,a'}^{C} = \sqrt{\overline{g}_{a,a} + \overline{g}_{a',a'} - 2\overline{g}_{a,a'}}$ (Anderson, 2017).

However, from Elaboration 6.2 we realize that for semi-metric dissimilarity measures that disobey the triangular inequality, such as the Bray–Curtis dissimilarity, the group means are not representative for the group. Gower matrices calculated from such dissimilarities usually contain negative eigenvalues, due to the triangular inequality disobedience. Thus for Gower matrices with negative eigenvalues, we conjecture that the calculation of effect sizes is still not possible unless the negative eigenvalues are corrected for (see Sections 6.2.1 and 7.2.3).

6.2.4 Permutations in PERMANOVA

In Section 3.7.3, statistical validation of univariate models using permutations was introduced. In PERMANOVA, the multivariate data is summarized in a single between-sample dissimilarity matrix. The dissimilarities between samples of different groups are compared to the dissimilarities of samples in the same group. Section 3.7.3 described how permutation tests should be applied in the case of a single-factor design and in designs with multiple factors in nested or crossed fashion. For exact permutation tests, the values in the distance matrix of the samples that are exchangeable under the H_0 hypothesis should be permuted. After these permutations, the PERMANOVA of the permuted data can be performed, and the statistic of interest can be calculated from the permuted distance matrix. The statistic of interest in PERMANOVA is usually the pseudo-F-value of the ratio between the between-group variation over the within-group variation (see Equation 6.13). Thus the distances between the samples do not have to be recalculated; only the position of the samples in the distance matrix is permuted. For the permutation, *e.g.*, $d_{1,2}$ switches position with $d_{3,6}$ and $d_{2,1}$ switches position with $d_{6,3}$ because the distance matrix $\mathbf{D}$ is symmetric.

In exact permutation tests of two or more factors, the permutation of location in the distance matrix should take the level of the other factor into account, just as was described in Section 3.7.3. For approximate permutation tests, *e.g.*, to test interactions in a two-factor design, the effect of the two main factors first has to be removed from the distances. In PERMANOVA the variation described by both factors can be obtained by projecting the Gower matrix $\mathbf{G}$ on the hat matrix of the design factors. If $\mathbf{X}_{12} = [\mathbf{X}_1|\mathbf{X}_2]$ contains the design matrices of both factors, then the hat matrix for both factors is calculated as: $\mathbf{H}_{12} = \mathbf{X}_{12}(\mathbf{X}_{12}^t\mathbf{X}_{12})^{-1}\mathbf{X}_{12}^t$. The part of the variation in $\mathbf{G}$ not explained by factors 1 and 2 equals $\mathbf{E}_{12} = (\mathbf{I} - \mathbf{H}_{12})\mathbf{G}$. The explained variation of this residual when projected on the model matrix for the interaction $\mathbf{H}_{Int}\mathbf{E}_{12}$ provides the variation explained solely by the interaction. Here, $\mathbf{H}_{Int}$ is the hat matrix based on the design matrix for the interaction. This variation is compared to the same variation estimated from permuted residual matrices $\mathbf{E}_{12}$ in an approximate permutation test.

6.2.5 Assumptions for PERMANOVA

Different from ANOVA, PERMANOVA does not assume normally distributed residuals with equal variance, nor does it have assumptions on the distribution of the measured variables. Because PERMANOVA uses permutations for significance testing, the only assumption is that the experimental units are exchangeable under a true null hypothesis. Related to the exchangeability requirement is the requirement of independence, as any dependency between experimental units would be destroyed by the permutation. PERMANOVA can be sensitive to differences in spread among groups, especially for unbalanced designs. Thus, if the permutation test detects a significant difference between the groups, this could be due to the levels of the treatment, but also to differences in spread, or a combination of the two. Anderson (2005), therefore, suggests performing an additional test on homogeneity, to find the real cause of the significant permutation test.

6.3 ANOSIM

Analysis of similarities (ANOSIM) is a nonparametric statistical method related to PERMANOVA that tests whether the similarity between groups is larger than the similarity within the groups (Clarke, 1993). ANOSIM is applied to ranks of the distances. Thus, the distances are first ranked before analysis is performed. The main idea of ANOSIM is to test whether the average rank distance between samples of different groups is larger than the average rank distance of samples in the same group. Such an analysis of ranks is more common in ecological analyses where the focus is not on the actual distances between samples but on dissimilarities being larger or smaller. Rank-based analyses are commonly more robust (less affected by outliers due to specific choice of distance measure) than analyses on actual distances but can have a reduced power (*i.e.*, small group differences might not be detected). ANOSIM was found to be less robust to heterogeneity of the groups than PERMANOVA, leading to more false positives for increasing sample size (Anderson and Walsh, 2013).

7

Reviews and Reflections

In this chapter, we will try to summarize the many HD-ANOVA methods presented in Chapters 4 and 5 by giving reviews of these methods as they are used in different fields. These reviews are not exhaustive but give a realistic view of which methods are used in which types of applications. We will also put the methods in context by reflecting on the usefulness of the different methods, discussing their (dis-)advantages and giving some recommendations. We will end by presenting some unsolved issues and remaining challenges. These may point to future research and development.

7.1 Reviews

In this section, we will give an overview of HD-ANOVA applications in different fields of the life sciences with a focus on the methods as explained in Chapters 4 and 5. Hence, ANOVA methods possibly combined with multiple testing corrections are not covered. The overview will be presented in tables for the different fields where a dashed line will separate the use of ASCA (+) versus other types of analysis methods to improve readability. A brief introduction to the separate fields is already given in Chapter 1. Some methods in this book, such as RDA and PRC, are not covered in these reviews since they are (almost) exclusively used in ecology; a field which is only touched upon briefly in this book.

The review was performed by crossing keywords of the application areas presented below with words like ASCA, APCA, ASCA+, LiMM-PCA, Analysis of Variance, and ANOVA. We do not claim that this review is exhaustive, but we think that in this way we covered a substantial part of the papers using HD-ANOVA methods.

7.1.1 Metabolomics

The use of analysis of variance-type models as discussed in this book has really taken off in metabolomics. To a large extent, this may be due to the availability of software (see Chapter 8). A summary of applications is given in the different subsections related to different fields.

7.1.1.1 Plant Science

In plant science, metabolomics is used in very different ways. A summary is given in Table 7.1 which shows that indeed the HD-ANOVA methods are starting to get used. This is not surprising given the fact that experimental designs already have a long history in plant science.

Analysis of Variance for High-Dimensional Data: Applications in Life, Food and Chemical Sciences,
First Edition. Age K. Smilde, Federico Marini, Johan A. Westerhuis, and Kristian H. Liland.

Table 7.1 Overview of metabolomics applications using HD-ANOVA in the plant sciences.

Type	Samples	Design	Method	Reference
R. glutinosa	Roots	Treatments	ASCA	Chang *et al.* (2006)
H. perforatum	Roots	Treatment/time	ASCA	Brasili *et al.* (2014)
H. perforatum	Roots	Treatment/time	ASCA	Brasili *et al.* (2016)
Lettuce	Leaves	Treatment/time	ASCA	Feizi *et al.* (2020)
White grape	Exocarp, seed	Treatments	ASCA	Garrido *et al.* (2021)
Populus trichocarpa	Wood	Genetics/blocking	ASCA	Guerra *et al.* (2016)
Maize	Grains	Location/trait/group	ASCA	Harrigan *et al.* (2016)
Arabidopsis	Roots	Treatment/time	ASCA	Hildreth *et al.* (2020)
Rice	Grain	Cultivar/time	ASCA	Hu *et al.* (2016)
Rice	Flag leaves	Varieties/time	ASCA	Hu *et al.* (2020)
Barley	Seeds	Strains/time	ASCA	Khakimov *et al.* (2017)
R. coreanus	Fruit	Groupings	ASCA	Lee *et al.* (2013)
C. arabica	Beans	Treatments	ASCA	Marcheafave *et al.* (2020)
Rice	Aerial part	Treatment/time	ASCA	Navarro-Reig *et al.* (2017)
Rice	Roots, aerial	Treatments	ASCA	Pérez-Cova *et al.* (2022)
Lentil	Leaves	Genotype/treatment/time	ASCA	Adobor *et al.* (2023)
Yerba-mate	Leaves	Treatment/season	ASCA	Tormena *et al.* (2020)
Potato	Leaves	Cultivar/strain	ASCA	Manasseh *et al.* (2023)
Yerba-mate	Leave extracts	Gender/cultivation	ASCA	Marcheafave *et al.* (2024)
Lettuce	Leaves	Varieties/conditions/time	APCA	Sun *et al.* (2018)
Strawberry	Fruit	UV-radiation/time	APCA	Sun *et al.* (2020)
A. thaliana/ D. melanogaster/ O. cincta	Whole-body tissue	Treatments	PC-MANOVA	Noer *et al.* (2020)
Arabidopsis	Whole rosettes	Treatments	PC-MANOVA	Johnson *et al.* (2007)
M. truncatula	Leaves, roots	Treatments	PC-MANOVA	Rusilowicz *et al.* (2016)

Table 7.1 (Continued)

Type	Samples	Design	Method	Reference
A. capillaris, R. acris	Leaves	Treatment/location	PC-ANOVA, ASCA	Gargallo-Garriga *et al.* (2017)
L. usitatissimum	Roots/shoots	Treatment	AMOPLS	Elboutachfaiti *et al.* (2022)
Potato	Skin	Cultivar/inoculation	AMOPLS	Massana-Codina *et al.* (2020)
L. sativum	Leaves and roots	Treatment/leaves, roots	AMOPLS	Salehi *et al.* (2024)
Lettuce	Leaves and roots	Treatments	AMOPLS	Zhang *et al.* (2024)

What is obvious from Table 7.1 is that ASCA is the most used method. Yet, some other methods (PC-MANOVA, AMOPLS) are also starting to be used. The Factor *Time* is sometimes used in the design. Depending on the specific design of the study (are the same plants harvested at each time point or not), this could call for repeated-measures type approaches (*e.g.*, RM-ASCA+), but these methods are relatively new and not yet used.

7.1.1.2 Microbiology and Biotechnology

In the field of microbiology and biotechnology, the HD-ANOVA methods for metabolomics are not used that much yet, see Table 7.2. ASCA is again the most used method, and *Time* is often a factor in the design calling for specific methods (see Section 7.1.1.1).

Table 7.2 Overview of metabolomics applications using HD-ANOVA in microbiology and biotechnology.

Type	Samples	Design	Method	Reference
Lactic acid bacterium	Fermentation	Treatment/time	ASCA	Ebrahimi *et al.* (2016)
Microcystis aeruginosa	Whole cells	Treatment/time	ASCA	Georges des Aulnois *et al.* (2020)
S. thermophilus	Fermentation	Strains/batches/time	ASCA	Khakimov *et al.* (2017)
Yeast	Cells	Conditions/time	ASCA	Puig-Castellví *et al.* (2016)
Yeast	Cells	Cultures/time	ASCA	Puig-Castellví *et al.* (2018)
Wine fermentation	Process samples	Time/treatment	ASCA	Schorn-García *et al.* (2023)
T. brucei	Culture	Time/treatment	ASCA	Fall *et al.* (2024)
Microorganisms	Volatiles	Different types	APCA	Boots *et al.* (2014)
Beer	Fermentation samples	Yeast type/temperature/time	ASCA, i-ASCA	Grassi *et al.* (2017)
C. crescentus	Cells	Sample handling	AMOPLS	Pezzatti *et al.* (2019)
Fungi	Cells	Species/time	AMOPLS	Azzollini *et al.* (2018)

7.1.1.3 Animal Science

In the animal science metabolomics field, a variety of animals is used, see Table 7.3. ASCA is again the most used method, and also in this case the experimental design contains often the Factor *Time*. Hence, the same comments as in the previous sections in this respect apply here. In some cases, multiple methods are used and compared.

Table 7.3 Overview of metabolomics applications using multivariate ANOVA in animal science.

Type	Samples	Design	Method	Reference
Steers	Urine	Group/time	ASCA	Artegoitia *et al.* (2022)
Salmon	Plasma	Group/time	ASCA	Aru *et al.* (2021)
D. melanogaster	Whole body	Group/treatment/time	ASCA	Colinet and Renault (2018)
Rats	Urine	Treatment/time	ASCA	Dall'Acqua *et al.* (2014)
Rats	Serum	Grouping/time	ASCA	Dai *et al.* (2016)
Rats	Serum, brain tissue	Treatment/time	ASCA	Du *et al.* (2017)
D. magna	Whole body	Treatments	ASCA	Garreta-Lara *et al.* (2018)
D. magna	Whole body	Reproduction stage/time	ASCA	Jeong and Simpson (2019)
Cows	Serum	Treatment/time	ASCA	Ghaffari *et al.* (2019)
Zebrafish	Muscle tissue	Treatment/time	ASCA	Gómez-Canela *et al.* (2017)
Zebrafish	Muscle tissue	Treatment/time	ASCA	Gómez-Canela *et al.* (2018)
Pigs	Blood	Treatment/time	ASCA	Hernández-Mesa *et al.* (2022)
Cows	Plasma	Genetic strains/time	ASCA	Jorge-Smeding *et al.* (2021)
Rats	Urine/serum	Treatment/time	ASCA	Kim *et al.* (2019)
Golden hamster	Plasma	Treatment/time	ASCA	Kokova *et al.* (2020)
Rats	Plasma	Treatment/time	ASCA	Li *et al.* (2018)
Mice	Serum	Treatment/time	ASCA	Li *et al.* (2021)
Cows	Serum	Grouping	ASCA	Luke *et al.* (2020)
Cows	Serum	Grouping	ASCA	Luke *et al.* (2020)
Rats	Liver tissue	Treatments	ASCA	Lv *et al.* (2017)
Mice	Liver tissue	Treatment/time	ASCA	Ly-Verdú *et al.* (2015)
Zebrafish	Tissue	Phenotype/time	ASCA	Olmos *et al.* (2019)
Zebrafish	Tissue	Tissue type/treatment	ASCA	Pérez *et al.* (2020)
Rats	Brain tissue	Treatment/location	ASCA	Quansah *et al.* (2017)
Rats	Blood	Diet/treatment	ASCA	Rago *et al.* (2013)
Sheep	Urine	Treatment/time	ASCA	Rosales-Nieto *et al.* (2021)
Mice	Liver tissue	Mutants/time	ASCA	Ruiz-Rodado *et al.* (2016)
Horses	Plasma	Conditions/time	ASCA	Salle *et al.* (2021)
G. pulex	Body tissue	Treatments/time	ASCA	Sheikholeslami *et al.* (2020)
Mice	Urine/feces	Diets/time	ASCA	Su *et al.* (2021)

Table 7.3 (Continued)

Type	Samples	Design	Method	Reference
Rats	Urine	Intervention/time	ASCA	Tomassini *et al.* (2014)
Rats	Serum, liver	Treatment/time	ASCA	Tremblay-Franco *et al.* (2015)
Minks	Plasma, urine	Diet/time	ASCA	Trimigno *et al.* (2022)
Rats	Urine	Treatment/time	ASCA	Wang *et al.* (2009)
Rats	Plasma	Treatment/age	ASCA	Xu *et al.* (2019)
Pigs	Plasma	Diet/time	ASCA	Yde *et al.* (2012)
Mice	Organ tissue	Gender/disease	ASCA	Zhang *et al.* (2020)
Mice	Organ tissues	Transgenic/time	ASCA	Zheng *et al.* (2018)
Rats	Gut-content	Dose/time	ASCA	Zhou *et al.* (2020)
Glass eels	Homogenized eels	Treatment	ASCA	Alvarez-Mora *et al.* (2023)
Rats	Urine	Diet/time	ASCA	Zhou *et al.* (2021)
Mice	Urine, Kidney, Plasma	Control/treated	ASCA	Barla *et al.* (2023)
Rats	Urine	Sex and sexual maturity/time	ASCA	Lee *et al.* (2023)
Rats	Lung tissue	Treatment/time	ASCA	Liang *et al.* (2024)
Rats	Plasma, Amniotic fluid, Placental tissue	Time/compartment	ASCA	Galineau *et al.* (2024)
Cows	Milk	Feeding regimes	ASCA	Tsermoula *et al.* (2024)
Piglets	Brain tissue	Group/treatment	PE-ASCA	Alinaghi *et al.* (2020)
Parasites	Extracts	Extraction method	ASCA+, APCA+	Mamede *et al.* (2024)
G. fossarum	Whole body	Group/treatment	APCA+	Bonnefoy *et al.* (2019)
D. magna	Tissue	subpopulations/ treatment	PC-ANOVA, ASCA	Zhang *et al.* (2021)
Rat	Urine	Treatment/time	PC-MANOVA	Gu *et al.* (2007)
Earthworm	Tissue	Treatments	PC-MANOVA	McKelvie *et al.* (2011)
Marine mussels	Tissue	Gender/exposure	AMOPLS	Dumas *et al.* (2020)

7.1.1.4 Human Science

Also in human science, there is a large variety of metabolomics applications using HD-ANOVA approaches, see Table 7.4. ASCA is again used mostly and, in many cases, there is a Factor *Time* involved calling again for special methods. In the cases where MANOVA is used, it is not always clear how the problem of having more variables than samples is handled. In many cases, also alternative methods are used to analyze the data, predominantly (O)PLS or (O)PLS-DA types of approaches.

Table 7.4 Overview of metabolomics applications using HD-ANOVA in human science.

Samples	Design	Method	Reference
Tissue/wound headspace	Site/time	ASCA	Ashrafi *et al.* (2020)
Plasma	Treatment/time	ASCA	Caellas *et al.* (2013)
Urine	Intervention/time	ASCA	Combrink *et al.* (2019)
Fecal	Sample treatment	ASCA	Cui *et al.* (2020)
Fecal	Grouping	ASCA	Cui *et al.* (2021)
Cancer tissue	Grouping/time	ASCA	Díaz *et al.* (2022)
Muscle tissue, plasma	Grouping/time	ASCA	Fazelzadeh *et al.* (2016)
Plasma, urine	Nutri-intervention/time	ASCA	Gürdeniz *et al.* (2016)
Breath	Treatment/time	ASCA	Henderson *et al.* (2022)
Plasma	Disease/grouping	ASCA	Horta *et al.* (2020)
Urine	Intervention/time	ASCA	Irwin *et al.* (2016)
Urine	Intervention/time	ASCA	Irwin *et al.* (2016)
Plasma	Diet/time	ASCA	Khakimov *et al.* (2016)
Urine/serum	Groupings	ASCA	Khan *et al.* (2017)
Serum	Intervention/time	ASCA	Khan *et al.* (2019)
Plasma	Grouping/time	ASCA	Lamichhane *et al.* (2018)
Plasma	Grouping/time	ASCA	Lamichhane *et al.* (2019)
Saliva	Treatment/time	ASCA	Lemanska *et al.* (2012)
HepG2 cells	Treatment/time	ASCA	Liu *et al.* (2017)
HepG2 cells	Treatment/time	ASCA	Lv *et al.* (2018)
Urine	Intervention	ASCA	Miccheli *et al.* (2015)
Hep2G	Handling conditions	ASCA	Moreno-Torres *et al.* (2021)
Hepatocytes	Cell lines/treatments/time	ASCA	Moreno-Torres *et al.* (2022)
Saliva	Grouping/time	ASCA	Neyraud *et al.* (2020)
Urine	Treatment/time	ASCA	Opperman *et al.* (2021)
Peripheral blood lymphocytes	case-control/time	ASCA	Perl *et al.* (2015)
Chondrocyte cells	Donor/time	ASCA	Salinas *et al.* (2019)
Serum	Intervention/time	ASCA	Silva *et al.* (2019)
Serum	Grouping/time	ASCA	Sonn *et al.* (2022)
Serum	Time	ASCA	Stander *et al.* (2020)
Fecal	Cross-over intervention/gender	ASCA	Trimigno *et al.* (2017)
Urine	Cohort-grouping	ASCA	Trimigno *et al.* (2019)
Urine	Diet/grouping/time	ASCA	Trimigno *et al.* (2020)
Breath	Grouping/disease	ASCA	van de Kant *et al.* (2013)
Urine	Intervention/time	ASCA	Xi *et al.* (2021)

Table 7.4 (Continued)

Samples	Design	Method	Reference
Urine, blood, feces	Diet/time	ASCA	Zheng *et al.* (2016)
Blood	Time	ASCA	Bester *et al.* (2023)
Stool	Gender/coupling/time	ASCA	Trimigno *et al.* (2024)
Serum	Treatment/time	RM-ASCA+	Torkildsen *et al.* (2023)
Fecal	Intervention/Cross-over	MSCA	Jacobs *et al.* (2008)
Blood	Nutri-intervention/time	MSCA, ASCA	Bondia-Pons *et al.* (2013)
Blood/serum	Case/control	MANOVA	Adam *et al.* (2021)
Breath	Grouping	MANOVA	Blanchet *et al.* (2017)
Blood	Sample freezing protocols	MANOVA	La Frano *et al.* (2018)
Blood	Grouping	MANOVA	Martínez-Arranz *et al.* (2015)
Breath	Healthy/diseased	PC-ANOVA	Maiti *et al.* (2020)
Serum	Cross-over intervention	ANOVA-PLS	Rådjursöga *et al.* (2019)
Serum	Time	ANOVA-TP	Stanimirova *et al.* (2020)
Serum	Patients/time	AMOPLS	Gagnebin *et al.* (2020)
Serum	Patient grouping	AMOPLS	Gagnebin *et al.* (2020)
Neural cells	Treatment/time	AMOPLS	(González-Ruiz *et al.*, 2017)
Neural cells	Treatment/time	AMOPLS	(González-Ruiz *et al.*, 2019)
Plasma	Nutrition/time	AMOPLS	Rangel-Huerta *et al.* (2022)
Cell lines	Treatments	AMOPLS	Boccard *et al.* (2019)

7.1.2 Microbiome

The use of HD-ANOVA methods in the microbiome field is dominated by methods like PERMANOVA based on different types of distances. We omit these types of applications because they are very abundant and good reviews already exist (see, *e.g.*, Kleine Bardenhorst *et al.*, 2021). The use of HD-MANOVA methods directly on the OTUs or ASVs is very limited; see Table 7.5. The problem of zero inflation in the microbiome data is usually tackled by ad-hoc methods and not modeled explicitly.

Table 7.5 Overview of applications using HD-ANOVA (except PERMANOVA) in the microbiome field.

Type	Samples	Design	Method	Reference
In-vitro gut	In-vitro metaproteome	Treatment/growth-stage	ASCA, MANOVA	Hao *et al.* (2020)
Mice-gut	Feces	Interventions/diet	ASCA, FFMANOVA	Khomich *et al.* (2021)
Cultures	Fermentation samples	Conditions/time	(Fused-)ASCA	Raimondi *et al.* (2021)

7.1.3 Genomics

The HD-ANOVA methods are also being used in genomics, see Table 7.6.

In most cases, the measurements are (still) performed using micro-arrays, but in the later publications, RNAseq is being used.

Table 7.6 Overview of applications using HD-ANOVA in genomics applications.

Type	Samples	Design	Method	Reference
A. fumigatus	RNA-seq of in vitro infection	Batch effects	ASCA	Irmer *et al.* (2015)
Honeycrisp apples	RNA-seq of fruits	Orchards/storage time	ASCA	Leisso *et al.* (2016)
Rats	Micro-array of liver tissue	Treatment/time	ASCA	Nueda *et al.* (2007)
NSF Potato	Micro-array	Stress/time	ASCA	Nueda *et al.* (2009)
NSF Potato	Micro-array	Noise characteristic	ASCA	Nueda *et al.* (2012)
Human embryonic stem cells	Micro-array	Conditions	ASCA	Tarazona *et al.* (2012)
Maize	RNA-seq of Maize kernels	Cultivars/time	(fused-)ASCA	Hu *et al.* (2016)
DNA oligonucleotides	DNA sequences	Conditions/time	APCA	Fritzsch *et al.* (2018)

7.1.4 Proteomics

The use of HD-ANOVA methods in proteomics is still limited, see Table 7.7.

Table 7.7 Overview of applications using HD-ANOVA in proteomics.

Type	Samples	Design	Method	Reference
Human	Serum	Sample handling factors	ASCA	Mitra *et al.* (2016)
Milk	In-vitro samples	Types of milk/time	ASCA	Aalaei *et al.* (2021)
Women	Amniotic fluid	Grouping/time/sample handling	APCA	Harrington *et al.* (2006)
E. coli	Cells	Strains/growth ages	APCA	Chen *et al.* (2008)
Cows	Liver	Treatment/time	RM-ASCA	Veshkini *et al.* (2022)

7.1.5 Food Science

An overview of the use of HD-ANOVA methods in food science is presented in Table 7.8. Clearly, HD-ANOVA methods are getting used in food science and with a variety of methods.

Table 7.8 Overview of applications using HD-ANOVA in food science.

Type	Samples	Design	Method	Reference
Pasta	Pasta samples	Composition/temperature/time	ASCA	Bevilacqua *et al.* (2013)
Cranberry	Fruit	Fruit type/extraction solvents	ASCA	Turi *et al.* (2015)
Coffee	Beans	Species/roasting time	ASCA	De Luca *et al.* (2016)
Cheddar cheese	Cheese slices	Storage temperature/time	ASCA	Firmani *et al.* (2020)
Mango	Dried slices	Batches/drying treatments	ASCA	Cheng *et al.* (2019)
Bread	Baked bread	Treatment/site/time	ASCA	Amigo *et al.* (2019)
Beer	Beer samples	Types of beer	ASCA	Dysvik *et al.* (2019)
Milk	Rennet	Process Factor Design	ASCA	Strani *et al.* (2019)
Milk	Milk samples	Instruments/time	ASCA	Nieuwoudt *et al.* (2021)
Palm oil	Leaves	Sample handling	ASCA	Vargas *et al.* (2016)
Duck breast	Meat	Breed/age	ASCA	Wang *et al.* (2020)
S. nigra	Elderberries	Cultivar/time	ASCA	Salvador *et al.* (2016)
Basil	Volatiles	Variety/cut/year	ASCA	D'Alessandro *et al.* (2021)
Wine	Grapes	Varieties	ASCA	Khakimov *et al.* (2022)
Wine	Grapes	Terroir/enzyme treatment	ASCA	Apolinar-Valiente *et al.* (2014)
Cheese	Cheese samples	Dairy factory/time	ASCA	Ricci *et al.* (2023)
Grapes	Crushed grapes	Time/position of the grape	ASCA	Schorn-García *et al.* (2023)
Wine fermentation	Process samples	Initial condition/treatment/time	ASCA	Schorn-García *et al.* (2023)
Cocoa	Ground beans	Drying/varieties	ASCA	Erazo Solorzano *et al.* (2023)
Wine	Wine sample	Temperature/time	ASCA	García-Aguilera *et al.* (2023)
External reference material	Wheat/sunflower kernels	Storage conditions	APCA	Sarembaud *et al.* (2007)
Milk	Milk sample	Class/production site/temperature/time	APCA	Harnly *et al.* (2014)
Broccoli	Powdered composites	Cultivars/treatment	APCA	Luthria *et al.* (2008)
Grapefruit	Juice	Conditions/time	APCA	Chen *et al.* (2010)
Wine	Grapes	Maturity/treatment/aging	AMOPLS	Zhang *et al.* (2020)

(Continued)

Table 7.8 (Continued)

Type	Samples	Design	Method	Reference
Wine	Grape	Cultivar/sites/time	AMOPLS	Schmidtke *et al.* (2020)
Parmigiano Cheese	Cheese samples	Time/site/treatment	AMOPLS	Becchi *et al.* (2024)
Rooibos tea	Extracts	Batches/treatment	ASCA, ANOVA-TP, rMANOVA	Marini *et al.* (2017)
Bread	Volatiles	Types of flour/treatment	MANOVA	Nissen *et al.* (2020)
Mango	Fruit	Storage condition/time	MANOVA	Uarrota *et al.* (2019)
Grana Pardano cheese	Cheese wheels	Season/factory/position	LMM-ASCA	Ricci *et al.* (2022)

7.1.6 Sensory Science

An overview of the use of HD-ANOVA methods in food science is presented in Table 7.9 and shows that these methods are not used very much in this field.

Table 7.9 Overview of applications using HD-ANOVA in sensory science.

Type	Samples	Design	Method	Reference
Tea	Shoot samples	Treatment/species/batches	ASCA	Joubert *et al.* (2024)
Tea	Prepared tea	Tasters/growing regions	MANOVA	Fernando *et al.* (2021)
Cheese	Aroma samples	Types	MANOVA	Haddi *et al.* (2010)
Candies	Products	Types/Assessors	ASCA, PC-ANOVA	Luciano and Næs (2009)

7.1.7 Chemistry

An overview of the use of HD-ANOVA methods in chemistry is presented in Table 7.10. Also in this field, the use is relatively minor.

Table 7.10 Overview of applications using HD-ANOVA in chemistry.

Type	Samples	Design	Method	Reference
Beer	Aroma samples	Ingredients/conditions	ASCA	Bevilacqua *et al.* (2017)
Streams/rivers	Water	Location/seasonal	ASCA	Hohrenk-Danzouma *et al.* (2022)
X. leavis kidney	Epithelial cells	Conditions	ASCA	Gorrochategui *et al.* (2016)
Waste water	Water	Location/time	ASCA	Hohrenk *et al.* (2019)

Table 7.10 (Continued)

Type	Samples	Design	Method	Reference
LC-Stationary phases	Standard mixture	Experimental conditions	ASCA	Navarro-Reig *et al.* (2017)
LC-Stationary phases	Standard mixture	Experimental conditions	ASCA	Ortiz-Villanueva *et al.* (2017)
Lipid extraction solvent system	Human LDL standard	Solvents/lipid class	ASCA	Reis *et al.* (2013)
Wastewater	Water	Location/time	GASCA	Khatoonabadi *et al.* (2021)
Carragehan gels	Gels	Concentration/ temperature	APCA	Pinto *et al.* (2008)
Biodiesel	Diesel samples	Experimental conditions	APCA	Pinto and Lemos (2013)
Gasoline	Gasoline samples	Brands/time	APCA	Barnett and Zhang (2018)

7.2 Reflections

7.2.1 Summary of Reviews

An overview of the various reviews is given in Table 7.11. In this table, the column with ASCA means publications where exclusively the ASCA method was used as an HD-ANOVA method. Papers that used various other HD-ANOVA methods are collected in the column *Other*. As mentioned before, PERMANOVA was excluded from this list.

Table 7.11 Overview of the reviews.

	MX		Micro		GX		PX		Food		Sens		Chem		
Years	ASCA	Other	ASCA	Other	ASCA	Other	ASCA	Other	ASCA	Other	ASCA	Other	ASCA	Other	Total
2005–2008	1	3	0	0	1	0	0	2	0	2	0	0	0	3	12
2009–2012	2	1	0	0	3	0	0	0	0	1	0	2	0	0	9
2013–2016	26	4	0	0	2	1	1	0	5	1	0	0	2	0	43
2017–2020	47	21	0	1	0	1	0	0	6	5	0	0	4	0	84
2021–2024	33	8	0	2	0	1	1	1	8	2	1	1	1	1	60
Total	109	37	0	3	6	3	2	3	19	11	1	3	7	4	208

MX is metabolomics; Micro is microbiome; GX is genomics; PX is proteomics; Sens is sensory science; Chem is chemistry. Note that the year 2024 was not finished yet while writing this book.

Several remarks are appropriate:

- By far the most used method is the simplest version of ASCA. This may have to do with the availability of software. For the more advanced ASCA-type methods, the software is not easily available.

- ASCA and related methods are mostly used in metabolomics; the field in which ASCA was introduced in 2005.
- The HD-ANOVA methods are hardly used stand-alone. They are often preceded by PCA and accompanied by other methods, notably (univariate) ANOVA or (O)PLS-type analyses. To some extent, this is problematic since univariate and multivariate methods may give different results (Saccenti *et al.*, 2014). Also, OPLS-DA and ASCA are not directly comparable.
- Over the years, ASCA and related methods have been used increasingly. In some fields, it has not taken off that much.
- Which method to apply also depends on the (publishing) culture in a field. In genomics, *e.g.*, most researchers use univariate methods such as differential expression combined with multiple testing corrections. In the microbiome field, PERMANOVA is very popular. This is not easily changed apart from the question of whether that is desirable in terms of the quality of the data analysis methods.
- There also seems to be another cultural aspect toward which methods to use. There are a number of groups that consistently use ASCA (and not only their developers). Others use their own developed methods (*e.g.*, AMOPLS). The large majority in the metabolomics community is using (O)PLS-type models. This may have to do with the availability of commercial software in which these tools are implemented, but it clearly also has to do with the culture in a community, see Levin (2014).

7.2.2 Overview of Methods

We will present an overview of methods (and also a mild critique) that are frequently used (see the tables in the previous sections of this chapter) in the form of a table (see Table 7.12) showing the following properties:

F/M: The method is only dealing with fixed effects models (F) or can also handle mixed models (M).
B/UB: The property of being able to deal with unbalanced (UB) data or only balanced (B) data.
Versatility: The range of applicability of a method: can it handle complex designs?
Metaparameters: This is a listing of the metaparameters of the method that has to be chosen a priori. These can be latent variables (LVs) or tuning parameters.
Validation: The validation and statistical inference tools attached to the method.
Interpretation: Can importance of variables be calculated? Visualization using plots?

For all applications, it is important that a good description of the statistical design of the discussed data is given and how this is accommodated by the used method. Statistical design is not discussed in detail in this book, but good references are available (Casella, 2008; Welham *et al.*, 2014).

There is a lot to say about all these methods, and a detailed discussion of all advantages and disadvantages would require another chapter. Moreover, some of these methods are still very recently developed and not used yet a lot. So, it is difficult at this point to evaluate their relative merits and usefulness. Some general conclusions, however, can be given.

Some methods use established building blocks such as ANOVA and PCA (ASCA, ASCA+, APCA, RM-ASCA+, LiMM-PCA, and PC-ANOVA). They require some understanding of (univariate) ANOVA models and coding schemes. This has some advantages in terms of interpretation and software development/use. The methods LiMM-PCA, ASCA+, and RM-ASCA+ are the most versatile since they can deal with complex designs (ASCA+) and mixed models and repeated-measures (LiMM-PCA and RM-ASCA+). To some extent, this

Table 7.12 Properties of methods.

Method	F/M	B/UB	Versatility	Metaparameters	Validation	Interpretation
ASCA	F	B	□	LVs	+	++
ASCA+	F	B/UB	+	LVs	+	++
APCA	F	B	□	LVs	+	++
ANOVA-TP	F	B/UB	□	LVs	+	□
rMANOVA	F	B	□	Tuning parameters	+	+
AComDim	F	B/UB	□	LVs	□	□
AMOPLS	F	B/UB	□	LVs	+	□
PC-ANOVA	F	B	□	LVs	□	+
PERMANOVA	M	B/UB	+	Type of distance	+	-
LiMM-PCA	M	B/UB	++	LVs, tuning parameter	+	+
RM-ASCA+	M	B/UB	+	Splines, LVs	+	+

Dashed line distinguishes F from M-type methods. ++ means very good; + means good, □ means moderate and – means poor.

leads to a decrease in ease of use, *e.g.*, setting metaparameters in LiMM-PCA (LVs before and after the regression step, and the α parameter) and reduced interpretability. Also the software requirements for these methods are more demanding than, *e.g.*, ASCA, although LiMM-PCA can be used with standard software for LMMs and PCA.

Methods such as ANOVA-TP, AComDim, and AMOPLS require extra steps in terms of PLS (including multiple target projections in ANOVA-TP) or ComDim and thus are less easy to use and interpret. The method rMANOVA requires selecting a target covariance matrix and a tuning parameter; it is also rather limited in the type of designs it can handle. All methods rely on resampling tools for statistical validation (see Section 7.2.3). PERMANOVA has the drawback that it can be affected by different within-group correlations (see Section 7.2.3) and it does not readily give the importance of variables/features, hampering interpretation.

7.2.3 Remaining Challenges

At the end of this book, we would like to address some remaining challenges. There are many unsolved issues in the field of HD-ANOVA, and we will address some of them.

7.2.3.1 ASCA+ and Partial RDA

While using ASCA+ for unbalanced data, the effect matrices are not mean-centered. When performing a PCA on those effect matrices including a centering step, the resulting scores and loadings do not represent an LS solution anymore (ter Braak, 2023). One way to solve this is to use partial RDA which postulates a model and finds the parameters using an LS method. This approach, however, is not without difficulties. The orthogonalization is performed by removing from $\mathbf{X}_2$ the part overlapping with $\mathbf{X}_1$ using the multiplication $(\mathbf{I} - \mathbf{X}_1\mathbf{X}_1^+)\mathbf{X}_2$. Then the column-space of this resulting matrix $(\mathbf{I} - \mathbf{X}_1\mathbf{X}_1^+)\mathbf{X}_2$ is not in the range of $\mathbf{X}_2$ anymore since $\mathbf{X}_2$ is projected onto $R(\mathbf{X}_1)^\perp$. Hence, there is new variation added to $\mathbf{X}_2$ by this step which may hamper interpretation.

This may give problems in practice and an example of a related problem in the context of deflating a matrix with variables outside its range is reported in Westerhuis *et al.* (2001). A method that resembles partial RDA is called Sequentially Orthogonalized PLS (SO-PLS) which also has this problem (Næs *et al.*, 2011). The problem may be mitigated in different ways (Westerhuis and Smilde, 2001; Smilde *et al.*, 2022).

7.2.3.2 Permutations: Correlations and Unbalancedness

Permutation approaches for high-dimensional data seem to be sensitive to heterogeneity in multivariate dispersions. Anderson (2017) proposed an extension using the PERMANOVA pseudo-F-statistic to deal with unequal dispersions among groups. The effect of unequal correlation structures among groups has not received much attention, although initial work appeared (Hemerik *et al.*, 2021). More research is needed to explore the effect of unequal correlation structures between groups on permutation results.

The final topic regarding permutations for high-dimensional data is related to how to deal with unbalanced designs. The unbalancedness leads to correlated effect matrices. Some work based on Type III sum-of-squares has been done in this respect, but as of yet it is not completely clear how to perform permutations in case of unbalanced designs (Kherad-Pajouh and Renaud, 2010; Rasmussen *et al.*, 2024).

7.2.3.3 PERMANOVA and Effect Sizes

Effect size calculations in PERMANOVA approaches are discussed in Section 6.2.3. There it was stated that for calculating effect sizes, it may be required that the Gower matrix is semi-positive definite. This is a conjecture and not yet proven.

7.2.3.4 Back-Projection Approach

In LiMM-PCA it is suggested to use the factor associated with the denominator of the F-test as the error term for back-projection and to scale the error term using an F-distribution and effective dimensions. It is unclear if the correction should also be used for fixed effect models and how large the scaling effect is in practice. Further, effective dimensions in mixed models seem to be highly sensitive to the parameters used in software for the model fitting process. This begs the question if the use of effective dimensions should be recommended in this case. An alternative could be to use degrees of freedom with a possible cost of biased estimates. More investigation is needed.

7.2.3.5 Inferential Statistics

For linear models, rigorous inferential statistics based on the assumption of normally distributed variables is very well developed. In the high-dimensional case, this only holds for MANOVA but that method can only handle a very limited number of variables. For HD-ANOVA methods, rigorous inferential statistics based on distributions are poorly developed. Statistical inference relies mostly on approximations and resampling methods which, when performed properly, is a powerful alternative. Yet, it would be nice to develop some more rigorous inferential statistics for HD-ANOVA methods.

7.2.3.6 Advanced HD-ANOVA Methods

As mentioned above, LiMM-PCA and RM-ASCA+ are the most advanced HD-ANOVA methods in terms of enabling complex designs. However, both methods have their disadvantages considering their use for analyzing omics data. Consider genomics data based on RNAseq measurements. These can easily contain thousands of variables. In such a case RM-ASCA+ is not feasible since it should start with building thousands of linear mixed models (one for each variable). Recent unpublished results have shown that the performance of LiMM-PCA depends crucially on the initial number of principal components that should be used as a first step in that method. For data of low intrinsic rank (such as chemical data), LiMM-PCA seems to work fine, but for more complex data without an obvious low-rank structure it may run into difficulties due to the selection of the initial number of PCs.

8

Software

8.1 HD-ANOVA Software

Many of the HD-ANOVA methods found in this book are available in software implementations, of which a large but non-exhaustive selection is found in Section 8.8. As of 2024, the largest collection of methods is found in the programming language R – scattered over several packages. Many methods are also found in various open-source and commercial MATLAB toolboxes.

Chapter Structure

The rest of this chapter contains five sections describing the R package `HDANOVA` accompanying this book, followed by an overview of many HD-ANOVA methods implemented in software. The `HDANOVA` sections include *installing and starting* the package, *data handling*, *basic methods*, and *alternative methods*.

8.2 R Package `HDANOVA`

For easy application of many of the methods presented in this book, we have developed an R package simply called `HDANOVA`, freely available from The Comprehensive R Archive Network (CRAN, (R Core Team, 2024)). This package includes methods that are unique to the package and imports a large range of methods from other packages. We have attempted to make the interface to the methods as unified as possible for easy testing of many methods without large changes to R scripts. Code examples with short explanations are found in Sections 8.3–8.7. A basic proficiency in R programming is a prerequisite for benefiting from this section. Overviews of basic package usage and included methods are split into the following subsections, largely following the chapters in this book:

- **Installing and starting the package**
  Section 8.3
- **Data handling**
  Section 8.4
- **Analysis of variance**
  Section 8.5, Chapter 3

Analysis of Variance for High-Dimensional Data: Applications in Life, Food and Chemical Sciences,
First Edition. Age K. Smilde, Federico Marini, Johan A. Westerhuis, and Kristian H. Liland.

- **ASCA family**
  Section 8.6, Chapters 4 and 5.3
- **Alternative methods**
  Section 8.7, Chapter 6

8.3 Installing and Starting the Package

After starting the R (https://cran.r-project.org/) interface of choice, *e.g.*, a terminal, a form of R GUI, or Integrated Development Environment (IDE) like R Studio (https://rstudio.com), the HDANOVA package can be installed and started from the R terminal using the following commands.

```
# One-time installation
install.packages("HDANOVA")

# Starting the package
library(HDANOVA)
```

The installation of the HDANOVA package also invokes the installation of several other packages that it depends on. Among these are the mixlm package enabling user-friendly mixed modeling. This package is used in the ANOVA examples below.

8.4 Data Handling

This section briefly shows how to read data from selected data formats, perform basic preprocessing and prepare blocks of data for multiblock analysis.

8.4.1 Read from File

Data are stored in many different file formats. The following three examples cover two types of CSV-files and generic flat files.

```
# Find directory extdata from the HDANOVA package
mbdir <- system.file('extdata/', package = "HDANOVA")

# Comma-separated values, row names in first column
meta_data <- read.csv(paste0(mbdir, "/meta_data.csv"), row.names = 1)
# If working directory matches file location:
# meta_data <- read.csv('meta_data.csv', row.names = 1)
meta_data
```

```
        temperature colour
John           38.0   blue
Julia          37.0  green
James          37.5   blue
Jacob          37.6    red
```

```
Jane             37.2    red
Johanna          37.9  green
```

```
# Semi-colon separated values(locales where the decimal point is comma),
# no row names
proteins <- read.csv2(paste0(mbdir, "/proteins.csv"))
proteins
```

```
        prot1       prot2      prot3
1  0.46532048  0.30183300 -1.4654414
2 -1.79802081 -0.22812232 -0.4639203
3 -1.92962434 -0.40513080  0.1767796
4  0.87437138  0.79843798  0.1234731
5 -0.62445278 -0.07975479 -1.1126332
6 -0.07493721  1.09576027  1.2656596
```

```
# Blank space separated data without labels
genes <- read.table(paste0(mbdir, "/genes.dat"))
genes
```

```
           V1         V2         V3
1  0.39033106 -0.5720390  1.9147573
2  0.55352785  0.0948703 -0.2239755
3  0.09872346 -0.1029385  0.9047138
4 -0.59213740 -0.6027739  0.6177083
5 -0.02350148  0.3572809 -0.5168416
6  0.76644845  1.2863428  1.8239298
```

8.4.2 Data Pre-processing

Before analysis, various types of pre-processing may be needed. Centering and standardizing/scaling may be considered the most basic. In R, these operations are performed column-wise by default with the *scale* function, leading to autoscaling. If these operations are performed on the rows, we perform the standard normal variate (SNV) instead.

```
# Column-centring
genes_centred <- scale(genes, scale=FALSE)
colMeans(genes_centred) # Check mean values
```

```
          V1           V2           V3
1.850372e-17 0.000000e+00 7.401487e-17
```

```
# Autoscaling
genes_scaled <- scale(genes)
apply(genes_scaled, 2, sd) # Check standard deviations
```

```
V1 V2 V3
 1  1  1
```

```
# SNV (transpose, autoscale, re-transpose)
genes_snv <- t(scale(t(genes)))
apply(genes_snv, 1, sd) # Check standard deviations
```

```
[1] 1 1 1 1 1 1
```

8.4.2.1 Re-coding Categorical data

Most analysis methods require continuous input data. The **meta_data data.frame** contains a character vector (a factor in older R versions) of categories. This package has a function **dummycode** for converting categorical data to various dummy formats.

```
# Default is sum-to-zero coding
dummycode(meta_data$colour)
```

```
  x1 x2
1  1  0
2  0  1
3  1  0
4 -1 -1
5 -1 -1
6  0  1
```

```
# Reference-coding (called "treatment coding" in R)
dummycode(meta_data$colour, "contr.treatment")
```

```
  xgreen xred
1      0    0
2      1    0
3      0    0
4      0    1
5      0    1
6      1    0
```

```
# Full-coding (rank deficient)
dummycode(meta_data$colour, drop = FALSE)
```

```
  xblue xgreen xred
1     1      0    0
2     0      1    0
3     1      0    0
4     0      0    1
5     0      0    1
6     0      1    0
```

```
# Replace categorical with dummy-coded, use I() to index by common name
meta_data2 <- meta_data
```

```
meta_data2$colour <- I(dummycode(meta_data$colour, drop = FALSE))
meta_data2

         temperature colour.xblue colour.xgreen colour.xred
John            38.0            1             0           0
Julia           37.0            0             1           0
James           37.5            1             0           0
Jacob           37.6            0             0           1
Jane            37.2            0             0           1
Johanna         37.9            0             1           0

meta_data2$colour
  xblue xgreen xred
1     1      0    0
2     0      1    0
3     1      0    0
4     0      0    1
5     0      0    1
6     0      1    0
```

8.4.3 Data Structures for Analysis Including Blocks

The block terminology is borrowed from multiblock analysis. For HD-ANOVA analyses, the primary reason we might want to include blocks in a data structure is if different sets of measurements have been made in connection to the same design of experiments. This simplifies handling as a common object can be referred to across multiple analyses.

8.4.3.1 Create List of Blocks

A simple list of blocks can be created using the `list()` function. Naming of the blocks can be done directly or after creation.

```
# Direct approach
blocks1 <- list(meta = meta_data2, proteins = proteins, genes = genes)

# Two-step approach
blocks2 <- list(meta_data2, proteins, genes)
names(blocks2) <- c('meta', 'proteins', 'genes')

# Same result
identical(blocks1, blocks2)
[1] TRUE

# Access by name or number

blocks1[['meta']]
         temperature colour.xblue colour.xgreen colour.xred
John            38.0            1             0           0
Julia           37.0            0             1           0
```

```
James              37.5            1            0          0
Jacob              37.6            0            0          1
Jane               37.2            0            0          1
Johanna            37.9            0            1          0
blocks2[[1]]
        temperature colour.xblue colour.xgreen colour.xred
John               38.0            1            0          0
Julia              37.0            0            1          0
James              37.5            1            0          0
Jacob              37.6            0            0          1
Jane               37.2            0            0          1
Johanna            37.9            0            1          0
```

8.4.3.2 Create data.frame of Blocks

A **data.frame** is a convenient storage format for data in R and can handle many types of variables, *e.g.*, numeric, logical, character, factor, or matrices. The latter is useful for analyses of data with shared sample mode.

```
# Construct block data.frame from list
df1 <- block.data.frame(blocks1)

# Construct block data.frame from data.frame:
# First merge blocks into data.frame
my_data <- cbind(meta_data2, proteins, genes)
# Then construct block data.frame using named
# list of indexes
df2 <- block.data.frame(my_data, block_inds =
        list(meta = 1:2, proteins = 3:5, genes = 6:8))

# Same result
identical(df1,df2)
[1] TRUE

# Access by name or number

df1[[2]]
              prot1       prot2      prot3
John     0.46532048  0.30183300 -1.4654414
Julia   -1.79802081 -0.22812232 -0.4639203
James   -1.92962434 -0.40513080  0.1767796
Jacob    0.87437138  0.79843798  0.1234731
Jane    -0.62445278 -0.07975479 -1.1126332
Johanna -0.07493721  1.09576027  1.2656596

df2[['proteins']]
              prot1       prot2      prot3
John     0.46532048  0.30183300 -1.4654414
Julia   -1.79802081 -0.22812232 -0.4639203
James   -1.92962434 -0.40513080  0.1767796
```

```
Jacob     0.87437138  0.79843798  0.1234731
Jane     -0.62445278 -0.07975479 -1.1126332
Johanna  -0.07493721  1.09576027  1.2656596

df1[c(1,3)]
$meta
        temperature colour.xblue colour.xgreen colour.xred
John           38.0            1             0           0
Julia          37.0            0             1           0
James          37.5            1             0           0
Jacob          37.6            0             0           1
Jane           37.2            0             0           1
Johanna        37.9            0             1           0

$genes
                V1         V2         V3
John     0.39033106 -0.5720390  1.9147573
Julia    0.55352785  0.0948703 -0.2239755
James    0.09872346 -0.1029385  0.9047138
Jacob   -0.59213740 -0.6027739  0.6177083
Jane    -0.02350148  0.3572809 -0.5168416
Johanna  0.76644845  1.2863428  1.8239298

df1[-2]
$meta
        temperature colour.xblue colour.xgreen colour.xred
John           38.0            1             0           0
Julia          37.0            0             1           0
James          37.5            1             0           0
Jacob          37.6            0             0           1
Jane           37.2            0             0           1
Johanna        37.9            0             1           0

$genes
                V1         V2         V3
John     0.39033106 -0.5720390  1.9147573
Julia    0.55352785  0.0948703 -0.2239755
James    0.09872346 -0.1029385  0.9047138
Jacob   -0.59213740 -0.6027739  0.6177083
Jane    -0.02350148  0.3572809 -0.5168416
Johanna  0.76644845  1.2863428  1.8239298

df2[c('proteins','genes')]
$proteins
             prot1       prot2      prot3
John     0.46532048  0.30183300 -1.4654414
Julia   -1.79802081 -0.22812232 -0.4639203
```

```
James    -1.92962434 -0.40513080  0.1767796
Jacob     0.87437138  0.79843798  0.1234731
Jane     -0.62445278 -0.07975479 -1.1126332
Johanna  -0.07493721  1.09576027  1.2656596

$genes
                  V1         V2         V3
John      0.39033106 -0.5720390  1.9147573
Julia     0.55352785  0.0948703 -0.2239755
James     0.09872346 -0.1029385  0.9047138
Jacob    -0.59213740 -0.6027739  0.6177083
Jane     -0.02350148  0.3572809 -0.5168416
Johanna   0.76644845  1.2863428  1.8239298
```

```
# Use with formula interface (see other vignettes)
# sopls(meta ~ proteins + genes, data = df1)

# Use with single list interface (see other vignettes)
# mfa(df1[c(1,3)], ncomp = 3)
```

8.5 Analysis of Variance (ANOVA)

This example will focus on univariate ANOVA in various designs including fixed and mixed effects, and briefly introduce MANOVA. Models using random effects will be run through the `mixlm` package. The corresponding theory is found in Chapter 3. The following single-response ANOVA models will be demonstrated:

- Fixed effect models
  - One-way – Section 3.2.1
  - Two-way – Section 3.2.2
  - Fixed effect nested – Section 3.2.2.2
- Linear mixed models
  - Basic – Section 3.5.2
  - Repeated measures – Section 3.5.2.3

```
# Start the mixlm R package
library(mixlm, warn.conflicts = FALSE)
```

8.5.1 Simulated Data

We will start by simulating some data to use in the examples below. In this fictitious setup, milk *yield* is measured as a function of *feed* type (low/high protein content), *cow* breed, *bull* identity, *daughter*, and *age* (months). The three first factors are crossed and balanced, while *daughter* is nested under *bull* and *age* varies freely.

The *yield* is generated with a linear model with some noise.

```
dat <- data.frame(
  feed     = factor(rep(rep(c("low","high"), each=6), 4)),
  breed    = factor(rep(c("NRF","Hereford","Angus"), 16)),
  bull     = factor(rep(LETTERS[1:4], each = 12)),
  daughter = factor(c(rep(letters[1:4], 3), rep(letters[5:8], 3),
                   rep(letters[9:12], 3), rep(letters[13:16], 3))),
  age      = round(rnorm(48, mean = 36, sd = 5))
)
dat$yield <- 150*with(dat, 10 + 3 * as.numeric(feed) + as.numeric(breed)
    + 2 * as.numeric(bull) + 1 * as.numeric(sample(dat$daughter, 48))
    + 0.5 * age + rnorm(48, sd = 2))
head(dat)
```

```
  feed    breed bull daughter age    yield
1  low      NRF    A        a  33 5541.301
2  low Hereford    A        b  35 5662.560
3  low    Angus    A        c  44 6354.182
4  low      NRF    A        d  36 6325.363
5  low Hereford    A        a  37 6144.458
6  low    Angus    A        b  45 6187.227
```

8.5.2 Fixed Effect Models

The simplest form of ANOVA is the fixed effect model. This model assumes that the levels of the factor are fixed and that the only source of variation is the factor itself.

8.5.2.1 One-Way ANOVA

Here we assess only the *feed* effect on *yield*, *i.e.*, the following model:

$$yield_{an} = \mu + feed_a + \epsilon_{an}$$

where a is the *feed* level and n is the observation within the feed level.

```
mod <- lm(yield ~ feed, data = dat)
print(anova(mod))
```

```
Analysis of Variance Table

Response: yield
          Df   Sum Sq Mean Sq F value  Pr(>F)
feed       1   957882  957882  3.1072 0.08459.
Residuals 46 14180730  308277
---
Signif. codes:  0 '***' 0.001 '**' 0.01 '*' 0.05 '.' 0.1 ' ' 1
```

In the ANOVA table, we can look at $Pr(> F)$ to see if the *feed* factor has a significant effect on *yield*. The summary function can be used to get more information about the underlying regression model.

```
summary(mod)
```

```
Call:
lm(formula = yield ~ feed, data = dat)

Residuals:
    Min      1Q  Median      3Q     Max
-954.25 -480.25   75.11  470.14 1240.91

Coefficients:
             Estimate Std. Error t value Pr(>|t|)
(Intercept)  6354.28      80.14  79.290   <2e-16 ***
feed(low)    -141.27      80.14  -1.763   0.0846.
---
Signif. codes:  0 '***' 0.001 '**' 0.01 '*' 0.05 '.' 0.1 ' ' 1

s: 555.2 on 46 degrees of freedom
Multiple R-squared: 0.06327,
Adjusted R-squared: 0.04291
F-statistic: 3.107 on 1 and 46 DF,  p-value: 0.08459
```

Basic model assessment can be done using the plot function (see Figure 8.1).

```
old.par <- par(mfrow=c(2,1), mar=c(4,4,2,0.5))
plot(mod, which = 1:2, ask=FALSE)
par(old.par)
```

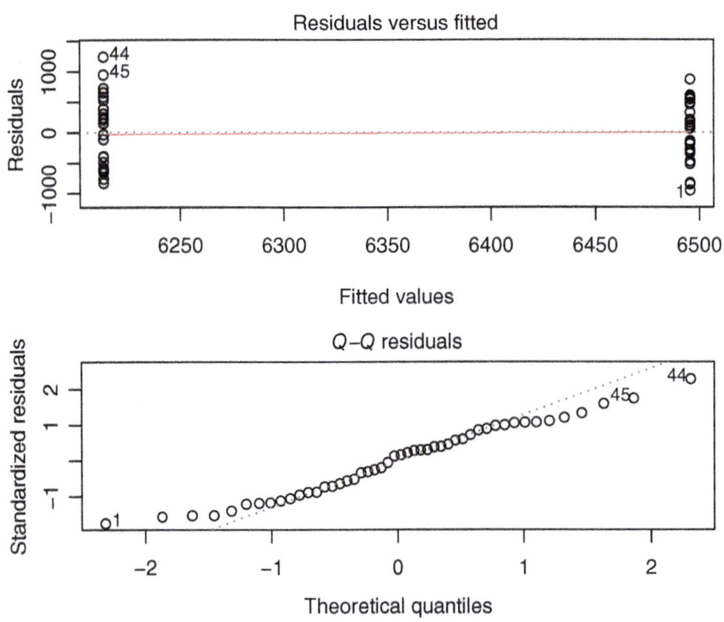

Figure 8.1 Model assessment plot for assessing homogeneity and normality.

8.5.2.2 Two-Way Crossed Effects ANOVA

Here we assess the *feed* and *breed* effects and their interaction effect on *yield*, *i.e.*, the following model:

$$yield_{abn} = \mu + feed_a + breed_b + (feed : breed)_{ab} + \epsilon_{abn}$$

where a is the feed level, b is the breed level, and n is the observation within the feed: breed-combination.

```
mod <- lm(yield ~ feed*breed, data = dat)
print(anova(mod))
```

```
Analysis of Variance Table

Response: yield
           Df   Sum Sq Mean Sq F value Pr(>F)
feed        1   957882  957882  2.9904 0.0911.
breed       2   284231  142116  0.4437 0.6447
feed:breed  2   443052  221526  0.6916 0.5064
Residuals  42 13453447  320320
---
Signif. codes:  0 '***' 0.001 '**' 0.01 '*' 0.05 '.' 0.1 ' ' 1
```

If the interaction effect is not significant, we can simplify the model to:

$$yield_{abn} = \mu + feed_a + breed_b + \epsilon_{abn}$$

```
mod <- lm(yield ~ feed+breed, data = dat)
print(anova(mod))
```

```
Analysis of Variance Table

Response: yield
          Df   Sum Sq Mean Sq F value  Pr(>F)
feed       1   957882  957882  3.0329 0.08858.
breed      2   284231  142116  0.4500 0.64055
Residuals 44 13896499  315830
---
Signif. codes:  0 '***' 0.001 '**' 0.01 '*' 0.05 '.' 0.1 ' ' 1
```

8.5.2.3 Types of Sums of Squares

The classical way of defining sums-of-squares are Type I, Type II, and Type III, as described in the documentation of the `Anova()` function in the `car` package.

```
# Type I - Sequential testing, including one and one effect
print(anova(mod))
```

```
Analysis of Variance Table
```

```
Response: yield
          Df   Sum Sq Mean Sq F value  Pr(>F)
feed       1   957882  957882  3.0329 0.08858.
breed      2   284231  142116  0.4500 0.64055
Residuals 44 13896499  315830
---
Signif. codes:  0 '***' 0.001 '**' 0.01 '*' 0.05 '.' 0.1 ' ' 1

# Type II - Testing each term after all others,
# except ignoring the term's higher-order relatives
print(Anova(mod, type="II"))


Anova Table (Type II tests)

Response: yield
            Sum Sq Df F value  Pr(>F)
feed        957882  1  3.0329 0.08858.
breed       284231  2  0.4500 0.64055
Residuals 13896499 44
---
Signif. codes:  0 '***' 0.001 '**' 0.01 '*' 0.05 '.' 0.1 ' ' 1

# Type III - Testing each term after all others,
# including the term's higher-order relatives
print(Anova(mod, type="III"))


Anova Table (Type III tests)

Response: yield
                Sum Sq Df   F value  Pr(>F)
(Intercept) 1938092247  1 6136.5139 < 2e-16 ***
feed            957882  1    3.0329 0.08858.
breed           284231  2    0.4500 0.64055
Residuals     13896499 44
---
Signif. codes:  0 '***' 0.001 '**' 0.01 '*' 0.05 '.' 0.1 ' ' 1
```

For the two-way ANOVA model, the Type I and Type II sums-of-squares are the same, while Type III can differ. With balanced data, this only happens when the contrast coding is of the treatment/reference type (sum coding above).

8.5.2.4 Coding Schemes

The coding schemes can be specified for each factor in the model. The default is reference-coding, but other codings can be specified using the `contrasts` variable. Since we are running `lm()` through the `mixlm` package, we can use the `contrasts` argument to specify the coding for all effects simultaneously.

```
# Sum-coding, i.e., the sum of all levels is zero and all effects
# are orthogonal in the balanced case.
mod <- lm(yield ~ feed*breed, data = dat, contrasts="contr.sum")
print(Anova(mod, type="III"))

Anova Table (Type III tests)

Response: yield
                Sum Sq Df   F value Pr(>F)
(Intercept) 1938092247  1 6050.4845 <2e-16 ***
feed            957882  1    2.9904 0.0911.
breed           284231  2    0.4437 0.6447
feed:breed      443052  2    0.6916 0.5064
Residuals     13453447 42
---
Signif. codes:  0 '***' 0.001 '**' 0.01 '*' 0.05 '.' 0.1 ' ' 1
```

```
# Weighted-coding, i.e., the sum of all levels is zero and the effects
# are weighted by the number of levels, effect-wise.
mod <- lm(yield ~ feed*breed, data = dat, contrasts="contr.weighted")
print(Anova(mod, type="III"))

Anova Table (Type III tests)

Response: yield
                Sum Sq Df   F value Pr(>F)
(Intercept) 1938092247  1 6050.4845 <2e-16 ***
feed            957882  1    2.9904 0.0911.
breed           284231  2    0.4437 0.6447
feed:breed      443052  2    0.6916 0.5064
Residuals     13453447 42
---
Signif. codes:  0 '***' 0.001 '**' 0.01 '*' 0.05 '.' 0.1 ' ' 1
```

Instead of specifying the contrasts in a specific model, it is also possible to set the contrasts globally for the session. This means that all subsequent models, unless specified otherwise, will use the specified contrasts.

```
options(contrasts = c("unordered"="contr.sum", "ordered"="contr.poly"))
```

8.5.2.5 Fixed Effect Nested ANOVA

In the case of nested factors, we can specify this in the model. In the current model, we assume that bulls are fixed effects that we are interested in and that daughters are nested under bulls. In this case, the daughters do not have any special attributes that would interfere with the estimation of the bull effect, so we do not have to assume that they are random effects. The model becomes:

$$yield_{abn} = \mu + bull_a + daugter_{b(a)} + \epsilon_{abn}$$

```
mod <- lm(yield ~ bull + daughter%in%bull, data = dat)
print(Anova(mod, type="II")) # See Section 3.A.3

Anova Table (Type II tests)

Response: yield
                Sum Sq Df F value    Pr(>F)
bull           6293005  3  8.9924 0.0001822 ***
bull:daughter  1380934 12  0.4933 0.9034247
Residuals      7464672 32
---
Signif. codes:  0 '***' 0.001 '**' 0.01 '*' 0.05 '.' 0.1 ' ' 1
```

8.5.3 Linear Mixed Models

Adding random effects to a model can be done either using least squares modeling through the `mixlm` package or using ML/REML estimation through the lme4 package (or similar).

8.5.3.1 Least Squares – `mixlm`

Using the `mixlm` package, we specify the random effects using the `r()` function. If we assume that the bull is a random selection from the population of bulls, we can specify this as a random effect when focusing on feed. The model looks like a fixed effect model, but the error structure is different:

$$yield_{abn} = \mu + feed_a + bull_b + (feed : bull)_{ab} + \epsilon_{abn}$$

```
mod <- lm(yield ~ feed*r(bull), data = dat)
print(Anova(mod, type="II"))

Analysis of variance (unrestricted model)
Response: yield
              Mean Sq     Sum Sq Df F value Pr(>F)
feed        957881.78  957881.78  1    3.75 0.1481
bull       2097668.39 6293005.18  3    8.22 0.0587
feed:bull   255297.76  765893.29  3    1.43 0.2472
Residuals   178045.79 7121831.56 40       -      -

             Err.term(s) Err.df VC(SS)
1 feed               (3)      3  fixed
2 bull               (3)      3 153531
3 feed:bull          (4)     40  12875
4 Residuals            -      - 178046
(VC = variance component)
```

```
            Expected mean squares
feed        (4) + 6 (3) + 24 Q[1]
bull        (4) + 6 (3) + 12 (2)
feed:bull   (4) + 6 (3)
Residuals   (4)
```

In addition to the ordinary ANOVA table, an overview of variance components and expected mean squares is printed.

8.5.3.2 Restrictions

The `mixlm` package has unrestricted models as default, but it is possible to turn on restrictions that affect how interactions between fixed and random effects are perceived (unrestricted means completely random interactions).

```
mod <- lm(yield ~ feed*r(bull), data = dat, unrestricted = FALSE)
print(Anova(mod, type="II"))
```

```
Analysis of variance (restricted model)
Response: yield
              Mean Sq      Sum Sq Df F value Pr(>F)
feed         957881.78   957881.78  1    3.75 0.1481
bull        2097668.39  6293005.18  3   11.78 0.0000
feed:bull    255297.76   765893.29  3    1.43 0.2472
Residuals    178045.79  7121831.56 40       -      -

              Err.term(s) Err.df VC(SS)
1 feed                (3)      3  fixed
2 bull                (4)     40 159969
3 feed:bull           (4)     40  12875
4 Residuals             -      - 178046
(VC = variance component)

            Expected mean squares
feed        (4) + 6 (3) + 24 Q[1]
bull        (4) + 12 (2)
feed:bull   (4) + 6 (3)
Residuals   (4)
```

This affects which tests are performed and how the variance components are estimated.

8.5.3.3 Repeated Measures

A repeated measures model can be a mixed model with a random effect for the repeated measures, where the repeated measures are nested under the subjects. Longitudinal data is a common example of repeated measures data, where the replicates are repetitions over *time* within *subject*. We extract a subset of the simulated data and add a longitudinal effect to the

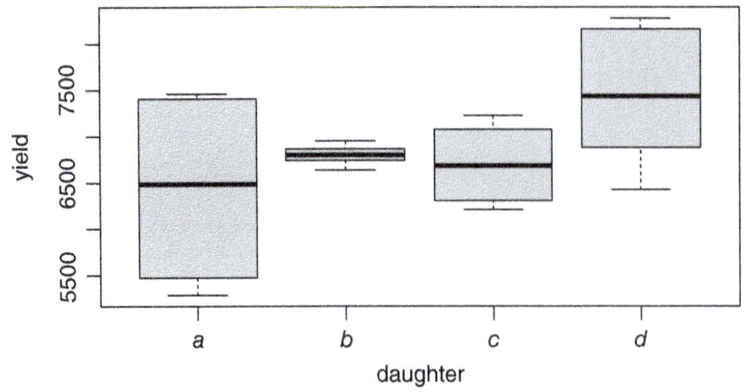

Figure 8.2 Boxplots over repetitions within daughters.

model (see Figure 8.2). This is chosen to be a random variation over three time points. The time effect does not necessarily need to be random.

```
set.seed(123)
long <- dat[c(1:4,9:12), c("feed", "daughter", "yield")]
long <- rbind(long, long, long)
long$daughter <- factor(long$daughter) # Remove redundant daughter
long$time  <- factor(rep(1:3, each=8))
long$yield <- long$yield + rnorm(24, sd = 100)
              + rep(c(-200,0,200), each=8)
plot(yield~daughter, data=long)
```

Now we have a *feed* effect, *individuals* (daughters), and a *time* effect repeated inside *daughters*, with the model:

$$yield_{ait} = \mu + daughter_i + feed_a + time_t + (feed : time)_{at} + \epsilon_{ait}$$

```
mod <- lm(yield ~ r(daughter) + feed*r(time), data = long,
          unrestricted=FALSE)
print(Anova(mod, type="II"))
```

```
Analysis of variance (restricted model)
Response: yield
             Mean Sq      Sum Sq Df F value Pr(>F)
daughter  1085642.42 3256927.25  3   1.854 0.1808
feed        21759.92   21759.92  1   7.079 0.1170
time       213465.79  426931.58  2   0.365 0.7005
feed:time    3073.71    6147.41  2   0.005 0.9948
Residuals  585554.07 8783311.09 15       -      -

             Err.term(s) Err.df   VC(SS)
1 daughter           (5)     15   83348
2 feed               (4)      2   fixed
```

```
3 time              (5)     15  -46511
4 feed:time         (5)     15 -145620
5 Residuals           -      -  585554
(VC = variance component)

             Expected mean squares
daughter  (5) + 6 (1)
feed      (5) + 4 (4) + 12 Q[2]
time      (5) + 8 (3)
feed:time (5) + 4 (4)
Residuals (5)
```

8.5.3.4 REML

REML estimation can be done directly with the lme4 package, but we can also do this through the `mixlm` package, leveraging the r() function.

```
mod <- lm(yield ~ feed*r(bull), data = dat, REML = TRUE)
print(Anova(mod, type="III"))

Analysis of Deviance Table (Type III Wald chisquare tests)

Response: yield
                Chisq Df Pr(>Chisq)
(Intercept) 3350.9550  1     <2e-16 ***
feed           2.3294  1     0.1269
---
Signif. codes:  0 '***' 0.001 '**' 0.01 '*' 0.05 '.' 0.1 ' ' 1
```

To see how `mixlm` transforms the model to lme4, we can print the model.

```
print(mod)

Linear mixed model fit by REML ['lmerMod']
Formula: yield ~ feed + (1 | bull) + (1 | feed:bull)
   Data: dat
REML criterion at convergence: 760.7979
Random effects:
 Groups    Name        Std.Dev.
 feed:bull (Intercept)   0.0
 bull      (Intercept)   0.0
 Residual              868.2
Number of obs: 48, groups:  feed:bull, 8; bull, 4
Fixed Effects:
(Intercept)        feed1
     7254.3       -191.3
```

```
optimizer (nloptwrap) convergence code: 0 (OK); 0 optimizer
     warnings; 1 lme4 warnings
```

We observe that (1 | bull) and (1 | feed:bull) are added to the model, which means that random intercepts are added for both bull and the interaction.

8.5.4 Multivariate ANOVA (MANOVA)

Basic multivariate ANOVA can be done using the `lm()` function, if we create a matrix of responses. In this case, we add a mastitis response to the model.

```
dat$mastitis <- as.numeric(dat$breed) + as.numeric(dat$feed)
     + rnorm(48, sd = 1)
```

The model becomes:

$$[yield_{abn}|mastitis_{abn}] = \mu + feed_a + breed_b + (feed\ :\ breed)_{ab} + \epsilon_{abn}$$

where each of the model terms now is a vector matching the number of responses.

```
mod <- lm(cbind(yield,mastitis) ~ feed*breed, data = dat)
print(Anova(mod, type="II"))


Type II MANOVA Tests: Pillai test statistic
           Df test stat approx F num Df den Df   Pr(>F)
feed        1   0.16426   4.0292      2     41 0.025260 *
breed       2   0.47415   6.5256      4     84 0.000127 ***
feed:breed  2   0.14929   1.6939      4     84 0.159025
---
Signif. codes:  0 '***' 0.001 '**' 0.01 '*' 0.05 '.' 0.1 ' ' 1
```

The test statistics are joint for all responses, here in the form of the default Pillai's test statistics. Other statistics can be produced as follows:

```
print(Anova(mod, type="II", test="Wilks"))


Type II MANOVA Tests: Wilks test statistic
           Df test stat approx F num Df den Df   Pr(>F)
feed        1   0.83574   4.0292      2     41  0.02526 *
breed       2   0.52761   7.7226      4     82 2.48e-05 ***
feed:breed  2   0.85375   1.6865      4     82  0.16100
---
Signif. codes:  0 '***' 0.001 '**' 0.01 '*' 0.05 '.' 0.1 ' ' 1
print(Anova(mod, type="II", test="Hotelling-Lawley"))
```

```
Type II MANOVA Tests: Hotelling-Lawley test statistic
           Df test stat approx F num Df den Df    Pr(>F)
feed        1   0.19655   4.0292      2     41   0.02526 *
breed       2   0.89201   8.9201      4     80 5.227e-06 ***
feed:breed  2   0.16775   1.6775      4     80   0.16336
---
Signif. codes:  0 '***' 0.001 '**' 0.01 '*' 0.05 '.' 0.1 ' ' 1

print(Anova(mod, type="II", test="Roy"))

Type II MANOVA Tests: Roy test statistic
           Df test stat approx F num Df den Df    Pr(>F)
feed        1   0.19655   4.0292      2     41   0.02526 *
breed       2   0.88826  18.6535      2     42 1.595e-06 ***
feed:breed  2   0.14285   2.9999      2     42   0.06056.
---
Signif. codes:  0 '***' 0.001 '**' 0.01 '*' 0.05 '.' 0.1 ' ' 1
```

The summary function can be used to get more information about the underlying regression model, here revealing the regressions are performed separately for each response.

```
summary(mod)

Response yield:

Call:
lm(formula = yield ~ feed * breed, data = dat)

Residuals:
    Min      1Q  Median      3Q     Max
-2123.5  -506.2  -109.0   625.8  1762.0

Coefficients:
                           Estimate Std. Error t value Pr(>|t|)
(Intercept)                 7254.28     129.27  56.118   <2e-16 ***
feed(high)                  -191.27     129.27  -1.480    0.146
breed(Angus)                  40.78     182.81   0.223    0.825
breed(Hereford)              -72.52     182.81  -0.397    0.694
feed(high):breed(Angus)      116.25     182.81   0.636    0.528
feed(high):breed(Hereford)    71.26     182.81   0.390    0.699
---
Signif. codes:  0 '***' 0.001 '**' 0.01 '*' 0.05 '.' 0.1 ' ' 1
```

```
Residual standard error: 895.6 on 42 degrees of freedom
Multiple R-squared:  0.07529,    Adjusted R-squared:  -0.03479
F-statistic: 0.6839 on 5 and 42 DF,  p-value: 0.6382


Response mastitis:

Call:
lm(formula = mastitis ~ feed * breed, data = dat)

Residuals:
     Min       1Q   Median       3Q      Max
-2.25984 -0.60241  0.07629  0.60307  1.82843

Coefficients:
                           Estimate Std. Error t value Pr(>|t|)
(Intercept)                  3.5537     0.1316  26.994  < 2e-16 ***
feed(high)                  -0.3506     0.1316  -2.663   0.0109 *
breed(Angus)                -0.8802     0.1862  -4.728 2.56e-05 ***
breed(Hereford)             -0.1653     0.1862  -0.888   0.3795
feed(high):breed(Angus)     -0.3722     0.1862  -1.999   0.0521 .
feed(high):breed(Hereford)   0.3991     0.1862   2.144   0.0379 *
---
Signif. codes:  0 '***' 0.001 '**' 0.01 '*' 0.05 '.' 0.1 ' ' 1

Residual standard error: 0.9121 on 42 degrees of freedom
Multiple R-squared:  0.5399, Adjusted R-squared:  0.4852
F-statistic: 9.858 on 5 and 42 DF,  p-value: 2.755e-06
```

8.6 Basic ASCA Family

The ANOVA part of ASCA includes all the possible variations of ANOVA demonstrated in the ANOVA section and more. Also, generalized linear models (GLM) can be used, though this is not handled in this book. The corresponding theory of the remaining methods is found in Chapters 4 and 5. The following theory will be exemplified:

- Basic ASCA – Section 4.1.1
  - Permutation testing – Section 4.1.3
  - Random effects – Section 5.3
  - Scores and loadings – Section 4.1.4
  - Data and confidence ellipsoids – Section 4.3.1
  - Combined effects – Section 4.2
  - Numeric effects

- ANOVA–PCA (APCA) – Section 4.2.1
- PC-ANOVA – Section 5.4.1
- MSCA – Section 4.7
- LiMM-PCA – Section 5.3.3
- RM-ASCA – Section 5.3.1

We start by demonstrating ASCA with a fixed effect model of two factors with interactions and ordinary PCA on the effect matrices.

```
# Load Candy data
data(candies)
```

8.6.1 Fit ASCA Model

We begin with analyzing the candy data, also used earlier in this book. It contains 9 response columns (assessment), 5 candies, 11 assessors, and 3 replicates.

```
mod <- asca(assessment ~ candy*assessor, data=candies)
summary(mod)
```

```
Anova Simultaneous Component Analysis fitted using 'lm' (Linear Model)
- SS type II, sum-coding, restricted model, least squares estimation
                Sum.Sq.  Expl.var.(%)
candy          33416.66         74.48
assessor        1961.37          4.37
assessor:candy  3445.73          7.68
Residuals       6043.52         13.47
```

The summary shows that the candy effect is the largest by far.

8.6.1.1 Permutation Testing

To get more insight, we can perform permutation testing of the factors.

```
# Permutation testing (default = 1000 permutations, if not specified)
mod <- asca(assessment ~ candy*assessor, data=candies, permute=TRUE)
summary(mod)
```

```
Anova Simultaneous Component Analysis fitted using 'lm' (Linear Model)
- SS type II, sum-coding, restricted model, least squares estimation,
1000 permutations
                Sum.Sq.  Expl.var.(%) P-value
candy          33416.66         74.48       0
assessor        1961.37          4.37       0
assessor:candy  3445.73          7.68       0
Residuals       6043.52         13.47      NA
```

Here we see that all effects are significant, where the Candy effect is the dominating one. (The P-values are rounded from 0.001 to 0 in the print.) This can also be visualized by looking at

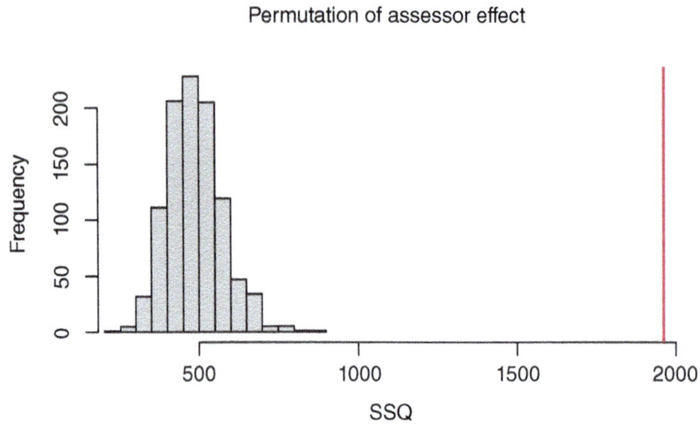

Figure 8.3 Histogram of SSQ values for assessor effect. The estimated effect is indicated in red.

the sums-of-squares values obtained under permutation compared to the original value (see Figure 8.3).

```
permutationplot(mod, factor = "assessor")
```

8.6.1.2 Random Effects

We can argue that the assessors are random effects and thus should be handled as such in the model. We can do this by adding *r*() around the assessor term.

```
# Fit ASCA model with random assessor
mod.mixed <- asca(assessment ~ candy*r(assessor), data=candies,
                  permute=TRUE)
summary(mod.mixed)
```

```
Anova Simultaneous Component Analysis fitted using 'lmm' (Linear Mixed Model)
- SS type II, sum-coding, restricted model, least squares estimation,
1000 permutations
                  Sum.Sq.  Expl.var.(%) P-value
candy            33416.66         74.48       0
assessor          1961.37          4.37       0
assessor:candy    3445.73          7.68       0
Residuals         6043.52         13.47      NA
```

8.6.1.3 Scores and Loadings

The effects can be visualized through, *e.g.*, loading and score plots to assess the relations between variables, objects, and factors. If a factor is not specified, the first factor is plotted (see Figure 8.4).

```
par.old <- par(mfrow=c(2,1), mar=c(4,4,2,1))
loadingplot(mod, scatter=TRUE, labels="names",
            main="Candy_loadings")
scoreplot(mod, main="Candy_scores")
par(par.old)
```

Figure 8.4 Loading plot and score plot for Candy effect.

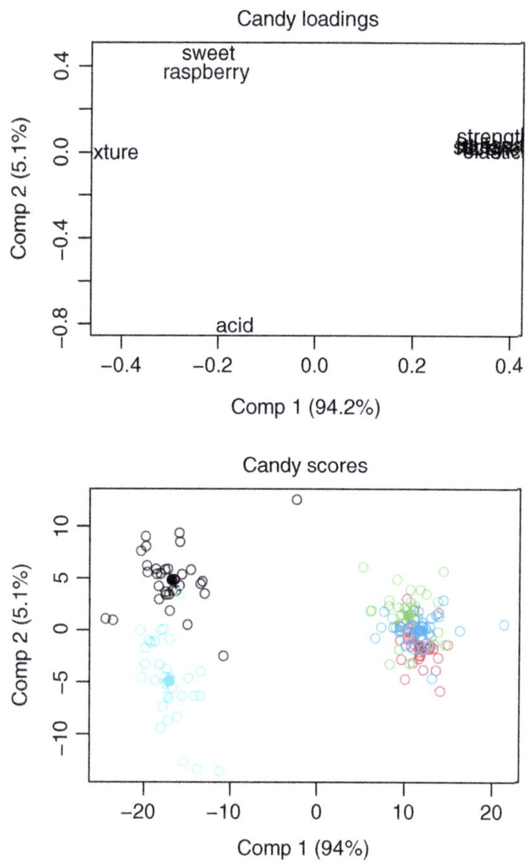

A specific factor can be plotted by specifying the factor name or number (see Figure 8.5).

```
par.old <- par(mfrow=c(2,1), mar=c(4,4,2,1))
loadingplot(mod, factor="assessor", scatter=TRUE,
            labels="names", main="Assessor_loadings")
scoreplot(mod, factor="assessor", main="Assessor_scores")
par(par.old)
```

Score plots can be modified, *e.g.*, omitting backprojections or adding spider plots (see Figure 8.6).

```
par.old <- par(mfrow=c(2,1), mar=c(4,4,2,1))
scoreplot(mod, factor="assessor", main="Assessor_scores",
     projections=FALSE)
scoreplot(mod, factor="assessor", main="Assessor_scores", spider=TRUE)
par(par.old)
```

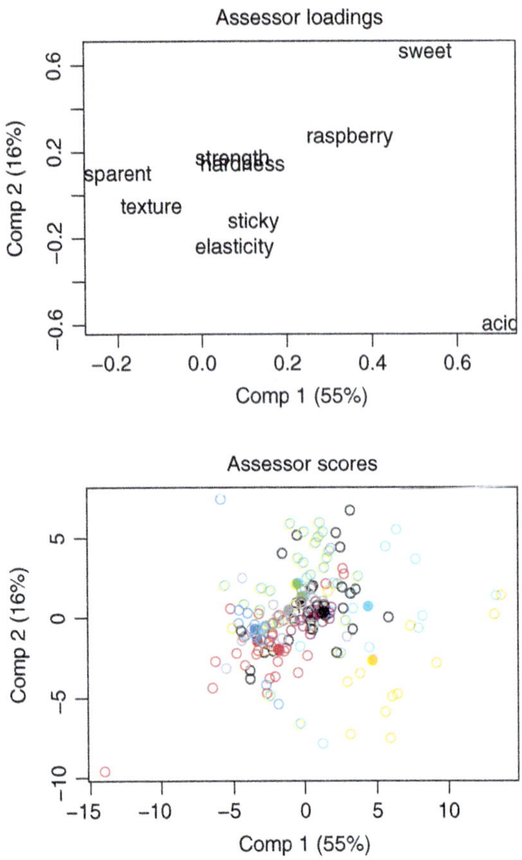

Figure 8.5 Loading plot and score plot for Assessor effect.

Scores and loadings can be extracted for further analysis.

```
L <- loadings(mod, factor="candy")
head(L)
```

```
                Comp 1        Comp 2      Comp 3      Comp 4
transparent  0.3927649  0.0279168248 -0.30598369  0.77398098
acid        -0.1649285 -0.8042012208  0.47650301  0.24192158
sweet       -0.2235032  0.4607791525  0.52330000 -0.01751194
raspberry   -0.2277610  0.3663430297  0.35862959  0.50808527
texture     -0.4302019 -0.0004154248 -0.13000106  0.22107937
strength     0.3653312  0.0694885591  0.09828578 -0.09582978
```

```
S <- scores(mod, factor="candy")
head(S)
```

```
     Comp 1   Comp 2     Comp 3    Comp 4
1 -16.81343 4.844756 -0.2914198 0.1811588
2 -16.81343 4.844756 -0.2914198 0.1811588
3 -16.81343 4.844756 -0.2914198 0.1811588
4 -16.81343 4.844756 -0.2914198 0.1811588
5 -16.81343 4.844756 -0.2914198 0.1811588
6 -16.81343 4.844756 -0.2914198 0.1811588
```

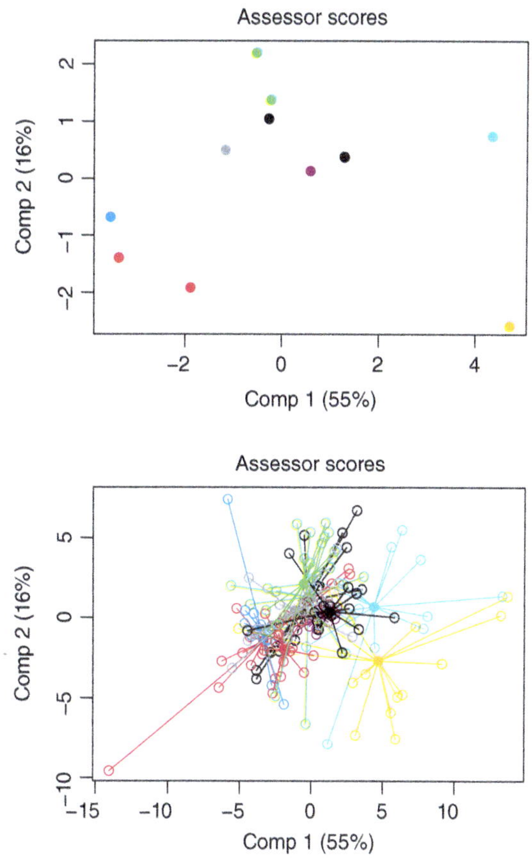

Figure 8.6 Score plot for Assessor effect without backprojection and as spider plot.

8.6.1.4 Data Ellipsoids and Confidence Ellipsoids

To emphasize factor levels or assess factor level differences, we can add data ellipsoids or confidence (model) ellipsoids to the score plot (see Figure 8.7). The confidence ellipsoids are built on the assumption of balanced data, and their theories are built around crossed designs.

```
par.old <- par(mfrow=c(2,1), mar=c(4,4,2,1))
scoreplot(mod, ellipsoids="data", factor="candy",
          main="Data_ellipsoids")
scoreplot(mod, ellipsoids="confidence", factor="candy",
          main="Confidence_ellipsoids")
par(par.old)
```

If we repeat this for the mixed model, we see that both types of ellipsoids change together with the change in the denominator term in the underlying ANOVA model (see Figure 8.8). It should be noted that the theory for confidence ellipsoids in mixed models is not fully developed, so interpretation should be done with caution.

```
par.old <- par(mfrow=c(2,1), mar=c(4,4,2,1))
scoreplot(mod.mixed, ellipsoids="data", factor="candy",
          main="Data_ellipsoids")
scoreplot(mod.mixed, ellipsoids="confidence", factor="candy",
          main="Confidence_ellipsoids")
par(par.old)
```

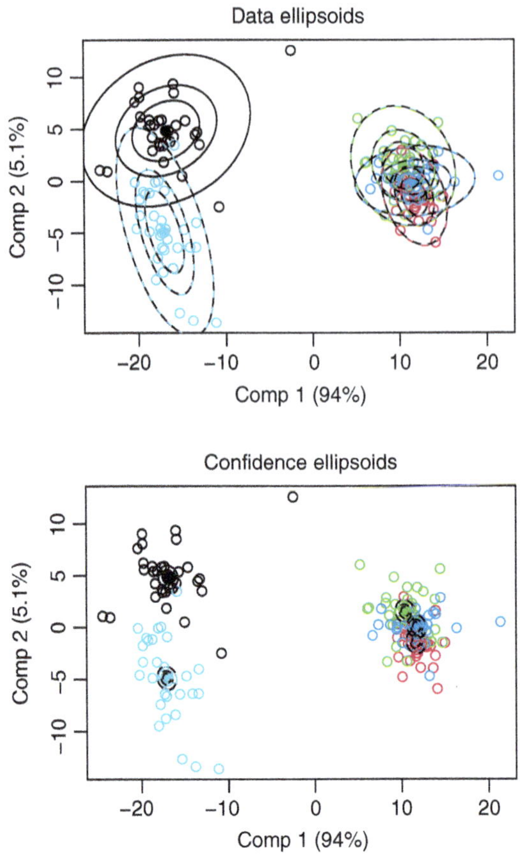

Figure 8.7 Data ellipsoids and confidence ellipsoids for the *Candy* effect.

8.6.1.5 Combined Effects

In some cases, it can be of interest to combine effects in ASCA. Here, we use an example with the Caldana data where we combine the *Light* effect with the *Time:Light* interaction using the *comb()* function.

```
# Load Caldana data
data(caldana)

mod.comb <- asca(compounds ~ time + comb(light + light:time),
                 data=caldana)
summary(mod.comb)
```

```
Anova Simultaneous Component Analysis fitted using 'lm' (Linear Model)
- SS type II, sum-coding, restricted model, least squares estimation
                  Sum.Sq.  Expl.var.(%)
time               154.58          9.69
light+light:time   349.64         21.92
Residuals         1091.14         68.39
```

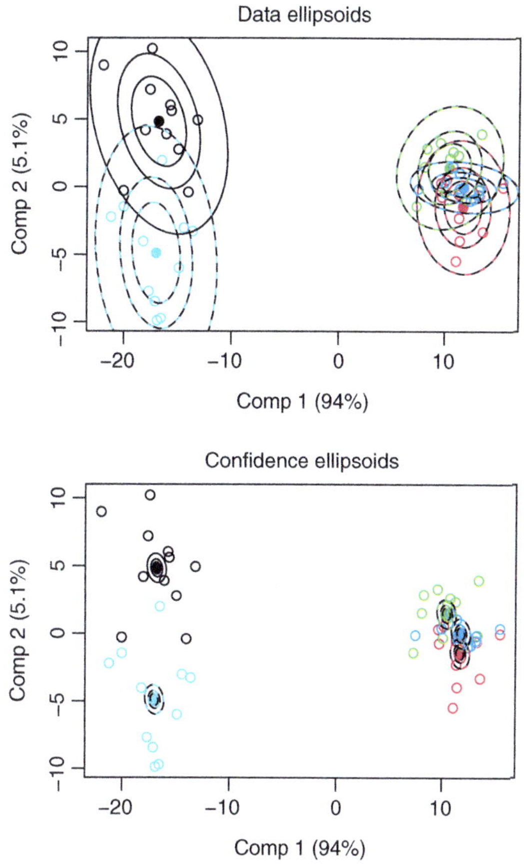

Figure 8.8 Data ellipsoids and confidence ellipsoids for the *Candy* effect in a mixed model.

When combined effects have a time factor, the scores can be plotted against time (see Figure 8.9).

```
# Scores plotted as a function of time
par.old <- par(mfrow=c(2,1), mar = c(4,4,1,1))
timeplot(mod.comb, factor="light", time="time", comb=2, comp=1,
     x_time=TRUE)
timeplot(mod.comb, factor="light", time="time", comb=2, comp=2,
     x_time=TRUE)
par(par.old)
```

8.6.1.6 Quantitative Effects

Quantitative effects, so-called covariates, can also be included in a model, though their use in ASCA is limited to ANOVA estimation and explained variance, not being used in subsequent PCA or permutation testing. We demonstrate this using the Caldana data again, but now recode the time effect to a quantitative effect, meaning it will be handled as a linear continuous effect.

```
caldanaNum <- caldana
caldanaNum$time <- as.numeric(as.character(caldanaNum$time))
mod.num <- asca(compounds ~ time*light, data = caldanaNum)
summary(mod.num)
```

```
Anova Simultaneous Component Analysis fitted using 'lm' (Linear Model)
- SS type II, sum-coding, restricted model, least squares estimation
              Sum.Sq.  Expl.var.(%)
time            55.42          3.47
light          102.49          6.42
light:time     109.56          6.87
Residuals     1327.90         83.23
```

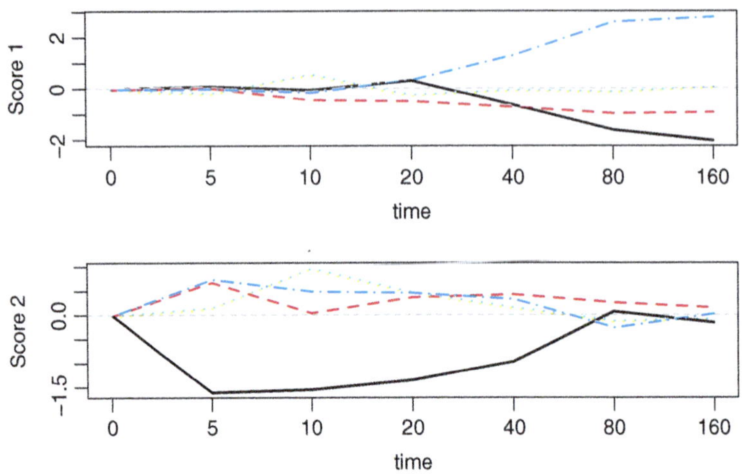

Figure 8.9 Scoreplot as a function of time for combined effects in ASCA.

8.6.2 ANOVA-PCA (APCA)

APCA differs from ASCA by adding the error term to the model before performing PCA instead of backprojecting errors afterward (see Figure 8.10).

```
# Fit APCA model
modp <- apca(assessment ~ candy*assessor, data=candies)
summary(modp)
```

```
Anova Principal Component Analysis fitted using 'lm' (Linear Model)
- SS type II, sum-coding, restricted model, least squares estimation
                  Sum.Sq.  Expl.var.(%)
candy            33416.66         74.48
assessor          1961.37          4.37
assessor:candy    3445.73          7.68
Residuals         6043.52         13.47
```

Plot scores and loadings (see Figure 8.10).

```
par.old <- par(mfrow=c(2,1), mar=c(4,4,2,1))
loadingplot(modp, scatter=TRUE, labels="names",
            main="Candy_loadings")
scoreplot(modp, main="Candy_scores")
par(par.old)
```

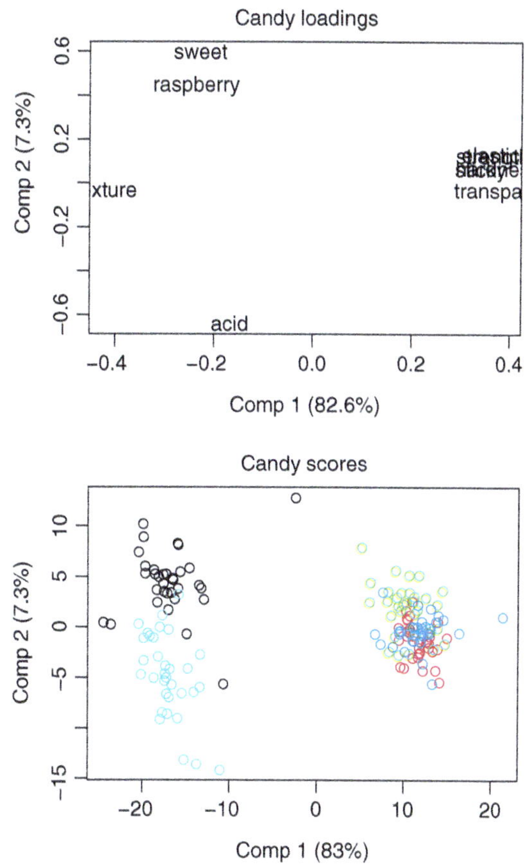

Figure 8.10 Loading plot and score plot of Candy effect in APCA model.

8.6.3 PC-ANOVA

In PC-ANOVA, a PCA is first applied to the data before the scores are subjected to ANOVA, effectively reversing the roles in ASCA. This means there will be one or more ANOVA tables in the summary of the output. In this example, we have chosen to use the number of components that explain at least 90% of the variation of the data.

```
mod <- pcanova(assessment ~ candy * assessor, data = candies,
                ncomp = 0.9)
print(mod)
```

```
PC-ANOVA - Principal Components Analysis of Variance

Call:
pcanova(formula = assessment ~ candy * assessor, data = candies,
ncomp = 0.9)
```

```
summary(mod)
```

```
PC-ANOVA fitted using 'lm' (Linear Model)
$`Comp. 1 (74.69%)`
Anova Table (Type II tests)

Response: assessment
                 Sum Sq  Df  F value  Pr(>F)
candy           31470.6   4 780.1762 < 2e-16 ***
assessor          224.9  10   2.2304 0.02089 *
candy:assessor    707.7  40   1.7545 0.01158 *
Residuals        1109.3 110
---
Signif. codes:  0 '***' 0.001 '**' 0.01 '*' 0.05 '.' 0.1 ' ' 1

$`Comp. 2 (9.38%)`
Anova Table (Type II tests)

Response: assessment
                Sum Sq  Df F value    Pr(>F)
candy           1573.8   4 33.1604 < 2.2e-16 ***
assessor         278.3  10  2.3455 0.0150274 *
candy:assessor 1053.3  40  2.2193 0.0005888 ***
Residuals       1305.2 110
---
Signif. codes:  0 '***' 0.001 '**' 0.01 '*' 0.05 '.' 0.1 ' ' 1

$`Comp. 3 (6.47%)`
Anova Table (Type II tests)

Response: assessment
                 Sum Sq  Df F value    Pr(>F)
candy            307.12   4   7.646 1.790e-05 ***
assessor        1006.62  10  10.024 8.574e-12 ***
candy:assessor   484.02  40   1.205    0.2229
Residuals       1104.61 110
---
Signif. codes:  0 '***' 0.001 '**' 0.01 '*' 0.05 '.' 0.1 ' ' 1
```

When creating score and loading plots for PC-ANOVA, the "global" scores and loadings will be shown, but the factors can still be used for coloring the symbols (see Figure 8.11).

```
par.old <- par(mfrow=c(2,1), mar=c(4,4,2,1))
loadingplot(mod, scatter=TRUE, labels="names",
            main="Global␣loadings")
scoreplot(mod, main="Global␣scores")
par(par.old)
```

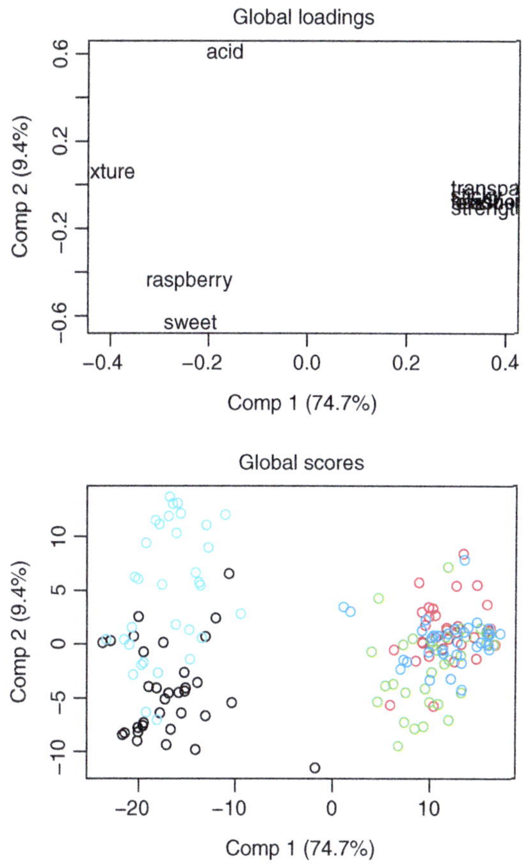

Figure 8.11 Loading plot and score plot for global PCA model in PC-ANOVA.

8.6.4 MSCA

Multilevel Simultaneous Component Analysis (MSCA) is a version of ASCA that assumes a single factor to be used as a between-individuals factor, while the within-individuals is assumed implicitly.

```
# Default MSCA model with a single factor
mod.msca <- msca(assessment ~ candy, data=candies)
summary(mod.msca)
```

```
Multilevel Simultaneous Component Analysis fitted using 'lm' (Linear Model)
- SS type II, sum-coding, restricted model, least squares estimation
         Sum.Sq.  Expl.var.(%)
Between 33416.66         74.48
Within  11450.62         25.52
```

Score plots can be created for the between-individuals factor and the within-individuals factor, and for each level of the within-individuals factor (see Figures 8.12–8.14, respectively).

```
# Scoreplot for between-individuals factor
scoreplot(mod.msca)

# Scoreplot of within-individuals factor
scoreplot(mod.msca, factor="within")
```

```
#.. per factor level
par.old <- par(mfrow=c(3,2), mar=c(4,4,2,1), mgp=c(2,0.7,0))
for(i in 1:length(mod.msca$scores.within))
 scoreplot(mod.msca, factor="within", within_level=i,
           main=paste0("Level: ", names(mod.msca$scores.within)[i]),
           panel.first=abline(v=0,h=0,col="gray",lty=2))
par(par.old)
```

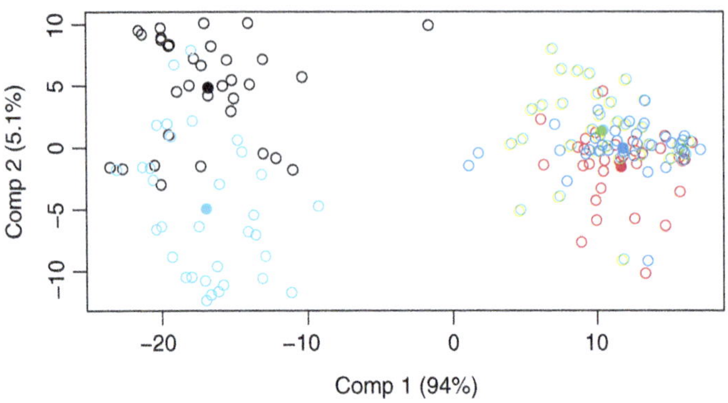

Figure 8.12 Score plot for between-individuals effect in MSCA.

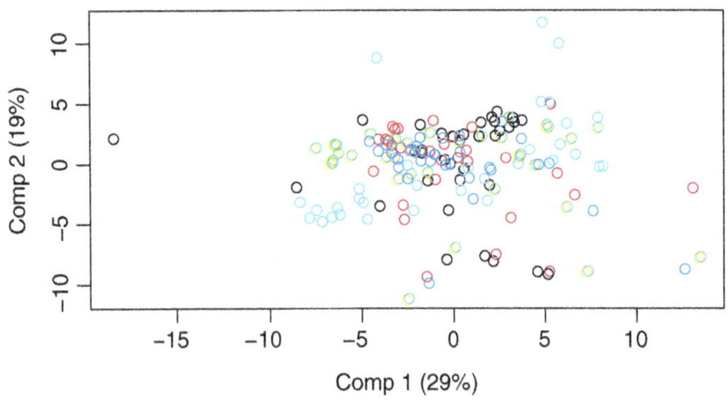

Figure 8.13 Score plot for within-individuals effect in MSCA.

8.6.5 LiMM-PCA

A version of LiMM-PCA is also implemented in HDANOVA. It combines REML-estimated mixed models with PCA and scales the back-projected errors according to degrees of freedom or effective dimensions (user choice) (see Figure 8.15).

```
# Default LiMM-PCA model with two factors and interaction,
     8 PCA components
mod.reml <- limmpca(assessment ~ candy*r(assessor), data=candies,
                            pca.in=8)
summary(mod.reml)
scoreplot(mod.reml, factor="candy")
```

```
LiMM-PCA fitted using 'lmm' (Linear Mixed Model)
- SS type II, sum-coding, restricted model, REML estimation
                    Sum.Sq.  Expl.var.(%)
candy              33415.98         74.73
assessor:candy       697.41          1.56
assessor             874.25          1.96
Residuals           7619.78         17.04
```

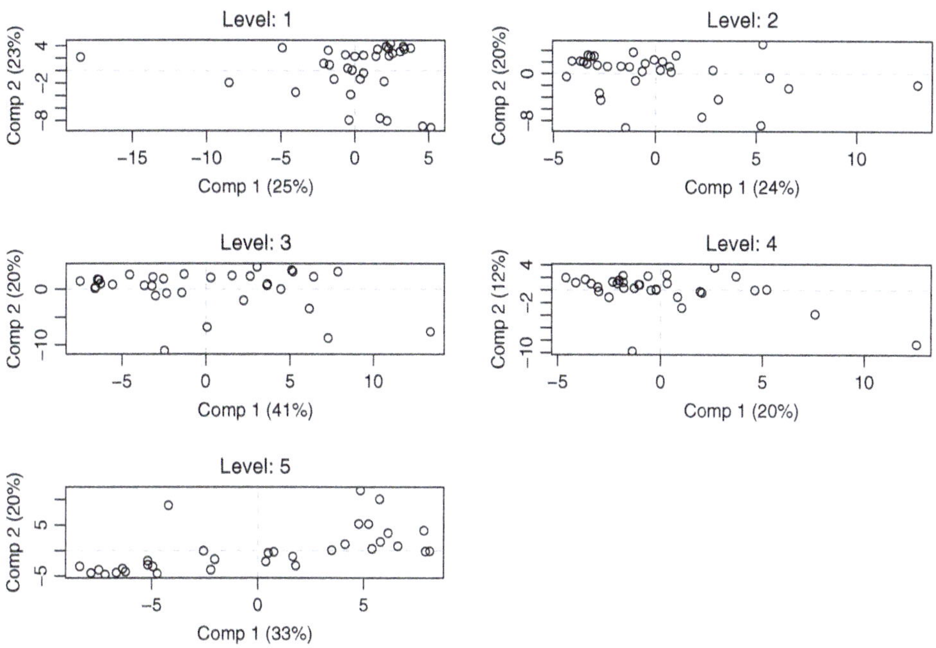

Figure 8.14 Score plot for within-individuals effect – one factor level at a time in MSCA.

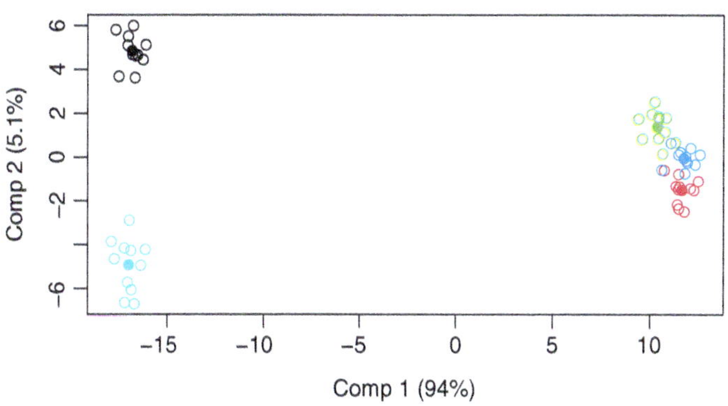

Figure 8.15 Score plot for fixed effect in LiMM-PCA.

We can also use least squares estimation without REML. This affects the random effects and scaling of backprojections (see Figure 8.16).

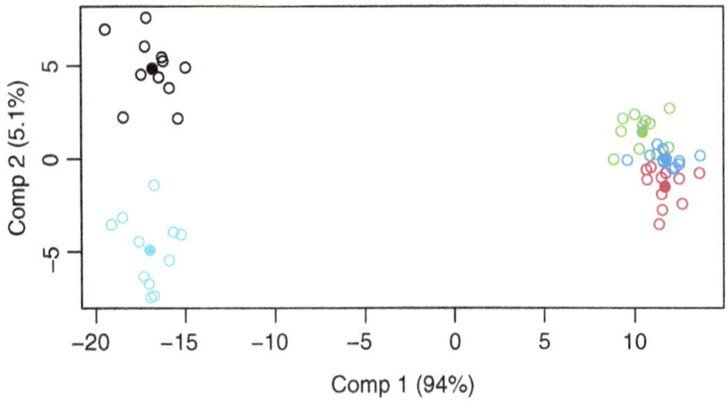

Figure 8.16 Score plot for fixed effect in LiMM-PCA when using least squares instead of REML estimation.

```
# LiMM-PCA with least squares estimation and 8 PCA components
mod.ls <- limmpca(assessment ~ candy*r(assessor), data=candies,
                  REML=NULL, pca.in=8)
summary(mod.ls)
scoreplot(mod.ls)
```

```
LiMM-PCA fitted using 'lmm' (Linear Mixed Model)
- SS type II, sum-coding, restricted model, least squares estimation
                Sum.Sq.  Expl.var.(%)
candy          33415.98         74.73
assessor        1948.75          4.36
assessor:candy  3419.04          7.65
Residuals       5934.46         13.27
```

8.6.6 Repeated Measures ASCA

We revisit the simulated data from the ANOVA vignette to demonstrate ASCA with repeated measures. The data are subset, a time effect is added, and the response is extended to the multivariate case.

```
set.seed(123)
# Original simulation
dat <- data.frame(
  feed   = factor(rep(rep(c("low","high"), each=6), 4)),
  breed  = factor(rep(c("NRF","Hereford","Angus"), 16)),
  bull   = factor(rep(LETTERS[1:4], each = 12)),
  daughter = factor(c(rep(letters[1:4], 3), rep(letters[5:8], 3),
                      rep(letters[9:12], 3), rep(letters[13:16], 3))),
  age    = round(rnorm(48, mean = 36, sd = 5))
)
dat$yield <- 150*with(dat, 10 + 3 * as.numeric(feed) + as.numeric(breed)
                      + 2 * as.numeric(bull) + 1 * as.numeric
                      (sample(dat$daughter, 48)) + 0.5 * age
                      + rnorm(48, sd = 2))
```

```
# Extended to repeated measures
long <- dat[c(1:4,9:12), c("feed", "daughter", "yield")]
long <- rbind(long, long, long)
long$daughter <- factor(long$daughter) # Remove redundant daughters
long$time  <- factor(rep(1:3, each=8))
long$yield <- long$yield + rnorm(24, sd = 100) + rep(c(-200,0,200),
                                                    each=8)
# Made multiresponse (no added structure, only noise)
long$yield <- I(matrix(rep(long$yield,10),nrow=length(long$yield),
               ncol=10)) + rnorm(length(long$yield)*10)
```

Analyzing the data with ASCA using the least squares approach.

```
# Least squares mixed model ASCA
mod.rm.asca <- asca(yield ~ r(daughter) + feed*r(time), data = long)
summary(mod.rm.asca)
```

```
Anova Simultaneous Component Analysis fitted using 'lmm' (Linear Mixed Model)
- SS type II, coding, restricted model, least squares estimation
                Sum.Sq. Expl.var.(%)
daughter   31997318.91        23.01
feed         306570.25         0.22
time        4309830.00         3.10
feed:time     57834.68         0.04
Residuals 102393907.07        73.63
```

Corresponding analysis using the LiMM-PCA approach with REML estimation.

```
# REML mixed model LiMM-PCA
mod.rm.limmpca <- limmpca(yield ~ r(daughter) + feed*r(time),
                  data = long, pca.in=2)
summary(mod.rm.limmpca)
```

```
LiMM-PCA fitted using 'lmm' (Linear Mixed Model)
- SS type II, coding, restricted model, REML estimation
                Sum.Sq. Expl.var.(%)
feed          306570.25         0.22
feed:time          0.86         0.00
daughter     7163945.08         5.15
time               0.15         0.00
Residuals 115642343.08        83.16
```

8.7 Alternative Methods

The examples shown here are HD-ANOVA methods that share aspects with ASCA. The corresponding theory is found in Chapters 4 and 6.

- Principal Response Curves (PRC) – Section 4.4
- Permutation-Based MANOVA (PERMANOVA) – Section 6.2.2

8.7.1 Principal Response Curves (PRC)

The PRC implementation we have wrapped is from the vegan package. A single treatment factor and a time factor are accepted together with a response matrix. The first level of the treatment factor is used as a reference level, effectively set to zero. As the vegan package is aimed at community ecology, the responses are called species. In our case, the responses are compounds from the Caldana data.

```
# Load Caldana data
data(caldana)

prc.cal <- prc(compounds ~ light * time, caldana)
summary(prc.cal)
```

```
Call:
prc(formula = compounds ~ light * time, data = caldana)
Species scores:
                      Alanine                      Valine
                     0.486693                    0.094963
                      Leucine                  Isoleucine
                    -0.355262                   -0.143003
                      Proline                      Serine
                     0.616952                   -0.027900
                    Threonine                beta-alanine
                     0.040614                   -0.392493
                Hydroxyproline                       GABA
                    -0.003806                   -0.587758
... [truncated output]

Coefficients for treatment + time:treatment interaction
which are contrasts to treatment Dark
rows are treatment, columns are time
                   0         5        10        20        40
Low Light  1.035e-15 -0.002369 -0.035096 -0.078319 -0.004197
Light      2.028e-15 -0.024598  0.066526 -0.054370  0.063304
High Light 1.041e-15 -0.007038 -0.006533  0.005982  0.200749
                 80     160
Low Light  0.07058 0.1179
Light      0.15538 0.2166
High Light 0.43379 0.4962
```

The default plot for PRC is a plot of *Treatment* + *Time:Treatment*. As we can see in Figure 8.17, the "Dark" level is the reference level from which the other levels are contrasted.

```
plot(prc.cal, species = FALSE, axis = 2, lwd = 4, legpos =
    "bottomright")
```

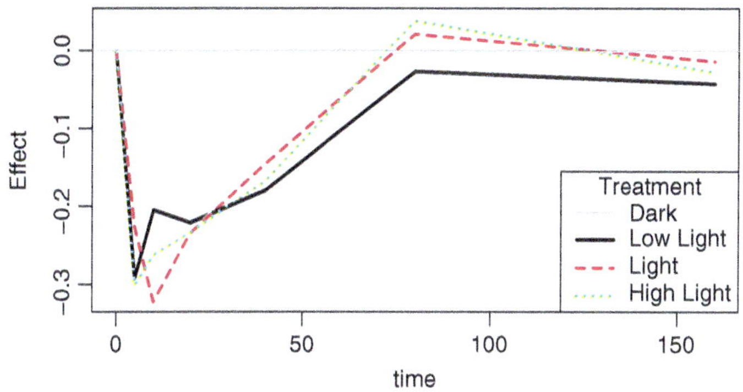

Figure 8.17 Plot of *Treatment* + *Time:Treatment* for PRC.

8.7.2 Permutation-Based MANOVA (PERMANOVA)

The PERMANOVA implementation we have wrapped is from the vegan package. Our wrapper takes care of the specialized formatting needed for the inputs. By default, 999 permutations are performed for the factors, and a standard Multivariate ANOVA is returned with permutation-based p-values.

```
permanova.cal <- permanova(compounds ~ light * time, caldana)
permanova.cal
```

```
Permutation test for adonis under reduced model
Terms added sequentially (first to last)
Permutation: free
Number of permutations: 999

vegan::adonis2(formula = formula, data = data)
             Df SumOfSqs      R2      F Pr(>F)
light         3  0.16005 0.07145 4.1266  0.001 ***
time          6  0.27300 0.12188 3.5194  0.001 ***
light:time   18  0.35885 0.16021 1.5421  0.001 ***
Residual    112  1.44793 0.64645
Total       139  2.23982 1.00000
---
Signif. codes:  0 '***' 0.001 '**' 0.01 '*' 0.05 '.' 0.1 ' ' 1
```

8.8 Software Packages

8.8.1 R Packages

In Table 8.1 we have collected methods related to this book that we could find in the Comprehensive R Archive Network (CRAN, (R Core Team, 2024)) and GitHub using simple search criteria. Only relevant methods are mentioned, though some packages have many

Table 8.1 R packages on CRAN having one or more HD-ANOVA methods.

Package name	Author	HD-ANOVA methods
ALASCA (GitHub)	Anders H. Jarmund	Longitudinal and cross-sectional analysis, ASCA+, LiMM-PCA, RM-ASCA+.
ASCATCATA	Michele Ricci	ASCA for Dynamic Sensory Analysis
ffmanova	Øyvind Langsrud	50-50 MANOVA
limpca (bioconductor)	Manon Martin	ASCA+ and APCA+
MANOVA.RM	Sarah Friedrich et al.	Resampling-Based MANOVA and Repeated Measures
MetaboAnalystR	Jianguo Xia et al.	ASCA with permutations
multiblock	Kristian Hovde Liland	ASCA with ellipsoids (ASCA parts will be superseded by HDANOVA)
vegan	Jari Oksanen et al.	PERMANOVA, Analysis of similarities ANOSIM, PRC

Table 8.2 MATLAB toolboxes and functions having one or more HD-ANOVA methods.

Package name	Author	HD-ANOVA methods
ASCA (GitHub)	Gooitzen Zwanenburg	ASCA
ASCA and ANOVA-TP	Federico Marini	ASCA and ANOVA-TP
MEDA toolbox	José Camacho Páez et al.	ASCA, APCA, VASCA, GASCA
PLS_toolbox	Eigenvector Research	ASCA+
rMANOVA (GitHub)	Jasper Engel	Regularized MANOVA

other methods in addition. Every package on CRAN is available directly using the canonical URL: https://CRAN.R-project.org/package=packagename, where "packagename" should be exchanged with the package of interest.

8.8.2 MATLAB Toolboxes

Most of the HD-ANOVA code available for MATLAB (MATLAB, 2024) comes in the form of collections of functions rather than full MATLAB toolboxes. This means that documentation is sometimes sparse. A non-exhaustive list of MATLAB code is found in Table 8.2.

8.8.3 Python

Very few relevant packages for Python (Rossum and Drake, 2009) have been found. A non-exhaustive list is found in Table 8.3.

Table 8.3 Python packages having one or more HD-ANOVA methods.

Package name HD-ANOVA methods	Author
ASCA (GitHub) ASCA	Sin Yong Teng
statsmodels MANOVA	Skipper Seabold et al.

References

Aalaei, K., Khakimov, B., De Gobba, C., and Ahrné, L. (2021) Digestion patterns of proteins in pasteurized and ultra-high temperature milk using in vitro gastric models of adult and elderly. *Journal of Food Engineering*, 292, 110305.

Abegaz, F., Abedini, D., White, F., Guerrieri, A., Zancarini, A., Dong, L., Westerhuis, J.A., van Eeuwijk, F., Bouwmeester, H., and Smilde, A.K. (2024) A strategy for differential abundance analysis of sparse microbiome data with group-wise structured zeros. *Scientific Reports*, 14(1), 12433.

Adam, M., Beyer, G., Christiansen, N., Kamlage, B., Pilarsky, C., Distler, M., Fahlbusch, T., Chromik, A., Klein, F., Bahra, M. *et al.* (2021) Identification and validation of a multivariable prediction model based on blood plasma and serum metabolomics for the distinction of chronic pancreatitis subjects from non-pancreas disease control subjects. *Gut*, 70(11), 2150–2158.

Adobor, S., Banniza, S., Vandenberg, A., and Purves, R.W. (2023) Untargeted profiling of secondary metabolites and phytotoxins associated with stemphylium blight of lentil. *Planta*, 257(4), 73.

Aitchison, J. (1986) *The Statistical Analysis of Compositional Data*, Chapman and Hall.

Aitchison, J. (1994) Principles of Compositional Data Analysis. *Lecture Notes-Monograph Series*, pp. 73–81.

Ali, N., Jansen, J., van den Doel, A., Tinnevelt, G.H., and Bocklitz, T. (2020) WE-ASCA: The weighted-effect ASCA for analyzing unbalanced multifactorial designs—a Raman spectra-based example. *Molecules*, 26(1), 66.

Alinaghi, M., Bertram, H.C., Brunse, A., Smilde, A.K., and Westerhuis, J.A. (2020) Common and distinct variation in data fusion of designed experimental data. *Metabolomics*, 16(1), 2.

Alvarez-Mora, I., Bolliet, V., Lopez-Herguedas, N., Olivares, M., Monperrus, M., and Etxebarria, N. (2023) Metabolomics to study the sublethal effects of diazepam and irbesartan on glass eels (*Anguilla anguilla*). *Aquatic Toxicology*, 259, 106547.

Amigo, J.M., Del Olmo, A., Engelsen, M.M., Lundkvist, H., and Engelsen, S.B. (2019) Staling of white wheat bread crumb and effect of maltogenic α-amylases. Part 2: Monitoring the staling process by using near infrared spectroscopy and chemometrics. *Food Chemistry*, 297, 124946.

Anderson, M.J. (2001) A new method for non-parametric multivariate analysis of variance. *Austral Ecology*, 26, 32–46.

Anderson, M.J. (2001) Permutation tests for univariate or multivariate analysis of variance and regression. *Canadian Journal of Fisheries and Aquatic Sciences*, 58, 626–639.

Anderson, M.J. (2005) Permutational multivariate analysis of variance, a computer program. *Department of Statistics, University of Auckland, Auckland*, 26, 32–46.

Anderson, M.J. (2017) *Permutational Multivariate Analysis of Variance (PERMANOVA)*, John Wiley and Sons, Ltd, pp. 1–15.

Analysis of Variance for High-Dimensional Data: Applications in Life, Food and Chemical Sciences, First Edition. Age K. Smilde, Federico Marini, Johan A. Westerhuis, and Kristian H. Liland.

Anderson, M.J. and Ter Braak, C. (2003) Permutation tests for multi-factorial analysis of variance. *Journal of Statistical Computation and Simulation*, 73(2), 85–113.

Anderson, M.J. and Walsh, D.C. (2013) PERMANOVA, ANOSIM, and the Mantel test in the face of heterogeneous dispersions: What null hypothesis are you testing? *Ecological Monographs*, 83(4), 557–574.

Apolinar-Valiente, R., Williams, P., Mazerolles, G., Romero-Cascales, I., Gómez-Plaza, E., López-Roca, J.M., Ros-García, J.M., and Doco, T. (2014) Effect of enzyme additions on the oligosaccharide composition of Monastrell red wines from four different wine-growing origins in Spain. *Food Chemistry*, 156, 151–159.

Armagan, A., Dunson, D.B., and Lee, J. (2013) Generalized double Pareto shrinkage. *Statistica Sinica*, 23(1), 119.

Artegoitia, V.M., Newman, J.W., Foote, A.P., Shackelford, S.D., King, D.A., Wheeler, T.L., Lewis, R.M., and Freetly, H.C. (2022) Non-invasive metabolomics biomarkers of production efficiency and beef carcass quality traits. *Scientific Reports*, 12(1), 231.

Aru, V., Khakimov, B., Sørensen, K.M., Chikwati, E.M., Kortner, T.M., Midtlyng, P., Krogdahl, Å., and Engelsen, S.B. (2021) The plasma metabolome of Atlantic salmon as studied by ^{1}H NMR spectroscopy using standard operating procedures: Effect of aquaculture location and growth stage. *Metabolomics*, 17(6), 50.

Ashrafi, M., Xu, Y., Muhamadali, H., White, I., Wilkinson, M., Hollywood, K., Baguneid, M., Goodacre, R., and Bayat, A. (2020) A microbiome and metabolomic signature of phases of cutaneous healing identified by profiling sequential acute wounds of human skin: An exploratory study. *PLoS One*, 15(2), e0229545.

Azzollini, A., Boggia, L., Boccard, J., Sgorbini, B., Lecoultre, N., Allard, P.M., Rubiolo, P., Rudaz, S., Gindro, K., Bicchi, C. *et al.* (2018) Dynamics of metabolite induction in fungal co-cultures by metabolomics at both volatile and non-volatile levels. *Frontiers in Microbiology*, 9, 72.

Barla, I., Efentakis, P., Lamprou, S., Gavriatopoulou, M., Dimopoulos, M.A., Terpos, E., Andreadou, I., Thomaidis, N., and Gikas, E. (2023) Metabolomics point out the effects of carfilzomib on aromatic amino acid biosynthesis and degradation. *International Journal of Molecular Sciences*, 24(18), 13966.

Barnes, R.J., Dhanoa, M.S., and Lister, S.J. (1989) Standard normal variate transformation and de-trending of near-infrared diffuse reflectance spectra. *Applied Spectroscopy*, 43(5), 772–777.

Barnett, I. and Zhang, M. (2018) Discrimination of brands of gasoline by using DART-MS and chemometrics. *Forensic Chemistry*, 10, 58–66.

Becchi, P.P., Rocchetti, G., García-Pérez, P., Michelini, S., Pizzamiglio, V., and Lucini, L. (2024) Untargeted metabolomics and machine learning unveil quality and authenticity interactions in grated Parmigiano Reggiano PDO cheese. *Food Chemistry*, 447, 138938.

Bester, R., Stander, Z., Mason, S., Howatson, G., Clifford, T., Stevenson, E.J., and Loots, D.T. (2023) The metabolic recovery of marathon runners: An untargeted ^{1}H-NMR metabolomics perspective. *Frontiers in Physiology*, 14, 1117687.

Bevilacqua, M., Bucci, R., Materazzi, S., and Marini, F. (2013) Application of near infrared (NIR) spectroscopy coupled to chemometrics for dried egg-pasta characterization and egg content quantification. *Food Chemistry*, 140(4), 726–734.

Bevilacqua, M., Bro, R., Marini, F., Rinnan, Å., Rasmussen, M.A., and Skov, T. (2017) Recent chemometrics advances for foodomics. *TrAC Trends in Analytical Chemistry*, 96, 42–51.

Blanchet, L., Smolinska, A., Baranska, A., Tigchelaar, E., Swertz, M., Zhernakova, A., Dallinga, J.W., Wijmenga, C., and van Schooten, F.J. (2017) Factors that influence the volatile organic compound content in human breath. *Journal of Breath Research*, 11(1), 016013.

Blyth, C.R. (1972) On Simpson's paradox and the sure-thing principle. *Journal of the American Statistical Association*, 67(338), 364–366.

Boccard, J. and Rudaz, S. (2016) Exploring omics data from designed experiments using analysis of variance multiblock orthogonal partial least squares. *Analytica Chimica Acta*, 920, 18–28.

Boccard, J., Tonoli, D., Strajhar, P., Jeanneret, F., Odermatt, A., and Rudaz, S. (2019) Removal of batch effects using stratified subsampling of metabolomic data for in vitro endocrine disruptors screening. *Talanta*, 195, 77–86.

Bondia-Pons, I., Caellas, N., Abete, I., Rodríguez, M.Á., Perez-Cornago, A., Navas-Carretero, S., Zulet, M.Á., Correig, X., and Martínez, J.A. (2013) Nutri-metabolomics: Subtle serum metabolic differences in healthy subjects by NMR-based metabolomics after a short-term nutritional intervention with two tomato sauces. *Omics: A Journal of Integrative Biology*, 17(12), 611–618.

Bonnefoy, C., Fildier, A., Buleté, A., Bordes, C., Garric, J., and Vulliet, E. (2019) Untargeted analysis of nanoLC-HRMS data by ANOVA-PCA to highlight metabolites in *Gammarus fossarum* after in vivo exposure to pharmaceuticals. *Talanta*, 202, 221–229.

Boots, A.W., Smolinska, A., Van Berkel, J.J.B.N., Fijten, R.R.R., Stobberingh, E.E., Boumans, M.L.L., Moonen, E.J., Wouters, E.F.M., Dallinga, J.W., and Van Schooten, F.J. (2014) Identification of microorganisms based on headspace analysis of volatile organic compounds by gas chromatography–mass spectrometry. *Journal of Breath Research*, 8(2), 027106.

Borg, I. and Groenen, J. (2010) *Modern Multidimensional Scaling*, Springer, New York, NY.

Boshuizen, H.C. and Te Beest, D.E. (2023) Pitfalls in the statistical analysis of microbiome amplicon sequencing data. *Molecular Ecology Resources*, 23(3), 539–548.

Bouveresse, D.J.R., Pinto, R.C., Schmidtke, L.M., Locquet, N., and Rutledge, D.N. (2011) Identification of significant factors by an extension of ANOVA–PCA based on multi-block analysis. *Chemometrics and Intelligent Laboratory Systems*, 106(2), 173–182.

ter Braak, C.J. (2023) Redundancy analysis includes analysis of variance-simultaneous component analysis (ASCA) and outperforms its extensions. *Chemometrics and Intelligent Laboratory Systems*, 240, 104898.

Brasili, E., Pratico, G., Marini, F., Valletta, A., Capuani, G., Sciubba, F., Miccheli, A., and Pasqua, G. (2014) A non-targeted metabolomics approach to evaluate the effects of biomass growth and chitosan elicitation on primary and secondary metabolism of *Hypericum perforatum* in vitro roots. *Metabolomics*, 10(6), 1186–1196.

Brasili, E., Miccheli, A., Marini, F., Praticò, G., Sciubba, F., Di Cocco, M.E., Cechinel Filho, V., Tocci, N., Valletta, A., and Pasqua, G. (2016) Metabolic profile and root development of *Hypericum perforatum* L. in vitro roots under stress conditions due to chitosan treatment and culture time. *Frontiers in Plant Science*, 7, 507.

Bro, R., Kjeldahl, K., Smilde, A., and Kiers, H. (2008) Cross-validation of component models: A critical look at current methods. *Analytical and Bioanalytical Chemistry*, 390(5), 1241–1251.

Caellas, N., Solà-Alberich, R., Brezmes, J., Mallol, R., Valls, R.M., Rodriguez, M.A., Vinaixa, M., Anguera, A., and Correig, X. (2013) Use of multivariate chemometric algorithms on ^{1}H NMR data to assess a soluble fiber (*Plantago ovata husk*) nutritional intervention. *Chemometrics and Intelligent Laboratory Systems*, 121, 1–8.

Caldana, C., Degenkolbe, T., Cuadros-Inostroza, A., Klie, S., Sulpice, R., Leisse, A., Steinhauser, D., Fernie, A., Willmitzer, L., and Hannah, M. (2011) High-density kinetic analysis of the metabolomic and transcriptomic response of Arabidopsis to eight environmental conditions. *The Plant Journal*, 67(5), 869–884.

Camacho, J. and Ferrer, A. (2012) Cross-validation in PCA models with the element-wise *k*-fold (*ekf*) algorithm: Theoretical aspects. *Journal of Chemometrics*, 26(7), 361–373.

Camacho, J. and Ferrer, A. (2014) Cross-validation in PCA models with the element-wise *k*-fold (*ekf*) algorithm: Practical aspects. *Chemometrics and Intelligent Laboratory Systems*, 131, 37–50.

Camacho, J., Rodríguez-Gómez, R.A., and Saccenti, E. (2017) Group-wise principal component analysis for exploratory data analysis. *Journal of Computational and Graphical Statistics*, 26(3), 501–512.

Camacho, J., Díaz, C., and Sánchez-Rovira, P. (2023) Permutation tests for ASCA in multivariate longitudinal intervention studies. *Journal of Chemometrics*, 37(7), e3398.

Camacho, J., Vitale, R., Morales-Jimenez, D., and Gómez-Llorente, C. (2023) Variable-selection ANOVA simultaneous component analysis (VASCA). *Bioinformatics*, 39(1), 795.

Casella, G. (2008) *Statistical Design*, Springer, New York.

Chang, W.T., Thissen, U., Ehlert, K.A., Koek, M.M., Jellema, R.H., Hankemeier, T., van der Greef, J., and Wang, M. (2006) Effects of growth conditions and processing on *Rehmannia glutinosa* using fingerprint strategy. *Planta Medica*, 72(05), 458–467.

Chen, P., Lu, Y., and Harrington, P.B. (2008) Biomarker profiling and reproducibility study of MALDI-MS measurements of *Escherichia coli* by analysis of variance- principal component analysis. *Analytical Chemistry*, 80(5), 1474–1481.

Chen, P., Harnly, J.M., and Lester, G.E. (2010) Flow injection mass spectral fingerprints demonstrate chemical differences in Rio Red grapefruit with respect to year, harvest time, and conventional versus organic farming. *Journal of Agricultural and Food Chemistry*, 58(8), 4545–4553.

Cheng, W., Sørensen, K.M., Mongi, R.J., Ndabikunze, B.K., Chove, B.E., Sun, D.W., and Engelsen, S.B. (2019) A comparative study of mango solar drying methods by visible and near-infrared spectroscopy coupled with ANOVA-simultaneous component analysis (ASCA). *LWT- Food Science and Technology*, 112, 108214.

Christian, G.D. and O'Reilly, J.E. (1988) *Instrumental Analysis*, Pearson Education (US), Upper Saddle River.

Clarke, K. (1993) Non-parametric multivariate analyses of changes in community structure. *Australian Journal of Ecology*, 18, 117–143.

Climaco-Pinto, R., Barros, A., Locquet, N., Schmidtke, L., and Rutledge, D. (2009) Improving the detection of significant factors using ANOVA-PCA by selective reduction of residual variability. *Analytica Chimica Acta*, 653(2), 131–142.

Colinet, H. and Renault, D. (2018) Similar post-stress metabolic trajectories in young and old flies. *Experimental Gerontology*, 102, 43–50.

Combrink, M., Du Preez, I., Ronacher, K., Walzl, G., and Loots, D.T. (2019) Time-dependent changes in urinary metabolome before and after intensive phase tuberculosis therapy: A pharmacometabolomics study. *Omics: A Journal of Integrative Biology*, 23(11), 560–572.

Coombs, C.H. (1964) *A Theory of Data*, John Wiley & Sons, New York.

Cornfield, J. and Tukey, J.W. (1956) Average values of mean squares in factorials. *The Annals of Mathematical Statistics*, 27(4), 907–949.

Cui, M., Trimigno, A., Aru, V., Khakimov, B., and Engelsen, S.B. (2020) Human Faecal ^{1}H NMR metabolomics: Evaluation of solvent and sample processing on coverage and reproducibility of signature metabolites. *Analytical Chemistry*, 92(14), 9546–9555.

Cui, M., Trimigno, A., Aru, V., Rasmussen, M.A., Khakimov, B., and Engelsen, S.B. (2021) Influence of age, sex, and diet on the human fecal metabolome investigated by ^{1}H NMR spectroscopy. *Journal of Proteome Research*, 20(7), 3642–3653.

Dai, D., Gao, Y., Chen, J., Huang, Y., Zhang, Z., and Xu, F. (2016) Time-resolved metabolomics analysis of individual differences during the early stage of lipopolysaccharide-treated rats. *Scientific Reports*, 6(1), 34136.

D'Alessandro, A., Ballestrieri, D., Strani, L., Cocchi, M., and Durante, C. (2021) Characterization of basil volatile fraction and study of its agronomic variation by ASCA. *Molecules*, 26(13), 3842.

Dall'Acqua, S., Stocchero, M., Clauser, M., Boschiero, I., Ndoum, E., Schiavon, M., Mammi, S., and Schievano, E. (2014) Changes in urinary metabolic profile after oral administration of curcuma extract in rats. *Journal of Pharmaceutical and Biomedical Analysis*, 100, 348–356.

Davies, P.T. and Tso, M.S. (1982) Procedures for reduced-rank regression. *Journal of the Royal Statistical Society: Series C (Applied Statistics)*, 31(3), 244–255.

De Jong, S. (1995) PLS shrinks. *Journal of Chemometrics*, 9(4), 323–326.

De Luca, S., De Filippis, M., Bucci, R., Magrì, A.D., Magrì, A.L., and Marini, F. (2016) Characterization of the effects of different roasting conditions on coffee samples of different geographical origins by HPLC-DAD, NIR and chemometrics. *Microchemical Journal*, 129, 348–361.

Díaz, C., González-Olmedo, C., Díaz-Beltrán, L., Camacho, J., Mena García, P., Martín-Blázquez, A., Fernández-Navarro, M., Ortega-Granados, A.L., Gálvez-Montosa, F., Marchal, J.A. *et al.* (2022) Predicting dynamic response to neoadjuvant chemotherapy in breast cancer: A novel metabolomics approach. *Molecular Oncology*, 16(14), 2658–2671.

Dieterle, F., Ross, A., Schlotterbeck, G., and Senn, H. (2006) Probabilistic quotient normalization as robust method to account for dilution of complex biological mixtures. Application in ^{1}H NMR metabonomics. *Analytical Chemistry*, 78(13), 4281–4290.

Dobson, A.J. and Barnett, A.G. (2018) *An Introduction to Generalized Linear Models*, CRC Press, Boca Raton.

Draper, N.R. and Smith, H. (1998) *Applied Regression Analysis*, vol. 326, John Wiley & Sons, Inc., New York, 3rd edn.

Du, Y., Zheng, H., Xia, H., Zhao, L., Hu, W., Bai, G., Yan, Z., and Gao, H. (2017) Early effect of amyloid β-peptide on hippocampal and serum metabolism in rats studied by an integrated method of NMR-based metabolomics and ANOVA-simultaneous component analysis. *BioMed Research International*, 2017(1), 3262495.

Dumas, T., Boccard, J., Gomez, E., Fenet, H., and Courant, F. (2020) Multifactorial analysis of environmental metabolomic data in ecotoxicology: Wild marine mussel exposed to WWTP effluent as a case study. *Metabolites*, 10(7), 269.

Durbin, B.P., Hardin, J.S., Hawkins, D.M., and Rocke, D.M. (2002) A variance-stabilizing transformation for gene-expression microarray data. *Bioinformatics*, 18(suppl_1), S105–S110.

Dysvik, A., La Rosa, S.L., Buffetto, F., Liland, K.H., Myhrer, K.S., Rukke, E.O., Wicklund, T., and Westereng, B. (2019) Secondary lactic acid bacteria fermentation with wood-derived xylooligosaccharides as a tool to expedite sour beer production. *Journal of Agricultural and Food Chemistry*, 68(1), 301–314.

Ebrahimi, P., Larsen, F.H., Jensen, H.M., Vogensen, F.K., and Engelsen, S.B. (2016) Real-time metabolomic analysis of lactic acid bacteria as monitored by in vitro NMR and chemometrics. *Metabolomics*, 12(4), 77.

Eckart, C. and Young, G. (1936) The approximation of one matrix by another of lower rank. *Psychometrika*, 1, 211–218.

Efron, B. (1979) Bootstrap methods: Another look at the jackknife. *The Annals of Statistics*, 7(1), 1–26.

Elboutachfaiti, R., Molinié, R., Mathiron, D., Maillot, Y., Fontaine, J.X., Pilard, S., Quéro, A., Brasselet, C., Dols-Lafargue, M., Delattre, C. *et al.* (2022) Secondary metabolism rearrangements in *Linum usitatissimum* L. After biostimulation of roots with COS oligosaccharides from fungal cell wall. *Molecules*, 27(7), 2372.

Ellekjær, M.R., Ilseng, M.A., and Næs, T. (1996) A case study of the use of experimental design and multivariate analysis in product improvement. *Food Quality and Preference*, 7(1), 29–36.

Endrizzi, I., Gasperi, F., Rodbotten, M., and Næs, T. (2013) Permutation testing for validating PCA, in *SIS 2013 Statistical Conference Advances in Latent Variables; Methods, Models and Applications*.

Engel, J., Blanchet, L., Bloemen, B., Van den Heuvel, L.P., Engelke, U.H.F., Wevers, R.A., and Buydens, L.M.C. (2015) Regularized MANOVA (rMANOVA) in untargeted metabolomics. *Analytica Chimica Acta*, 899, 1–12.

Engel, J., Postma, G., van Peufflik, I., Blanchet, L., and Buydens, L. (2015) Pseudo-sample trajectories for variable interaction detection in dissimilarity partial least squares. *Chemometrics and Intelligent Laboratory Systems*, 146, 89–101.

Engel, J., Houthuijs, K.J., Vasiliou, V., and Charkoftaki, G. (2020) Regularized multivariate analysis of variance, in *Comprehensive Chemometrics (Second Edition)* (eds S. Brown, R. Tauler, and B. Walczak), Elsevier, Oxford, pp. 479–494.

Erazo Solorzano, C.Y., Disca, V., Muoz-Redondo, J.M., Tuárez García, D.A., Sánchez-Parra, M., Carrilo Zenteno, M.D., Moreno-Rojas, J.M., and Rodríguez-Solana, R. (2023) Effect of drying technique on the volatile content of Ecuadorian bulk and fine-flavor cocoa. *Foods*, 12(5), 1065.

Erdős, B., Westerhuis, J.A., Adriaens, M.E., O'Donovan, S.D., Xie, R., Singh-Povel, C.M., Smilde, A.K., and Arts, I.C. (2023) Analysis of high-dimensional metabolomics data with complex temporal dynamics using RM-ASCA+. *PLoS Computational Biology*, 19(6), e1011221.

Evans, C., Hardin, J., and Stoebel, D.M. (2018) Selecting between-sample RNA-Seq normalization methods from the perspective of their assumptions. *Briefings in Bioinformatics*, 19(5), 776–792.

Fall, F., Mamede, L., Vast, M., De Tullio, P., Hayette, M.P., Michels, P.A.M., Frédérich, M., Govaerts, B., and Quetin-Leclercq, J. (2024) First comprehensive untargeted metabolomics study of suramin-treated *Trypanosoma brucei*: An integrated data analysis workflow from multifactor data modelling to functional analysis. *Metabolomics*, 20(2), 25.

Fazelzadeh, P., Hangelbroek, R.W.J., Tieland, M., de Groot, L.C.P.G.M., Verdijk, L.B., Van Loon, L.J.C., Smilde, A.K., Alves, R.D.A.M., Vervoort, J., Müller, M. *et al.* (2016) The muscle metabolome differs between healthy and frail older adults. *Journal of Proteome Research*, 15(2), 499–509.

Feizi, N., Seraj, M., Tajali, R., Shavandi, S.R., and Parastar, H. (2020) Chemometric techniques coupled with NMR for matabolic profiling of lettuce exposed to polycyclic aromatic hydrocarbones. *Analytical Biochemistry*, 611, 113945.

Fernando, D., Samita, S., and Peiris, T. (2021) Identifying panel inconsistency in sensory profiles using multivariate analysis of variance (MANOVA) and follow–up canonical variate Analysis (CVA). *Tropical Agricultural Research*, 33(1), 80–91.

Filzmoser, P. and Hron, K. (2009) Correlation analysis for compositional data. *Mathematical Geosciences*, 41, 905–919.

Filzmoser, P., Hron, K., and Templ, M. (2018) *Applied Compositional Data Analysis*, Springer, Cham.

Firmani, P., Vitale, R., Ruckebusch, C., and Marini, F. (2020) ANOVA-simultaneous component analysis modelling of low-level-fused spectroscopic data: A food chemistry case-study. *Analytica Chimica Acta*, 1125, 308–314.

Fisher, R.A. (1935) *The Design of Experiments*, Oliver & Boyd, Edinburgh & London, 2nd edn.

Fisher, R.A. and Mackenzie, W.A. (1923) Studies in crop variation. II. The manurial response of different potato varieties. *The Journal of Agricultural Science*, 13(3), 311–320.

Friedman, J.H. and Meulman, J.J. (2004) Clustering objects on subsets of attributes (with discussion). *Journal of the Royal Statistical Society: Series B (Statistical Methodology)*, 66(4), 815–849.

Friedman, J., Hastie, T., and Tibshirani, R. (2008) Sparse inverse covariance estimation with the graphical lasso. *Biostatistics*, 9(3), 432–441.

Friendly, M., Monette, G., and Fox, J. (2013) Elliptical insights: Understanding statistical methods through elliptical geometry. *Statistical Science*, 28(1), 1–39.

Fritzsch, R., Donaldson, P.M., Greetham, G.M., Towrie, M., Parker, A.W., Baker, M.J., and Hunt, N.T. (2018) Rapid screening of DNA–ligand complexes via 2D-IR spectroscopy and ANOVA–PCA. *Analytical Chemistry*, 90(4), 2732–2740.

Fu, W.J. (1998) Penalized regressions: The bridge versus the lasso. *Journal of Computational and Graphical Statistics*, 7(3), 397–416.

Gagnebin, Y., Jaques, D.A., Rudaz, S., de Seigneux, S., Boccard, J., and Ponte, B. (2020) Exploring blood alterations in chronic kidney disease and haemodialysis using metabolomics. *Scientific Reports*, 10(1), 19502.

Gagnebin, Y., Pezzatti, J., Lescuyer, P., Boccard, J., Ponte, B., and Rudaz, S. (2020) Combining the advantages of multilevel and orthogonal partial least squares data analysis for longitudinal metabolomics: Application to kidney transplantation. *Analytica Chimica Acta*, 1099, 26–38.

Galineau, L., Bourdin-Pintueles, A., Bodard, S., Busson, J., Nadal-Desbarats, L., Lefèvre, A., Emond, P., and Mavel, S. (2024) Temporal metabolomics state in pregnant rat: Analysis of amniotic fluid, placenta, and maternal plasma at embryonic and fetal time points. *Placenta*, 150, 22–30.

García-Aguilera, M.E., Delgado-Altamirano, R., Villalón, N., Ruiz-Terán, F., García-Garnica, M.M., Ocaa-Ríos, I., Rodríguez de San Miguel, E., and Esturau-Escofet, N. (2023) Study of the stability of wine samples for ^{1}H-NMR metabolomic profile analysis through chemometrics methods. *Molecules*, 28(16), 5962.

Gargallo-Garriga, A., Ayala-Roque, M., Sardans, J., Bartrons, M., Granda, V., Sigurdsson, B.D., Leblans, N.W., Oravec, M., Urban, O., Janssens, I.A. *et al.* (2017) Impact of soil warming on the plant metabolome of icelandic grasslands. *Metabolites*, 7(3), 44.

Garreta-Lara, E., Campos, B., Barata, C., Lacorte, S., and Tauler, R. (2018) Combined effects of salinity, temperature and hypoxia on *Daphnia magna* metabolism. *Science of the Total Environment*, 610, 602–612.

Garrido, A., Engel, J., Mumm, R., Conde, A., Cunha, A., and De Vos, R.C.H. (2021) Metabolomics of photosynthetically active tissues in white grapes: Effects of light microclimate and stress mitigation strategies. *Metabolites*, 11(4), 205.

Georges des Aulnois, M., Réveillon, D., Robert, E., Caruana, A., Briand, E., Guljamow, A., Dittmann, E., Amzil, Z., and Bormans, M. (2020) Salt shock responses of *Microcystis* revealed through physiological, transcript, and metabolomic analyses. *Toxins*, 12(3), 192.

Ghaffari, M.H., Jahanbekam, A., Sadri, H., Schuh, K., Dusel, G., Prehn, C., Adamski, J., Koch, C., and Sauerwein, H. (2019) Metabolomics meets machine learning: Longitudinal metabolite profiling in serum of normal versus overconditioned cows and pathway analysis. *Journal of Dairy Science*, 102(12), 11561–11585.

Gloor, G.B., Macklaim, J.M., Pawlowsky-Glahn, V., and Egozcue, J.J. (2017) Microbiome datasets are compositional: And this is not optional. *Frontiers in Microbiology*, 8, 2224.

Gómez-Canela, C., Prats, E., Pia, B., and Tauler, R. (2017) Assessment of chlorpyrifos toxic effects in zebrafish (*Danio rerio*) metabolism. *Environmental Pollution*, 220, 1231–1243.

Gómez-Canela, C., Prats, E., Lacorte, S., Raldúa, D., Pia, B., and Tauler, R. (2018) Metabolomic changes induced by nicotine in adult zebrafish skeletal muscle. *Ecotoxicology and Environmental Safety*, 164, 388–397.

González-Ruiz, V., Pezzatti, J., Roux, A., Stoppini, L., Boccard, J., and Rudaz, S. (2017) Unravelling the effects of multiple experimental factors in metabolomics, analysis of human neural cells with hydrophilic interaction liquid chromatography hyphenated to high resolution mass spectrometry. *Journal of Chromatography A*, 1527, 53–60.

González-Ruiz, V., Schvartz, D., Sandström, J., Pezzatti, J., Jeanneret, F., Tonoli, D., Boccard, J., Monnet-Tschudi, F., Sanchez, J.C., and Rudaz, S. (2019) An integrative multi-omics workflow to address multifactorial toxicology experiments. *Metabolites*, 9(4), 79.

Gorrochategui, E., Lacorte, S., Tauler, R., and Martin, F.L. (2016) Perfluoroalkylated substance effects in *Xenopus laevis* A6 kidney epithelial cells determined by ATR-FTIR spectroscopy and chemometric analysis. *Chemical Research in Toxicology*, 29(5), 924–932.

Gower, J. and Krzanowski, W. (1999) Analysis of distance for structured multivariate data and extensions to multivariate analysis of variance. *Journal of the Royal Statistical Society: Series C (Applied Statistics)*, 4, 505–519.

Gower, J.C. and Legendre, P. (1986) Metric and Euclidean properties of dissimilarity coefficients. *Journal of Classification*, 3, 5–48.

Gralka, E., Luchinat, C., Tenori, L., Ernst, B., Thurnheer, M., and Schultes, B. (2015) Metabolomic fingerprint of severe obesity is dynamically affected by bariatric surgery in a procedure-dependent manner. *The American Journal of Clinical Nutrition*, 102(6), 1313–1322.

Grassi, S., Lyndgaard, C.B., Rasmussen, M.A., and Amigo, J.M. (2017) Interval ANOVA simultaneous component analysis (i-ASCA) applied to spectroscopic data to study the effect of fundamental fermentation variables in beer fermentation metabolites. *Chemometrics and Intelligent Laboratory Systems*, 163, 86–93.

Gu, H., Chen, H., Pan, Z., Jackson, A., Talaty, N., Xi, B., Kissinger, C., Duda, C., Mann, D., Raftery, D. *et al.* (2007) Monitoring diet effects via biofluids and their implications for metabolomics studies. *Analytical Chemistry*, 79(1), 89–97.

Guerra, F.P., Richards, J.H., Fiehn, O., Famula, R., Stanton, B.J., Shuren, R., Sykes, R., Davis, M.F., and Neale, D.B. (2016) Analysis of the genetic variation in growth, ecophysiology, and chemical and metabolomic composition of wood of *Populus trichocarpa* provenances. *Tree Genetics & Genomes*, 12(1), 1–16.

Gürdeniz, G., Jensen, M.G., Meier, S., Bech, L., Lund, E., and Dragsted, L.O. (2016) Detecting beer intake by unique metabolite patterns. *Journal of Proteome Research*, 15(12), 4544–4556.

Haddi, Z., Annanouch, F., Amari, A., Hadoune, A., Bouchikhi, B., and ElBari, N. (2010) Application of a portable electronic nose device to discriminate and identify cheeses with known percentages of cow's and goat's milk, in *SENSORS, 2010 IEEE*, IEEE, pp. 771–774.

Hao, Z., Li, L., Ning, Z., Zhang, X., Mayne, J., Cheng, K., Walker, K., Liu, H., and Figeys, D. (2020) Metaproteomics reveals growth phase-dependent responses of an in vitro gut microbiota to metformin. *Journal of the American Society for Mass Spectrometry*, 31(7), 1448–1458.

Harnly, J.M., Harrington, P.B., Botros, L.L., Jablonski, J., Chang, C., Bergana, M.M., Wehling, P., Downey, G., Potts, A.R., and Moore, J.C. (2014) Characterization of near-infrared spectral variance in the authentication of skim and nonfat dry milk powder collection using ANOVA-PCA, pooled-ANOVA, and partial least-squares regression. *Journal of Agricultural and Food Chemistry*, 62(32), 8060–8067.

Harrigan, G.G., Venkatesh, T.V., Leibman, M., Blankenship, J., Perez, T., Halls, S., Chassy, A.W., Fiehn, O., Xu, Y., and Goodacre, R. (2016) Evaluation of metabolomics profiles of grain from maize hybrids derived from near-isogenic GM positive and negative segregant inbreds demonstrates that observed differences cannot be attributed unequivocally to the GM trait. *Metabolomics*, 12(5), 1–14.

Harrington, P.d.B., Vieira, N.E., Espinoza, J., Nien, J.K., Romero, R., and Yergey, A.L. (2005) Analysis of variance–principal component analysis: A soft tool for proteomic discovery. *Analytica Chimica Acta*, 544(1-2), 118–127.

Harrington, P.B., Vieira, N.E., Chen, P., Espinoza, J., Nien, J.K., Romero, R., and Yergey, A.L. (2006) Proteomic analysis of amniotic fluids using analysis of variance-principal component analysis and fuzzy rule-building expert systems applied to matrix-assisted laser desorption/ionization mass spectrometry. *Chemometrics and Intelligent Laboratory Systems*, 82(1-2), 283–293.

Hasdemir, D., van den Berg, R.A., van Kampen, A., and Smilde, A.K. (2020) Modeling adaptive response profiles in a vaccine clinical trial. *BMC Medical Research Methodology*, 20(1), 191.

Hastie, T., Tibshirani, R., and Friedman, J. (2009) *The Elements of Statistical Learning: Data Mining, Inference, and Prediction*, Springer Science & Business Media, New York.

Hayes, A.F. (1996) Permutation test is not distribution-free: Testing H_0: $\rho = 0$. *Psychological Methods*, 1(2), 184.

Hemerik, J., Thoresen, M., and Finos, L. (2021) Permutation testing in high-dimensional linear models: An empirical investigation. *Journal of Statistical Computation and Simulation*, 91(5), 897–914.

Henderson, B., Meurs, J., Lamers, C.R., Batista, G.L., Materić, D., Bertinetto, C.G., Bongers, C.C.W.G., Holzinger, R., Harren, F.J.M., Jansen, J.J. *et al.* (2022) Non-invasive monitoring of inflammation in inflammatory bowel disease patients during prolonged exercise via exhaled breath volatile organic compounds. *Metabolites*, 12(3), 224.

Hernández-Mesa, M., Narduzzi, L., Ouzia, S., Soetart, N., Jaillardon, L., Guitton, Y., Le Bizec, B., and Dervilly, G. (2022) Metabolomics and lipidomics to identify biomarkers of effect related to exposure to non-dioxin-like polychlorinated biphenyls in pigs. *Chemosphere*, 296, 133957.

Hildreth, S.B., Foley, E.E., Muday, G.K., Helm, R.F., and Winkel, B.S.J. (2020) The dynamic response of the Arabidopsis root metabolome to auxin and ethylene is not predicted by changes in the transcriptome. *Scientific Reports*, 10(1), 679.

Hoefsloot, H.C.J., Verouden, M.P.H., Westerhuis, J.A., and Smilde, A.K. (2006) Maximum likelihood scaling (MALS). *Journal of Chemometrics*, 20(3-4), 120–127.

Hoerl, A.E. and Kennard, R.W. (1970) Ridge regression: Biased estimation for nonorthogonal problems. *Technometrics*, 12(1), 55–67.

Hohrenk, L.L., Vosough, M., and Schmidt, T.C. (2019) Implementation of chemometric tools to improve data mining and prioritization in LC-HRMS for nontarget screening of organic micropollutants in complex water matrixes. *Analytical Chemistry*, 91(14), 9213–9220.

Hohrenk-Danzouma, L.L., Vosough, M., Merkus, V.I., Drees, F., and Schmidt, T.C. (2022) Non-target analysis and chemometric evaluation of a passive sampler monitoring of small streams. *Environmental Science & Technology*, 56(9), 5466–5477.

Horta, D., Moreno-Torres, M., Ramírez-Lázaro, M.J., Lario, S., Kuligowski, J., Sanjuan-Herráez, J.D., Quintas, G., Villoria, A., and Calvet, X. (2020) Analysis of the association between fatigue and the plasma lipidomic profile of inflammatory bowel disease patients. *Journal of Proteome Research*, 20(1), 381–392.

Hotelling, H. (1933) Analysis of a complex of statistical variables into principal components. *Journal of Educational Psychology*, 24, 417–441.

Hu, C., Li, Q., Shen, X., Quan, S., Lin, H., Duan, L., Wang, Y., Luo, Q., Qu, G., Han, Q. *et al.* (2016) Characterization of factors underlying the metabolic shifts in developing kernels of colored maize. *Scientific Reports*, 6(1), 35479.

Hu, C., Tohge, T., Chan, S.A., Song, Y., Rao, J., Cui, B., Lin, H., Wang, L., Fernie, A.R., Zhang, D. *et al.* (2016) Identification of conserved and diverse metabolic shifts during rice grain development. *Scientific Reports*, 6(1), 20942.

Hu, C., Rao, J., Song, Y., Chan, S.A., Tohge, T., Cui, B., Lin, H., Fernie, A.R., Zhang, D., and Shi, J. (2020) Dissection of flag leaf metabolic shifts and their relationship with those occurring simultaneously in developing seed by application of non-targeted metabolomics. *PLoS One*, 15(1), e0227577.

Huber, W., Von Heydebreck, A., Sültmann, H., Poustka, A., and Vingron, M. (2002) Variance stabilization applied to microarray data calibration and to the quantification of differential expression. *Bioinformatics*, 18(suppl_1), S96–S104.

Irmer, H., Tarazona, S., Sasse, C., Olbermann, P., Loeffler, J., Krappmann, S., Conesa, A., and Braus, G. (2015) RNAseq analysis of aspergillus fumigatus in blood reveals a just wait and see resting stage behavior. *BMC Genomics*, 16(1), 640.

Irwin, C., Van Reenen, M., Mason, S., Mienie, L.J., Westerhuis, J.A., and Reinecke, C.J. (2016) Contribution towards a metabolite profile of the detoxification of benzoic acid through glycine conjugation: An intervention study. *PLoS One*, 11(12), e0167309.

Jacobs, D., Deltimple, N., van Velzen, E., van Dorsten, F.A., Bingham, M., Vaughan, E.E., and van Duynhoven, J. (2008) ^{1}H NMR metabolite profiling of feces as a tool to assess the impact of nutrition on the human microbiome. *NMR in Biomedicine: An International Journal Devoted to the Development and Application of Magnetic Resonance In vivo*, 21(6), 615–626.

Janda, J.M. and Abbott, S.L. (2007) 16S rRNA gene sequencing for bacterial identification in the diagnostic laboratory: Pluses, perils, and pitfalls. *Journal of Clinical Microbiology*, 45(9), 2761–2764.

Jansen, J.J., Hoefsloot, H.C.J., van der Greef, J., Timmerman, M.E., and Smilde, A.K. (2005) Multilevel component analysis of time-resolved metabolic fingerprinting data. *Analytica Chimica Acta*, 530(2), 173–183.

Jansen, J.J., Hoefsloot, H.C.J., van der Greef, J., Timmerman, M.E., Westerhuis, J.A., and Smilde, A.K. (2005) ASCA: Analysis of multivariate data obtained from an experimental design. *Journal of Chemometrics*, 19(9), 469–481.

Jansen, J.J., Bro, R., Hoefsloot, H.C.J., van den Berg, F.W.J., Westerhuis, J.A., and Smilde, A.K. (2008) PARAFASCA: ASCA combined with PARAFAC for the analysis of metabolic fingerprinting data. *Journal of Chemometrics*, 22(1-2), 114–121.

Jeong, T.Y. and Simpson, M.J. (2019) Reproduction stage differentiates the time-course regulation of metabolites in daphnia magna. *Environmental Science & Technology*, 53(21), 12764–12773.

Jiang, R., Sun, T., Song, D., and Li, J.J. (2022) Statistics or biology: The zero-inflation controversy about scRNA-seq data. *Genome Biology*, 23(1), 31.

Johnson, H.E., Lloyd, A.J., Mur, L.A.J., Smith, A.R., and Causton, D.R. (2007) The application of MANOVA to analyse arabidopsis thaliana metabolomic data from factorially designed experiments. *Metabolomics*, 3(4), 517–530.

Johnston, J. (1963) *Econometric Methods*, McGraw Hill Book Company, New York.

Jolliffe, I.T. (2010) *Principal Component Analysis*, Springer, Berlin.

Jorge-Smeding, E., Carriquiry, M., Cantalapiedra-Hijar, G., Mendoza, A., and Astessiano, A.L. (2021) Plasma concentrations of branched-chain amino acids differ with Holstein genetic strain in pasture-based dairy systems. *Scientific Reports*, 11(1), 22414.

Joubert, E., Petrus, A., de Beer, D., Human, C., van der Rijst, M., Walczak, B., du Preez, B.V.P., Moelich, E., and Muller, M. (2024) Enhancing fruity and sweet-associated aroma attributes of honeybush tea (*Cyclopia* spp.) by ultra-high temperature oxidation. *Applied Food Research*, 4(1), 100398.

van de Kant, K.D.G., van Berkel, J.J.B.N., Jöbsis, Q., Passos, V.L., Klaassen, E.M.M., van der Sande, L., van Schayck, O.C.P., de Jongste, J.C., van Schooten, F.J., Derks, E. *et al.* (2013) Exhaled breath profiling in diagnosing wheezy preschool children. *European Respiratory Journal*, 41(1), 183–188.

Kaplan, D. (2008) *Structural Equation Modeling: Foundations and Extensions, Advanced Quantitative Techniques in the Social Sciences*, vol. 10, SAGE Publications, Inc, Thousand Oaks.

Keun, H.C., Ebbels, T.M.D., Bollard, M.E., Beckonert, O., Antti, H., Holmes, E., Lindon, J.C., and Nicholson, J.K. (2004) Geometric trajectory analysis of metabolic responses to toxicity can define treatment specific profiles. *Chemical Research in Toxicology*, 17(5), 579–587.

Khakimov, B., Poulsen, S.K., Savorani, F., Acar, E., Gürdeniz, G., Larsen, T.M., Astrup, A., Dragsted, L.O., and Engelsen, S.B. (2016) New Nordic diet versus average Danish diet: A randomized controlled trial revealed healthy long-term effects of the new Nordic diet by GC–MS blood plasma metabolomics. *Journal of Proteome Research*, 15(6), 1939–1954.

Khakimov, B., Christiansen, L.D., Heins, A.L., Sørensen, K.M., Schöller, C., Clausen, A., Skov, T., Gernaey, K.V., and Engelsen, S.B. (2017) Untargeted GC-MS metabolomics reveals changes in the metabolite dynamics of industrial scale batch fermentations of *Streptoccoccus thermophilus* broth. *Biotechnology Journal*, 12(10), 1700400.

Khakimov, B., Rasmussen, M.A., Kannangara, R.M., Jespersen, B.M., Munck, L., and Engelsen, S.B. (2017) From metabolome to phenotype: GC-MS metabolomics of developing mutant barley seeds reveals effects of growth, temperature and genotype. *Scientific Reports*, 7(1), 8195.

Khakimov, B., Bakhytkyzy, I., Fauhl-Hassek, C., and Engelsen, S.B. (2022) Non-volatile molecular composition and discrimination of single grape white of Chardonnay, Riesling, Sauvignon Blanc and Silvaner using untargeted GC–MS analysis. *Food Chemistry*, 369, 130878.

Khan, A., Park, H., Lee, H.A., Park, B., Gwak, H.S., Lee, H.R., Jee, S.H., and Park, Y.H. (2017) Elevated metabolites of steroidogenesis and amino acid metabolism in preadolescent female children with high urinary bisphenol a levels: A high-resolution metabolomics study. *Toxicological Sciences*, 160(2), 371–385.

Khan, A., Shin, O.S., Na, J., Kim, J.K., Seong, R.K., Park, M.S., Noh, J.Y., Song, J.Y., Cheong, H.J., Park, Y.H. *et al.* (2019) A systems vaccinology approach reveals the mechanisms of immunogenic responses to Hantavax vaccination in humans. *Scientific Reports*, 9(1), 4760.

Khatoonabadi, R.L., Vosough, M., Hohrenk, L.L., and Schmidt, T.C. (2021) Employing complementary multivariate methods for a designed nontarget LC-HRMS screening of a wastewater-influenced river. *Microchemical Journal*, 160, 105 641.

Kherad-Pajouh, S. and Renaud, O. (2010) An exact permutation method for testing any effect in balanced and unbalanced fixed effect ANOVA. *Computational Statistics & Data Analysis*, 54(7), 1881–1893.

Khomich, M., Måge, I., Rud, I., and Berget, I. (2021) Analysing microbiome intervention design studies: Comparison of alternative multivariate statistical methods. *PLoS One*, 16(11), e0259973.

Kim, J.K., Khan, A., Cho, S., Na, J., Lee, Y., Bang, G., Yu, W.J., Jeong, J.S., Jee, S.H., and Park, Y.H. (2019) Effect of developmental exposure to bisphenol A on steroid hormone and vitamin D3 metabolism. *Chemosphere*, 237, 124469.

Kleine Bardenhorst, S., Berger, T., Klawonn, F., Vital, M., Karch, A., and Rübsamen, N. (2021) Data analysis strategies for microbiome studies in human populations—a systematic review of current practice. *Msystems*, 6(1), 10–1128.

Kokova, D., Verhoeven, A., Perina, E.A., Ivanov, V.V., Knyazeva, E.M., Saltykova, I.V., and Mayboroda, O.A. (2020) Plasma metabolomics of the time resolved response to *Opisthorchis felineus* infection in an animal model (golden hamster, *Mesocricetus auratus*). *PLoS Neglected Tropical Diseases*, 14(1), e0008015.

Kroonenberg, P. (2008) *Applied Multiway Data Analysis*, John Wiley & Sons.

Krzanowski, W. (2000) *Principles of Multivariate Analysis*, vol. 23, OUP Oxford.

Kvalheim, O.M. (2010) Interpretation of partial least squares regression models by means of target projection and selectivity ratio plots. *Journal of Chemometrics*, 24(7-8), 496–504.

Kvalheim, O.M. and Karstang, T.V. (1989) Interpretation of latent-variable regression models. *Chemometrics and Intelligent Laboratory Systems*, 7(1), 39–51.

La Frano, M., Carmichael, S., Ma, C., Hardley, M., Shen, T., Wong, R., Rosales, L., Borkowski, K., Pedersen, T., Shaw, G. *et al.* (2018) Impact of post-collection freezing delay on the reliability of serum metabolomics in samples reflecting the California mid-term pregnancy biobank. *Metabolomics*, 14(11), 151.

Lamichhane, S., Ahonen, L., Dyrlund, T.S., Kemppainen, E., Siljander, H., Hyöty, H., Ilonen, J., Toppari, J., Veijola, R., Hyötyläinen, T. *et al.* (2018) Dynamics of plasma lipidome in progression to islet autoimmunity and type 1 diabetes–type 1 diabetes prediction and prevention study (DIPP). *Scientific Reports*, 8(1), 1–12.

Lamichhane, S., Kemppainen, E., Trošt, K., Siljander, H., Hyöty, H., Ilonen, J., Toppari, J., Veijola, R., Hyötyläinen, T., Knip, M. *et al.* (2019) Circulating metabolites in progression to islet autoimmunity and type 1 diabetes. *Diabetologia*, 62(12), 2287–2297.

Langsrud, Ø. (2002) 50–50 multivariate analysis of variance for collinear responses. *Journal of the Royal Statistical Society: Series D (The Statistician)*, 51(3), 305–317.

Langsrud, Ø. (2003) ANOVA for unbalanced data: Use Type II instead of Type III sums of squares. *Statistics and Computing*, 13(2), 163–167.

Ledoit, O. and Wolf, M. (2004) A well-conditioned estimator for large-dimensional covariance matrices. *Journal of Multivariate Analysis*, 88(2), 365–411.

Lee, D.Y., Heo, S., Kim, S.G., Choi, H.K., Lee, H.J., Cho, S.M., and Auh, J.H. (2013) Metabolomic characterization of the region-and maturity-specificity of *Rubus coreanus* Miquel (*Bokbunja*). *Food Research International*, 54(1), 508–515.

Lee, H.J., Jin, J., Seo, Y., Kang, I., Son, J., Yi, E.C., and Min, H. (2023) Untargeted metabolomics analysis reveals toxicity based on the sex and sexual maturity of single low-dose DEHP exposure. *Toxics*, 11(9), 794.

Legendre, P. and Anderson, M.J. (1999) Distance-based redundancy analysis: Testing multispecies responses in multifactorial ecological experiments. *Ecological Monographs*, 69(1), 1–24.

Legendre, P. and Legendre, L. (2012) *Numerical Ecology*, Elsevier, Oxford.

Leisso, R., Gapper, N., Mattheis, J., Sullivan, N., Watkins, C., Giovannoni, J., Schaffer, R., Johnston, J., Hanrahan, I., Hertog, M. *et al.* (2016) Gene expression and metabolism preceding soft scald, a chilling injury of 'Honeycrisp'apple fruit. *BMC Genomics*, 17(1), 798.

Lemanska, A., Grootveld, M., Silwood, C.J.L., and Brereton, R.G. (2012) Chemometric variance analysis of ^{1}H NMR metabolomics data on the effects of oral rinse on saliva. *Metabolomics*, 8(1), 64–80.

Levin, N. (2014) Multivariate statistics and the enactment of metabolic complexity. *Social Studies of Science*, 44(4), 555–578.

Li, J., Hou, L., Wang, C., Jia, X., Qin, X., and Wu, C. (2018) Short term intrarectal administration of sodium propionate induces antidepressant-like effects in rats exposed to chronic unpredictable mild stress. *Frontiers in Psychiatry*, 9, 454.

Li, Y., Xue, H., Fang, S., Wang, G., Wang, Y., Wang, T., Shi, R., Wu, J., and Ma, Y. (2021) Time-series metabolomics insights into the progressive characteristics of 3,5-diethoxycarbonyl-1,4-dihydrocollidine-induced cholestatic liver fibrosis in mice. *Journal of Pharmaceutical and Biomedical Analysis*, 198, 113986.

Liang, C., Lin, L., Hu, J., Ma, Y., Li, Y., and Sun, Z. (2024) Comprehensive pulmonary metabolic responses to silica nanoparticles exposure in Fisher 344 rats. *Ecotoxicology and Environmental Safety*, 275, 116256.

Liland, K.H., Smilde, A.K., Marini, F., and Næs, T. (2018) Confidence ellipsoids for ASCA models based on multivariate regression theory. *Journal of Chemometrics*, 32(5), e2990.

Lin, H. and Peddada, S.D. (2020) Analysis of microbial compositions: A review of normalization and differential abundance analysis. *NPJ Biofilms and Microbiomes*, 6(1), 60.

Little, R.J.A. and Rubin, D.B. (2019) *Statistical Analysis with Missing Data*, vol. 793, John Wiley & Sons, New York.

Liu, H., Cui, B., Xu, Y., Hu, C.and Liu, Y., Qu, G., Li, D., Wu, Y., Zhang, D., Quan, S. *et al.* (2017) Ethyl carbamate induces cell death through its effects on multiple metabolic pathways. *Chemico-Biological Interactions*, 277, 21–32.

Lohr, S.L. (1995) Hasse diagrams in statistical consulting and teaching. *The American Statistician*, 49(4), 376–381.

Love, M.I., Huber, W., and Anders, S. (2014) Moderated estimation of fold change and dispersion for RNA-seq data with DESeq2. *Genome Biology*, 15, 1–21.

Lu, C. and King, R.D. (2009) An investigation into the population abundance distribution of mRNAs, proteins, and metabolites in biological systems. *Bioinformatics*, 25(16), 2020–2027.

Lubbe, S., Filzmoser, P., and Templ, M. (2021) Comparison of zero replacement strategies for compositional data with large numbers of zeros. *Chemometrics and Intelligent Laboratory Systems*, 210, 104248.

Luciano, G. and Næs, T. (2009) Interpreting sensory data by combining principal component analysis and analysis of variance. *Food Quality and Preference*, 20(3), 167–175.

Luke, T.D.W., Pryce, J.E., Elkins, A.C., Wales, W., and Rochfort, S.J. (2020) Use of large and diverse datasets for ^{1}H NMR serum metabolic profiling of early lactation dairy cows. *Metabolites*, 10(5), 180.

Luke, T.D.W., Pryce, J.E., Wales, W.J., and Rochfort, S.J. (2020) A tale of two biomarkers: Untargeted ^{1}H NMR metabolomic fingerprinting of BHBA and NEFA in early lactation dairy cows. *Metabolites*, 10(6), 247.

Luthria, D.L., Lin, L.Z., Robbins, R.J., Finley, J.W., Banuelos, G.S., and Harnly, J.M. (2008) Discriminating between cultivars and treatments of broccoli using mass spectral fingerprinting and analysis of variance- principal component analysis. *Journal of Agricultural and Food Chemistry*, 56(21), 9819–9827.

Lutz, K.C., Jiang, S., Neugent, M.L., De Nisco, N.J., Zhan, X., and Li, Q. (2022) A survey of statistical methods for microbiome data analysis. *Frontiers in Applied Mathematics and Statistics*, 8, 884810.

Lv, Y., Wang, J., Xu, D., Liao, S., Li, P., Zhang, Q., Yang, M., and Kong, L. (2017) Comparative study of single/combination use of Huang-Lian-Jie-Du decoction and berberine on their protection on sepsis induced acute liver injury by NMR metabolic profiling. *Journal of Pharmaceutical and Biomedical Analysis*, 145, 794–804.

Lv, Y., Tian, N., Wang, J., Yang, M., and Kong, L. (2018) Metabolic switching in the hypoglycemic and antitumor effects of metformin on high glucose induced HepG2 cells. *Journal of Pharmaceutical and Biomedical Analysis*, 156, 153–162.

Ly-Verdú, S., Gröger, T.M., Arteaga-Salas, J.M., Brandmaier, S., Kahle, M., Neschen, S., Harbě de Angelis, M., and Zimmermann, R. (2015) Combining metabolomic non-targeted GC× GC–TOF–MS analysis and chemometric ASCA-based study of variances to assess dietary influence on type 2 diabetes development in a mouse model. *Analytical and Bioanalytical Chemistry*, 407(1), 343–354.

Madssen, T.S., Giskeødegård, G.F., Smilde, A.K., and Westerhuis, J.A. (2021) Repeated measures ASCA+ for analysis of longitudinal intervention studies with multivariate outcome data. *PLoS Computational Biology*, 17(11), e1009585.

Maiti, K.S., Roy, S., Lampe, R., and Apolonski, A. (2020) Breath indeed carries significant information about a disease: Potential biomarkers of cerebral palsy. *Journal of Biophotonics*, 13(11), e202000125.

Mamede, L., Fall, F., Schoumacher, M., Ledoux, A., Bugli, C., De Tullio, P., Quetin-Leclercq, J., Govaerts, B., and Frédérich, M. (2024) Comparison of extraction methods in vitro plasmodium falciparum: A ^{1}H NMR and LC-MS joined approach. *Biochemical and Biophysical Research Communications*, 703, 149684.

Manasseh, R., Berim, A., Kappagantu, M., Moyo, L., Gang, D.R., and Pappu, H.R. (2023) Pathogen-triggered metabolic adjustments to potato virus Y infection in potato. *Frontiers in Plant Science*, 13, 1031629.

Marcheafave, G.G., Tormena, C.D., Mattos, L.E., Liberatti, V.R., Ferrari, A.B.S., Rakocevic, M., Bruns, R.E., Scarminio, I.S., and Pauli, E.D. (2020) The main effects of elevated CO_2 and soil-water deficiency on ^{1}H NMR-based metabolic fingerprints of *Coffea arabica* beans by factorial and mixture design. *Science of the Total Environment*, 749, 142350.

Marcheafave, G., Pauli, E.D., Rakocevic, M., Wendling, I., Scarminio, I.S., Bruns, R.E., and Tauler, R. (2024) Uv–Vis molecular spectral resolution of multi-level mixture design-fingerprints of yerba mate leaves by chemometric methods. *Microchemical Journal*, 199, 110021.

Mardia, K.V., Kent, J.T., and Bibby, J.M. (1979) *Multivariate Analysis*, Academic Press, New York.

Marini, F. and Walczak, B. (2020) ANOVA-target projection (ANOVA-TP), in Brown, S.D., Tauler, R., and Walczak, B., *Comprehensive Chemometrics Chemical and Biochemical Data Analysis*, 2nd ed. *Volume 1*, Elsevier, Amsterdam, pp. 495–520.

Marini, F., de Beer, D., Joubert, E., and Walczak, B. (2015) Analysis of variance of designed chromatographic data sets: The analysis of variance-target projection approach. *Journal of Chromatography A*, 1405, 94–102.

Marini, F., de Beer, D., Walters, N.A., de Villiers, A., Joubert, E., and Walczak, B. (2017) Multivariate analysis of variance of designed chromatographic data. a case study involving fermentation of rooibos tea. *Journal of Chromatography A*, 1489, 115–125.

Martens, H. and Martens, M. (2000) Modified Jack-knife estimation of parameter uncertainty in bilinear modelling by partial least squares regression (PLSR). *Food Quality and Preference*, 11(1–2), 5–16.

Martin, M. and Govaerts, B. (2020) LiMM-PCA: Combining ASCA+ and linear mixed models to analyse high-dimensional designed data. *Journal of Chemometrics*, 34(6), e3232.

Martínez-Arranz, I., Mayo, R., Pérez-Cormenzana, M., Mincholé, I., Salazar, L., Alonso, C., and Mato, J. (2015) Enhancing metabolomics research through data mining. *Journal of Proteomics*, 127, 275–288.

Massana-Codina, J., Schnee, S., Allard, P.M., Rutz, A., Boccard, J., Michellod, E., Cléroux, M., Schürch, S., Gindro, K., and Wolfender, J.L. (2020) Insights on the structural and metabolic resistance of potato (*Solanum tuberosum*) cultivars to tuber black dot (*Colletotrichum coccodes*). *Frontiers in Plant Science*, 11, 1287.

MATLAB (2024) *R2024a*, The MathWorks Inc., Natick, MA.

McArdle, B.H. and Anderson, M.J. (2001) Fitting multivariate models to community data: A comment on distance-based redundancy analysis. *Ecology*, 82(1), 290–297.

McInnes, L., Healy, J., and Melville, J. (2018) UMAP: Uniform manifold approximation and projection for dimension reduction. *arXiv preprint arXiv:1802.03426*.

McKelvie, J., Wolfe, D., Celejewski, M., Alaee, M., Simpson, A., and Simpson, M. (2011) Metabolic responses of *Eisenia fetida* after sub-lethal exposure to organic contaminants with different toxic modes of action. *Environmental Pollution*, 159(12), 3620–3626.

Miccheli, A., Capuani, G., Marini, F., Tomassini, A., Praticò, G., Ceccarelli, S., Gnani, D., Baviera, G., Alisi, A., Putignani, L. *et al.* (2015) Urinary ^{1}H-NMR-based metabolic profiling of children with NAFLD undergoing VSL# 3 treatment. *International Journal of Obesity*, 39(7), 1118–1125.

Miller Jr, R.G. (1997) *Beyond ANOVA: Basics of Applied Statistics*, CRC Press, Boca Raton.

Mitra, V., Govorukhina, N., Zwanenburg, G., Hoefsloot, H., Westra, I., Smilde, A.K., Reijmers, T., Van Der Zee, A.G.J., Suits, F., Bischoff, R. *et al.* (2016) Identification of analytical factors affecting complex proteomics profiles acquired in a factorial design study with analysis of variance: Simultaneous component analysis. *Analytical Chemistry*, 88(8), 4229–4238.

Moreno-Torres, M., García-Llorens, G., Moro, E., Méndez, R., Quintás, G., and Castell, J.V. (2021) Factors that influence the quality of metabolomics data in in vitro cell toxicity studies: A systematic survey. *Scientific Reports*, 11(1), 1–13.

Moreno-Torres, M., Kumar, M., García-Llorens, G., Quintás, G., Tricot, T., Boon, R., Tolosa, L., Toprakhisar, B., Chesnais, F., Verfaillie, C. *et al.* (2022) A novel UPLC-MS metabolomic analysis-based strategy to monitor the course and extent of iPSC differentiation to hepatocytes. *Journal of Proteome Research*, 21(3), 702–712.

Morton, J.T., Toran, L., Edlund, A., Metcalf, J.L., Lauber, C., and Knight, R. (2017) Uncovering the horseshoe effect in microbial analyses. *mSystems*, 2(1), 10–1128.

Mosleth, E.F. and Liland, K.H. (2024) General Effect Modelling (GEM)-Part 1. Method description. *arXiv.2404.03024.*

Mosleth, E.F., McLeod, A., Rud, I., Axelsson, L., Solberg, L.E., Moen, B., Gilman, K.M.E., Færgestad, E.M., Lysenko, A., Rawlings, C. *et al.* (2020) Analysis of megavariate data in functional omics, in Brown, S.D., Tauler, R., and Walczak, B. *Comprehensive Chemometrics: Chemical and Biochemical Data Analysis*, 2nd ed., Volume 4, Elsevier, Amsterdam, p. 516–567.

Mosleth, E.F., Vedeler, C.A., Liland, K.H., McLeod, A., Bringeland, G.H., Kroondijk, L., Berven, F.S., Lysenko, A., Rawlings, C.J., Eid, K.E.H. *et al.* (2021) Cerebrospinal fluid proteome shows disrupted neuronal development in multiple sclerosis. *Scientific Reports*, 11(1), 4087.

Mosleth, E.F., Dankel, S.E.N., Mellgren, G., Olmos, F.M.B., Orozco, L.S., Lysenko, A., Ofstad, R., Begum, M.C., Martens, H., and Liland, K.H. (2024a) General Effect Modelling (GEM)-Part 2. Multivariate GEM applied to gene expression data of type 2 diabetes detects information that is lost by univariate validation. *arXiv.2404.03029.*

Mosleth, E.F., Myhr, K.M., Vedeler, C.A., Berven, F.S., Lysenko, A., Gavasso, S., and Liland, K.H. (2024b) General Effect Modelling (GEM)-Part 3. GEM applied on proteome data of cerebrospinal fluid of multiple sclerosis and clinically isolated syndrome. *arXiv.2404.03034.*

Murphy, K. (2012) *Machine Learning: A Probabilistic Perspective*, MIT Press, Cambridge.

Næs, T. and Langsrud, Ø. (1998) Fixed or random assessors in sensory profiling? *Food Quality and Preference*, 9(3), 145–152.

Næs, T., Tomic, O., Mevik, B.H., and Martens, H. (2011) Path modelling by sequential PLS regression. *Journal of Chemometrics*, 25(1), 28–40.

Navarro-Reig, M., Jaumot, J., Baglai, A., Vivó-Truyols, G., Schoenmakers, P.J., and Tauler, R. (2017) Untargeted comprehensive two-dimensional liquid chromatography coupled with high-resolution mass spectrometry analysis of rice metabolome using multivariate curve resolution. *Analytical Chemistry*, 89(14), 7675–7683.

Navarro-Reig, M., Ortiz-Villanueva, E., Tauler, R., and Jaumot, J. (2017) Modelling of hydrophilic interaction liquid chromatography stationary phases using chemometric approaches. *Metabolites*, 7(4), 54.

Nelder, J.A. (1994) The statistics of linear models: Back to basics. *Statistics and Computing*, 4, 221–234.

Neyraud, E., Schwartz, C., Brignot, H., Jouanin, I., Tremblay-Franco, M., Canlet, C., and Tournier, C. (2020) Longitudinal analysis of the salivary metabolome of breast-fed and formula-fed infants over the first year of life. *Metabolomics*, 16(3), 37.

Nieuwenhuis, R., te Grotenhuis, H.F., and Pelzer, B.J. (2017) Weighted effect coding for observational data with wec. *The R Journal*, 9, 477–485.

Nieuwoudt, M.K., Giglio, C., Marini, F., Scott, G., and Holroyd, S.E. (2021) Routine monitoring of instrument stability in a milk testing laboratory with ASCA: A pilot study. *Frontiers in Chemistry*, 9, 733331.

Nissen, L., Bordoni, A., and Gianotti, A. (2020) Shift of volatile organic compounds (VOCs) in gluten-free hemp-enriched sourdough bread: A metabolomic approach. *Nutrients*, 12(4), 1050.

Noer, N.K., Pagter, M., Bahrndorff, S., Malmendal, A., and Kristensen, T.N. (2020) Impacts of thermal fluctuations on heat tolerance and its metabolomic basis in *Arabidopsis thaliana, Drosophila melanogaster*, and *Orchesella cincta*. *PLoS One*, 15(10), e0237201.

de Noord, O.E. and Theobald, E.H. (2005) Multilevel component analysis and multilevel PLS of chemical process data. *Journal of Chemometrics*, 19(5–7), 301–307.

Nørgaard, S.K., Linder-Steinlein, K., Eliasen, A.U., Stokholm, J., Chawes, B.L., Bønnelykke, K., Bisgaard, H., Smilde, A.K., and Rasmussen, M.A. (2021) On using kernel integration by graphical LASSO to study partial correlations between heterogeneous data sets. *Journal of Chemometrics*, 35(10), e3324.

Nueda, M., Conesa, A., Westerhuis, J., Hoefsloot, H., Smilde, A., Talón, M., and Ferrer, A. (2007) Discovering gene expression patterns in time course microarray experiments by ANOVA–SCA. *Bioinformatics*, 23(14), 1792–1800.

Nueda, M., Sebastián, P., Tarazona, S., García-García, F., Dopazo, J., Ferrer, A., and Conesa, A. (2009) Functional assessment of time course microarray data. *BMC Bioinformatics*, 10, S9.

Nueda, M.J., Ferrer, A., and Conesa, A. (2012) ARSyN: A method for the identification and removal of systematic noise in multifactorial time course microarray experiments. *Biostatistics*, 13(3), 553–566.

Oehlert, G.W. (2000) *A First Course in Design and Analysis of Experiments*, W.H. Freeman, New York.

Olmos, V., Marro, M., Loza-Alvarez, P., Raldúa, D., Prats, E., Pia, B., Tauler, R., and De Juan, A. (2019) Assessment of tissue-specific multifactor effects in environmental–omics studies of heterogeneous biological samples: Combining hyperspectral image information and chemometrics. *Talanta*, 194, 390–398.

Olson, C.L. (1976) On choosing a test statistic in multivariate analysis of variance. *Psychological Bulletin*, 83(4), 579.

Opperman, M., Loots, D.T., van Reenen, M., Ronacher, K., Walzl, G., and du Preez, I. (2021) Chronological metabolic response to intensive phase TB therapy in patients with cured and failed treatment outcomes. *ACS Infectious Diseases*, 7(6), 1859–1869.

Ortiz-Villanueva, E., Navarro-Reig, M., Jaumot, J., and Tauler, R. (2017) Chemometric evaluation of hydrophilic interaction liquid chromatography stationary phases: Resolving complex mixtures of metabolites. *Analytical Methods*, 9(5), 774–785.

Ott, R.L. and Longnecker, M.T. (2015) *An Introduction to Statistical Methods and Data Analysis*, Cengage Learning, Boston.

Parsons, L., Haque, E., and Liu, H. (2004) Subspace clustering for high dimensional data: A review. *ACM SIGKDD Explorations Newsletter*, 6(1), 90–105.

Pearson, K. (1901) On lines and planes of closest fit to points in space. *Philosophical Magazine*, 2, 559–572.

Pérez, Y., Casado, M., Raldúa, D., Prats, E., Pia, B., Tauler, R., Alfonso, I., and Puig-Castellví, F. (2020) MCR-ALS analysis of ^{1}H NMR spectra by segments to study the zebrafish exposure to acrylamide. *Analytical and Bioanalytical Chemistry*, 412(23), 5695–5706.

Pérez-Cova, M., Tauler, R., and Jaumot, J. (2022) Adverse effects of arsenic uptake in rice metabolome and lipidome revealed by untargeted liquid chromatography coupled to mass spectrometry (LC-MS) and regions of interest multivariate curve resolution. *Separations*, 9(3), 79.

Perl, A., Hanczko, R., Lai, Z.W., Oaks, Z., Kelly, R., Borsuk, R., Asara, J.M., and Phillips, P.E. (2015) Comprehensive metabolome analyses reveal *N*-acetylcysteine-responsive accumulation of kynurenine in systemic lupus erythematosus: Implications for activation of the mechanistic target of rapamycin. *Metabolomics*, 11(5), 1157–1174.

Pezzatti, J., Bergé, M., Boccard, J., Codesido, S., Gagnebin, Y., Viollier, P.H., González-Ruiz, V., and Rudaz, S. (2019) Choosing an optimal sample preparation in *Caulobacter crescentus* for untargeted metabolomics approaches. *Metabolites*, 9(10), 193.

Phatak, A. and De Jong, S. (1997) The geometry of partial least squares. *Journal of Chemometrics*, 11(4), 311–338.

Pinto, L. and Lemos, S.G. (2013) Multivariate optimization of the voltammetric determination of Cd, Cu, Pb and Zn at bismuth film. Application to analysis of biodiesel. *Microchemical Journal*, 110, 417–424.

Pinto, R.C., Bosc, V., Nocairi, H., Barros, A.S., and Rutledge, D.N. (2008) Using ANOVA-PCA for discriminant analysis: Application to the study of mid-infrared spectra of carraghenan gels as a function of concentration and temperature. *Analytica Chimica Acta*, 629(1-2), 47–55.

Puig-Castellví, F., Alfonso, I., Pia, B., and Tauler, R. (2016) ^{1}H NMR metabolomic study of auxotrophic starvation in yeast using multivariate curve resolution-alternating least squares for pathway analysis. *Scientific Reports*, 6(1), 30982.

Puig-Castellví, F., Pérez, Y., Pia, B., Tauler, R., and Alfonso, I. (2018) Comparative analysis of ^{1}H NMR and ^{1}H–^{13}C HSQC NMR metabolomics to understand the effects of medium composition in yeast growth. *Analytical Chemistry*, 90(21), 12422–12430.

Qannari, E., Wakeling, I., Courcoux, P., and MacFie, H. (2000) Defining the underlying sensory dimensions. *Food Quality and Preference*, 11(1-2), 151–154.

Quansah, E., Ruiz-Rodado, V., Grootveld, M., Probert, F., and Zetterström, T.S.C. (2017) ^{1}H NMR-based metabolomics reveals neurochemical alterations in the brain of adolescent rats following acute methylphenidate administration. *Neurochemistry International*, 108, 109–120.

R Core Team (2024) *R: A Language and Environment for Statistical Computing*, R Foundation for Statistical Computing, Vienna, Austria.

Rådjursöga, M., Lindqvist, H.M., Pedersen, A., Karlsson, G.B., Malmodin, D., Brunius, C., Ellegård, L., and Winkvist, A. (2019) The ^{1}H NMR serum metabolomics response to a two meal challenge: A cross-over dietary intervention study in healthy human volunteers. *Nutrition Journal*, 18(1), 1–12.

Rago, D., Mette, K., Gürdeniz, G., Marini, F., Poulsen, M., and Dragsted, L.O. (2013) A LC–MS metabolomics approach to investigate the effect of raw apple intake in the rat plasma metabolome. *Metabolomics*, 9(6), 1202–1215.

Raimondi, S., Calvini, R., Candeliere, F., Leonardi, A., Ulrici, A., Rossi, M., and Amaretti, A. (2021) Multivariate analysis in microbiome description: Correlation of human gut protein degraders, metabolites, and predicted metabolic functions. *Frontiers in Microbiology*, 12, 723479.

Rajilić-Stojanović, M., Heilig, H.G.H.J., Molenaar, D., Kajander, K., Surakka, A., Smidt, H., and De Vos, W.M. (2009) Development and application of the human intestinal tract chip, a phylogenetic microarray: Analysis of universally conserved phylotypes in the abundant microbiota of young and elderly adults. *Environmental Microbiology*, 11(7), 1736–1751.

Rangel-Huerta, O.D., de la Torre-Aguilar, M.J., Mesa, M.D., Flores-Rojas, K., Pérez-Navero, J.L., Baena-Gómez, M.A., Gil, A., and Gil-Campos, M. (2022) The metabolic impact of two different parenteral nutrition lipid emulsions in children after hematopoietic stem cell transplantation: A lipidomics investigation. *International Journal of Molecular Sciences*, 23(7), 3667.

Rasmussen, M.A., Khakimov, B., Engel, J., and Jansen, J. (2024) Permutation strategies for inference in ANOVA-based models for nonorthogonal designs including continuous covariates. *Journal of Chemometrics*, 38(10), e3580.

Reis, A., Rudnitskaya, A., Blackburn, G.J., Fauzi, N.M., Pitt, A.R., and Spickett, C.M. (2013) A comparison of five lipid extraction solvent systems for lipidomic studies of human LDL [S]. *Journal of Lipid Research*, 54(7), 1812–1824.

Ricci, M., Gasperi, F., Endrizzi, I., Menghi, L., Cliceri, D., Franceschi, P., and Aprea, E. (2022) Effect of dairy, season, and sampling position on physical properties of Trentingrana cheese: Application of an LMM-ASCA model. *Foods*, 11(1), 127.

Ricci, M., Gasperi, F., Betta, E., Menghi, L., Endrizzi, I., Cliceri, D., Franceschi, P., and Aprea, E. (2023) Multivariate data analysis strategy to monitor Trentingrana cheese real-scale production through volatile organic compounds profiling. *LWT*, 173, 114364.

Risvik, E., McEwan, J.A., Colwill, J.S., Rogers, R., and Lyon, D.H. (1994) Projective mapping: A tool for sensory analysis and consumer research. *Food Quality and Preference*, 5(4), 263–269.

Robinson, M.D., McCarthy, D.J., and Smyth, G.K. (2010) edgeR: A Bioconductor package for differential expression analysis of digital gene expression data. *Bioinformatics*, 26(1), 139–140.

Roger, J.M., Chauchard, F., and Bellon-Maurel, V. (2003) EPO–PLS external parameter orthogonalisation of PLS application to temperature-independent measurement of sugar content of intact fruits. *Chemometrics and Intelligent Laboratory Systems*, 66(2), 191–204.

Rosales-Nieto, C.A., Rodríguez-Aguilar, M., Santiago-Hernandez, F., Cuevas-Reyes, V., Flores-Najera, M.J., Vázquez-García, J.M., Urrutia-Morales, J., Ghaffari, M.H., Meza-Herrera, C.A., González-Bulnes, A. *et al.* (2021) Periconceptional nutrition with spineless cactus (*Opuntia ficus-indica*) improves metabolomic profiles and pregnancy outcomes in sheep. *Scientific Reports*, 11(1), 7214.

Rossum, G.V. and Drake, F.L. (2009) *Python 3 Reference Manual*, CreateSpace, Scotts Valley, CA.

Rubingh, C.M., van Erk, M.J., Wopereis, S., van Vliet, T., Verheij, E.R., Cnubben, N.H.P., van Ommen, B., van der Greef, J., Hendriks, H.F.J., and Smilde, A.K. (2011) Discovery of subtle effects in a human intervention trial through multilevel modeling. *Chemometrics and Intelligent Laboratory Systems*, 106(1), 108–114.

Ruiz-Rodado, V., Nicoli, E.R., Probert, F., Smith, D.A., Morris, L., Wassif, C.A., Platt, F.M., and Grootveld, M. (2016) ^{1}H NMR-linked metabolomics analysis of liver from a mouse model of NP-C1 disease. *Journal of Proteome Research*, 15(10), 3511–3527.

Rusilowicz, M., Dickinson, M., Charlton, A., O'Keefe, S., and Wilson, J. (2016) A batch correction method for liquid chromatography–mass spectrometry data that does not depend on quality control samples. *Metabolomics*, 12(3), 56.

Ryckewaert, M., Gorretta, N., Henriot, F., Marini, F., and Roger, J.M. (2020) Reduction of repeatability error for analysis of variance-simultaneous component analysis (REP-ASCA): Application to nir spectroscopy on coffee sample. *Analytica Chimica Acta*, 1101, 23–31.

Saccenti, E. (2023) What can go wrong when observations are not independently and identically distributed: A cautionary note on calculating correlations on combined data sets from different experiments or conditions. *Frontiers in Systems Biology*, 3, 3.

Saccenti, E., Hoefsloot, H.C.J., Smilde, A.K., Westerhuis, J.A., and Hendriks, M.M.W.B. (2014) Reflections on univariate and multivariate analysis of metabolomics data. *Metabolomics*, 10(3), 361–374.

Saccenti, E., Tenori, L., Verbruggen, P., Timmerman, M.E., Bouwman, J., van der Greef, J., Luchinat, C., and Smilde, A.K. (2014) Of monkeys and men: A metabolomic analysis of static and dynamic urinary metabolic phenotypes in two species. *PLoS One*, 9(9), e106077.

Saccenti, E., Smilde, A.K., and Camacho, J. (2018) Group-wise ANOVA simultaneous component analysis for designed omics experiments. *Metabolomics*, 14, 73.

Saccenti, E., Hendriks, M.M.W.B., and Smilde, A.K. (2020) Corruption of the Pearson correlation coefficient by measurement error and its estimation, bias, and correction under different error models. *Scientific Reports*, 10(1), 438.

Salehi, H., Zhang, L., Alp-Turgut, F.N., Arikan, B., Elbasan, F., Ozfidan-Konakci, C., Balcı, M., Zengin, G., Yildiztugay, E., and Lucini, L. (2024) The exogenous application of naringenin and rosmarinic acid modulates functional traits in *Lepidium sativum*. *Journal of the Science of Food and Agriculture*, 104(5), 2761–2771.

Salinas, D., Mumey, B.M., and June, R.K. (2019) Physiological dynamic compression regulates central energy metabolism in primary human chondrocytes. *Biomechanics and Modeling in Mechanobiology*, 18(1), 69–77.

Salle, G., Canlet, C., Cortet, J., Koch, C., Malsa, J., Reigner, F., Riou, M., Perrot, N., Blanchard, A., and Mach, N. (2021) Integrative biology defines novel biomarkers of resistance to strongylid infection in horses. *Scientific Reports*, 11(1), 14278.

Salvador, A.C., Rudnitskaya, A., Silvestre, A.J.D., and Rocha, S.M. (2016) Metabolomic-based strategy for fingerprinting of *Sambucus nigra* L. berry volatile terpenoids and norisoprenoids: Influence of ripening and cultivar. *Journal of Agricultural and Food Chemistry*, 64(26), 5428–5438.

Sammel, M., Lin, X., and Ryan, L. (1999) Multivariate linear mixed models for multiple outcomes. *Statistics in Medicine*, 18(17-18), 2479–2492.

Sarembaud, J., Pinto, R., Rutledge, D.N., and Feinberg, M. (2007) Application of the ANOVA-PCA method to stability studies of reference materials. *Analytica Chimica Acta*, 603(2), 147–154.

Schäfer, J. and Strimmer, K. (2005) A shrinkage approach to large-scale covariance matrix estimation and implications for functional genomics. *Statistical Applications in Genetics and Molecular Biology*, 4(1), 1–32.

Schmidtke, L.M., Antalick, G., Šuklje, K., Blackman, J.W., Boccard, J., and Deloire, A. (2020) Cultivar, site or harvest date: The gordian knot of wine terroir. *Metabolomics*, 16(5), 52.

Schorn-García, D., Ezenarro, J., Busto, O., Acea, L., Boqué, R., Mestres, M., and Giussani, B. (2023) A new index to detect process deviations using IR spectroscopy and chemometrics process tools. *Food and Bioprocess Technology*, 17, 2782–2792.

Schorn-García, D., Giussani, B., Busto, O., Acea, L., Mestres, M., and Boqué, R. (2023) Methodologies based on ASCA to elucidate the influence of a subprocess: Vinification as a case of study. *Journal of Chemometrics*, 37(7), e3465.

Schorn-García, D., Giussani, B., García-Casas, M.J., Rico, D., Martin-Diana, A.B., Acea, L., Busto, O., Boqué, R., and Mestres, M. (2023) Assessment of variability sources in grape ripening parameters by using FTIR and multivariate modelling. *Foods*, 12(5), 962.

Schott, J.R. (2016) *Matrix Analysis for Statistics*, John Wiley & Sons, New York.

Schwarz, C.J. (1993) The mixed-model ANOVA: The truth, the computer packages, the books. *The American Statistician*, 47(1), 48–59.

Searle, S.R. (1971) *Linear Models*, John Wiley & Sons, New York.

Searle, S.R. (1982) *Matrix Algebra Useful for Statistics*, John Wiley & Sons, New York.

Searle, S.R. (2006) *Linear Models for Unbalanced Data*, John Wiley & Sons, New York.

Searle, S.R., Casella, G., and McCulloch, C. (1992) *Variance Components*, John Wiley & Sons, New York.

Sheikholeslami, M.N., Gómez-Canela, C., Barron, L.P., Barata, C., Vosough, M., and Tauler, R. (2020) Untargeted metabolomics changes on *Gammarus pulex* induced by propranolol, triclosan, and nimesulide pharmaceutical drugs. *Chemosphere*, 260, 127479.

Silva, D., Arend, E., Rocha, S.M., Rudnitskaya, A., Delgado, L., Moreira, A., and Carvalho, J. (2019) The impact of exercise training on the lipid peroxidation metabolomic profile and respiratory infection risk in older adults. *European Journal of Sport Science*, 19(3), 384–393.

Silverman, J.D., Roche, K., Mukherjee, S., and David, L.A. (2020) Naught all zeros in sequence count data are the same. *Computational and Structural Biotechnology Journal*, 18, 2789–2798.

Smilde, A.K., Geladi, P., and Bro, R. (2005) *Multi-Way Analysis: Applications in the Chemical Sciences*, John Wiley & Sons, New York.

Smilde, A.K., Hoefsloot, H.C.J., and Westerhuis, J.A. (2008) The geometry of ASCA. *Journal of Chemometrics*, 22(8), 464–471.

Smilde, A.K., Timmerman, M.E., Hendriks, M.M.W.B., Jansen, J.J., and Hoefsloot, H.C.J. (2012) Generic framework for high-dimensional fixed-effects ANOVA. *Briefings in Bioinformatics*, 13(5), 524–535.

Smilde, A.K., Næs, T., and Liland, K.H. (2022) *Multiblock Data Fusion in Statistics and Machine Learning: Applications in the Natural and Life Sciences*, John Wiley & Sons, New York.

Snee, R.D. (1974) "Computation and use of expected mean squares in analysis of variance." *Journal of Quality Technology*, 6, 128–137.

Song, Y., Westerhuis, J., and Smilde, A. (2020) Separating common (global and local) and distinct variation in multiple mixed types data sets. *Journal of Chemometrics*, 34, e3197.

Song, Y., Westerhuis, J.A., Aben, N., Wessels, L.F., Groenen, P.J., and Smilde, A.K. (2021) Generalized simultaneous component analysis of binary and quantitative data. *Journal of Chemometrics*, 35(3), e3312.

Sonn, B.J., Heard, K.J., Heard, S.M., D'Alessandro, A., Reynolds, K.M., Dart, R.C., Rumack, B.H., and Monte, A.A. (2022) Metabolomic markers predictive of hepatic adaptation to therapeutic dosing of acetaminophen. *Clinical Toxicology*, 60(2), 221–230.

Sprocati, A.R., Alisi, C., Pinto, V., Montereali, M.R., Marconi, P., Tasso, F., Turnau, K., De Giudici, G., Goralska, K., Bevilacqua, M., Marini, F., and Cremisini, C. (2014) Assessment of the applicability of a "toolbox" designed for microbially assisted phytoremediation: The case study at Ingurtosu mining site (Italy). *Environmental Science and Pollution Research*, 21(11), 6939–6951.

Stahle, L. and Wold, S. (1990) Multivariate analysis of variance (MANOVA). *Chemometrics and Intelligent Laboratory Systems*, 9(2), 127–141.

Stander, Z., Luies, L., Mienie, L.J., Van Reenen, M., Howatson, G., Keane, K.M., Clifford, T., Stevenson, E.J., and Loots, D.T. (2020) The unaided recovery of marathon-induced serum metabolome alterations. *Scientific Reports*, 10(1), 11060.

Stanimirova, I., Banasik, M., Zabek, A., Dawiskiba, T., Kościelska-Kasprzak, K., Wojtowicz, W., Krajewska, M., Janczak, D., and Młynarz, P. (2020) Serum metabolomics approach to monitor the changes in metabolite profiles following renal transplantation. *Scientific Reports*, 10(1), 17223.

Stevens, S. (1946) On the theory of scales of measurement. *Science*, 103(2684), 677–680.

Stevens, J. (1979) Comment on Olson: Choosing a test statistic in multivariate analysis of variance. *Psychological Bulletin*, 86(2), 355–360.

Strani, L., Grassi, S., Casiraghi, E., Alamprese, C., and Marini, F. (2019) Milk renneting: Study of process factor influences by FT-NIR spectroscopy and chemometrics. *Food and Bioprocess Technology*, 12(6), 954–963.

Stuart, A., Ord, J.K., and Arnold, S.F. (2004) *Kendall's Advanced Theory of Statistics: Classical Inference and the Linear Model. Volume* 2A, John Wiley & Sons, New York.

Su, X., Yu, J., Wang, N., Zhao, S., Han, W., Chen, D., Li, L., and Li, L. (2021) High-coverage metabolome analysis reveals significant diet effects of autoclaved and irradiated feed on mouse fecal and urine metabolomics. *Molecular Nutrition & Food Research*, 65(12), 2100110.

Sun, J., Zhang, M., Kubzdela, N., Luo, Y., Harnly, J.M., and Chen, P. (2018) Determination of variance of secondary metabolites in lettuces grown under different light sources by flow injection mass spectrometric (FIMS) fingerprinting and ANOVA–PCA. *Journal of Analysis and Testing*, 2(4), 312–321.

Sun, J., Janisiewicz, W.J., Takeda, F., Evans, B., Jurick II, W.M., Zhang, M., Yu, L., and Chen, P. (2020) Effect of nighttime UV-C irradiation of strawberry plants on phenolic content of fruit: Targeted and non-targeted metabolomic analysis. *Journal of Berry Research*, 10(3), 365–380.

Takeuchi, K., Yanai, H., and Mukherjee, B.N. (1982) *Foundations of Multivariate Analysis: A Unified Approach by Means of Projection Onto Linear Subspaces*, John Wiley & Sons, Inc, New York.

Tarazona, S., Prado-López, S., Dopazo, J., Ferrer, A., and Conesa, A. (2012) Variable selection for multifactorial genomic data. *Chemometrics and Intelligent Laboratory Systems*, 110(1), 113–122.

Te Grotenhuis, M., Pelzer, B., Eisinga, R., Nieuwenhuis, R., Schmidt-Catran, A., and Konig, R. (2017) When size matters: Advantages of weighted effect coding in observational studies. *International Journal of Public Health*, 62, 163–167.

Ten Berge, J. (1993) *Least Squares Optimization in Multivariate Analysis*, DSWO Press, Leiden.

Thiel, M., Féraud, B., and Govaerts, B. (2017) ASCA+ and APCA+: Extensions of ASCA and APCA in the analysis of unbalanced multifactorial designs. *Journal of Chemometrics*, 31(6), e2895.

Tibshirani, R. (1996) Regression shrinkage and selection via the lasso. *Journal of the Royal Statistical Society: Series B (Statistical Methodology)*, 58(1), 267–288.

Timmerman, M.E. (2006) Multilevel component analysis. *British Journal of Mathematical and Statistical Psychology*, 59(Part 2), 301–320.

Timmerman, M.E. and Kiers, H.A.L. (2003) Four simultaneous component models for the analysis of multivariate time series from more than one subject to model intraindividual and interindividual differences. *Psychometrika*, 68(1), 105–121.

Timmerman, M.E., Kiers, H.A.L., and Smilde, A.K. (2007) Estimating confidence intervals for principal component loadings: A comparison between the bootstrap and asymptotic results. *British Journal of Mathematical and Statistical Psychology*, 60(2), 295–314.

Timmerman, M.E., Hoefsloot, H.C.J., Smilde, A.K., and Ceulemans, E. (2015) Scaling in ANOVA-simultaneous component analysis. *Metabolomics*, 11(5), 1265–1276.

Tomassini, A., Vitalone, A., Marini, F., Praticò, G., Sciubba, F., Bevilacqua, M., Delfini, M., Di Sotto, A., Di Giacomo, S., Mariani, P. *et al.* (2014) ^{1}H NMR-based urinary metabolic profiling reveals changes in nicotinamide pathway intermediates due to postnatal stress model in rat. *Journal of Proteome Research*, 13(12), 5848–5859.

Torkildsen, C.F., Austdal, M., Iversen, A.C., Bathen, T.F., Giskeødegård, G.F., Nilsen, E.B., Iversen, G.A., Sande, R.K., Bjørge, L., and Thomsen, L.C.V. (2023) Primary treatment effects for high-grade serous ovarian carcinoma evaluated by changes in serum metabolites and lipoproteins. *Metabolites*, 13(3), 417.

Tormena, C.D., Pauli, E.D., Marcheafave, G.G., Scheel, G.L., Rakocevic, M., Bruns, R.E., and Scarminio, I.S. (2020) FT-IR biomarkers of sexual dimorphism in yerba-mate plants: Seasonal and light accessibility effects. *Microchemical Journal*, 158, 105329.

Tremblay-Franco, M., Cabaton, N.J., Canlet, C., Gautier, R., Schaeberle, C.M., Jourdan, F., Sonnenschein, C., Vinson, F., Soto, A.M., and Zalko, D. (2015) Dynamic metabolic disruption in rats perinatally exposed to low doses of bisphenol-a. *PLoS One*, 10(10), e0141698.

Trimigno, A., Khakimov, B., Mejia, J.L.C., Mikkelsen, M.S., Kristensen, M., Jespersen, B.M., and Engelsen, S.B. (2017) Identification of weak and gender specific effects in a short 3 weeks intervention study using barley and oat mixed linkage β-glucan dietary supplements: A human fecal metabolome study by GC-MS. *Metabolomics*, 13(10), 108.

Trimigno, A., Khakimov, B., Savorani, F., Tenori, L., Hendrixson, V., Čivilis, A., Glibetic, M., Gurinovic, M., Pentikäinen, S., Sallinen, J. *et al.* (2019) Investigation of variations in the human urine metabolome amongst European populations: An exploratory search for biomarkers of people at risk-of-poverty. *Molecular Nutrition & Food Research*, 63(1), 1800 216.

Trimigno, A., Khakimov, B., Savorani, F., Poulsen, S.K., Astrup, A., Dragsted, L.O., and Engelsen, S.B. (2020) Human urine ^{1}H NMR metabolomics reveals alterations of protein and carbohydrate metabolism when comparing habitual Average Danish diet vs. healthy New Nordic diet. *Nutrition*, 79, 110867.

Trimigno, A., Khakimov, B., Lauge Quaade M., Honoré, O.L., Clausen, T., Blaabjerg, K., Engelsen, S.B., and Vedsted Hammer, A.S. (2022) Urinary and plasma metabolome of farm mink (*Neovison vison*) after an intervention with raw or cooked poultry offal: A ^{1}H NMR investigation. *Archives of Animal Nutrition*, 76(1), 74–91.

Trimigno, A., Łoniewska, B., Skonieczna-Żydecka, K., Kaczmarczyk, M., Łoniewski, I., and Picone, G. (2024) The application of high-resolution nuclear magnetic resonance (HR NMR) in metabolomic analyses of meconium and stool in newborns. A preliminary pilot study of MABEL project: Metabolomics approach for the assessment of baby-mother enteric microbiota legacy. *PharmaNutrition*, 27, 100378.

Tsermoula, P., Kristensen, N.B., Mobaraki, N., Engelsen, S.B., and Khakimov, B. (2024) Efficient quantification of milk metabolites from ^{1}H NMR spectra using the signature mapping (SigMa) approach: Chemical shift library development for cows' milk and colostrum. *Analytical Chemistry*, 96(5), 1861–1871.

Turi, C.E., Finley, J., Shipley, P.R., Murch, S.J., and Brown, P.N. (2015) Metabolomics for phytochemical discovery: Development of statistical approaches using a cranberry model system. *Journal of Natural Products*, 78(4), 953–966.

Twisk, J., Bosman, L., Hoekstra, T., Rijnhart, J., Welten, M., and Heymans, M. (2018) Different ways to estimate treatment effects in randomised controlled trials. *Contemporary Clinical Trials Communications*, 10, 80–85.

Uarrota, V.G., Fuentealba, C., Hernández, I., Defilippi-Bruzzone, B., Meneses, C., Campos-Vargas, R., Lurie, S., Hertog, M., Carpentier, S., Poblete-Echeverría, C. *et al.* (2019) Integration of proteomics and metabolomics data of early and middle season Hass avocados under heat treatment. *Food Chemistry*, 289, 512–521.

Uh, H., Klarić, L., Ugrina, I., Lauc, G., Smilde, A., and Houwing-Duistermaat, J. (2020) Choosing proper normalization is essential for discovery of sparse glycan biomarkers. *Molecular Omics*, 16(3), 231–242.

Van Batenburg, M.F., Coulier, L., van Eeuwijk, F., Smilde, A.K., and Westerhuis, J.A. (2011) New figures of merit for comprehensive functional genomics data: The metabolomics case. *Analytical Chemistry*, 83(9), 3267–3274.

Van den Brink, P.J. and Ter Braak, C.J.F. (1999) Principal response curves: Analysis of time-dependent multivariate responses of biological community to stress. *Environmental Toxicology and Chemistry*, 18(2), 138–148.

Van Den Wollenberg, A.L. (1977) Redundancy analysis an alternative for canonical correlation analysis. *Psychometrika*, 42(2), 207–219.

Van der Maaten, L. and Hinton, G. (2008) Visualizing data using t-SNE. *Journal of Machine Learning Research*, 9(11), 2579–2605.

Van Mechelen, I. and Smilde, A.K. (2011) Comparability problems in the analysis of multiway data. *Chemometrics and Intelligent Laboratory Systems*, 106(1), 2–11.

Vargas, L.H.G., Neto, J.C.R., de Aquino Ribeiro, J.A., Ricci-Silva, M.E., Souza, M.T., Rodrigues, C.M., de Oliveira, A.E., and Abdelnur, P.V. (2016) Metabolomics analysis of oil palm (*Elaeis guineensis*) leaf: Evaluation of sample preparation steps using UHPLC–MS/MS. *Metabolomics*, 12(10), 153.

van Velzen, E.J.J., Westerhuis, J.A., Grün, C.H., Jacobs, D.M., Eilers, P.H.C., Mulder, T.P., Foltz, M., Garczarek, U., Kemperman, R., Vaughan, E.E. *et al.* (2014) Population-based nutrikinetic modeling of polyphenol exposure. *Metabolomics*, 10, 1059–1073.

Verbeke, G., Molenberghs, G., and Verbeke, G. (2009) *Linear Mixed Models for Longitudinal Data*, Springer, New York.

Veshkini, A., Hammon, H.M., Vogel, L., Delosière, M., Viala, D., Dèjean, S., Tröscher, A., Ceciliani, F., Sauerwein, H., and Bonnet, M. (2022) Liver proteome profiling in dairy cows during the transition from gestation to lactation: Effects of supplementation with essential fatty acids and conjugated linoleic acids as explored by PLS-DA. *Journal of Proteomics*, 252, 104436.

Vis, D.J., Westerhuis, J.A., Smilde, A.K., and van der Greef, J. (2007) Statistical validation of megavariate effects in ASCA. *BMC Bioinformatics*, 8(1), 322.

Vis, D.J., Westerhuis, J.A., Jacobs, D.M., van Duynhoven, J.P.M., Wopereis, S., van Ommen, B., Hendriks, M.M.W.B., and Smilde, A.K. (2015) Analyzing metabolomics-based challenge tests. *Metabolomics*, 11, 50–63.

Vitale, R., Westerhuis, J., Næs, T., Smilde, A., de Noord, O., and Ferrer, A. (2017) Selecting the number of factors in principal component analysis by permutation testing—numerical and practical aspects. *Journal of Chemometrics*, 31(12), e2937.

Waaijenborg, S., Korobko, O., Willems van Dijk, K., Lips, M., Hankemeier, T., Wilderjans, T.F., Smilde, A.K., and Westerhuis, J.A. (2018) Fusing metabolomics data sets with heterogeneous measurement errors. *PLoS One*, 13(4), e0195939.

Wang, J., Reijmers, T., Chen, L., Van Der Heijden, R., Wang, M., Peng, S., Hankemeier, T., Xu, G., and Van Der Greef, J. (2009) Systems toxicology study of doxorubicin on rats using ultra performance liquid chromatography coupled with mass spectrometry based metabolomics. *Metabolomics*, 5(4), 407–418.

Wang, X., Jiang, G., Kebreab, E., Li, J., Feng, X., Li, C., Zhang, X., Huang, X., Fang, C., Fang, R. *et al.* (2020) ^{1}H NMR-based metabolomics study of breast meat from Pekin and Linwu duck of different ages and relation to meat quality. *Food Research International*, 133, 109126.

Welham, S.J., Gezan, S.A., Clark, S.J., and Mead, A. (2014) *Statistical Methods in Biology: Design and Analysis of Experiments and Regression*, CRC Press, Boca Raton.

Wentzell, P.D., Andrews, D.T., Hamilton, D.C., Faber, K., and Kowalski, B.R. (1997) Maximum likelihood principal component analysis. *Journal of Chemometrics*, 11(4), 339–366.

Westerhuis, J.A. and Smilde, A.K. (2001) Deflation in multiblock PLS. *Journal of Chemometrics*, 15(5), 485–493.

Westerhuis, J.A., De Jong, S., and Smilde, A.K. (2001) Direct orthogonal signal correction. *Chemometrics and Intelligent Laboratory Systems*, 56(1), 13–25.

Westerhuis, J.A., van Velzen, E.J.J., Hoefsloot, H.C.J., and Smilde, A.K. (2010) Multivariate paired data analysis: Multilevel PLSDA versus OPLSDA. *Metabolomics*, 6, 119–128.

White, F.T., Heintz-Buschart, A., Dong, L., Bouwmeester, H.J., Westerhuis, J.A., and Smilde, A.K. (2024) MASCARA: Coexpression analysis in data from designed experiments. *bioRxiv: 2024.02.29.582876*.

Wishart, D.S. (2008) Quantitative metabolomics using NMR. *TrAC Trends in Analytical Chemistry*, 27(3), 228–237.

Wold, S., Ruhe, A., Wold, H., and Dunn, III, W.J. (1984) The collinearity problem in linear regression. The partial least squares (PLS) approach to generalized inverses. *SIAM Journal on Scientific and Statistical Computing*, 5(3), 735–743.

Xi, M., Berendsen, A.A.M., Ernst, M., Hu, T., Vázquez-Manjarrez, N., Feskens, E.J.M., Dragsted, L.O., and La Barbera, G. (2021) Combined urinary biomarkers to assess coffee intake using untargeted metabolomics: Discovery in three pilot human intervention studies and validation in cross-sectional studies. *Journal of Agricultural and Food Chemistry*, 69(25), 7230–7242.

Xu, D., Liao, S., Li, P., Zhang, Q., Lv, Y., Fu, X., Yang, M., Wang, J., and Kong, L. (2019) Metabolomics coupled with transcriptomics approach deciphering age relevance in sepsis. *Aging and Disease*, 10(4), 854.

Yau, K.K., Wang, K., and Lee, A.H. (2003) Zero-inflated negative binomial mixed regression modeling of over-dispersed count data with extra zeros. *Biometrical Journal: Journal of Mathematical Methods in Biosciences*, 45(4), 437–452.

Yde, C.C., Jansen, J.J., Theil, P.K., Bertram, H.C., and Knudsen, K.E.B. (2012) Different metabolic and absorption patterns of betaine in response to dietary intake of whole-wheat grain, wheat aleurone or rye aleurone in catheterized pigs. *European Food Research and Technology*, 235(5), 939–949.

Yuan, M. and Lin, Y. (2006) Model selection and estimation in regression with grouped variables. *Journal of the Royal Statistical Society: Series B (Statistical Methodology)*, 68(1), 49–67.

Zancarini, A., Echenique-Subiabre, I., Debroas, D., Taïb, N., Quiblier, C., and Humbert, J.F. (2017) Deciphering biodiversity and interactions between bacteria and microeukaryotes within epilithic biofilms from the Loue River, France. *Scientific Reports*, 7(1), 4344.

Zhang, X., Kontoudakis, N., Šuklje, K., Antalick, G., Blackman, J.W., Rutledge, D.N., Schmidtke, L.M., and Clark, A.C. (2020) Changes in red wine composition during bottle aging: Impacts of grape variety, vineyard location, maturity, and oxygen availability during aging. *Journal of Agricultural and Food Chemistry*, 68(47), 13331–13343.

Zhang, X., Xu, H., Ning, J., Ji, H., Yan, J., Zheng, Y., Xu, Q., Li, C., Zhao, L., Zheng, H. *et al.* (2020) Sex-specific metabolic changes in peripheral organs of diabetic mice. *Journal of Proteome Research*, 19(8), 3011–3021.

Zhang, C., Jones, M., Govaert, L., Viant, M., De Meester, L., and Stoks, R. (2021) Resurrecting the metabolome: Rapid evolution magnifies the metabolomic plasticity to predation in a natural daphnia population. *Molecular Ecology*, 30(10), 2285–2297.

Zhang, L., Vaccari, F., Ardenti, F., Fiorini, A., Tabaglio, V., Puglisi, E., Trevisan, M., and Lucini, L. (2024) The dosage-and size-dependent effects of micro-and nanoplastics in lettuce roots and leaves at the growth, photosynthetic, and metabolomics levels. *Plant Physiology and Biochemistry*, 208, 108531.

Zheng, H., Lorenzen, J.K., Astrup, A., Larsen, L.H., Yde, C.C., Clausen, M.R., and Bertram, H.C. (2016) Metabolic effects of a 24-week energy-restricted intervention combined with low or high dairy intake in overweight women: An NMR-based metabolomics investigation. *Nutrients*, 8(3), 108.

Zheng, H., Cai, A., Shu, Q., Niu, Y., Xu, P., Li, C., Lin, L., and Gao, H. (2018) Tissue-specific metabolomics analysis identifies the liver as a major organ of metabolic disorders in amyloid precursor protein/presenilin 1 mice of Alzheimer's disease. *Journal of Proteome Research*, 18(3), 1218–1227.

Zhou, X. and Stephens, M. (2014) Efficient multivariate linear mixed model algorithms for genome-wide association studies. *Nature Methods*, 11(4), 407–409.

Zhou, J., Tang, L., Shen, C.L., and Wang, J.S. (2020) Green tea polyphenols boost gut-microbiota-dependent mitochondrial TCA and urea cycles in Sprague–Dawley rats. *The Journal of Nutritional Biochemistry*, 81, 108395.

Zhou, X., Ulaszewska, M.M., De Gobba, C., Rinnan, Å., Poulsen, M.W., Chen, J., Mattivi, F., Hedegaard, R.V., Skibsted, L.H., and Dragsted, L.O. (2021) New advanced glycation end products observed in rat urine by untargeted metabolomics after feeding with heat-treated skimmed milk powder. *Molecular Nutrition & Food Research*, 65(7), 2001049.

Zimmerman, D.L. (2020) *Linear Model Theory*, Springer, New York.

Zou, H. and Hastie, T. (2003) Regression shrinkage and selection via the elastic net, with applications to microarrays. *Journal of the Royal Statistical Society: Series B (Statistical Methodology)*, 67, 301–320.

Zuur, A.F., Ieno, E.N., and Smith, G.M. (2007) *Analysing Ecological Data*, vol. 680, Springer, New York.

Zwanenburg, G., Hoefsloot, H.C.J., Westerhuis, J.A., Jansen, J.J., and Smilde, A.K. (2011) ANOVA–principal component analysis and ANOVA–simultaneous component analysis: A comparison. *Journal of Chemometrics*, 25(10), 561–567.

Index

Analysis of Variance for High-Dimensional Data: Applications in Life, Food and Chemical Sciences,
First Edition. Age K. Smilde, Federico Marini, Johan A. Westerhuis, and Kristian H. Liland.